THIRD EDITION

An Introduction to Numerical Methods

A MATLAB® Approach

THIRD EDITION

An Introduction to Numerical Methods
A MATLAB® Approach

Abdelwahab Kharab
Abu Dhabi University

Ronald B. Guenther
Oregon State University

CRC Press
Taylor & Francis Group
Boca Raton   London   New York

CRC Press is an imprint of the
Taylor & Francis Group, an **informa** business

A CHAPMAN & HALL BOOK

CRC Press
Taylor & Francis Group
6000 Broken Sound Parkway NW, Suite 300
Boca Raton, FL 33487-2742 *100654146* 3

© 2012 by Taylor & Francis Group, LLC
CRC Press is an imprint of Taylor & Francis Group, an Informa business

No claim to original U.S. Government works

Printed in the United States of America on acid-free paper
Version Date: 2011909

International Standard Book Number: 978-1-4398-6899-7 (Hardback)

Visit the Taylor & Francis Web site at
http://www.taylorandfrancis.com

and the CRC Press Web site at
http://www.crcpress.com

Contents

Preface

This is a textbook designed for an introductory course in numerical methods. It deals with the theory and application of the most commonly used numerical methods for solving numerical problems on microcomputers. It is intended for students in mathematics, science, and engineering who have completed the introductory calculus sequence. In addition, the reader is assumed to have taken a structured programming course. The thrust of this text is to assist the students to become familiar with the most common numerical methods encountered in science and engineering. The content material of this book has been designed to be compatible with any introductory numerical textbook that exists in the market. Students will be able to examine and solve many numerical problems, using MATLAB®[1] in a short period of time.

Due to the rapid advancement of computer technology and software developments, we have used MATLAB as the computing environment throughout all the chapters of the book. Each numerical method discussed in this book is demonstrated through the use of MATLAB which is easy to use and has many features such as:

1. Powerful matrix structure,

2. Powerful two- and three-dimensional graphing facilities,

3. A vast number of powerful built-in functions,

4. MATLAB's structured programming style that resembles FORTRAN and BASIC.

The goal of this present third edition is the same as the previous one. The book introduces students to a wide range of useful and important algorithms. Computer results are presented with full details so that the main steps of the algorithm of each

---

[1]MATLAB® is a registered trademark of the MathWorks, Inc.
For product information, please contact:
The MathWorks, Inc.
3 Apple Hill Drive
Natick, MA 01760-2098 USA
Tel: 508-647-7000   Fax: 508-647-7001
E-mail: info@mathworks.com   Web: www.mathworks.com

numerical method are visualized and interpreted. For this reason, a supplementary CD-ROM, attached at the back of the book, has been developed for students' use with MATLAB. The CD-ROM contains simple MATLAB functions that give a clear step-by-step explanation of the mechanism behind the algorithm of each numerical method covered. Emphasis is placed on understanding how the methods work. These functions guide the student through the calculations necessary to understand the algorithm(s). The main feature of this book, beside the use of MATLAB as its computing environment, is that the style of the book is easy, direct, and simple. This is likely to boost students' confidence in their ability to master the elements of the subject.

The book is organized in a fairly standard manner. Topics that are simpler, both theoretically and computationally, come first; for example, the root of equations is covered in Chapter 3. Chapter 2 contains an introduction to computer floating point arithmetic, errors, and interval arithmetic.

Both direct and iterative methods are presented in Chapter 4 for solving systems of linear equations.

Interpolation, spline functions, concepts of least squares data fitting, and numerical optimization are the subjects of Chapters 5, 6, 7, and 8. Interpolation forms the theoretical basis for much of numerical analysis.

Chapters 9 and 10 are devoted to numerical differentiation and integration. Several efficient integration techniques are presented.

In Chapters 11 and 12 a wide variety of numerical techniques is presented for solving linear integral equations and ordinary differential equations. An introduction for solving boundary value problems is presented in Chapter 13. Chapter 14 is devoted to some numerical techniques for computing the eigenvalues and eigenvectors of a matrix.

The last Chapter 15 provides a basic introduction to numerical techniques for solving partial differential equations.

In each chapter we have attempted to present clear examples in every section followed by a good number of related exercises at the end of each section with answers to some exercises.

It is the purpose of this book to implement various important numerical methods on a personal computer and to provide not only a source of theoretical information on the methods covered, but also to allow the student to easily interact with the microcomputer and the algorithm for each method using MATLAB.

This text should provide an excellent tool suitable for teaching numerical method courses at colleges and universities. It is also suitable for self-study purposes.

In this third edition all suggestions from the expert reviewers were addressed. They all deserve our sincere thanks and appreciation.

Features in the Third Edition

There have been some minor changes in some sections. Major new features are as follows:

- A new chapter on numerical solution of integral equations.

- A new section on nonlinear PDEs has been added in Chapter 15.

- MATLAB GUI has been incorporated as an integral part of the text.

Acknowledgments

We wish to thank the many persons who have helped us in the creation of this book. They are: R. Baker Kearfoot of the University of Southwestern Louisiana, Rachid Kharab of the University of Rouen, A. Laradji of the King Fahd University of Petroleum & Minerals, A.R. Kashani of the Sahand University of Technology, Z. Khiari of King Fahd University of Petroleum & Minerals, and H. Cheded of the King Fahd University of Petroleum & Minerals, who encouraged us to submit a book proposal.

Special thanks is due to Dr. B. Khoshandam of Semnan University for his keen interest and suggestions in this edition and careful work on the use of MATLAB GUI. His work was extraordinarily detailed and helpful to us.

We also thank the Anonymous Reviewers who made useful recommendations for the third edition.

The authors remain very grateful to the editorial and production staff of Chapman & Hall/CRC who have been helpful and available at all stages. Among them are Sarah Morris, Project Editor, who was in communication with us during the preparation of the final version of this third edition, and Sunil Nair, mathematics and statistics publisher, for his assistance and encouragement.

Finally, the authors are grateful for the financial and facilities support provided by Mr. H. Hassnaoui, General Director of the GFP in Muscat.

Suggestions for improvements to the book are always welcome and can be made by e-mail at: awkharab@yahoo.com.

Abdelwahab Kharab
Ronald B. Guenther

Chapter 1

Introduction

The Taylor Series is one of the most important tools in numerical analysis. It constitutes the foundation of numerical methods and will be used in most of the chapters of this text. From the Taylor Series, we can derive the formulas and error estimates of the many numerical techniques used. This chapter contains a review of the Taylor Series, and a brief introduction to MATLAB®.

1.1 ABOUT MATLAB and MATLAB GUI (Graphical User Interface)

MATLAB (MATrix LABoratory) is a powerful interactive system for matrix-based computation designed for scientific and engineering use. It is good for many forms of numeric computation and visualization. MATLAB language is a high-level matrix/array language with control flow statements, functions, data structures, input/output, and object-oriented programming features. To fully use the power and computing capabilities of this software program in classrooms and laboratories by teachers and students in science and engineering, part of this text is intended to introduce the computational power of MATLAB to modern numerical methods.

MATLAB has several advantages. There are three major elements that have contributed to its immense popularity. First, it is extremely easy to use since data can be easily entered, especially for algorithms that are adaptable to a table format. This is an important feature because it allows students to experiment with many numerical problems in a short period of time. Second, it includes high-level commands for two-dimensional and three-dimensional data visualization, and presentation graphics. Plots are easily obtained from within a script or in command mode. Third, the most evident power of a MATLAB is its speed of calculation. The program gives instantaneous feedback. Because of their popularity, MATLAB and other software such as MAPLE and Mathematica are now available in most university microcomputer laboratories.

One of the primary objectives of this text is to give students a clear step-by-step explanation of the algorithm corresponding to each numerical method used. To accomplish this objective, we have developed MATLAB M-functions contained in a supplementary CD-ROM at the back of the book. These M-files can be used for the application or illustration of all numerical methods discussed in this text. Each M-function will enable students to solve numerical problems simply by entering data and executing the M-functions.

It is well known that the best way to learn computer programming is to write computer programs. Therefore, we believe that by understanding the basic theory underlying each numerical method and the algorithm behind it, students will have the necessary tools to write their own programs in a high-level computer language such as C or FORTRAN.

Another future of MATLAB is that it provides the Graphical User Interface (GUI). It is a pictorial interface to a program intended to provide students with a familiar environment in which to work. This environment contains push buttons, toggle buttons, lists, menus, text boxes, and so forth, all of which are already familiar to the user, so that they can concentrate on using the application rather than on the mechanics involved in doing things. They do not have to type commands at the command line to run the MATLAB functions. For this reason, we introduced GUI in this edition to make it easier for the students to run the M-functions of the book. A readme file named "SUN Package readme contained in the directory NMETH" of the CD attached at the back cover of the book, explains in details the steps to follow to run the MATLAB functions using GUI. In addition, Appendix D shows the use of GUI for solving several examples.

1.2   AN INTRODUCTION TO MATLAB

In this section we give to the reader a brief tutorial introduction to MATLAB. For additional information we urge the reader to use the reference and user's guides of MATLAB.

1.2.1   Matrices and matrix computation

MATLAB treats all variables as matrices. They are assigned to expressions by using an equal sign and their names are case-sensitive. For example,

```
>> A = [4 -2 5; 6 1 7; -1 0 6]
A =
       4    -2     5
       6     1     7
      -1     0     6
```

New rows may be indicated by a new line or by a semicolon. A column vector may be given as

```
>> x = [2; 8; 9] or x = [2 8 9]'
x =
```

$$2$$
$$8$$
$$9$$

Elements of the matrix can be a number or an expression, like

```
>> x = [2    1+2    12/4    2^3]
x =
      2    3    3    8
```

One can define an array with a particular structure by using the command

$$x = a : step : b$$

As an example

```
>> y = [0: 0.2 : 1]
y =
         0   0.2000   0.4000   0.6000   0.8000   1.0000
```

```
>> y = [0: pi/3 : pi]
y =
         0   1.0472   2.0944   3.1416
```

```
>> y = [20: -5 : 2]
y =
        20   15   10   5
```

MATLAB has a number of special matrices that can be generated by built-in functions

```
>> ones(2)
ans =
        1     1
        1     1
```
a matrix of all 1's.

```
>> zeros(2,4)
ans =
        0   0   0   0
        0   0   0   0
```
a 2 × 4 matrix of zeros.

```
>> rand(2,4)
ans =
       0.9501    0.6068    0.8913    0.4565
       0.2311    0.4860    0.7621    0.0185
```

a 2×4 random matrix with uniformly distributed random elements.

```
>> eyes(3)
ans =
        1    0    0
        0    1    0
        0    0    1
```
the 3×3 identity matrix.

The diag function either creates a matrix with specified values on a diagonal or extracts the diagonal entries. For example,

```
>> v = [3 2 5];
>> D=diag(v)
D =
        3    0    0
        0    2    0
        0    0    5
```
a 3×3 diagonal matrix with v on the main diagonal.

To extract the diagonal entries of an existing matrix, the diag function is used:

```
>> C=[2 -4 7; 3 1 8; -1 5 6];
>> u=diag(C)
u =
      2
      1
      6
```

The linspace function generates row vectors with equally spaced elements. The form of the function is

linspace(firstValue, lastValue, numValues)

where firstValue and lastValue are the starting and ending values in the sequence of elements and NumValues is the number of elements to be created. For example,

```
>> evens = linspace(0,10,6)
evens =
        0    2    4    6    8    10
```

The most useful matrix functions are

eig	eigenvalues and eigenvectors
inv	inverse
lu	LU decomposition
qr	QR factorization
rank	rank

A component-by-component addition of two vectors of the same dimensions is indicated as

```
>> x = [1 5 8];
>> y = [2 4 3];
>> x + y

ans =
        3    9    11
```

Multiplying a vector or matrix by a scalar will scale each element of the vector or matrix by the value of the scalar.

```
>> C = 2*[1 3; 4 2]
C =
        2    6
        8    4
>> v = 3*[1   4   -5   7]
v =
        3    12   -15   21
```

The component-by-component multiplication of the vectors x and y is indicated as

```
>> x. * y
ans =
        2    20    24
```

The inner, or dot, product of two vectors is a scaler and can be obtained by multiplying a row vector and a column vector. For example,

```
>> u=[5 7 -1 2]; v=[2; 3; 10; 5];
>> u*v

ans =
        31
```

The transpose of a matrix or a vector is denoted by a prime symbol. For example

```
>> x'
ans =
        1
        5
        8
```

The matrix operations of multiplication, power, and division are indicated as

```
>> B = [1 3; 4 2];

>> A = [2 5; 0 6];

>> A*B

ans =
        22      16
        24      12

>> A^2
ans =
        4       40
        0       36

>> A/B
ans =
        1.6000        0.1000
        2.4000       -0.6000

>> A.*B
ans =
        2       15
        0       12
```

Note that the three operations of multiplication, division, and power can operate elementwise by preceding them with a period.

1.2.2   Polynomials

MATLAB provides a number of useful built-in functions that operates on polynomials. The coefficients of a polynomial are specified by the entries of a vector. For example, the polynomial $p(x) = 2x^3 - 5x^2 + 8$ is specified by [2 -5 0 8]. To evaluate the polynomial at a given x we use the polyval function.

```
>> c = [2 -5 0 8];
>> polyval(c,2)
ans =
```

evaluates the polynomials $p(x) = 2x^3 - 5x^2 + 8$ at $x = 2$.

The function roots finds the n roots of a polynomial of degree n. For example, to find the roots of the polynomial $p(x) = x^3 - 5x^2 - 17x + 21$, enter

```
>> a = [1 -5 -17 21];
>> roots(a)
ans =
          7.0000
         -3.0000
          1.0000
```

1.2.3   Output format

While all computations in MATLAB are performed in double precision, the format of the displayed output can be controlled by the following commands:

format short	fixed point with 4 decimal places (the default)
format long	fixed point with 14 decimal places
format short e	scientific notation with 4 decimal places
format long e	scientific notation with 15 decimal places
format rat	approximation by ratio of small integers

Once invoked, the chosen format remains in effect until changed.

It is possible to make the printed output from a MATLAB function look good by using the disp and fprintf functions.

disp(X) displays the array X, without printing the array name. If X is a string, the text is displayed. For example,

disp('The input matrix A is')

will display the text placed in single quotes. The function disp also allows us to print numerical values placed in text strings. For example,

disp(['Newton's method converges after ',num2str(iter),' iterations'])

will write on the screen

Newton's method converges after 9 iterations.

Here the function num2str(iter) converts a number to a string.

The other function fprintf is more flexible and has the form

$$\text{fprintf}('\text{filename}','\text{format}',\text{list})$$

where

filename is optional; if it is omitted, output goes to the screen.

format is a format control string containing conversion specifications or any optional text. The conversion specifications control the output of array elements.

Common conversion specifications include:

%P.Qe	for exponential
%P.Qf	fixed point
%P.Qg	to automatically select the shorter of %P.Qe or %P.Qf

where P and Q are integers that set the field width and the number of decimal places, respectively.

list is a list of variable names separated by commas, and the special format \n produces a new line.

For example, the statements

```
>> x = pi ; y = 4.679 ; z = 9 ;
>> fprintf('\n x = %8.7f\n y = %4.3f\n z = %1.0f\n',x,y,z)
```

prints on the screen

```
x = 3.1415927
y = 4.679
z = 9
```

1.2.4   Planar plots

Plotting functions is very easy in MATLAB. Graphs of two-dimensional functions can be generated by using the MATLAB command plot. You can, for example, plot the function $y = x^2 + 1$ over the interval $[-1, 2]$ with the following commands:

```
>> x = [-1: 0.02: 2];
>> y = x.^2 +1;
>> plot(x,y)
```

The plot is shown in Figure 1.1. The command plot(x,y,'r+:') will result in r=red with + for points and dotted line. Use help plot for many options.

To plot a function $f1$ defined by an M-file (see Subsection 1.2.7) in an interval $[a, b]$, we use the command

```
>> fplot('f1',[a,b]).
```

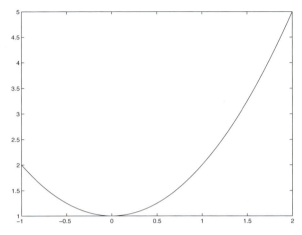

FIGURE 1.1
Plot of the function $f(x) = x^2 + 1$ **in** $[-1, 2]$.

Other features can be added to a given graph using commands such as grid, xlabel, ylabel, title, etc.... For example, the command grid on will create grid lines on the graph.

Plots of parametrically defined curves can also be made. Try, for example,

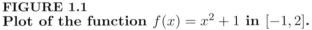

Multiple plots on a single graph are illustrated by

```
>> x=[0:0.1:2*pi];
>> y1=cos(x); y2=sin(x); y3=cos(2*x);
>> plot(x,y1,x,y2,x,y3)
```

1.2.5   3-D mesh plots

Three-dimensional or 3-D mesh surface plots are drawn with the function mesh. The command mesh(z) creates a three-dimensional perspective plot of the elements of the matrix z.

The following example will generate the plot shown in Figure 1.2 of the function $z = e^{-x^2-y^2}$ over the square $[-4, 4] \times [-4, 4]$.

```
>> [x y]=meshgrid(-4.0:0.2:4.0,-4.0:0.2:4.0);
>> z=exp(-x.^2-y.^2);
>> mesh(x,y,z)
```

The first command creates a matrix whose entries are the points of a grid in

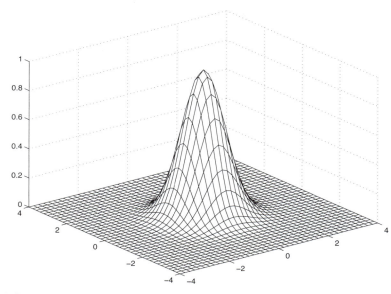

FIGURE 1.2
Three-dimensional surface.

the square $-4 \leq x \leq 4$, $-4 \leq y \leq 4$ with grid 0.2 units wide and 0.2 units tall. The second command creates a matrix whose entries are the values of the function $z(x, y)$ at the grid points. The third command uses this information to construct the graph.

Other commands related to 3-D graphics are plot3 and surf.

1.2.6 Function files

MATLAB contains a vast collection of built-in functions from elementary functions like sine and cosine, to more sophisticated functions like matrix inversions, matrix eigenvalues, and fast Fourier transforms. Table 1.1 contains a small selection of elementary mathematical MATLAB functions and Appendix B contains additional lists of functions grouped by subject area.

Note that all these function names must be in lower cases. MATLAB functions may have single or multiple arguments. For example,

[U,D]=eig(A)

produces a matrix U whose columns are the eigenvectors of A and a diagonal matrix D with the eigenvalues of A on its diagonal.

1.2.7 Defining functions

MATLAB allows users to define their own functions by constructing an M-file in the M-file Editor/Debugger. The first line of a function has the form

cos(x)	cosine of x
sin(x)	sine of x
tan(x)	tangent of x
sqrt(x)	square root of x
abs(x)	absolute value of x
exp(x)	exponential of x
log(x)	log to the base e of x
log10(x)	log to the base 10 of x
cosh(x)	hyperbolic cosine of x
tanh(x)	hyperbolic tangent of x
asin(x)	inverse sine function of x
expm(A)	exponential of matrix A

Table 1.1   Elementary mathematical MATLAB functions.

$$\text{function } y = \text{function\_name(input arguments)}$$

As an example, let us define the function $f1(x) = 2x^3 - 3x + 1$ as an m.file. Using the Editor/Debugger or any text editor like *emacs*, or *vi*, we enter

```
function y=f1(x)
y=2*x.^3-3*x+1;
```

Once this function is saved as an M-file named f1.m, we can use the MATLAB command window to compute, for example,

```
>> f1(2)
ans =
        11
```

User defined functions can themselves have functions as input parameters. They can be evaluated internally by using the function feval. For example,

```
>> x1=fzero('f1',0)
x1=
      0.3949
```
gives the zero of f1 closest to 0.

We can pass one function f1.m to another function f2.m in MATLAB by using the @ symbol:

```
>> f2(@f1)
```

For example, the built-in function quad(f,a,b) in MATLAB approximately integrates the function $y = f(x)$ from a to b. To use this built-in function to approximately integrate $y = \cos(\pi x) + 2$ from $x = -1$ to $x = 1$, we begin by writing a

MATLAB function called f1.m:

```
function y=f1(x)
    y=cos(pi*x)+2;
```

If we are in the correct directory, we type in MATLAB:

```
>> quad(@f1,-1,1)
ans =
            4.00000001667477
```

1.2.8   Relations and loops

Like most computer languages, MATLAB offers a variety of flow control statements like for, while, and if. The statements, that we use to control the flow are called relations.

The *Relational operators* in MATLAB are

==	equal
<=	less than or equal
>=	greater than or equal
˜ =	not equal
<	less than
>	greater than

Note that "=" is used in an assignment statement while "==" is used in a relation.

Logical operators

&	and
|	or
˜	not

if, for, while statements

The general form of an if statement is

```
if <if _expression>
    <statements>
elseif <if _expression>
    <statements>
        ...
else
```

```
        <statements>
    end
```

Here is a simple example. If $i, j = 1, ..., 10$, then

```
if i==j
        A(i,j)=1;
else
        A(i,j)=0;
end
```

will result in a diagonal matrix A with 1's in its diagonal.

The general form of a **for** loop is

```
for <loop_variable= loop_expression>
        <statements>
end
```

For example,

```
S=0;
for i=1:100
    S=S+1;
end
S
```
results in the output

```
    S =
        100
```

You can also nest **for** statements. The following statements create an $m \times n$ Hilbert matrix, H.

```
for i=1:m
    for j=1:n
        H(i,j)=1/(i+j-1);
    end
end
```

The **switch** command is ideal for decision making. The general form of the **switch** command is

```
switch switch expression
    case case-expression
        commands
```

. . . .

    otherwise

        commands

end

Here is a code that finds the number of days in a month. For simplicity we will ignore the effect of leap years on the number of days in February and assume that there are always 28 days in February. The switch command is ideal for this kind of code. We implement it as follows:

```
switch month
     case {1,3,5,7,8,10,12}
          days = 31;
     case {4,6,9,11}
          days = 30;
     case 2
          days = 28;
end
```

The variable month carries the number of the month, and depending on its value the first case, second case, or third case will be executed depending on the outcome of comparing the switch expression (month) to the members of the various case expressions. The same result could also have been accomplished with an if-structure, but in this case the switch-command resulted in a more compact and elegant code.

Finally, MATLAB also has its own version of the while loop, which repeatedly executes a group of statements while a condition remains true. Its form is

```
while < while _expression >
     < condition statements>
end
```

Given a positive number n the following statements compute and display the even powers of 3 less than n.

```
k=0;
while 3^k < n
     if rem(k,2)==0
          3^k
          k=k+1;
     else
          k=k+1;
     end
end
```

1.3 TAYLOR SERIES

THEOREM 1.1 (Taylor)
Suppose f has $n + 1$ continuous derivatives on $[a, b]$. Let $x, x_0 \in [a, b]$. Then

$$f(x) = \sum_{k=0}^{n} \frac{f^{(k)}(x_0)}{k!}(x - x_0)^k + R_{n+1}(x) \tag{1.1}$$

where

$$R_{n+1}(x) = \frac{f^{(n+1)}(\xi)}{(n+1)!}(x - x_0)^{n+1} \tag{1.2}$$

for some $\xi = \xi(x)$ between x_0 and x.

Eqn. (1.1) is called the **Taylor series** expansion of $f(x)$ about x_0 and $R_{n+1}(x)$ is called the **remainder** or the **truncation error**. Since in practical computation with the Taylor series only a finite number of terms can be carried out, the term **truncation error** generally refers to the error involved in using a finite summation to approximate the sum of an infinite series.

EXAMPLE 1.1
Derive the Taylor series for $f(x) = e^x$ near $x_0 = 0$.

Since $f^{(k)}(0) = e^0 = 1$ for $k = 0, 1, \ldots$, then from Eqn. (1.1) we have

$$e^x = 1 + x + \frac{x^2}{2!} + \frac{x^3}{3!} + \cdots + \frac{x^n}{n!} + \frac{e^\xi}{(n+1)!}x^{n+1}$$

$$= \sum_{k=0}^{n} \frac{x^k}{k!} + \frac{e^\xi}{(n+1)!}x^{n+1}. \tag{1.3}$$

EXAMPLE 1.2
Using Example 1.1, approximate e with $n = 3$.

Set $x = 1$ and $n = 3$ in (1.3) to get

$$e \approx 1 + \frac{1}{1!} + \frac{1}{2!} + \frac{1}{3!} = 2.66667.$$

Now using (1.2), one can get an estimate of the truncation error. We have the result

$$R_4 = \frac{e^\xi}{4!} \quad \text{with} \quad 0 < \xi < 1.$$

Since $\left| e^\xi \right| \le e \le 2.8$, the error is no greater in magnitude than

$$(1/4!)(2.8) = 0.11667.$$

If we let $x_0 = 0$ in Eqn. (1.1), we obtain the series known as **Maclaurin series**

$$f(x) = \sum_{k=0}^{n} \frac{f^k(0)}{k!} x^k + \frac{f^{(n+1)}(\xi)}{(n+1)!} x^{n+1} \tag{1.4}$$

with $\xi(x)$ between 0 and x.

EXAMPLE 1.3
Use the Taylor series to approximate e to three decimal places.

Using (1.1) we need to choose n so large that

$$\frac{e^\xi}{(n+1)!} < \frac{2.8}{(n+1)!} < 5 \times 10^{-4}$$

which yields $n = 7$ for the smallest n that approximates e to three decimal places. Setting $n = 7$ in the Taylor series we obtain

$$e \approx 1 + \frac{1}{1!} + \frac{1}{2!} + \frac{1}{3!} + \frac{1}{4!} + \frac{1}{5!} + \frac{1}{6!} + \frac{1}{7!} = 2.718253968254.$$

The calculator value is 2.718281828459.

EXAMPLE 1.4
Derive the Maclaurin series for $f(x) = \sin x$.

The derivatives of $\sin x$ are

$$\begin{array}{rclcrcl}
f'(x) &=& \cos x, & \qquad & f'(0) &=& 1, \\
f''(x) &=& -\sin x, & \qquad & f''(0) &=& 0, \\
f'''(x) &=& -\cos x, & \qquad & f'''(0) &=& -1, \\
\cdots &=& \cdots & & \cdots &=& \cdots.
\end{array}$$

Therefore,

$$\sin x = \sum_{k=0}^{n} (-1)^k \frac{x^{2k+1}}{(2k+1)!} + (-1)^n \frac{x^{2n+3}}{(2n+3)!} \cos(\xi).$$

Another useful form of the Taylor series is obtained by replacing in Eqn. (1.1) x by $x + h$ and x_0 by x. The result is

$$f(x + h) = \sum_{k=0}^{n} \frac{f^{(k)}(x)}{k!} h^k + \frac{f^{(n+1)}(\xi)}{(n+1)!} h^{n+1} \tag{1.5}$$

with ξ between x and $x + h$.

Here we introduce a commonly used notation for the error term. We say that $F(h)$ *is of the order of* h^n and write

$$F(h) = O(h^n)$$

if there exists a real constant $C > 0$ and a positive integer n such that

$$|F(h)| \leq C |h^n| \qquad \text{for sufficiently small } h.$$

EXAMPLE 1.5

In Example 1.4 we found that for $n = 2$.

$$\sin x = x - \frac{1}{3!} x^3 + \frac{1}{5!} x^5 - \frac{1}{7!} x^7 \cos(\xi), \qquad 0 < \xi < x.$$

Since

$$\left| \frac{\sin x - x + \frac{1}{3!} x^3 - \frac{1}{5!} x^5}{x^7} \right| = \left| \frac{1}{7!} \cos(\xi) \right| \leq \frac{1}{7!} = C$$

then

$$\sin x - x + \frac{1}{3!} x^3 - \frac{1}{5!} x^5 = O(x^7).$$

The following theorem is a special case of Taylor's Theorem ($n = 0$).

THEOREM 1.2 (Mean Value Theorem)
 If f is continuous on $[a, b]$ and is differentiable on (a, b), then there exists a number $c \in (a, b)$ such that (see Figure 1.3)

$$f(b) - f(a) = f'(c)(b - a).$$

The next theorem can be found in any calculus text and is an important tool for determining the existence of the solution of certain problems.

THEOREM 1.3 (Intermediate Value Theorem)
 Suppose f is continuous on a closed interval $[a, b]$. If w is any number between $f(a)$ and $f(b)$, then there exists at least one c in (a, b) such that $f(c) = w$ (See Figure 1.4).

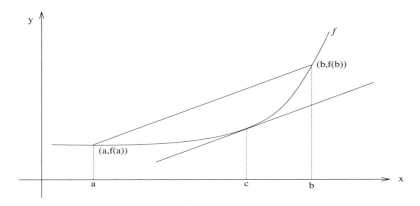

FIGURE 1.3
Mean Value Theorem.

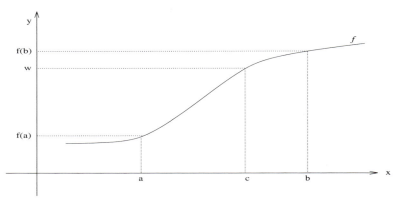

FIGURE 1.4
Intermediate Value Theorem.

EXAMPLE 1.6
Show that the equation $x^5 - 5x^4 - 7x^3 - x + 1 = 0$ has a solution in the interval $[0, 1]$.

Consider the function $f(x) = x^5 - 5x^4 - 7x^3 - x + 1$. It is easy to see that f is continuous on $[0, 1]$ and $f(0) = 1, f(1) = -11$. Since $f(1) < 0 < f(0)$, then, from the intermediate value theorem (IVT), there exists x in $(0, 1)$ such that

$$x^5 + 5x^4 - 7x^3 - x + 1 = 0.$$

EXERCISE SET 1.3

1. Find the Maclaurin series for $f(x) = x^2 e^x$.

2. Determine whether the following functions satisfy the hypotheses of the Mean Value Theorem. If so, find a number c satisfying the conclusion of the Mean Value Theorem.

   (a) $f(x) = x^{1/3}$, on $[-1, 2]$,
   (b) $f(x) = x^{1/3}$, on $[0, 2]$,
   (c) $f(x) = |x|$, on $[-2, 2]$.

3. Find all numbers c in $[1, 3]$ that satisfy the conclusion of the Mean Value Theorem for
$$f(x) = 2x + \frac{3}{x}.$$

4. Does the Mean Value Theorem tell us anything about the function
$$f(x) = \left| x^2 - 1 \right| \quad \text{in } [0, 2]?$$

5. Show that $f(x) = 2e^{2x} + 4x - 3$ has at least one zero in $[0, 1]$.

6. Find the Taylor series expansion of $f(x) = e^{-x}$ about $x_0 = 0$ and approximate $f(-0.88)$ with $n = 3$.

7. Find the Taylor series expansion for $f(x) = \ln(1 - x)$ about $x_0 = 0$.

8. Approximate $f(x) = e^x \sin x$ at $x = 0.2$ by using the Taylor series.

9. Find the Taylor series expansion of $\cos x$ about $x = \pi/3$.

10. What is the third-order Taylor polynomial for $f(x) = \sqrt{x + 1}$ about x_0?

11. Use a Taylor polynomial to find an approximate value for $\sqrt{e}$.

12. Use Taylor's Theorem to show that
$$\frac{1 - \cos x}{x} = \frac{1}{2}x + O(x^2)$$

   for x is sufficiently small.

13. Use Taylor's Theorem to show that
$$(1 + x)^{-1} = 1 - x + x^2 + O(x^3)$$

   for x is sufficiently small.

14. What is the error in the Taylor polynomial of degree 5 for $f(x) = 1/x$, using $x_0 = 3/4$ for $x \in [1/2, 1]$?

15. How many terms must be taken in the previous exercise to get an error of less than 10^{-2}? 10^{-4}?

16. Construct a Taylor polynomial approximation to $f(x) = e^{-x}$ with $x_0 = 0$ that is accurate to within 10^{-3} on the interval $[0, 1]$. Find a value M such that $|f(x_1) - f(x_2)| \leq M|x_1 - x_2|$ for all x_1, x_2 on the interval.

17. Find the Taylor polynomial of order 2 around $x_0 = 1$ for $f(x) = x^3 + x$.

18. Find intervals containing solutions to the following equations

   (a) $x - 3^{-x} = 0$,

   (b) $4x^2 - e^x = 0$.

19. Use the identity
$$\cos^2(x) = \frac{1 + \cos(2x)}{2}$$
   to find the Taylor series representation of $\cos^2(x)$ about 0.

20. Use the previous exercise and the identity
$$\sin^2(x) = 1 - \cos^2(x)$$
   to find the Taylor series representation of $\sin^2(x)$ about 0.

21. Evaluate to five decimal places the integrals

   (a) $\int_0^1 \frac{\sin x}{x} dx$,

   (b) $\int_0^1 \frac{1 - \cos x}{x} dx$,

   (c) $\int_0^1 \frac{e^x - 1}{x} dx$,

   (d) $\int_1^2 \frac{e^x}{x} dx$

   and prove your answers are correct by estimating the error.

22. Find to four decimal places the length of

   (a) the ellipse $(\frac{x}{4})^2 + (\frac{y}{2})^2 = 1$,

   (b) the parabola $y = x^2$ for $-2 \leq x \leq 2$,

   (c) the hyperbola $y^2 = x^2 + 1$ for $y > 0$ and $-1 \leq x \leq 1$.

COMPUTER PROBLEM SET 1.3

1. Consider the so-called cosine integral
$$Cin(x) = \int_0^x \frac{1 - \cos t}{t} dt.$$

(a) Find the degree $2n$ Taylor polynomial with remainder $R(x)$ about $a = 0$. Give an estimate for $|R(x)|$ that still depends on x.

(b) Write a MATLAB function $Cin(x, tol, ca, n)$ with input parameters x, tol computing an approximate value ca for $Cin(X)$ with an error less than tol. The program should use the Taylor polynomial of degree n to compute ca. n should be determined by the program as small as possible, depending on x and tol (use the results of part (a)).
Test your program for $x = 1$ and $tol = 0.25$, 0.06, 10^{-4}. The exact value is $Cin(1) = 0.239811742$. Print in every case the error $ca - Cin(x)$, and n. Check if the actual errors are really $\leq tol$. Print your results ca and the value of n for $x = 0.5$, $tol = 10^{-4}$.

2. The following MATLAB algorithm computes e^{-x} using a Taylor series expansion:

```
s=1;t=1;i=1;
while (abs(t) > eps*abs(s))
        .... stop iterating when adding t to s does not change s
        t=-t*x/i;
        s=s+t;
        i=i+1;
end
result=s
```

Run this algorithm for x=1:20.

Chapter 2

Number System and Errors

The way in which numbers are stored and manipulated when arithmetic operations are performed on microcomputers is different from the way we, humans, do our arithmetic. We use the so-called decimal number system, while in microcomputers, internal calculations are done in the binary system. In this chapter we consider methods for representing numbers on computers and the errors involved.

2.1 FLOATING-POINT ARITHMETIC

In order to understand the major sources of errors involved in numerical solutions, we need to examine the manner in which numbers are stored in the computer. Since only numbers with a finite number of digits can be represented in the computer, *floating-point* form is used for the representation of real numbers.

A k-floating-point number in base β has the form

$$x = \pm (.b_1 b_2 \ldots b_k) \times \beta^e \tag{2.1}$$

where $\beta, b_1, \ldots, b_k$ are integers such that $\beta \geq 2$ and $0 \leq b_i \leq \beta - 1$. Here $(b_1 b_2 \ldots b_k)$ is called the **mantissa** and e the **exponent**. If $b_1 \neq 0$, or else $b_1 = b_2 = \cdots = b_k = 0$ such a floating-point number is said to be **normalized**.

β varies from computer to computer, but since most computers use the binary system ($\beta = 2$), we consider the normalized binary form

$$x = \pm (.1 b_2 \ldots b_k) \times 2^e \tag{2.2}$$

where the mantissa is expressed as a sequence of zeros and ones.

As an abbreviation, the two binary digits 0 and 1 are usually called **bits** and the fixed-length group of binary bits is generally called a **computer word**. As an example, let us consider the floating-point number system of a 32-bit word length microcomputer. The internal representation of a word is shown in Figure 2.1. The leftmost bit is used for the sign of the mantissa, where the zero bit corresponds to a plus sign, and the unit bit to a negative sign. The next seven bits are used to represent the exponent with the first bit used as a sign to indicate whether the

	Exponent (7 bits)	normalized Mantissa (24 bits)

$\uparrow$
sign of Mantissa
(1 bit)

FIGURE 2.1
Internal representation of a 32-bit word length microcomputer.

exponent is positive or negative. The final 24 bits represent the normalized mantissa with $b_1 = 1$.

EXAMPLE 2.1
Consider the machine number.

1	0000010	110100000000000000000000

The number is negative since the leftmost bit is 1. The exponent is positive and is equal to

$$0 \times 2^0 + 1 \times 2^1 + 0 \times 2^2 + 0 \times 2^3 + 0 \times 2^4 + 0 \times 2^5 = (2)_{10}.$$

The final 24 bits indicate that the normalized mantissa is

$$1 \times 2^{-1} + 1 \times 2^{-2} + 1 \times 2^{-4} = (0.8125)_{10}.$$

Thus, this machine number represents the decimal number

$$-0.8125 \times 2^2 = (-3.25)_{10}.$$

EXAMPLE 2.2
Write the decimal number 42.78125 in the normalized binary form.

We have
$$(42.)_{10} = (101010.)_2 \quad \text{and} \quad (.78125)_{10} = (.110010)_2.$$

Now
$$(42.78125)_{10} = (101010.110010)_2 = (0.101010110010)_2 \times 2^6.$$

Since the exponent $(6)_{10} = (110)_2$, the internal representation of 42.78125 in normalized binary form is

$$(0000011010101011001000000000000)_2.$$

The largest normalized positive machine number is

0	0111111	11111111111111111111111	$\approx 2^{63} \approx 10^{18}$.	

Thus, the range of numbers that can be stored is from about 10^{-19} to almost 10^{18} in decimal magnitude.

Numbers occurring in calculations that have a magnitude outside the above computer range result in what is called an **underflow** or an **overflow**.

The internal representation of a number that we have just described is an example of how numbers can be stored in a 32-bit word length microcomputer. However, a widely used internal representation of numbers in almost all new computers is the **IEEE standard for binary floating-point arithmetic**. The single-precision format uses 32 bits and operates as follows:

$$\text{Floating-point number} = (-1)^s \times (1.f)_2 \times (2^{c-127})_{10}.$$

The first bit is reserved for the sign bit s where $s = 0$ corresponds to $+$ and $s = 1$

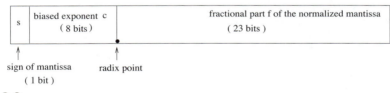

FIGURE 2.2
IEEE 32-bit format.

corresponds to $-$. The next 8 bits are reserved for the **biased** or **offset** exponent c. The value of c ranges from $(0000\,0000)_2 = (0)_{10}$ to $(1111\,1111)_2 = (255)_{10}$. Finally, the remaining 23 bits are used for the mantissa. Note that only the part of the mantissa denoted by f is stored because the first bit in the mantissa is always equal to 1 so that this bit does not need to be stored. The IEEE 32-bit format is illustrated in Figure 2.2.

EXAMPLE 2.3
Determine the IEEE format of the decimal number -45.8125.

We have

$$(-45.)_{10} = (-101101)_2 \quad \text{and} \quad (.8125)_{10} = (110100)_2.$$

Hence,
$$(-45.8125)_{10} = (-101101.110100)_2 = (-1.011011101)_2 \times 2^5.$$

The value of c is $c = 127 + 5 = (132)_{10}$ whose 8-bit form is $(10000100)_2$ and $s = 1$ since the number is negative. Thus, the IEEE format of -45.8125 is

1	1000 0100	0110 1110 1000 0000 0000 000

The largest number in the IEEE 32-bit format is obtained when the mantissa $(1.f)$ has a 1 in each bit position and the biased exponent $c = (1111\,1110)_2 = (254)_{10}$. The value of this number is therefore $(2-2^{-23})2^{127} \approx 3 \times 10^{38}$. The smallest positive number is $2^{-126} \approx 10^{-38}$.

The IEEE **double-precision** format employs 64 bits, of which 11 bits are reserved for the biased exponent, and 52 bits for the fractional part f of the normalized mantissa. The leading bit remains the sign bit. The double precision has the advantage of representing more numbers more accurately, while single precision has the advantage of requiring less space (which may be an important concern in storing huge amounts of data). Since hardware speeds are often the same for the two formats, the double precision format is the more used in practice. Intel processors also have an 80-bit extended format that is used within a computer's hardware arithmetic unit, but which is not always available to the programmer.

EXERCISE SET 2.1

1. Write the following decimal numbers in the normalized binary form:
   (a) $(3.4375)_{10}$,   (b) $(12.875)_{10}$.

2. Determine the representation of the following decimal numbers in the floating-point number system of a 32-bit word length microcomputer.
   (a) $(5.25)_{10}$,   (b) $(-3.84375)_{10}$.

3. On a computer, floating-point numbers are represented in the following way

t	e	u	m

t = sign of exponent, 1 bit
e = absolute value of exponent, 3 bits
u = sign of mantissa, 1 bit

m = mantissa, 5 bits
The base is 2 and the mantissa is normalized.

(a) Give the largest number that can be represented exactly.

(b) Determine the decimal number corresponding to the word

$$(0101010110)_2.$$

4. Determine the IEEE format of the following decimal numbers

   (a) 351.78215,   (b) −0.5625.

5. Convert the following 32-bit machine numbers to decimal

   (a) $a =$

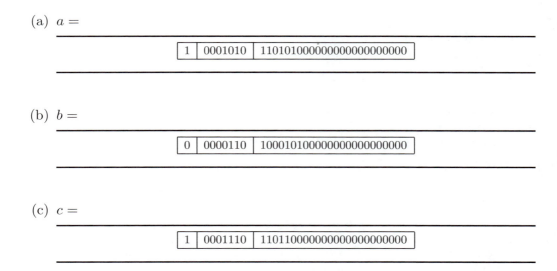

   (b) $b =$

   (c) $c =$

6. In the IEEE double-precision floating-point standard, 64 bits (binary digits) are used to represent a real number: 1 bit for the sign, 11 bits for the exponent, and 52 bits for the mantissa. A double-precision normalized nonzero number x can be written in this standard as $x = \pm(1.d_1 d_2 \cdots d_{52})_2 \times 2^{n-1023}$ with $1 \leq n \leq 2046$, and $0 \leq d_k \leq 1$ for $k = 1, 2, \ldots, 52$.

   (a) What is the smallest positive number in this system?

   (b) What is the smallest negative number in this system?

   (c) How many real numbers are in this system?

   Note: You may express your answers with formulas.

2.2   ROUND-OFF ERRORS

There are two commonly used ways of representing a given real number x by a floating-point machine number, denoted by $fl(x)$, **rounding** and **chopping**. Consider a positive real number x in the normalized decimal form

$$x = (0.b_1 b_2 \ldots b_k b_{k+1} \ldots)_{10} \times 10^e. \tag{2.3}$$

We say that the number x is chopped to k digits when all digits following the kth digits are discarded; that is, the digits $b_{k+1} b_{k+2} \ldots$ are chopped off to obtain

$$fl(x) = (0.b_1 b_2 \ldots b_k)_{10} \times 10^e.$$

Conversely, x is rounded to k digits when $fl(x)$ is obtained by choosing $fl(x)$ nearest to x; that is, adding one to b_k if $b_{k+1} \geq 5$ and chop off all but the first k digits if $b_{k+1} < 5$.

EXAMPLE 2.4
Consider the number e in the decimal system

$$e = 0.2718281828 \ldots \times 10^1.$$

If $k = 7$ and rounding is employed, then

$$fl(e) = (0.2718281 + 0.0000001) \times 10^1 = 2.718282.$$

If, on the other hand, chopping is employed, then

$$fl(e) = 0.2718281 \times 10^1 = 2.718281.$$

The error that results from replacing a number x with its floating-point form $fl(x)$ is called **round-off error**. It is equal to the difference between x and $fl(x)$. The round-off error depends on x and is therefore best measured relative to the size of x.

DEFINITION 2.1   *If $x \neq 0$, the **absolute error** is given by $|x - fl(x)|$ and the **relative error** is given by $|x - fl(x)| / |x|$.*

As an example, let us find the relative error for the machine representation of x, given by Eqn. (2.3), when chopping is used to k digits. We have

$$\left| \frac{x - fl(x)}{x} \right| = \left| \frac{0.b_1 b_2 \cdots b_k \cdots \times 10^n - 0.b_1 b_2 \cdots b_k \times 10^n}{0.b_1 b_2 \cdots b_k \cdots \times 10^n} \right|$$

$$= \left| \frac{0.b_{k+1} b_{k+2} \cdots}{0.b_1 b_2 \cdots} \right| \times 10^{-k} \leq \frac{1}{0.1} \times 10^{-k}.$$

Thus,

$$\left| \frac{x - fl(x)}{x} \right| \leq 10^{-k+1}.$$

In a similar manner when rounding is used, the relative error is

$$\left| \frac{x - fl(x)}{x} \right| \leq 0.5 \times 10^{-k+1}.$$

For a binary computer with k bits in the mantissa, the bound on the relative error is 2^{-k} for rounding and 2^{1-k} for chopping.

There are situations when the round-off errors are so large that the accuracy of the numerical solution becomes questionable. An example of such a situation is the evaluation of polynomials of high degree near their zeros. Just having 24 bits for the mantissa in this case is not enough. One way of improving the accuracy of the solution is to use **double precision arithmetic**. This is done by adding another computer word to the mantissa. The result will provide the mantissa with 56 bits, that is 17 decimal digits of accuracy. A disadvantage in using double precision is that it increases the memory size required by the computer program and it consumes more computer time.

To see how round-off can drastically reduce the accuracy in some cases, let us consider the two numbers:

$$x = 1.23456 \qquad \text{and} \qquad y = 1.22222.$$

These two numbers are nearly equal and both carry six decimal digits of precision. If x and y are replaced by their first three digits, then

$$x - y = 1.23 - 1.22 = 0.01.$$

Thus, the error is about 1% in each. But if we compute the difference

$$1.23456 - 1.22222 = 0.01234$$

we can see that 0.01 differs from 0.01234 by about 20%. Similarly, when numbers are accurate to one part in 10^8, subtraction of nearly equal numbers can result in a large percent error compared with their actual difference.

We end this section with the following definition:

DEFINITION 2.2 *If x^* is an approximation to x, then we say that x^* approximates x to k **significant digits** if k is the largest nonnegative integer for which the relative error*

$$\left| \frac{x - x^*}{x} \right| < 5 \times 10^{-k}.$$

Take for example, $x = 3.1415$. For x^* to approximate x to three significant digits, it has to satisfy

$$\left| \frac{3.1415 - x^*}{3.1415} \right| < 5 \times 10^{-3}.$$

That is,

$$3.13 < x^* < 3.15.$$

EXERCISE SET 2.2

1. Perform the following operations using
   (i) four-digits rounding arithmetic
   (ii) four-digits chopping arithmetic
   (a) $\frac{1}{3} + \frac{7}{4}$ (b) $\left(\frac{2}{3} - \frac{4}{7}\right) + \left(\frac{8}{5} - \frac{13}{7}\right)$ (c) $\left(\frac{5}{3} \times \frac{2}{7}\right) + \frac{2}{3}$

2. Find the roots of the quadratic equation $x^2 + 6x + 2 = 0$, using four digits and rounding.

3. Let $x = 0.86439868$ and $y = 0.86433221$. Compute the relative error for the following operations if the operations were done using five significant digits and rounding:
   (a) $fl(x) + fl(y)$ (b) $fl(x) \times fl(y)$.

4. If x_1, x_2 are approximations to X_1, X_2 with errors ϵ_1, ϵ_2, show that the relative error
   $$\frac{X_1 X_2 - x_1 x_2}{X_1 X_2}$$
   of the product $X_1 X_2$ is approximately the sum $\frac{\epsilon_1}{X_1} + \frac{\epsilon_2}{X_2}$.

5. Given the numbers
   $$a = 0.67323, \quad b = 12.751, \quad \text{and} \quad c = 12.687,$$
   write the quotient $\frac{a}{b-c}$ in the form $\frac{a(b+c)}{b^2-c^2}$ and compute it using the numbers as given, and then using three digits and rounding.

6. The number $\pi = 3.14159265358979\ldots$. If we use the approximation $\pi = 3.14$, what is the absolute error? Express your answer using chopping to a decimal normalized floating-point representation with 5 significant digits.

7. Evaluate the following polynomial for $x = 1.07$ using three-digits chopping after each operation:
   $$p(x) = 2.75x^3 - 2.95x^2 + 3.16x - 4.67.$$
   Find the absolute and relative error of your results.
   (a) Proceeding from left to right.
   (b) Proceeding from right to left.

8. Suppose $\tilde{p}$ must approximate p with relative error at most 10^{-3}. Find the largest interval in which $\tilde{p}$ must lie for each value of p
   (a) 150   (b) 1500.

9. Use 3-digits chopping after each addition for

$$\sum_{i=1}^{N} \frac{1}{i}.$$

   (a) Find N such that N is the smallest integer for which your sum is not the exact solution.

   (b) What is the absolute error for this value of N?

   (c) What is the relative error for this value of N?

10. The number e can be calculated as

$$e = \sum_{n=0}^{\infty} \frac{1}{n!}.$$

   Use 3-digits chopping arithmetic to compute the following approximations for e. Also compute the relative error using $e = 2.7182818$ as the exact value.

   (a) $S_1 = \sum_{n=0}^{5} \frac{1}{n!}$
   (b) $S_2 = \sum_{n=0}^{5} \frac{1}{(5-n!)}$

11. Assume we are using 3-digit decimal arithmetic. For $\epsilon = 0.0001$, $a_1 = 5$, compute

$$a_2 = a_0 + \left(\frac{1}{\epsilon}\right)a_1$$

   for a_0 equal to each of 1, 2, and 3. Comment.

12. Using the floating-point representation $x = 2^e(1+g)$ where $0 \le g < 1$, rewrite x as $x = 2^{2\bar{e}}(1+\bar{g})$ where $\bar{e} = [e/2]$ (i.e., $e/2$ rounded to the nearest integer) and $\bar{g}$ is modified approximately. What range can $1 + \bar{g}$ lie in?

COMPUTER PROBLEM SET 2.2

1. Write a MATLAB function to evaluate the accuracy of Stirling's famous approximation

$$n! \approx n^n e^{-n}\sqrt{2\pi n}.$$

   The output table should have the form

n	$n!$	Stirling's approximation	Absolute error	Relative error

Judging from the table, does the accuracy increase or decease with increasing n?

2. In this assignment, you will compute the number π using an iterative method. An equilateral regular polygon, inscribed in a circle of radius 1, has the perimeter nL_n, where n is the number of sides of the polygon, and L_n is the length of one side. This can serve as an approximation for the circle perimeter 2π. Therefore, $\pi \approx nL_n/2$. From trigonometry, a polygon with twice as many sides, inscribed in the same circle, has the side length

$$L_{2n} = \sqrt{2 - \sqrt{4 - L_n^2}}. \qquad (E1)$$

(a) Write a MATLAB function to iteratively compute approximations for n using equation (E1) and starting from $n = 6$ and $L_6 = 1$ (regular hexagon). You need to do the computations using double precision floating-point numbers. Output a table of the form

n	L_n	Absolute error in approximating π

for $n = 6, 6 \times 2, 6 \times 4, \ldots, 6 \times 2^{20}$.

(b) Use the formula $b - \sqrt{b^2 - a} = \frac{a}{b + \sqrt{b^2 - a}}$ to derive a different form of equation (E1).

(c) Modify the MATLAB function using the new equation and repeat the computation to produce a new table.

2.3 TRUNCATION ERROR

A round-off error arises in terminating the decimal expansion and rounding. In contrast, the truncation error terminates a process. So, the **truncation error** generally refers to the error involved in terminating an infinite sequence or series after a finite number of terms, that is, replacing noncomputable problems with computable ones. To make the idea precise, consider the familiar infinite Taylor series (Section 1.3)

$$\cos x = 1 - \frac{x^2}{2!} + \frac{x^4}{4!} - \frac{x^6}{6!} + \frac{x^8}{8!} - \cdots$$

which can be used to approximate the cosine of an angle x in radians.

Let us find, for instance, the value of $\cos(0.5)$ using three terms of the series. We have

$$\cos(0.5) \approx 1 - \frac{1}{8} + \frac{1}{384} \approx 0.8776041667 = x^*.$$

Since $|\cos(0.5) - x^*|/|\cos(0.5)| = 2.4619 \times 10^{-5} < 5 \times 10^{-5}$, x^* approximates $\cos(0.5)$ to five significant digits.

Note that one can use the expression of the remainder R_{n+1} of the Taylor series to get a bound on the truncation error. In this case

$$R_6(x) = \frac{f^{(6)}(z)x^6}{6!}.$$

Now $f^{(6)}(z) = -\cos z$, so the remainder term is $-x^6 \cos z/6!$. Since $|-\cos z| \le 1$ on the interval $(0, 0.5)$, so we see that the remainder is bounded by $(0.5)^6(1)/6! = 0.217 \times 10^{-4}$. The actual absolute error is 0.216×10^{-4}, versus 0.217×10^{-4} remainder-formula bound.

EXAMPLE 2.5

Given that

$$I = \int_0^{1/4} e^{-x^2}\,dx = 0.244887887180,$$

approximate the integral by replacing e^{-x^2} with the truncated Maclaurin series $p_6(x) = 1 - x^2 + \frac{1}{2}x^4 - \frac{1}{6}x^6$.

We have

$$\int_0^{1/4}\left(1 - x^2 + \frac{1}{2}x^4 - \frac{1}{6}x^6\right)dx = \left[x - \frac{1}{3}x^3 + \frac{1}{10}x^5 - \frac{1}{42}x^7\right]_0^{1/4}$$

$$= \frac{1}{4} - \frac{1}{192} + \frac{1}{10240} - \frac{1}{688130}$$

$$= 0.2447999705027 = I^*.$$

Since $\frac{|I - I^*|}{|I|} = 3.590 \times 10^{-4} < 5 \times 10^{-4}$, I^* approximates I to four significant digits.

EXERCISE SET 2.3

1. Find the Maclaurin series for $f(x) = e^x$. Use three terms of the series to find an approximation to $e^{0.5}$ and the truncation error.

2. Approximate $\ln 0.5$ using three terms of the Taylor series expansion of $\ln(1-x)$ about 0.

3. Using the Taylor series expansion for $\cos x$, approximate $\cos(1.7)$ to an accuracy of 10^{-4}.

2.4   INTERVAL ARITHMETIC

Most practical problems requiring numerical computation involve quantities determined experimentally by approximate measurements, which are often estimated to a certain number of significant decimal digits. Results computed from such inexact initial data will also be of a limited precision. One way of dealing with such a problem is to replace the inexact initial data with intervals known in advance to contain the desired exact results. Our main objective is then to narrow such intervals. Interval arithmetic is also used to construct mathematically rigorous error bounds around approximate solutions. The property that interval evaluations bound the range is also useful in various contexts. In this section we will introduce and discuss an arithmetic for intervals.

To begin, we define an interval number to be the ordered pair of real numbers, $[a_1, a_2]$, with $a_1 < a_2$. The interval $[a_1, a_2]$ is also a set of real numbers x such that $a_1 \le x \le a_2$. We can therefore write

$$[a_1, a_2] = \{\, x \mid a_1 \le x \le a_2 \,\},$$

$x \in [a_1, a_2]$ means x is a real number in the interval $[a_1, a_2]$ and $[b_1, b_2] \subset [a_1, a_2]$ means that the interval $[b_1, b_2]$ is contained as a set in the interval $[a_1, a_2]$, that is, $a_1 \le b_1 \le b_2 \le a_2$.

We now introduce some further notation that will be used in this section. We denote the set of all closed real intervals by $I(\Re)$ and the elements of $I(\Re)$ by $A, B, ..., Z$. Intervals of the form $[a_1, a_1]$ are equivalent to the real number a_1.

DEFINITION 2.3   *If $A = [a_1, a_2]$, $B = [b_1, b_2] \in I(\Re)$, then*

1. $A = B \quad \Longleftrightarrow \quad a_1 = b_1$ *and* $a_2 = b_2$.

2. *The absolute value or "magnitude" of A is defined as*

$$|A| = \max\{\, |a_1|, |a_2| \,\}.$$

3. *The width of A is defined to be*

$$d(A) = a_2 - a_1 \ge 0.$$

4. *A partial ordering of the elements of $I(\Re)$ is defined by*

$$A < B \quad \Longleftrightarrow \quad a_2 < b_1.$$

We now introduce arithmetic operations on elements from $I(\Re)$.

DEFINITION 2.4   *Let $*$ be one of the symbols $\{+, -, \cdot, \div\}$. If $A, B \in I(\Re)$, we define arithmetic operations on intervals by*

$$A * B = \{\, x * y \mid x \in A, \ y \in B \,\} \tag{2.4}$$

except that we do not define A/B if $0 \in B$.

The definition of the interval arithmetic operations given by Eqn. (2.4) emphasizes the fact that the sum, difference, product, or quotient of two intervals is the set of sums, differences, products, or quotients, respectively, of pairs of real numbers, one from each of the two intervals.

In general, the new set of possible values of x that result from the operation on intervals $A = [a_1, a_2]$ and $B = [b_1, b_2]$ may be found explicitly as

$$A + B = [a_1 + b_1, a_2 + b_2],$$

$$A - B = [a_1 - b_2, a_2 - b_1], \tag{2.5}$$

$$A \cdot B = [\min\{a_1b_1, a_1b_2, a_2b_1, a_2b_2\}, \max\{a_1b_1, a_1b_2, a_2b_1, a_2b_2\}],$$

$$A \div B = [a_1, a_2] \cdot [1/b_2, 1/b_1] \text{ provided that } 0 \notin [b_1, b_2].$$

EXAMPLE 2.6

Let $A = [-1, 2]$ and $B = [1, 3]$. The operations mentioned above are now illustrated for these two intervals.

$$[-1, 2] + [1, 3] = [0, 5]$$
$$[-1, 2] - [1, 3] = [-4, 1]$$
$$[-1, 2] \cdot [1, 3] = [\min\{-1, -3, 2, 6\}, \max\{-1, -3, 2, 6\}]$$
$$= [-3, 6]$$
$$[-1, 2] \div [1, 3] = [-1, 2] \cdot [1/3, 1] = [-1, 2].$$

From the definition given by 2.4 it is clear that interval arithmetic is an extension of real arithmetic. It follows directly from Definition 2.4 that the interval addition and interval multiplication are both commutative and associative. That is, if A, B, and $C \in I(\Re)$, then it follows that

$$A + B = B + A, \quad A \cdot B = B \cdot A, \qquad \text{(commutativity)}$$

$$(A + B) + C = A + (B + C), \quad (A \cdot B) \cdot C = A \cdot (B \cdot C). \qquad \text{(associativity)}$$

The real numbers 0 and 1 are identities for interval addition and multiplication, respectively. That is, for any $A \in I(\Re)$, we have

$$0 + A = A + 0 = A$$
$$1 \cdot A = A \cdot 1 = A.$$

In general, the distributive law does not hold for interval arithmetic. In order to show that, consider the example

$$[1, 3] \cdot ([1, 4] - [1, 4]) = [1, 3] \cdot ([-3, 3]) = [-9, 9]$$

whereas

$$[1,3] \cdot ([1,4] - [1,4]) \neq [1,3] \cdot [1,4] - [1,3] \cdot [1,4]$$
$$\neq [1,12] - [1,12] = [-11,11].$$

But the *subdistributivity* holds for any interval A, B, and C. That is,

$$A \cdot (B + C) \subset A \cdot B + A \cdot C, \qquad \text{subdistributivity}.$$

Proof: $\quad A \cdot (B + C) = \{w = x\,(y + z) \,|\, x \in A,\, y \in B,\, z \in C\}$
$$\subseteq \{v = x\,y + \bar{x}\,z \,|\, x,\, \bar{x} \in A,\, y \in B,\, z \in C\}$$
$$= AB + AC.$$

We now want to turn to an application of the interval arithmetic for computing the range of real functions. We have the following results:

If $f(x_1, x_2, ..., x_n)$ is a real rational expression in which each variable x_i appears only once and to the first power, then the corresponding interval expression $F(X_1, X_2, ..., X_n)$, *called interval evaluation for f*, will compute the actual range of values of f for $x_i \in X_i$ (see [2]). Here we assume that all occurring expressions are composed of finitely many operations for which the interval arithmetic operations in F are defined.

EXAMPLE 2.7

Consider the two expressions for the function $f(x, a) = 1 - a/x$, with $x \in [2,3]$

$$f_1(x, a) = 1 - \frac{a}{x}$$
$$f_2(x, a) = \frac{x - a}{x}.$$

Using interval arithmetic, the interval evaluations for f_1 and f_2 when $a \in [0,1]$ are

$$F_1([2,3], [0,1]) = 1 - \frac{[0,1]}{[2,3]} = \left[\frac{1}{2}, 1\right],$$

$$F_2([2,3], [0,1]) = \frac{[2,3] - [0,1]}{[2,3]} = \left[\frac{1}{3}, \frac{3}{2}\right].$$

The range of all values of f when $x \in [2,3]$ and $a \in [0,1]$ is $\left[\frac{1}{2}, 1\right]$.

From the above results, it is clear that the interval evaluation of a function f is dependent on the choice of expression for f. In this example, the range obtained by the expression F_2 is different from the actual range of f because of the appearance of the variable x twice in the expression. For a continuous function of a real variable, we have the following result:

Let f be a continuous function of the real variable x. Furthermore, we assume that the interval evaluation $f(Y)$ is defined for the interval Y. It follows that

$$\text{for } X \subseteq Y,$$

and

actual range of $f = \{f(x)|\, x \in X\} \subseteq f(X)$.

EXAMPLE 2.8

Consider the function

$$f(x) = \frac{x-3}{x+2}, \quad x \neq -2.$$

___

For $x = [2, 4]$, we have

$$\{f(x)|\, x \in [2, 4]\} = [-\frac{1}{4}, \frac{1}{6}] \subset f([2, 4]) = [-\frac{1}{4}, \frac{1}{4}].$$

Another important application of interval arithmetic is the solvability of the equation

$$AX = B \tag{2.6}$$

where $X \in I\,(\Re)$ and $A \neq [0, 0]$. The solution set of the equation is

$$\{\, x = b/a|\, a \in [a_1, a_2], \ b \in [b_1, b_2]\}.$$

EXAMPLE 2.9

Consider the equation

$$[1, 3]\, X = [-1, 4].$$

___

The solution set of the equations $ax = b$ with $a \in [1, 3]$ and $b \in [-1, 4]$ is given by

$$\{\, x = b/a|\, a \in [1, 3], \ b \in [-1, 4]\} = [-1, 4]/[1, 3] = [-1, 4]$$

which is different from the unique interval solution $X = [-\frac{1}{3}, \frac{4}{3}]$ of the equation $AX = B$. Note that $[-\frac{1}{3}, \frac{4}{3}] \subset [-1, 4]$. In general, one can show that

$$X \subseteq B/A$$

as follows:

if $z \in X$, then there exists $a \in A$ and $b \in B$ such that

$$az = b \Rightarrow \ z = b/a \in B/A.$$

The starting point for the application of interval analysis was, in retrospect, the desire in numerical mathematics to be able to execute algorithms on digital computers capturing all the round-off errors automatically and therefore to calculate strict error bounds automatically. Interval arithmetic, when practical, allows rigor in scientific computations and can provide tests of correctness of hardware, and function libraries for floating-point computations. A complete study of interval arithmetic can be found in [2]. In that reference, more topics such as interval matrix operations and interval systems of equations are presented.

EXERCISE SET 2.4

1. Compute a solution set for $[2, 4]X = [-2, 1]$.

2. Using interval arithmetic, compute bounds on the range of the following real rational functions:

   (a) $f(x) = x^2 + 2x + 1, \quad 0 \le x \le 1$.
   (b) $f(x, y, z) = xy + z, \quad -1 \le x \le 1, \ -1 \le y \le 1, \ 1 \le z \le 2$.
   (c) $f(x) = x/(1 + x), \quad 0 \le x \le 1$.

3. By testing the sign of the endpoints of the intervals $A = [a, b]$ and $B = [c, d]$, break the formula for interval multiplication into nine cases only one of which requires more than two multiplications.

4. Using the results of the previous exercise carry out the following interval arithmetic:

   (a) $[1, 2] \cdot [3, 4]$,
   (b) $[-1, 2] \cdot [-4, -3]$.

5. Carry out the following interval arithmetic:

   (a) $[0, 2] + [2, 3]$,
   (b) $[2.0, 2.2] - [1.0, 1.1]$,
   (c) $[-5, -3] \cdot [-7, 4]$,
   (d) $[-2, 1] \cdot [-4.4, 2.1]$,
   (e) $[-2, 2]/[-3/2, -1]$,
   (f) $3 \cdot [-2, 4]$,
   (g) $[-1, 0]/[1, 2] + [-2, 3] \cdot [4, 5]$.

Chapter 3

Roots of Equations

In this chapter we shall discuss one of the oldest approximation problems which consists of finding the roots of an equation. It is also one of the most commonly occurring problems in applied mathematics. The **root-finding problem** consists of the following: given a continuous function f, find the values of x that satisfy the equation

$$f(x) = 0. \tag{3.1}$$

The solutions of this equation are called the **zeros** of f or the **roots** of the equation. In general, Eqn. (3.1) is impossible to solve exactly. Therefore, one must rely on some numerical methods for an approximate solution. The methods we will discuss in this chapter are iterative and consist basically of two types: one in which the convergence is guaranteed and the other in which the convergence depends on the initial guess.

EXAMPLE 3.1 : Floating Sphere

Consider a sphere of solid material floating in water. Archimedes' principle states that the buoyancy force is equal to the weight of the replaced liquid. Let $V_s = (4/3)\pi r^3$ be the volume of the sphere, and let V_w be the volume of water it displaces when it is partially submerged. In static equilibrium, the weight of the sphere is balanced by the buoyancy force

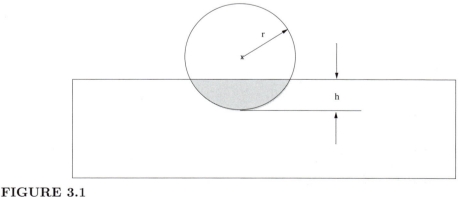

FIGURE 3.1
Floating Sphere.

$$\rho_s g V_s = \rho_w g V_w$$

where ρ_s is the density of the sphere material, g is the acceleration due to gravity, ρ_w, is the density of water. The volume V_h of water displaced when a sphere is submerged to a depth h is (see Figure 3.1)

$$V_h = \frac{\pi}{3}h^2(3r - h).$$

Applying Archimedes' principle produces the following equation in term of h

$$h^3 - 3rh^2 + 4\rho r^3 = 0. \tag{3.2}$$

Given values of r and the specific gravity of the sphere material $\rho = \frac{\rho_s}{\rho_w}$, the solutions h of the above equation can be obtained by one of the iterative methods described in this chapter.

EXAMPLE 3.2 : **Area of a Segment**

A segment of circle is the region enclosed by an arc and its chord (see Figure 3.2). If r is the radius of the circle and θ the angle subtended at the center of the circle, then it can be shown that the area A of the segment is

$$A = \frac{1}{2}r^2(\theta - \sin\theta) \tag{3.3}$$

where θ is in radian. Given A and r one can determine the value of θ by finding the zeros of $f(x) = \frac{1}{2}r^2(\theta - \sin\theta) - A$.

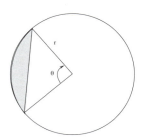

FIGURE 3.2
Area of a segment.

3.1   THE BISECTION METHOD

Let $f(x)$ be a given function, continuous on an interval $[a, b]$, such that

$$f(a)f(b) < 0. \tag{3.4}$$

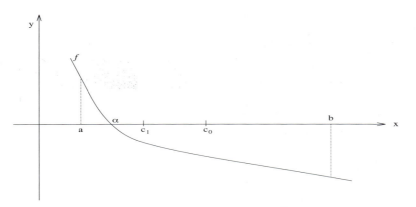

FIGURE 3.3
The bisection method and the first two approximations to its zero α.

Using the Intermediate Value Theorem, it follows that there exists at least one zero of f in (a, b). To simplify our discussion, we assume that f has exactly one root α. Such a function is shown in Figure 3.3.

The bisection method is based on halving the interval $[a, b]$ to determine a smaller and smaller interval within which α must lie. The procedure is carried out by first defining the midpoint of $[a, b]$, $c = (a + b)/2$ and then computing the product $f(c)f(b)$. If the product is negative, then the root is in the interval $[c, b]$. If the product is positive, then the root is in the interval $[a, c]$. Thus, a new interval containing α is obtained. The process of halving the new interval continues until the root is located as accurately as desired, that is

$$|a_n - b_n| < \epsilon \tag{3.5}$$

where a_n and b_n are the endpoints of the nth interval $[a_n, b_n]$ and ϵ is a specified tolerance value.

Some other stopping criteria that one can use, other than (3.5), are given by

$$\frac{|a_n - b_n|}{|a_n|} < \epsilon \tag{3.6}$$

or

$$|f(a_n)| < \epsilon. \tag{3.7}$$

An algorithm statement of this method is shown below. Suppose $f(a)f(b) \leq 0$. Let $a_0 = a$ and $b_0 = b$.

> **for** $n = 0, 1, \ldots, \text{ITMAX}$
> $\quad c \leftarrow \frac{a_n + b_n}{2}$
> $\quad$ **if** $\quad f(a_n)f(c) \leq 0$, set $a_{n+1} = a_n$, $b_{n+1} = c$
> $\quad$ **otherwise**, set $a_{n+1} = c$, $b_{n+1} = b_n$

EXAMPLE 3.3

The function $f(x) = x^3 - x^2 - 1$ has exactly one zero in $[1, 2]$. Use the bisection algorithm to approximate the zero of f to within 10^{-4}.

Since $f(1) = -1 < 0$ and $f(2) = 3 > 0$, then (3.4) is satisfied. Starting with $a_0 = 1$ and $b_0 = 2$, we compute

$$c_0 = \frac{a_0 + b_0}{2} = \frac{1 + 2}{2} = 1.5 \quad \text{and} \quad f(c_0) = 0.125.$$

Since $f(1.5)f(2) > 0$, the function changes sign on $[a_0, c_0] = [1, 1.5]$. To continue, we set $a_1 = a_0$ and $b_1 = c_0$; so

$$c_1 = \frac{a_1 + b_1}{2} = \frac{1 + 1.5}{2} = 1.25 \quad \text{and} \quad f(c_1) = -0.609375.$$

Again $f(1.25)f(1.5) < 0$ so the function changes sign on $[c_1, b_1] = [1.25, 1.5]$. Next we set $a_2 = c_1$ and $b_2 = b_1$. Continuing in this manner leads to the values in Table 3.1, which converge to $r = 1.465454$.

Note that, before calling the MATLAB function bisect, we must define the MATLAB function f1 as follows:

```
function f=f1(x)
f=x.^3-x.^2-1;
```

Now call the the function bisect to get the results shown in Table 3.1.

» bisect('f1',1,2,10^(-4),40)					
iter	a	b	c	f(c)	|b-a|/2
0	1.0000	2.0000	1.500000	0.125000	0.500000
1	1.0000	1.5000	1.250000	-0.609375	0.250000
2	1.2500	1.5000	1.375000	-0.291016	0.125000
3	1.3750	1.5000	1.437500	-0.095947	0.062500
4	1.4375	1.5000	1.468750	0.011200	0.031250
5	1.4375	1.4688	1.453125	-0.043194	0.015625
6	1.4531	1.4688	1.460938	-0.016203	0.007813
7	1.4609	1.4688	1.464844	-0.002554	0.003906
8	1.4648	1.4688	1.466797	0.004310	0.001953
9	1.4648	1.4668	1.465820	0.000875	0.000977
10	1.4648	1.4658	1.465332	-0.000840	0.000488
11	1.4653	1.4658	1.465576	0.000017	0.000244
12	1.4653	1.4656	1.465454	-0.000411	0.000122

Table 3.1   Solution of $x^3 - x^2 - 1 = 0$ in $[1, 2]$ using the bisection method.

Some of the features of MATLAB can be used to view, for example, step-by-step calculations involved in the function bisect. This can be done by using the pause function of the form pause(s). If we add the line

pause(1);

just before the end statement of the while loop, execution is held up for approximately 1 second at the end of each output line. MATLAB also provides a wide range of graphic facilities that may be used to graphically visualize the results. For example if the following commands

fplot('f1',[a0,b0])
grid

are added at the end of the function bisect.m, then a plot of the function $f(x) = x^3 - x^2 - 1$ in $[a, b]$, shown in Figure 3.4, will be given with the output.

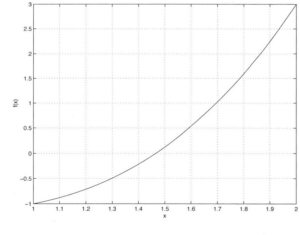

FIGURE 3.4
A plot of $f(x) = x^3 - x^2 - 1$ in $[1, 2]$.

Although the bisection method always converges and is easy to program, compared to other methods, its rate of convergence is rather slow. We state the following convergence result.

THEOREM 3.1

Let f be a continuous function on an interval $[a, b]$ where $f(a)f(b) < 0$. At the nth step, the bisection procedure approximates a root with an error of at most $(b - a)/2^{n+1}$.

Proof: Let $[a_0, b_0]$ be the original interval and α a zero of f in this interval. Define

$$c_0 = \frac{b_0 + a_0}{2} \qquad \text{and} \qquad \alpha \in [a_0, b_0]$$

then

$$|\alpha - c_0| < (b_1 - a_1) = \frac{b_0 - a_0}{2}$$

with a_1 and b_1 being the endpoints of the new subinterval containing α. If c_n denotes the nth value of c, then we have

$$|\alpha - c_n| < (b_{n+1} - a_{n+1}) = \frac{b_0 - a_0}{2^{n+1}}, \quad n = 0, 1, \ldots.$$

EXAMPLE 3.4

Determine the number of iterations required to find the zero of $f(x)$, defined in Example 3.3, with an absolute error of no more than 10^{-6}.

We want to find n such that

$$|\alpha - c_n| \le \frac{b - a}{2^{n+1}} = \frac{2 - 1}{2^{n+1}} < 10^{-6}.$$

This implies that

$$-(n+1)\log(2) < -6 \quad \text{or} \quad n > \frac{6}{\log 2} - 1 \approx 19.$$

So one needs at most 19 iterations to obtain an approximation accurate to 10^{-6}.

EXERCISE SET 3.1

1. Verify that the function $f(x) = x^2 \sin x + 2x - 3$ has exactly one root in $(0, 2)$. Find this root by using the bisection method with an error of no more than 10^{-5}. Count the number of iterations needed to find the root and compare your answer with the result obtained by using the error formula given by Theorem 1 of Section 3.1.

2. Find a root of $f(x) = x^3 + 2x - 3$ in the range $0 \le x \le 7/5$ using the bisection method.

3. The function $f(x) = x^2 - 2.6x - 2.31$ has one root in the interval $[3, 4]$. How many steps of the bisection method are needed to locate the root with an error of at most 10^{-5}.

4. The function $f(x) = x^5 - 3x^3 + 2.5x - 0.6$ has exactly one root in $(1, 2)$. Demonstrate it graphically and determine the root with an error of at most 10^{-5}?

M-function 3.1

The following MATLAB function **bisect.m** finds the solution of an equation using the Bisection method. INPUTS are a function *f(x)*; the endpoints *a* and *b*; a tolerance *tol*; the maximum number of iterations *n*. The input function *f(x)* should be defined as an M-file.

```
function bisect(f,a,b,tol,n)
% Bisection method for solving the nonlinear equation f(x)=0.
a0=a; b0=b;
iter=0;
u=feval(f,a);
v=feval(f,b);
c=(a+b)*0.5;
err=abs(b-a)*0.5;
disp('_____')
disp(' iter        a          b         c        f(c)      lb-al/2  ')
disp('_____')
fprintf('\n')
if (u*v<=0)
   while (err>tol)&(iter<=n)
     w=feval(f,c);
     fprintf('%2.0f %10.4f %10.4f %12.6f %10.6f %10.6f\n', iter,
a, b, c, w, err)
     if (w*u<0)
        b=c;v=w;
     end;
     if (w*u>0)
        a=c;u=w;
     end;
     iter=iter+1;
     c=(a+b)*0.5;
     err=abs(b-a)*0.5;
   end;
   if (iter>n)
     disp(' Method failed to converge')
   end;
else
   disp('  The method cannot be applied f(a)f(b)>0')
end;
% Plot f(x) in the interval [a,b].
fplot(f, [a0 b0])
xlabel('x'); ylabel('f(x)'); grid
```

5. Find a root of the equation $e^{2x} - 7x = 0$.

6. Consider the function

$$f(x) = x^4 - 5x^3 + \frac{22}{3}x^2 - \frac{116}{27}x + \frac{8}{9}.$$

   (a) Check that $f(x)$ has a root α_1 between 0 and 1 and another root α_2 between 1 and 4.

   (b) Compute both roots using the bisection method.

7. Sketch the graph of $f(x) = \tan x + \tanh x = 0$ and find an interval in which its smallest positive root lies. Determine the root correct to two decimal digits using the bisection method.

8. The function $f(x) = x^4 - 8.6x^3 - 35.51x^2 + 464.4x - 998.46$ has a simple root in the interval $[6, 8]$ and a double root in the interval $[4, 5]$. Use the bisection method to find both roots.

9. Find to four decimal places a root of $0.1x^2 - x \ln x = 0$ between 1 and 2.

10. Use the bisection method to find the root of the equation $x + \cos x = 0$ correct to two decimal places.

11. If the bisection method is applied on the interval from $a = 14$ to $b = 16$, how many iterations will be required to guarantee that a root is located to the maximum accuracy of the IEEE double-precision standard?

12. Let n be the number of iterations of the bisection method needed to ensure that a root α is bracketed in an interval $[a, b]$ of length less than ϵ. Express n in terms of ϵ, a, and b.

13. Consider the function
$$f(x) = (x - 2)^2 - \ln x$$
on the interval $[1, 2]$.

   (a) Prove that there is exactly one root of this equation in this interval.

   (b) Use the bisection method to approximate a root to 6 digits accuracy.

   (c) How many iterates of the Bisection method are needed to find an approximation to the root of $f(x) = 0$ in the interval to within an accuracy of 10^{-4}?

14. Find the points where the hyperbola $y = 1/x$, $x > 0$ intersect the curve $y = \cot x$.

COMPUTER PROBLEM SET 3.1

1. Write a computer program in a programming language of your choice that finds the zeros of a function $f(x)$ on an interval $[a, b]$ using the bisection method. Assume that there is only one simple root in this interval. Input data to the program should be a function $f(x)$, a, b, and the error tolerance ϵ. Use your program to find the zeros of the function

$$f(x) = 0.25x^4 - 1.5x^3 - x^2 + 3.1x - 3$$

in the interval $[-2.5, 0]$. What does your program do if the interval is reduced to $[-1, 0]$?

2. The roots of the polynomial

$$p(x) = (x - 1)(x - 2) \ldots (x - 20) - 10^{-8} x^{19}$$

are highly sensitive to small alterations in their coefficients.

   (a) Prove that p has a root in $[20, 22]$, and then use the MATLAB function bisect.m to find this root.

   (b) Find the number of iteration needed in a) to get the root to within an error $< 10^{-3}$; find also the number of iteration for accuracy $< 10^{-12}$.

3. The equation

$$x + 4\cos x = 0$$

has three solutions. Locate all of them, using the MATLAB function bisect.m, selecting appropriate initial interval for each.

4. Use the MATLAB function bisect.m to solve the following two problems.

   (a) To test the MATLAB function, start by computing something familiar. How about π? Solve the equation

$$\sin x = 0$$

   using the initial interval $a = 2$ and $b = 4$. Determine the number of iterations required to compute the number π with six significant decimal digits.

   (b) The equation

$$x + e^x = 0$$

   has the root around $x \approx -0.6$. Find a better approximation using the initial interval $[-1, 0]$.

5. Use the MATLAB function bisect.m to find the first eight roots for $f(x) = -e^{-x} + \sin x$ to three decimal places. Observe there are infinitely many roots. If x_n is the nth root, calculate the ratios $\frac{x_n}{n\pi}$, $n = 1, 2, \ldots$. Based on the evidence of these calculations, guess the value of

$$\lim_{n \to \infty} \frac{x_n}{n\pi}.$$

3.2   THE METHOD OF FALSE POSITION

The method of false position, also known as **regula falsi**, is similar to the Bisection method but has the advantage of being slightly faster than the latter. As in the bisection method, the function is assumed to be continuous over the interval $[x_1, x_2]$ with

$$f(x_1)f(x_2) < 0.$$

In this procedure, the first point x_3 is selected as point of intersection of the x-axis, and the straight line joining the points $(x_1, f(x_1))$ and $(x_2, f(x_2))$ as shown in Figure 3.5. From the equation of the secant line, it follows that

$$x_3 = x_2 - f(x_2)\frac{x_2 - x_1}{f(x_2) - f(x_1)}. \tag{3.8}$$

With x_3 now known, we then compute $f(x_3)$ and repeat the procedure between the

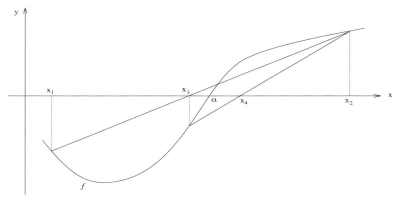

FIGURE 3.5
The method of false position and the first two approximations to its zero α**.**

values at which the function changes sign, that is, if $f(x_1)f(x_3) < 0$ set $x_2 = x_3$, otherwise set $x_1 = x_3$. At each step the method of false position produces a new interval that contains a root of f and the sequence of points generated by the method will eventually converge to the root. An algorithm of the method of *false position* is given below.

Given a continuous function $f(x)$ on $[a_0, b_0]$ with $f(a_0)f(b_0) < 0$,

> **for** $n = 0, 1, \ldots, \text{ITMAX}$
> $$c \leftarrow \frac{f(b_n)a_n - f(a_n)b_n}{f(b_n) - f(a_n)}$$
>
> **if** $f(a_n)f(c) < 0$, set $a_{n+1} = a_n$, $b_{n+1} = c$
> **otherwise**, set $a_{n+1} = c$, $b_{n+1} = b_n$.

Because in some cases the method of false position fails to give a small interval in which the zero is known to lie, we terminate the iterations if

$$|f(x_n)| \leq \epsilon$$

or

$$|x_n - x_{n-1}| \leq \epsilon$$

where ϵ is a specified tolerance value.

One of the main disadvantages of this method is that if the sequence of points generated by its algorithm is one-sided, it will tremendously slow the convergence of the method. This applies to a function whose graph is concave up or concave down on $[a_0, b_0]$ (see Figure 3.6).

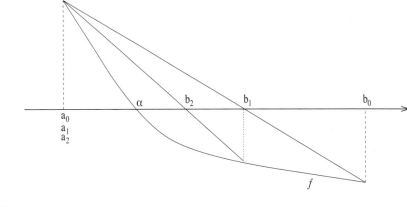

FIGURE 3.6
The method of false position on $[a_0, b_0]$.

EXAMPLE 3.5
The function $f(x) = x^3 - x^2 - 1$ has exactly one zero in $[1, 2]$. Use the method of false position to approximate the zero of f to within 10^{-4}.

A root lies in the interval $[1, 2]$ since $f(1) = -1$ and $f(2) = 3$. Starting with $a_0 = 1$ and $b_0 = 2$, we get using Eqn. (3.8)

$$c_0 = 2 - \frac{3(2 - 1)}{3 - (-1)} = 1.25 \quad \text{and} \quad f(c_0) = -0.609375.$$

Here, $f(c_0)$ has the same sign as $f(a_0)$ and so the root must lie on the interval $[c_0, b_0] = [1.25, 2]$. Next, we set $a_1 = c_0$ and $b_1 = b_0$ to get the next approximation

$$c_1 = 2 - \frac{3 - (2 - 1.25)}{3 - (-0.609375)} = 1.37662337 \quad \text{and} \quad f(c_1) = -0.2862640.$$

Now $f(x)$ changes sign on $[c_1, b_1] = [1.37662337, 2]$. Thus, we set $a_2 = c_1$ and $b_2 = b_1$. Continuing in this manner leads to the values in Table 3.2 that converge to $r = 1.465558$.

» falsep('f1',1,2,10^(-4),40)					
iter	a	b	c	f(c)	|b-a|
0	1	2	1.250000	-0.609375	1.000000
1	1.25	2	1.376623	-0.286264	0.750000
2	1.3766	2	1.430925	-0.117660	0.623377
3	1.4309	2	1.452402	-0.045671	0.569075
4	1.4524	2	1.460613	-0.017331	0.547598
5	1.4606	2	1.463712	-0.006520	0.539387
6	1.4637	2	1.464875	-0.002445	0.536288
7	1.4649	2	1.465310	-0.000916	0.535125
8	1.4653	2	1.465474	-0.000343	0.53469
9	1.4655	2	1.465535	-0.000128	0.534526
10	1.4655	2	1.465558	-0.000048	0.534465

Table 3.2   Solution of $x^3 - x^2 - 1 = 0$ in $[1, 2]$ using the method of false position.

The false position method, which sometimes keeps an older reference point to maintain an opposite sign bracket around the root, has a lower and uncertain convergence rate compared to the secant method (described in Section 3.4). The emphasis on bracketing the root may sometimes restrict the false position method in difficult situations while solving highly nonlinear equations.

EXERCISE SET 3.2

1. Verify that the function $f(x) = x^2 \sin x + 2x - 3$ has exactly one root in $(0, 2)$. Find this root using the method of false position with an error of no more than 10^{-3}.

2. Find a root of $f(x) = x^3 + 2x - 3$ in the range $0 \le x \le 7/5$ using the method of false position.

3. The function $f(x) = x^5 - 3x^3 + 2.5x - 0.6$ has exactly one root in $(1, 2)$. Demonstrate it graphically and determine the root with an error of at most 10^{-4} using the method of false position.

M-function 3.2

The following MATLAB function **falsep.m** finds the solution of an equation using the method of false position. INPUTS are a function *f*; the endpoints *a* and *b*; a tolerance *tol*; the maximum number of iteration *n*. The input function *f(x)* should be defined as an M-file.

```
function falsep(f,a,b,tol,n)
% False position method for solving the nonlinear equation f(x)=0.
a0=a; b0=b;
iter=0;
u=feval(f,a);
v=feval(f,b);
c=(v*a-u*b)/(v-u);
w=feval(f,c);
disp('_____')
disp(' iter       a        b        c          f(c)        lb-al   ')
disp('_____')
fprintf('\n')
if (u*v<=0)
   while (abs(w)>tol)&(abs(b-a)>tol)&(iter<=n)&((v-u)~=0)
      w=feval(f,c);
      fprintf('%2.0f %12.4f %12.4f  %12.6f  %10.6f %10.6f\n', iter, a,
b, c, w, abs(b-a))
      if (w*u<0)
         b=c;v=w;
      end;
      if (w*u>0)
         a=c;u=w;
      end;
      iter=iter+1;
      c=(v*a-u*b)/(v-u);
   end;
   if (iter>n)
      disp(' Method failed to converge')
   end;
   if (v-u==0)
      disp(' Division by zero')
   end;
else
   disp(' The method cannot be applied f(a)f(b)>0')
end;
fplot(f,[a0 b0])
xlabel('x');ylabel('f(x)'); grid
```

4. Show that formula (3.6) for the false position method is algebraically equivalent to

$$c_n = \frac{f(b_n)a_n - f(a_n)b_n}{f(b_n) - f(a_n)}.$$

5. Find a root of the equation $e^{2x} - 7x = 0$.

6. Find the interval in which the smallest positive root of the following equations lies:

   (a) $\tan x + \tanh x = 0$,
   (b) $x^3 - x - 4 = 0$.

   Determine the roots correct to three decimals using the false position method.

7. Find the real roots of $f(x) = x^3 - 2.56x^2 - 34.6x + 112.5$ using the method of false position.

8. The function $f(x) = x^4 - 8.6x^3 - 35.51x^2 + 464.4x - 998.46$ has a simple root in the interval $[6, 8]$ and a double root in the interval $[4, 5]$. Use the false position method to find both roots.

9. Find the points where the hyperbola $y = 1/x$, $x > 0$ intersect the curve $y = \cot x$.

10. Show that $f(x) = x^2 + 3x - 4$ has only one root and that it lies between 0 and 1. Find the root to five decimal places using the bisection and false position methods.

11. The function $f(x) = \cos x - 3x^3 + 2x - 5$ has one zero in the interval $[-2, -1]$. Show it using the intermediate value theorem. Find the root to four decimal places using the false position method.

COMPUTER PROBLEM SET 3.2

1. Write a computer program in a language of your choice that finds the zeros of a function $f(x)$ on an interval $[a, b]$ using the method of false position. Input data to the program should be a function $f(x)$, a, b, the error tolerance ϵ, and the maximum number of iterates.
   Use your program to find the zeros of the function

$$f(x) = 0.1x^8 - 1.9x^6 + 0.2x^5 - 1.2x^3 - 2.5x^2 - 19.6$$

   in the interval $[2.5, 5]$.

2. The roots of the polynomial

$$p(x) = (x-1)(x-2)\ldots(x-20) - 10^{-8}x^{19}$$

   are highly sensitive to small alterations in their coefficients.

   (a) Prove that p has a root in $[20, 22]$, and then use the MATLAB function falsep.m to find this root.

   (b) Find the number of iteration needed in (a) to get the root to within an error $< 10^{-3}$; find also the number of iteration for accuracy $< 10^{-12}$.

3. Use the MATLAB function falsep.m to solve the following two problems.

   (a) To test the MATLAB function, start by computing something familiar. How about π? Solve the equation

   $$\sin x = 0$$

   using the initial interval $a = 2$ and $b = 4$.

   (b) The equation
   $$x + e^x = 0$$

   has the root around $x \approx -0.6$. Find a better approximation using the initial interval $[-1, 0]$.

4. Use the MATLAB function falsep.m to find the first eight roots for $f(x) = -e^{-x} + \sin x$ to four decimal places. Observe there are infinitely many roots. If x_n is the nth root, calculate the ratios $\frac{x_n}{n\pi}$, $n = 1, 2, \ldots$. Based on the evidence of these calculations, guess the value of

$$\lim_{n\to\infty} \frac{x_n}{n\pi}.$$

3.3 FIXED POINT ITERATION

In this section we consider an important iterative method for the solution of equations of the form
$$g(x) = x. \tag{3.9}$$

A solution to this equation is called a **fixed point** of g. The fixed point iteration, which is used to determine the fixed point of g, starts from an initial guess p_0 to determine a sequence of values $\{p_n\}$ obtained from the rule

$$p_{n+1} = g(p_n), \qquad n = 0, 1, 2, \ldots. \tag{3.10}$$

If the process converges to a root α and g is continuous, then

$$\lim_{n \to \infty} p_n = \alpha$$

and a solution to Eqn. (3.9) is obtained.

Geometrically, a fixed point p of g can be regarded as the point of intersection of the functions $y = x$ and $y = g(x)$. Figures 3.7 and 3.8 give graphical interpretations of convergence and divergence iterations, respectively.

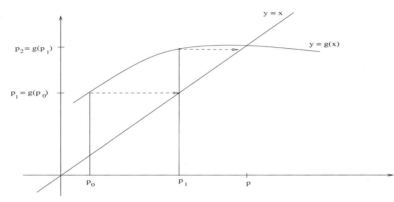

FIGURE 3.7
The fixed point iteration with convergence iteration to its zero p.

EXAMPLE 3.6
Solve the equation $0.4 - 0.1x^2 = 0$ with $x_0 = 1$ using the fixed point iteration and choosing

(F1) $g(x) = 0.4 + x - 0.1x^2$,

(F2) $g(x) = \frac{4}{x}$.

Table 3.3 gives the iterates x_n of the fixed point iteration method for both F1 and F2. The exact solution is 2.0. From the results given in Table 3.3, it is interesting to note that the choice of $g(x)$ in F1 led to convergence and the one in F2 to divergence.

The question that one can ask is how can we choose $g(x)$ so that convergence will occur? The following theorem gives an answer to that question.

THEOREM 3.2 (Fixed Point Theorem)
If g is continuous on $[a, b]$ and $a \le g(x) \le b$ for all x in $[a, b]$, then g has at least a fixed point in $[a, b]$.

Further, suppose $g'(x)$ is continuous on (a, b) and that a positive constant c exists with

$$|g'(x)| \le c < 1, \qquad \text{for all } x \text{ in } (a, b). \tag{3.11}$$

Then there is a unique fixed point α of g in $[a, b]$. Also, the iterates

$$x_{n+1} = g(x_n) \qquad n \ge 0$$

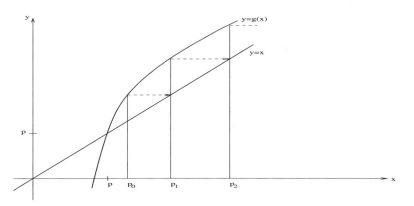

FIGURE 3.8
The fixed point iteration with divergence iteration from its zero p.

will converge to α for any choice of x_0 in $[a, b]$.

Proof: Define the function $f(x) = x - g(x)$. Then f is continuous on $[a, b]$, and $f(a) \leq 0$ and $f(b) \geq 0$. By the Intermediate Value Theorem there exists α in (a, b) for which $f(x) = 0$. Thus, α is a fixed point of g.

In addition, by the Mean Value Theorem, we have that for any two fixed points r and s in $[a, b]$ there exists ξ such that

$$|r - s| = |g(r) - g(s)| = |g'(\xi)||r - s| \leq c|r - s|.$$

Thus,

$$|r - s| \leq c|r - s| \quad \text{or} \quad (1 - c)|r - s| \leq 0.$$

Since $c < 1$, we must have $r = s$. Hence, g has a unique fixed point in $[a, b]$. For the convergence,

$$|\alpha - x_{n+1}| = |g(\alpha) - g(x_n)| = |g'(\xi_n)||\alpha - x_n| \leq c|\alpha - x_n|$$

with $\xi_n \in (\alpha, x_n)$. By induction

$$|\alpha - x_n| \leq c^n|\alpha - x_0| \qquad n \geq 0.$$

As $n \to \infty$, $c^n \to 0$; therefore, $x_n \to \alpha$.

EXAMPLE 3.7
Suppose we want to solve the equation $x = \pi + 0.5 \sin x$ using the fixed point iteration. If $g(x) = \pi + 0.5 \sin x$, let us find the interval $[a, b]$ satisfying the hypotheses of Theorem 3.2.

It is easy to see that g is continuous and

$$0 \leq g(x) \leq 2\pi \qquad \text{for all } x \text{ in } [0, 2\pi].$$

n	F1	F2
0	1.00000000	1
1	1.30000000	4
2	1.53100000	1
3	1.69660390	4
4	1.80875742	1
5	1.88159708	4
6	1.92755632	1
7	1.95600899	4
8	1.97341187	1
9	1.98397643	4
10	1.99036018	1
11	1.99420682	4
12	1.99652073	1
13	1.99791123	4
14	1.99874630	1
15	1.99924762	4
16	1.99954852	1
17	1.99972909	4
18	1.99983745	1

Table 3.3   The fixed point iteration for (F1) and (F2).

Moreover, g' is continuous and

$$|g'(x)| = |0.5 \cos x| < 1 \quad \text{in } [0, 2\pi];$$

thus, g satisfies the hypotheses of Theorem 3.2 and has a unique fixed point in the interval $[0, 2\pi]$.

EXAMPLE 3.8
The equation $x = \frac{5}{x^2} + 2$ has exactly one zero in $[2.5, 3]$. Use the fixed point method to find the root with $g(x) = 5/x^2 + 2$ and $x_0 = 2.5$ to within 10^{-4}.

Both $g(x)$ and $g'(x) = -10/x^3$ are continuous in $[2.5, 3]$. It is easy to check that $|g'(x)| \leq 1$ and $2.5 \leq g(x) \leq 3$ for all x in $[2.5, 3]$. Thus, g has a unique fixed point in $[2.5, 3]$.

The first three iterations are

$$p_1 = 5/p_0^2 - 2 = 2.8$$
$$p_2 = 5/p_1^2 - 2 = 2.637755$$
$$p_3 = 5/P_2^2 - 2 = 2.718623.$$

Continuing in this manner leads to the values in Table 3.4, which converge to $r = 2.69062937$. In Table 3.4 the MATLAB function g1 is defined as follows:

```
» fixed('g1',2.5,10^(-4),100)
```

iter	xn	g(xn)	| xn+1-xn |
0	2.500000	2.800000	
1	2.800000	2.637755	0.30000000
2	2.637755	2.718623	0.16224490
3	2.718623	2.676507	0.08086781
4	2.676507	2.697965	0.04211628
5	2.697965	2.686906	0.02145790
6	2.686906	2.692572	0.01105819
7	2.692572	2.689660	0.00566568
8	2.689660	2.691154	0.00291154
9	2.691154	2.690387	0.00149391
10	2.690387	2.690781	0.00076713
11	2.690781	2.690579	0.00039377
12	2.690579	2.690683	0.00020216
13	2.690683	2.690629	0.00010378
14	2.690629	2.690657	0.00005328

Table 3.4 Solution of $x = \frac{5}{x^2} + 2$ using the fixed point iteration method with $x_0 = 2.5$.

```
function f=g1(x)
    f=5/x.^2+2;
```

EXERCISE SET 3.3

1. Use the fixed point iteration to find an approximation to $\sqrt{2}$ with 10^{-3} accuracy.

2. The quadratic equation $x^2 - 2x - 3 = 0$ has two roots. Consider the following rearrangements to approximate the roots using the fixed point iteration:

   (a) $x = \sqrt{2x + 3}$,
   (b) $x = 3/(x - 2)$,
   (c) $x = (x^2 - 3)/2$,

   starting from $x_0 = 4$.

3. Solve the following equations using the fixed point iteration method:

   (a) $x = \sin x + x + 1$, in $[3.5, 5]$,

M-function 3.3

The following MATLAB function **fixed.m** finds the solution of an equation using the fixed-point iteration method. INPUTS are a function g; an initial approximation x_0; a tolerance *tol*; the maximum number of iterations n. The input function $g(x)$ should be defined as an M-file.

```
function fixed(g,x0,tol,n)
% The fixed-point iteration method for solving the nonlinear
%  equation f(x)=0.
iter=0;
u=feval(g,x0);
err=abs(u-x0);
disp('_____')
disp('  iter              x              g(x)          |xn+1-xnl ')
disp('_____')
fprintf('%2.0f %12.6f %12.6f\n',iter,x0,u)
while (err>tol)&(iter<=n)
    x1=u;
    err=abs(x1-x0);
    x0=x1;
    u=feval(g,x0);
    iter=iter+1;
    fprintf('%2.0f %12.6f %12.6f %12.8f\n',iter,x0,u,err)
end
if (iter>n)
    disp(' Method failed to converge')
end
```

(b) $x = \sqrt{x^2 + 1} - x + 1$, in $[0, 3]$,

(c) $x = \ln x^2 + x - 2$, in $[-4, -2]$.

4. For each of the following functions, locate an interval on which fixed point iteration will converge.

(a) $x = 0.2 \sin x + 1$,

(b) $x = 1 - x^2/4$.

5. Find the solution of the following equations using the fixed point iteration:

(a) $x = x^5 - 0.25$ starting from $x_0 = 0$,

(b) $x = 2 \sin x$ starting from $x_0 = 2$,

(c) $x = \sqrt{3x + 1}$ starting from $x_0 = 2$,

(d) $x = \frac{2 - e^x + x^2}{3}$ starting from $x_0 = 1$.

6. Let $g(x) = \frac{x+4}{2}$.

(a) Show that $\alpha = 4$ is a fixed point of $g(x)$,

(b) Let $x_0 = 5$, show that $|\alpha - x_n| = |\alpha - x_0|/2^n$ for $n = 1, 2, \ldots$.

7. Let x_0 be a real number. Define a sequence $\{x_n\}_{n=1}^{\infty}$ by

$$x_{n+1} = \sin x_n, \quad n = 0, 1, 2, \ldots.$$

Using the fact that $|\sin x| < |x|$ for every $x \neq 0$, show that the sequence x_n converges to a unique fixed point $x = 0$ of the function $g(x) = \sin x$.

8. Use the bisection method to the zero of the function

$$f(x) = t^3 + t - 9$$

correct to one decimal place.
Give the definition of the fixed point of the iteration

$$s_{n+1} = s_n + \lambda(s_n^3 + s_n - 9);$$

find an approximate range of values of the constant λ such that the iteration converges to the fixed point near 2.

9. Explain the different convergence behavior of the fixed point iterations given by

   (a) $x_{n+1} = \frac{1}{2} \tan x_n$,
   (b) $x_{n+1} = \arctan(2x_n)$

   for the solution of the equation $2x - \tan x = 0$.

10. Consider the fixed point iteration scheme: $x_{n+1} = 1 + e^{-x_n}$. Show that this scheme converges for any $x_0 \in [1, 2]$. How many iterations does the theory predict it will take to achieve 10^{-6} accuracy?

11. Show that $x = 4$ is a fixed point of the following iteration schemes:

   (a) $x_{n+1} = \frac{1}{4}(8x_n - x_n^2)$,
   (b) $x_{n+1} = \frac{1}{3}(x_n^2 - 4)$,
   (c) $x_{n+1} = \sqrt{3x_n + 4}$.

   Compute a few iterations for each scheme choosing an x_0 as a starting value. Determine which of these methods should converge to the root $x = 4$ given a starting guess close enough. Which method should converge fastest?

12. Consider the fixed point problem

$$x = \frac{1}{2} \tan \pi \frac{x}{2}$$

on the open interval $0 < x < 1$. Explain both graphically and analytically why fixed point iteration $x_{n+1} = (1/2) \tan(\pi x_n/2)$, $n = 0, 1, 2, \ldots$ does not converge to the unique fixed point $s = 1/2$ in the interval $(0, 1)$ for any starting point $x_0 \in (0, 1)$, $x_0 \neq 1/2$.

13. Show that $f(x) = x^2 + 3x - 4$ has only one root and that it lies between 0 and 1. Find the root to five decimal places using the fixed point iteration method. Use

$$x = \frac{4}{x^2 + 3} = g(x).$$

14. Consider the fixed point iteration $p_{n+1} = g(p_n)$ when the function $g(x) = 2(x - 1)^{1/2}$ for $x \geq 0$ is used. Plot the function and the line $y = x$ on the same plot and determine how many fixed points exist. Iterate with $p_0 = 1.5$ and with $p_0 = 2.5$. Explain the result in the light of the fixed point theorem.

COMPUTER PROBLEM SET 3.3

1. Use the MATLAB function fixed.m to examine the convergence of fixed point algorithms. Given the equation

$$x^3 = \arctan x$$

find the root of this equation that is close to 1. To do that convert the equation into the form $x = g(x)$ using four different forms of $g(x)$. Make your four choices for g so that at least once you have convergence and at least once you do not have convergence. Could you tell in advance, by studying g first, if convergence is expected?

3.4   THE SECANT METHOD

Because the bisection and the false position methods converge at a very slow speed, our next approach is an attempt to produce a method that is faster. One such method is the **secant method**. Similar to the false position method, it is based on approximating the function by a straight line connecting two points on the graph of the function f, but we do not require f to have opposite signs at the initial points. Figure 3.9 illustrates the method.

In this method, the first point, x_2, of the iteration is taken to be the point of intersection of the x-axis and the secant line connecting two starting points $(x_0, f(x_0))$ and $(x_1, f(x_1))$. The next point, x_3, is generated by the intersection of the new secant line, joining $(x_1, f(x_1))$ and $(x_2, f(x_2))$ with the x-axis. The new point, x_3, together with x_2, is used to generate the next point, x_4, and so on.

A formula for x_{n+1} is obtained by setting $x = x_{n+1}$ and $y = 0$ in the equation of the secant line from $(x_{n-1}, f(x_{n-1}))$ to $(x_n, f(x_n))$

$$x_{n+1} = x_n - f(x_n) \left[\frac{x_n - x_{n-1}}{f(x_n) - f(x_{n-1})} \right]. \tag{3.12}$$

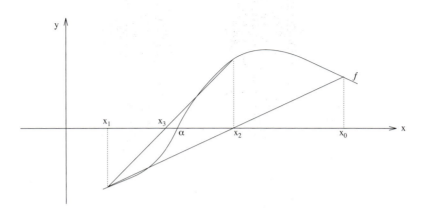

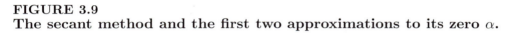

FIGURE 3.9
The secant method and the first two approximations to its zero α.

Note that x_{n+1} depends on the two previous elements of the sequence and therefore two initial guesses, x_0 and x_1, must be provided to generate $x_2, x_3, \ldots$.

An algorithmic statement of this method is shown below.

Let x_0 and x_1 be two initial approximations,

$$\textbf{for } n = 1, 2, \ldots, \text{ITMAX}$$
$$\lfloor \quad x_{n+1} \leftarrow x_n - f(x_n)\left[\frac{x_n - x_{n-1}}{f(x_n) - f(x_{n-1})}\right].$$

A suitable stopping criterion is

$$|f(x_n)| \leq \epsilon, \quad |x_{n+1} - x_n| \leq \epsilon \quad \text{or} \quad \frac{|x_{n+1} - x_n|}{|x_{n+1}|} \leq \epsilon$$

where ϵ is a specified tolerance value.

EXAMPLE 3.9
Use the secant method with $x_0 = 1$ and $x_1 = 2$ to solve $x^3 - x^2 - 1 = 0$ to within 10^{-4}.

With $x_0 = 1$, $f(x_0) = -1$ and $x_1 = 2$, $f(x_1) = 3$, we have

$$x_2 = 2 - \frac{(2 - 1)(3)}{3 - (-1)} = 1.25$$

from which $f(x_2) = f(1.25) = -0.609375$. The next iterate is

$$x_3 = 1.25 - \frac{(1.25 - 2)(-0.609375)}{-0.609375 - 3} = 1.3766234.$$

Continuing in this manner leads to the values in Table 3.5, which converge to $r = 1.4655713$.

```
» secant('f1',1,2,10^(-4),40)
```

n	xn	f(xn)	f(xn+1)-f(xn)	l xn+1-xn l
0	1.000000	-1.000000		
1	2.000000	3.000000	4.000000	1.000000
2	1.250000	-0.609375	-3.609375	0.750000
3	1.376623	-0.286264	0.323111	0.126623
4	1.488807	0.083463	0.369727	0.112184
5	1.463482	-0.007322	-0.090786	0.025325
6	1.465525	-0.000163	0.007160	0.002043
7	1.465571	3.20E-07	0.000163	0.000046

Table 3.5 Solution of $x^3 - x^2 - 1 = 0$ using the secant method with $x_0 = 1$, $x_1 = 2$.

EXERCISE SET 3.4

1. Approximate to within 10^{-6} the root of the equation $e^{-2x} - 7x = 0$ in $[1/9, 2/3]$ by the secant method.

2. The function $f(x) = x^5 - 3x^3 + 2.5x - 0.6$ has two zeros in the interval $[0, 1]$. Approximate these zeros by using the secant method with:

   (a) $x_0 = 0, x_1 = 1/2$,

   (b) $x_0 = 3/4, x_1 = 1$.

3. Solve the equation $x^3 - 4x^2 + 2x - 8 = 0$ with an accuracy of 10^{-4} by using the secant method with $x_0 = 3, x_1 = 1$.

4. Use the secant method to approximate the solution of the equation $x = x^2 - e^{-x}$ to within 10^{-5} with $x_0 = -1$ and $x_1 = 1$.

5. If the secant method is used to find the zeros of $f(x) = x^3 - 3x^2 + 2x - 6$ with $x_0 = 1$ and $x_1 = 2$, what is x_2?

6. Use the secant method to approximate the solution of the equation $x = -e^x \sin x - 5$ to within 10^{-10} in the interval $[3, 3.5]$.

7. Given the following equations:

   (a) $x^4 - x - 10 = 0$,

   (b) $x - e^{-x} = 0$.

   Determine the initial approximations. Use these to find the roots correct to four decimal places using the secant method.

M-function 3.4

The following MATLAB function **secant.m** finds the solution of an equation using the secant method. INPUTS are a function f; initial approximations x_0, x_1; a tolerance *tol*; the maximum number of iterations n. The input function $f(x)$ should be defined as an M-file.

```
function secant(f,x0,x1,tol,n)
% The secant method for solving the nonlinear equation f(x)=0.
iter=0;
u=feval(f,x0);
v=feval(f,x1);
err=abs(x1-x0);
disp('_____')
disp(' iter      xn       f(xn)      f(xn+1)-f(xn)    Ixn+1-xnI ')
disp('_____')
fprintf('%2.0f  %12.6f  %12.6f\n',iter,x0,u)
fprintf('%2.0f  %12.6f  %12.6f  %12.6f  %12.6f\n',iter,x1,v,v-u,err)
while (err>tol)&(iter<=n)&((v-u)~=0)
    x=x1-v*(x1-x0)/(v-u);
    x0=x1;
    u=v;
    x1=x;
    v=feval(f,x1);
    err=abs(x1-x0);
    iter=iter+1;
    fprintf('%2.0f  %12.6f  %12.6f  %12.6f  %12.6f\n',iter,x1,v,v-u,
err)
end
if ((v-u)==0)
    disp(' Division by zero')
end
if (iter>n)
    disp(' Method failed to converge')
end
```

8. Show that the formula for the secant method is algebraically equivalent to
$$x_{n+1} = \frac{f(x_n)x_{n-1} - f(x_{n-1})x_n}{f(x_n) - f(x_{n-1})}.$$

9. Use the secant method to compute the next two iterates x_2 and x_3 for the zeros of the following functions.

   (a) $f(x) = x^2 - 3x - 1$ with $x_0 = 3.0, x_1 = 3.1$,
   (b) $f(x) = x^3 - x - 1$ with $x_0 = 1.2, x_1 = 1.3$,
   (c) $f(x) = x^3 - 2x + 1$ with $x_0 = 0.6, x_1 = 0.5$.

10. Given the equation $f(x) = 0$, obtain an iterative method using the rational approximation
$$f(x) = \frac{x - a_0}{b_0 + b_1 x}$$

where the coefficient a_0, b_0, and b_1 are determined by evaluating $f(x)$ at x_k, x_{k-1}, and x_{k-2}. Carry out two iterations using this method for the equation $2x^3 - 3x^2 + 2x - 3 = 0$ with $x_0 = 0$, $x_1 = 1$, and $x_2 = 2$.

COMPUTER PROBLEM SET 3.4

1. Write a computer program in a language of your choice that finds the zeros of a function $f(x)$ using the secant method. Input data to the program should be a function $f(x)$, two initial guesses for the root and the error tolerance ϵ. Use your program to find a zero of the function

$$f(x) = x^4 - 6x^3 + 3x^2 - 2x + \cos(3x)$$

   using $x_0 = 0$ and $x_1 = 0.25$.

2. The Peng-Robinson equation of state is given by

$$P = \frac{RT}{V - b} - \frac{a}{V(V + b) + b(V - b)}$$

   where

   P = Pressure

   V = Molar Volume, volume of one mole of gas or liquid

   T = Temperature(K)

   R = ideal gas constant

   Find V at $P = 778$ kPa and $T = 350$ K with $a = 365$ m^{6}kPa/(kg mole)^2, $b = 0.3$ m^3/kg mole, and $R = 1.618$. Use the secant method with $V_0 = 1.0$ and $V_1 = 1.5$ as initial estimates.

3. The roots of the polynomial

$$p(x) = (x - 1)(x - 2)\ldots(x - 20) - 10^{-8}x^{19}$$

   are highly sensitive to small alterations in their coefficients.

   (a) Prove that p has a root in $[20, 22]$, and then use the MATLAB function secant.m to find this root.

   (b) Find the number of iteration needed in a) to get the root to within an error $< 10^{-3}$; find also the number of iteration for accuracy $< 10^{-12}$.

4. Use the MATLAB function secant.m to find the first eight roots for $f(x) = -e^{-x} + \sin x$ to ten decimal places. Observe there are infinitely many roots. If x_n is the nth root, calculate the ratios $\frac{x_n}{n\pi}$, $n = 1, 2, \ldots$. Based on the evidence of these calculations, guess the value of

$$\lim_{n \to \infty} \frac{x_n}{n\pi}.$$

3.5 NEWTON'S METHOD

Newton's method is one of the most widely used of all iterative techniques for solving equations. Rather than using a secant line, the method uses a tangent line to the curve. Figure 3.10 gives a graphical interpretation of the method. To use the method we begin with an initial guess x_0, sufficiently close to the root α. The next approximation x_1 is given by the point at which the tangent line to f at $f(x_0, f(x_0))$ crosses the x-axis. It is clear that the value x_1 is much closer to α than the original guess x_0. If x_{n+1} denotes the value obtained by the succeeding iterations, that is the x-intercept of the tangent line to f at $(x_n, f(x_n))$, then a formula relating x_n and x_{n+1}, known as **Newton's method**, is given by

$$x_{n+1} = x_n - \frac{f(x_n)}{f'(x_n)}, \qquad n \geq 0 \tag{3.13}$$

provided $f'(x_n)$ is not zero.

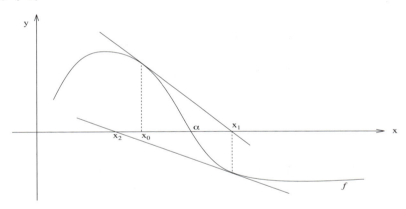

FIGURE 3.10
Newton's method and the first two approximations to its zero α.

To derive Eqn. (3.13), notice that the equation of the tangent line at $(x_0, f(x_0))$ is

$$y - f(x_0) = f'(x_0)(x - x_0).$$

If x_1 denotes the point where this line intersects the x-axis, then $y = 0$ at $x = x_1$, that is

$$-f(x_0) = f'(x_0)(x_1 - x_0),$$

and solving for x_1 gives

$$x_1 = x_0 - \frac{f(x_0)}{f'(x_0)}.$$

By repeating the process using the tangent line at $(x_1, f(x_1))$, we obtain for x_2

$$x_2 = x_1 - \frac{f(x_1)}{f'(x_1)}.$$

By writing the preceding equation in more general terms, one gets Eqn. (3.13). A suitable way of terminating the iterations is by using one of the stopping criteria suggested for the secant method.

An algorithmic statement of Newton's method is given below.

Given a continuously differentiable function f and an initial value x_0

$$\textbf{for } n = 0, 1, \ldots, \text{ITMAX}$$
$$\quad x_{n+1} \leftarrow x_n - \frac{f(x_n)}{f'(x_n)}$$

EXAMPLE 3.10

Use Newton's method to compute a root of

$$x^3 - x^2 - 1 = 0$$

to an accuracy of 10^{-4}. Use $x_0 = 1$.

The derivative of f is $f'(x) = 3x^2 - 2x$. Using $x_0 = 1$ gives $f(1) = -1$ and $f'(1) = 1$ and so the first Newton iterate is

$$x_1 = 1 - \frac{-1}{1} = 2 \quad \text{and} \quad f(2) = 3, \quad f'(2) = 8.$$

The next iterate is

$$x_2 = 2 - \frac{3}{8} = 1.625.$$

Continuing in this manner leads to the values in Table 3.6, which converge to $r = 1.465571$. In Table 3.6 the MATLAB function df1 is defined as follows:

```
function f=df1(x)
f=3*x.^2-2*x;
```

» newton('f1','df1',1,10^(-4),40)				
iter	xn	f(xn)	f'(xn)	| xn+1-xn |
0	1.000000	-1.000000	1.000000	
1	2.000000	3.000000	8.000000	1.000000
2	1.625000	0.650391	4.671875	0.375000
3	1.485786	0.072402	3.651108	0.139214
4	1.465956	0.001352	3.515168	0.019830
5	1.465571	0.000001	3.512556	0.000385
6	1.465571	6.93E-14	3.512555	1.43E-07

Table 3.6   Solution of $x^3 - x^2 - 1 = 0$ using Newton's method with $x_0 = 1$.

If we set $g(x_n) = x_n - \frac{f(x_n)}{f'(x_n)}$, then Eqn. (3.13) can be written as

$$x_n = g(x_n).$$

Thus, the Newton method for finding the root of the equation $f(x) = 0$ can be viewed as using repeated substitution to find a fixed point of the function $g(x)$.

Newton's method is a powerful technique, but has some difficulties. One of the difficulties is that if x_0 is not sufficiently close to the root, the method may not converge. It is evident that one should not also choose x_0 such that $f'(x_0)$ is close to zero, for then the tangent line is almost horizontal, as illustrated in Figure 3.11.

The following convergence theorem for Newton's method illustrates the importance of the choice of the initial estimate x_0.

THEOREM 3.3 (Newton's Theorem)
Assume that $f \in C^2[a, b]$. If $\alpha \in [a, b]$ is such that $f(\alpha) = 0$ and $f'(\alpha) \neq 0$, then there exists $\delta > 0$ such that Newton's method generates a sequence $\{x_n\}_{n=1}^{\infty}$ converging to α for any initial approximation $x_0 \in [\alpha - \delta, \alpha + \delta]$.

Proof. The proof consists of showing the hypotheses of the fixed point Theorem 3.2 with

$$g(x) = x - \frac{f(x)}{f'(x)}.$$

We have

$$g'(x) = 1 - \frac{f'(x)f'(x) - f(x)f''(x)}{[f'(x)]^2} = \frac{f(x)f''(x)}{[f'(x)]^2}.$$

By assumption $f(\alpha) = 0$ and $f'(\alpha) \neq 0$, so

$$g'(\alpha) = \frac{f(\alpha)f''(\alpha)}{[f'(\alpha)]^2} = 0.$$

Since $g'(\alpha) = 0$ and $g(x)$ is continuous, there exists $\delta > 0$ such that $|g'(x)| < 1$ on $(\alpha - \delta, \alpha + \delta)$. Therefore, if we choose δ such that

$$\frac{|f(x)f''(x)|}{|f'(x)|^2} < 1 \quad \text{for all} \quad x \in (\alpha - \delta, \alpha + \delta)$$

then a sufficient condition for the sequence $\{x_n\}_{n=1}^{\infty}$ to converge to a root of $f(x) = 0$ is to choose $x_0 \in [\alpha - \delta, \alpha + \delta]$.

MATLAB's Methods

MATLAB contains two commands to compute the roots of functions. The first command, **roots**, is specifically written to find the roots of a polynomial. The second command, **fzero**, is designed to find the roots of $f(x)$ provided the function actually crosses the x-axis at the root.

roots
The coefficients of the various powers are the critical data in a polynomial. For example, the coefficients of the cubic polynomial $3x^3 - 2x^2 + 4$ may be entered in a row (or column) vector [3 -2 0 4] where the coefficients are in descending order, beginning

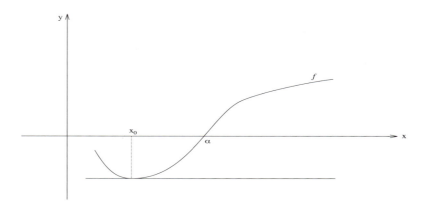

FIGURE 3.11
Failure of Newton's method due to a bad starting point.

with the highest power. The roots of this cubic polynomial may be found as follows:

>> roots([3 -2 0 4])
ans =
    $0.7921 + 0.9088i$
    $0.7921 - 0.9088i$
    -0.9174

One may verify an approximate x^* root by evaluating the polynomial at x^* using the MATLAB command **polyval**.

fzero
MATLAB's root finding procedure, **fzero**, may be used to compute the zeros of nonpolynomial functions. The usual syntax is

>> fzero('fun',initial value)
where the function, $f(x)$, is entered in an M-file. The function $f(x)$ may be entered directly as a string. As an example, the MATLAB result for finding a zero of $f(x) = x^2 - 3$ is:

>> fzero('x.^2-3', 1)
ans =
        1.7321

EXAMPLE 3.11
Use the function fzero to solve the area of a segment problem given in Example 3.2.

In this example, the fzero function is used to solve the Eqn. (3.3) to find θ given $r = 4$ and $A = 0.3243$. The MATLAB result is

```
>> format long
>> fzero('0.5*4^2*(x-sin(x))-0.3243',1)
ans =
       0.62834063441724 ≈ π/5
```

EXAMPLE 3.12

Use the function **roots** to solve the floating sphere problem given in Example 3.1 with $r = 12$ cm and $\rho = 0.64$.

In the case with $r = 12$ and $\rho = 0.64$ Eqn. 3.1 becomes

$$h^3 - 36h^2 + 4423.68 = 0.$$

The MATLAB result is

```
>> roots([1   -36   0   4423.68 ]
ans =
```

> 31.5582
> 14.2670
> − 9.8252

The root 31.5582 is rejected because it is larger than the diameter, −9.8252 is negative and therefore it is not a feasible solution, and the root 14.2670 is the proper solution since it lies in the interval $[0, 24]$.

EXERCISE SET 3.5

1. Find the roots of the following equations correct to six decimal places using Newton's method.

   (a) $x^5 - 3x^3 - x + 4 = 0$, $x_0 = 0$,

   (b) $3x^3 - x + 4 = 0$, $x_0 = -1$,

   (c) $x^3 - x^2 + e^x = 0$, $x_0 = -1$,

   (d) $e^x + x \sin x = 0$, $x_0 = 0$.

2. Calculate an approximate value for $\sqrt{5}$ using Newton's method with $x_0 = 5$.

3. Find the roots of the following equations in the interval indicated with an accuracy of 10^{-3} by using Newton's method.

   (a) $x^3 + 3x^2 + x - 4 = 0$, in $[0, 1]$,

M-function 3.5

The following MATLAB function **newton.m** approximates the solution of an equation using Newton's method. INPUTS are a function *f* and its derivative *f'*; an initial guess x_0; a tolerance *tol*; the maximum number of iterations *n*. The input functions *f(x)* and *f'(x)* should be defined as M-files.

```
function newton(f,df,x0,tol,n)
% Newton's method for solving the nonlinear equation f(x)=0.
iter=0;
u=feval(f,x0);
v=feval(df,x0);
err=abs(u/v);
disp('_____')
disp(' iter        x           f(x)        df(x)        |xn+1-xn|   ')
disp('_____')
fprintf('%2.0f %12.6f %12.6f %12.6f\n', iter, x0, u, v)
while (err>tol)&(iter<=n)&(v~=0)
    x1=x0-u/v;
    err=abs(x1-x0);
    x0=x1;
    u=feval(f,x0);
    v=feval(df,x0);
    iter=iter+1;
    fprintf('%2.0f %12.6f %12.6f %12.6f %12.6f\n',iter,x0,u,v,err)
end
if (v==0)
   disp(' division by zero')
end
if (iter>n)
   disp(' Method failed to converge')
end
```

(b) $x^4 + 2x^3 - 3x^2 - 4x - 2 = 0$, in $[1, 2]$,

(c) $x^3 - 3x^2 + 3 = 0$, in $[1, 2]$.

4. Use Newton's method to compute a zero of the function

$$f(x) = x^5 - 3x^3 - 5x + 4$$

to an accuracy of 10^{-6}. Use $x_0 = 1$.

5. Find an approximate root of $e^{-x} - x^2 - x = 0$ in $[-1, 1]$ using Newton's method with 10^{-5} accuracy.

6. The equation $e^x + x^3 \cos x + 2 = 0$ has two positive roots in $[2, 4]$. Find them by using Newton's method.

7. Find an approximate root of $x + \sin x - \cos 2x = 0$, by using:

(a) Newton's method with $x_0 = 0$,

(b) The secant method with $x_0 = 0$ and $x_1 = 1$.

8. Use Newton's method with $x_0 = 0$ to approximate the value of x that produces the point on the graph of $y = x^2$ that is closest to $(3, 2)$.

9. The equation $x - \cos x = 0$ has a root near $x = 0$ and $x = 1$. Compute the root with 10^{-4} accuracy first using the secant method and then Newton's method.

10. Derive the formula

$$x_n = \frac{x_{n-1}^2 - b}{2x_{n-1} - a}$$

for the approximate roots of the quadratic equation

$$x^2 - ax + b = 0$$

using Newton's method. Use the formula to find the positive root of $x^2 - 13x - 1 = 0$.

11. Show that the Newton method iterates for the function

$$f(x) = \frac{1}{x} - 5$$

is given by $x_{n+1} = 2x_n - 5x_n^2$. Approximate the zero of f using Newton's method with $x_0 = 1/4$.

12. Show that when Newton's method is used to approximate $\sqrt[3]{a}$, the sequence of iterates is given by

$$x_{n+1} = \frac{2x_n^3 + a}{3x_n^2}.$$

13. The iteration formula $x_{n+1} = x_n + ae^{-x_n} - 1$, where a is a positive constant, was obtained by applying Newton's method for some function $f(x)$. Find $f(x)$.

14. Experiment with Newton's method to solve

$$x_{n+1} = 2x_n - ax_n^2$$

using several different values of a. Make a conjecture about what this method computes.

15. Consider finding the real root of $(1 + \ln x)/x = 0$ by Newton's Method. Show that this leads to the equation

$$x_{n+1} = 2x_n + \frac{x_n}{\ln x_n}.$$

Apply the method with $x_1 = 1.21$. Next try it with $x_1 = 0.5$. Graph $y = (1 + \ln x)/x$ to understand your results.

16. Let
$$f(x) = -x^4 + x^2 + A$$

for constant A. What value of A should be chosen that guarantees that if $x_0 = 1/3$ as the initial approximation, the Newton method produces

$$x_1 = -x_0, x_2 = x_0, x_3 = -x_0, \ldots ?$$

17. In this problem we will find the roots of

$$\lambda x = \tan x.$$

This equation describes the states of a quantum-mechanical particle in a rectangular box with finite walls. This equation cannot be solved analytically and one needs to use numerical root-finding. You are asked to analyze this equation for three positive values of λ: one $\lambda < 1$, another $\lambda = 1$, and third $\lambda > 1$. (You choose the values of λ yourself)

(a) (Analytical) What happens to the roots as $\lambda \to 1$?

(b) Plot the functions λx and $\tan x$ on the same graph and find out approximately where the roots are.

(c) Find approximate formula for x_{n+1} using Newton's method.

(d) Use MATLAB's function Newton.m to compute a root of the equation on $(\pi/2, 3\pi/2)$.

18. Derive Newton's method using a Taylor series truncated to two terms.

19. Consider the function $f(x) = 1 + x^2 - \tan x$.

(a) Use Newton's method to approximate the first positive root. Begin with $x_0 = 0.1$ and compute x_4 using hand computations. Estimate the error in x_4.

(b) Repeat part a) with $x_0 = 1$.

20. The function $e^{x-1} - 5x^3 + 5$ has one root near $x = 1$. Using Newton's method to approximate this root, does the number of correct digits double with each iteration?

21. Consider the rootfinding problem

$$f(x) = x^2 \cos \frac{\pi x}{4} = 0. \tag{R}$$

Does Newton's method applied to (R) converge to some root of (R) for any starting point x_0? Justify your answer.

22. Find the points where the hyperbola $y = 1/x$, $x > 0$ intersect the cotangent curves $y = \cot x$.

23. Show that $f(x) = e^{-x^2} - 2x$ has only one root and that it lies between 0 and 1. Find the root to five decimal places using the bisection, false position, and Newton's methods. Prove that your approximations really are valid to five decimal places. Compare the methods to determine which method is the most efficient, that is, the number of iterations it takes each method to find the solution.

24. An ancient technique for extracting the square root of an integer $N > 1$ going back to the Babylonians involves using the iteration process

$$x_{n+1} = \frac{1}{2}\left(x_n + \frac{N}{x_n}\right)$$

where x_0 is the largest integer for which $x^2 \leq N$. Show that if by chance $x^2 = N$, the sequence is constant, i.e., $x_n = \sqrt{N}$ for all $n = 1, 2, \ldots$. Show that the sequence is simply Newton's method applied to

$$f(x) = x^2 - N.$$

How many iterations does it take to calculate $\sqrt{411}$ to three decimal places using $x_0 = 20$? $\sqrt{3}$ and $\sqrt{5}$ using, respectively, $x_0 = 1$ and $x_0 = 2$?

COMPUTER PROBLEM SET 3.5

1. Write a computer program in a language of your choice that finds the zeros of a function $f(x)$ using Newton's method. Input data to the program should be a function $f(x)$ and its derivative $f'(x)$, an initial guess x_0 for the root, and the error tolerance ϵ.
   Use your program to find the three zeros of the function

   $$f(x) = x^5 - 21x^2 - 8x^3 - 4x^4 - 28x + 60.$$

2. Let $f(x) = 2^x - 2x$. Show that f has exactly two real roots. Find for which starting points Newton's method leads to convergence to each of these roots.

3. Use MATLAB's function Newton.m to estimate the cubic root of 100 to 10 decimal places starting with $x_0 = 2$.

4. Use MATLAB's function Newton.m to estimate the square root of 7 to 12 decimal places.

5. The equation

   $$-210.7 + 79.1x - 67.2x^2 + 18.3x^3 - 5.3x^4 + x^5 = 0$$

   has only one real root. Use MATLAB's function Newton.m to estimate this root to 8 decimal places.

6. Use Newton's method to find a root of the equation

$$e^{-x^2} - \cos 2x - 1 = 0$$

   on $[0, 3]$. What happens if we start with $x_0 = 0$?

7. In an effort to determine the optimum damping ratio of a spring-mass-damper system designed to minimize the transmitted force when an impact is applied to the mass the following equation must be solved for

$$\cos[4x\sqrt{1 - x^2}] + 8x^4 - 8x^2 + 1 = 0.$$

   Use MATLAB's function Newton.m to compute the solution of this problem.

8. Use the MATLAB function newton.m to find the first eight roots for $f(x) = -e^{-x} + \sin x$ to ten decimal places. Observe there are infinitely many roots. If x_n is the nth root, calculate the ratios $\frac{x_n}{n\pi}$, $n = 1, 2, \ldots$. Based on the evidence of these calculations, guess the value of

$$\lim_{n \to \infty} \frac{x_n}{n\pi}.$$

9. In the year 1225 Leonardo of Pisa studied the equation

$$x^3 + 2x^2 + 10x - 20 = 0$$

   and produced $x = 1.368\,808\,107$. Nobody knows by what method Leonardo found this value but it is a good approximation for this time since it is accurate to nine decimal digits. Apply the MATLAB function newton.m to obtain this result.

3.6   CONVERGENCE OF THE NEWTON AND SECANT METHODS

In order to determine the order of convergence of the Newton and secant methods, we make the following definition:

DEFINITION 3.1   *Let $x_0, x_1, \ldots$ be a sequence that converges to a number α, and set $e_n = \alpha - x_n$. If there exists a number k and a positive constant C such that*

$$\lim_{n \to \infty} \frac{|e_{n+1}|}{|e_n|^k} = C$$

then k is called the order of convergence of the sequence and C the asymptotic error constant.

We now examine the rate of convergence to simple roots for both the Newton and secant methods. We consider first Newton's method.

We have from Eqn. (3.13)

$$e_{n+1} = \alpha - x_{n+1} = \alpha - x_n + \frac{f(x_n)}{f'(x_n)}. \tag{3.14}$$

If f is twice-differentiable, the Taylor series expansion of f about x_n gives

$$f(\alpha) = f(x_n) + (\alpha - x_n) f'(x_n) + \frac{(\alpha - x_n)^2}{2} f''(\xi_n) \tag{3.15}$$

where ξ_n lies between x_n and α.

Since $f(\alpha) = 0$ and $f'(\alpha) \neq 0$, then combining Eqns. (3.14) and (3.15) we obtain

$$e_{n+1} = -(x_n - \alpha)^2 \frac{f''(\xi_n)}{2f'(x_n)}$$

$$= -e_n^2 \frac{f''(\xi_n)}{2f'(x_n)}. \tag{3.16}$$

Therefore, if α is a simple root, that is $f'(\alpha) \neq 0$, Eqn. (3.16) shows that Newton's method converges quadratically. In other words, for a simple root, Newton's method will usually converge with the number of accurate decimal places approximately doubling in one Newton iteration.

Similarly, it can be shown that the error for the secant method satisfies the equation

$$e_{n+1} = \frac{f''(\xi_n)}{2f'(\zeta_n)} e_n e_{n-1} \tag{3.17}$$

where ξ_n and ζ_n are in the smallest interval containing α, x_n, and x_{n-1}. Because a complete error analysis of the secant method is too complicated, we omit the proof (see [11] for more details). In fact, it can also be shown that

$$|e_{n+1}| \leq C|e_n|^\lambda$$

for some constant C. Here

$$\lambda = \lim_{n \to \infty} \frac{\lambda_{n+1}}{\lambda_n} = \frac{1 + \sqrt{5}}{2}$$

where λ_n are the Fibonacci numbers $(1, 1, 2, 3, 5, 8, \ldots)$ obtained by adding the previous two numbers and given that $\lambda_0 = \lambda_1 = 1$. Since $(1 + \sqrt{5})/2 \approx 1.62$, the secant method is said to be **superlinearly** convergent.

EXERCISE SET 3.6

1. The function $f(x) = x^4 - x^3 - 1$ has one root in $[1, 2]$. Using Newton's method, find this zero to an accuracy of 10^{-5}. Calculate $|e_n|$ and $|e_{n+1}|/|e_n^2|$ for $n = 2, 3, 4$ to verify the error formula (3.16).

2. Find the real positive root of the equation
$$x^3 - x^2 - 1 = 0$$
with an accuracy of 10^{-4} by using the secant method with $x_0 = 1$ and $x_1 = 2$. Calculate $|e_n|$ and $|e_{n+1}|/|e_n e_{n-1}|$ for $n = 2, 3, 4$ to verify the error formula (3.17).

3. The equation $x^2 + ax + b = 0$ has two real roots, α and β. Show that the iteration method
$$x_{k+1} = -(ax_k + b)/x_k$$
is convergent near $x = \alpha$ if $|\alpha| > |\beta|$ and that
$$x_{k+1} = -b/(x_k + a)$$
is convergent near $x = \alpha$ if $|\alpha| < |\beta|$.

4. Prove that the equation
$$c_0 = 3 \qquad c_{n+1} = c_n - \tan c_n, \qquad n = 1, 2, \ldots$$
converges. Find the order of convergence.

5. Show that the sequence defined by
$$x_n = \frac{1}{2}x_{n-1} + \frac{3}{2x_{n-1}}, \qquad \text{for} \quad n \geq 1$$
converges to $\sqrt{3}$ whenever the starting point $x_0 > \sqrt{3}$.

6. If f is such that $|f''(x)| \leq 4$ for all x and $|f'(x)| \geq 2$ for all x and if the initial error in Newton's method is less than $1/3$, what is an upper bound on the error at each of the first 3 steps?

7. Show that $x^3 - 36 = 0$ has exactly one root on the interval $[3, 4]$ and call it α. Set up a quadratically convergent scheme to find α. Let $x_0 = 3$. Find x_2.

8. Determine whether the sequence $x_n = 1/2^{2^n}$ converges quadratically or not.

9. Consider the equation
$$e^x \sin^2 x - x^2 = 0$$
which has one root at $x = 0$ and another near $x = 2.3$. Use Newton's method to approximate these roots using $x_0 = 1$ and $x_0 = 2$ as starting values. Determine the order of convergence in each case.

10. Sometimes it is difficult or expensive to compute $f'(x)$ for use in Newton's method, so we use an approximation α instead. Find the condition on α to ensure that the iteration
$$x_{n+1} = x_n - \frac{f(x_n)}{\alpha}$$
will converge(at least) linearly to a zero of f if started near enough to the zero.

3.7 MULTIPLE ROOTS AND THE MODIFIED NEWTON METHOD

We say that α is a root of **multiplicity** k of f if

$$f(\alpha) = f'(\alpha) = \cdots = f^{(k-1)}(\alpha) = 0, \qquad \text{but} \qquad f^{(k)}(\alpha) \neq 0. \tag{3.18}$$

In Section 3.5, Newton's algorithm was derived under the assumption that $f'(\alpha) \neq 0$, that is, α was assumed to be a simple zero. As one can see from Example 3.13 below, difficulties might occur with Newton's method if α is not a simple root.

EXAMPLE 3.13

The function $f(x) = x^3 - 7x^2 + 11x - 5$ has a root of multiplicity two at $x = 1$. Use Newton's method with $x_0 = 0$ to approximate it.

A plot of $f(x)$ from MATLAB is shown in Figure 3.12.

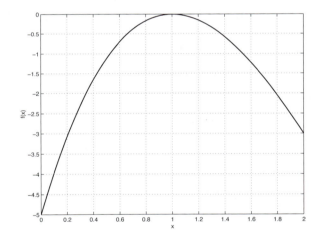

FIGURE 3.12
Plot of $f(x) = x^3 - 7x^2 + 11x - 5$ with a root of multiplicity two at $x = 1$.

From Definition 3.1 and the numerical results given in Table 3.7, it shows that Newton's method convergence linearly to $a = 1$ since the error in each successive iteration is proportional to the error in the previous iteration. That is,

$$|a - x_{n+1}|/|a - x_n| \approx C,$$

where $C \approx 1/2$ as shown in Table 3.7. The difficulty with multiple roots leads us to consider a modification of Newton's method.

As before, we wish to find a zero of a function f. We assume that f has a zero α of multiplicity m. Let us define the new function

$$u(x) = \frac{f(x)}{f'(x)}. \tag{3.19}$$

n	x_n	$\|x_{n+1}-x_n\|$	$\|a-x_n\|$	$\|e_{n+1}\| / \|e_n\|$
0	0.000000	0.454545	1.000000	0.545455
1	0.454545	0.257290	0.545455	0.528302
2	0.711835	0.139399	0.288165	0.516254
3	0.851234	0.073073	0.148766	0.508807
4	0.924307	0.037498	0.075693	0.504600
5	0.961805	0.019008	0.038195	0.502353
6	0.980813	0.009571	0.019187	0.501191
7	0.990384	0.004802	0.009616	0.500599
8	0.995186	0.002406	0.004814	0.500300
9	0.997592	0.001204	0.002408	0.500150
10	0.998795	0.000602	0.001205	0.500075
11	0.999398	0.000301	0.000602	0.500038
12	0.999699	0.000151	0.000301	0.500019
13	0.999849	0.000075	0.000151	0.500009
14	0.999925	0.000000	0.000075	

Table 3.7   Solution of $x^3 - 7x^2 + 11x - 5 = 0$ using Newton's method with $x_0 = 0$.

Using Eqn. (3.18), the Taylor series expansion of $f(x)$ about α becomes

$$f(x) = (x - \alpha)^m h(x) \tag{3.20}$$

with

$$h(x) = \frac{f^{(m)}(\alpha)}{m!} + \frac{f^{(m+1)}(\alpha)}{(m+1)!}(x - \alpha) + \cdots .$$

By differentiating Eqn. (3.20) we obtain

$$\begin{aligned} f'(x) &= (x - \alpha)^m h'(x) + m(x - \alpha)^{m-1} h(x) \\ &= (x - \alpha)^{m-1} \left[(x - \alpha)h'(x) + mh(x) \right]. \end{aligned} \tag{3.21}$$

Thus, from (3.20) and (3.21), (3.19) can be written in the form

$$u(x) = \frac{(x - \alpha)h(x)}{(x - \alpha)h'(x) + mh(x)} = (x - \alpha)\Psi(x) \tag{3.22}$$

where

$$\Psi(x) = \frac{h(x)}{(x - \alpha)h'(x) + mh(x)}.$$

Since $h(\alpha) = \frac{f^{(m)}(\alpha)}{m!}$, we have

$$\lim_{x \to \alpha} \Psi(x) = \frac{1}{m} \tag{3.23}$$

and

$$\lim_{x \to \alpha} u'(x) = \lim_{x \to \alpha} \left[(x - \alpha)\Psi'(x) + \Psi(x) \right] = \frac{1}{m}. \tag{3.24}$$

n	x_n	$|x_{n+1}-x_n|$	$|a-x_n|$	$|e_{n+1}|/|e_n|^2$
0	0.00000000	0.90909091	1.00000000	0.09090909
1	0.90909091	0.08991009	0.09090909	0.12087912
2	0.99900100	0.00099888	0.00099900	0.12495319
3	0.99999988	0.00000012	0.00000012	
4	1.00000000	0.00000000	0.00000000	

Table 3.8 Solution of $x^3 - 7x^2 + 11x - 5 = 0$ using the modified Newton's method with $x_0 = 0$ and $m = 2$.

In view of (3.22) and (3.24), we conclude that $u(x)$ has a simple root at $x = \alpha$. Applying now Newton's method to the function $u(x)$ and using (3.24) yields

$$x_{n+1} = x_n - m\frac{f(x_n)}{f'(x_n)} \qquad (3.25)$$

where m represents the multiplicity of the root α. Eqn. (3.25) is called the **modified Newton's method**.

EXAMPLE 3.14
Use the modified Newton's method with $x_0 = 0$ and $m = 2$ to approximate the double root $a = 1$ of the equation given in Example 3.13. Use the results to show that the modified Newton's method converges quadratically.

Using the modified Newton's method

$$x_{n+1} = x_n - 2\frac{x_n^3 - 7x_n^2 + 11x_n - 5}{3x_n^2 - 14x_n + 11}$$

with $m = 2$, we get the results shown in Table 3.8. From the numerical results given in Table 3.8, it shows that the modified Newton's method converges quadratically to $a = 1$ since the error in each successive iteration is proportional to the square of the error in the previous iteration. That is,

$$|a - x_{n+1}|/|a - x_n|^2 \approx C,$$

where $C \approx 0.12$.

Note that it took 14 iterations for Newton's method to converge and only 3 iterations for the modified Newton's method with the same tolerance.

EXAMPLE 3.15
The equation $x^3 - 4x^2 - 3x + 18 = 0$ has a double root at $x = 3$. Use the modified Newton's method with $x_0 = 0$ and $m = 2$ to approximate it.

```
» newton2('f3','df3',0,2,10^(-12),40)

iter          xn              f(xn)            f '(xn)         | xn+1-xn |

 0        0.000000         18.000000        -3.000000          ---
 1       12.000000       1134.000000       333.000000        12.000000
 2        5.189189         34.454544        36.269540         6.810811
 3        3.289273          0.442600         3.143766         1.899916
 4        3.007700          0.000297         0.077175         0.281573
 5        3.000006          1.70E-10         5.90E-05         0.007694
 6        3.000000          0                0                5.90E-06
 7        3.000000          0                0                0
```

Table 3.9   Solution of $x^3 - 4x^2 - 3x + 18 = 0$ using the modified Newton's method with $x_0 = 0$ and $m = 2$.

Define the MATLAB function f3 and its derivative df3 as follows:

```
function f=f3(x)
f=x.^3-4*x.^2-3*x+18;

function f=df3(x)
f=3*x.^2-8*x-3;
```

and then call the function newton2 to get the results shown in Table 3.9.

EXERCISE SET 3.7

1. Try to solve the equation $x^2 - 1 = 0$ using Newton's method with $x_0 = 0$. Use a graphic interpretation to explain what happened.

2. The function

$$f(x) = x^3 + 2x^2 - 7x + 4$$

has a zero of order 2 at $x = 1$. Start with $x_0 = 2$ and use Newton's method to find this zero. Show that in this case Newton's method converges linearly. Also show that if the modified Newton's method is used with $x_0 = 2$ and $m = 2$ to find this zero, the iteration does converge quadratically.

3. The function $f(x) = x^2 - 2e^{-x}x + e^{-2x}$ has one multiple real root. Use the modified Newton's method to approximate to within 10^{-4} this root with $x_0 = 1$. Also try to find the root using Newton's method with the same initial guess and compare the number of iterations needed to attain the same accuracy.

M-function 3.7

The following MATLAB function **newton2.m** finds the solution of an equation using the modified Newton's method. INPUTS are a function *f* and its derivative *f*; an initial approximation x_0; the number of multiplicity *m*; a tolerance *tol*; the maximum number of iterations *n*. The input functions *f(x)* and *f'(x)* should be defined as M-files.

```
function newton2(f,df,x0,m,tol,n)
% Modified Newton's method for solving the nonlinear equation
% f(x)=0.
iter=0;
u=feval(f,x0);
v=feval(df,x0);
err=abs(m*u/v);
disp('_____')
disp('  iter          x          f(x)          df(x)          |xn+1-xnl ')
disp('_____')
fprintf('%2.0f  %12.6f  %12.6f  %12.6f\n',iter,x0,u,v)
while (err>tol)&(iter<=n)&(v~=0)
   x1=x0-m*u/v;
   err=abs(x1-x0);
   x0=x1;
   u=feval(f,x0);
   v=feval(df,x0);
   iter=iter+1;
   fprintf('%2.0f  %12.6f  %12.6f  %12.6f  %12.7f\n',iter,x0,u,v,err)
   pause(1)
   y(iter)=err;
end
if (v==0)
   disp('  division by zero')
end
if (iter>n)
   disp('  Method failed to converge')
end
```

4. The function $f(x) = x^4 - 8.6x^3 - 35.51x^2 + 464.4x - 998.46$ has a double root in the interval $[4, 5]$. Use the modified Newton's method to find it.

5. Show that the function

$$f(x) = (4x - 8) + x^3 - 2x^2 - 4(x^2 - 2x)$$

has a zero of order 3. Use the modified Newton method to find this zero starting with $x_0 = 1$.

6. Use the modified Newton's method to find the root $\alpha = 1$ for the function $f(x) = 1 - xe^{1-x}$. Is the quadratic convergence recovered?

7. Consider the equation

$$e^x \sin^2 x - x^2 = 0$$

which has one root at $x = 0$ of multiplicity three. Use the modified Newton's method to approximate this root using $x_0 = 1$ as starting value. Determine the order of convergence.

COMPUTER PROBLEM SET 3.7

1. Write a computer program in a language of your choice that finds the zeros of a function $f(x)$ using the modified Newton's method. Input data to the program should be a function $f(x)$ and its derivative $f'(x)$, an initial guess x_0 for the root, and the error tolerance ϵ.

   Use your program to find the zero of multiplicity three of the function

$$f(x) = -126.9 + 153.9x - 96.3x^2 + 40.7x^3 - 10x^4 + x^5$$

2. Each of the following functions has a root or roots with multiplicity greater than one. Determine the multiplicity in each case. Explain your reasoning. Use the MATLAB function newton2.m to approximate the root or roots.

   (a) $f(x) = e^x - x - 1$,
   (b) $f(x) = \cos(x + \sqrt{2}) + x(x/2 + \sqrt{2})$,
   (c) $f(x) = x^3 + 3.3x^2 + 3.63x + 1.331$.

3.8   NEWTON'S METHOD FOR NONLINEAR SYSTEMS

Newton's method introduced in Section 3.5 for finding the zeros of a nonlinear function of a single variable can be extended to nonlinear functions in several variables. However, the problem of finding the zeros of nonlinear functions is much more difficult than for linear equations. Starting with an initial estimate of the solution, Newton's method consists of approximating each nonlinear function by its tangent plane; and the common root of the resulting linear equations provides the next approximation. To see how the algorithm works, we first consider a system of two nonlinear equations. That is,

$$f_1(x_1, x_2) = 0$$
$$f_2(x_1, x_2) = 0.$$

Let $(x_1^{(0)}, x_2^{(0)})$ be an initial estimate to the solution and f be differentiable at $\mathbf{x}^{(0)} = (x_1^{(0)}, x_2^{(0)})$. The equation of the tangent plane to the function $y = f_1(x_1, x_2)$ at $(x_1^{(0)}, x_2^{(0)})$ is

$$y - f_1(x_1^{(0)}, x_2^{(0)}) = \frac{\partial}{\partial x_1}\left[f_1(x_1^{(0)}, x_2^{(0)})\right](x_1 - x_1^{(0)})$$
$$+ \frac{\partial}{\partial x_2}\left[f_1(x_1^{(0)}, x_2^{(0)})\right](x_2 - x_2^{(0)}).$$

Similarly, the equation of the tangent plane to the function $y = f_2(x_1, x_2)$ at $(x_1^{(0)}, x_2^{(0)})$ is

$$y - f_2(x_1^{(0)}, x_2^{(0)}) = \frac{\partial}{\partial x_1}\left[f_2(x_1^{(0)}, x_2^{(0)})\right](x_1 - x_1^{(0)})$$
$$+ \frac{\partial}{\partial x_2}\left[f_2(x_1^{(0)}, x_2^{(0)})\right](x_2 - x_2^{(0)}).$$

The intersection of these two tangent planes with the xy-plane is obtained by solving the system

$$\frac{\partial}{\partial x_1}\left[f_1(x_1^{(0)}, x_2^{(0)})\right]\Delta x_1^{(0)} + \frac{\partial}{\partial x_2}\left[f_1(x_1^{(0)}, x_2^{(0)})\right]\Delta x_2^{(0)} = -f_1(x_1^{(0)}, x_2^{(0)})$$
$$\frac{\partial}{\partial x_1}\left[f_2(x_1^{(0)}, x_2^{(0)})\right]\Delta x_1^{(0)} + \frac{\partial}{\partial x_2}\left[f_2(x_1^{(0)}, x_2^{(0)})\right]\Delta x_2^{(0)} = -f_2(x_1^{(0)}, x_2^{(0)})$$

$$(3.26)$$

where $\Delta x_1^{(0)} = x_1 - x_1^{(0)}$ and $\Delta x_2^{(0)} = x_2 - x_2^{(0)}$.

We solve Eqn. (3.26) by Gaussian elimination (see Chapter 4) and the resulting solution $(x_1^{(0)}, x_2^{(0)})$ is used to get an improved estimate to the solution. That is,

$$x_1^{(1)} = x_1^{(0)} + \Delta x_1^{(0)}$$
$$x_2^{(1)} = x_2^{(0)} + \Delta x_2^{(0)}.$$

We repeat the process until f_1 and f_2 are close to zero or $\|\Delta \mathbf{x}\|_\infty = \max_i\{|\Delta x_i|\}$ is less than a given specified tolerance value, in which case convergence has been achieved.

Starting with an approximate solution $(x_1^{(i)}, x_2^{(i)})$, the vector-matrix notation of these equations is

$$\begin{bmatrix} \partial f_1(x_1^{(i)}, x_2^{(i)})/\partial x_1 & \partial f_1(x_1^{(i)}, x_2^{(i)})/\partial x_2 \\ \partial f_2(x_1^{(i)}, x_2^{(i)})/\partial x_1 & \partial f_2(x_1^{(i)}, x_2^{(i)})/\partial x_2 \end{bmatrix} \begin{bmatrix} \Delta x_1^{(i)} \\ \Delta x_2^{(i)} \end{bmatrix} = - \begin{bmatrix} f_1(x_1^{(i)}, x_2^{(i)}) \\ f_2(x_1^{(i)}, x_2^{(i)}) \end{bmatrix} \quad (3.27)$$

and

$$\begin{bmatrix} x_1^{(i+1)} \\ x_2^{(i+1)} \end{bmatrix} = \begin{bmatrix} x_1^{(i)} + \Delta x_1^{(i)} \\ x_2^{(i)} + \Delta x_2^{(i)} \end{bmatrix}, \quad i = 0, 1, 2, \ldots.$$

For a system of n equations, we can write Newton's method by generalizing Eqn. (3.27). The generalized system is given by

$$
\begin{bmatrix} \frac{\partial f_1}{\partial x_1} & \cdots & \frac{\partial f_1}{\partial x_n} \\ \vdots & \cdots & \vdots \\ \frac{\partial f_n}{\partial x_1} & \cdots & \frac{\partial f_n}{\partial x_n} \end{bmatrix} \begin{bmatrix} \Delta x_1^{(i)} \\ \vdots \\ \Delta x_n^{(i)} \end{bmatrix} = - \begin{bmatrix} f_1 \\ \vdots \\ f_n \end{bmatrix}
\tag{3.28}
$$

where the functions are evaluated at $\mathbf{x}^{(i)} = (x_1^{(i)}, x_2^{(i)}, ..., x_n^{(i)})$. Solve this system to compute

$$
\mathbf{x}^{(i+1)} = \mathbf{x}^{(i)} + \Delta\mathbf{x}^{(i)}, \quad i = 0, 1, 2, \ldots .
\tag{3.29}
$$

The coefficient matrix in Eqn. (3.28) is known as the Jacobian matrix $\mathbf{J}(x_1, \ldots, x_n)$. If the initial estimate is sufficiently close to the solution, Newton's method is generally expected to give quadratic convergence, provided the Jacobian matrix is non-singular.

We now illustrate the method on the following 2×2 system.

EXAMPLE 3.16

Solve the nonlinear system

$$
f_1(x_1, x_2) = x_1^3 + 3x_2^2 - 21 = 0
$$
$$
f_2(x_1, x_2) = x_1^2 + 2x_2 + 2 = 0
$$

by Newton's method starting, with the initial estimate $\mathbf{x}^{(0)} = (1, -1)$. Iterate until $\|\Delta\mathbf{x}\|_\infty = \max_i\{|\Delta x_i|\} < 10^{-6}$.

The Jacobian matrix is

$$
\mathbf{J}(x_1, x_2) = \begin{bmatrix} 3x_1^2 & 6x_2 \\ 2x_1 & 2 \end{bmatrix} .
$$

At the point $(1, -1)$ the function vector and the Jacobian matrix take on the values

$$
\mathbf{F}(1, -1) = \begin{bmatrix} -17 \\ 1 \end{bmatrix}, \quad \mathbf{J}(1, -1) = \begin{bmatrix} 3 & -6 \\ 2 & 2 \end{bmatrix} ;
$$

the differentials $\Delta x_1^{(0)}$ and $\Delta x_2^{(0)}$ are solutions of the system

$$
\begin{bmatrix} 3 & -6 \\ 2 & 2 \end{bmatrix} \begin{bmatrix} \Delta x_1^{(0)} \\ \Delta x_2^{(0)} \end{bmatrix} = - \begin{bmatrix} -17 \\ 1 \end{bmatrix} .
$$

Its solution is

$$
\Delta\mathbf{x}^{(0)} = \begin{bmatrix} \Delta x_1^{(0)} \\ \Delta x_2^{(0)} \end{bmatrix} = \begin{bmatrix} 1.555556 \\ -2.055560 \end{bmatrix} .
$$

Thus, the next point of iteration is

$$\mathbf{x}^{(1)} = \begin{bmatrix} x_1^{(1)} \\ x_2^{(1)} \end{bmatrix} = \begin{bmatrix} 1 \\ -1 \end{bmatrix} + \begin{bmatrix} 1.555556 \\ -2.055556 \end{bmatrix} = \begin{bmatrix} 2.555556 \\ -3.055556 \end{bmatrix}.$$

Similarly, the next three points are

$$\mathbf{x}^{(2)} = \begin{bmatrix} 1.865049 \\ -2.500801 \end{bmatrix}, \quad \mathbf{x}^{(3)} = \begin{bmatrix} 1.661337 \\ -2.359271 \end{bmatrix}, \quad \mathbf{x}^{(4)} = \begin{bmatrix} 1.643173 \\ -2.349844 \end{bmatrix}.$$

The results are summarized in Table 3.10.

Iteration i	$\mathbf{x}^{(i)}$	$\|\Delta\mathbf{x}\|_\infty$
0	$[1, -1]$	
1	$[2.555556, -3.055556]$	2.055556
2	$[1.865049, -2.500801]$	0.690 507
3	$[1.661337, -2.359271]$	0.203712
4	$[1.643173, -2.349844]$	0.018164
5	$[1.643038, -2.349787]$	0.000135
6	$[1.643038, -2.349787]$	7.3×10^{-9}

Table 3.10 Newton's method for Example 3.16.

This example shows that Newton's method can converge very rapidly once an initial estimate sufficiently close to the solution is obtained. A disadvantage of Newton's method is the necessity of solving the linear system (3.28) at each step.

To solve the above example using the MATLAB function newton_sys first requires saving the nonlinear system and its related Jacobian matrix, respectively, as M-files:

```
function f=f1(X)
x=X(1);y=X(2);
f=[x.^3+3*y.^2-21; x.^2+2*y+2];

function df=df1(X)
x=X(1);y=X(2);
df=[3*x.^2, 6*y; 2*x, 2];
```

Now call the MATLAB function

```
>> newton_sys('f1', 'df1', [1,-1]', 10^(8), 40)
```

to get the results shown in Table 3.10.

M-function 3.8

The following MATLAB function **newton_sys.m** finds the solution of a nonlinear system of equations using Newton's method. INPUTS are the system F given by an *(nx1)* vector; the Jacobian JF of F given by an *(nxn)* matrix; the initial vector *xx0*; a tolerance *tol*; the max. number of iterations   *maxit*. The system F and its Jacobian JF should be defined as M-files.

```
function newton_sys(F,JF,xx0,tol,maxit)
% Solve a nonlinear system of equations using Newton's method.
x0=xx0
iter=1;
while(iter<=maxit)
   y=-feval(JF,x0)\feval(F,x0);
   xn=x0+y;
   err= max(abs(xn-x0));
   if (err<=tol)
      x=xn;
      fprintf(' Newton's method converges after %3.0f iterations to
\n', iter)
      x
      return;
   else
      x0=xn;
   end
   iter=iter+1;
end
disp(' Newton's method does not converge')
x=xn;
```

EXERCISE SET 3.8

1. Find the Jacobian matrix $J(x,y)$ at the point $(-1,4)$ for the functions

   $f_1(x,y) = x^3 - y^2 + y = 0,$
   $f_2(x,y) = xy + x^2 = 0.$

2. Find the exact solution for the nonlinear system

   $x^2 + y^2 = 4,$
   $x^2 - y^2 = 1.$

3. Solve the nonlinear system of equations

   $$\ln(x^2 + y) - 1 + y = 0$$
   $$\sqrt{x} + xy = 0$$

   using Newton's method with $x_0 = 2.4$ and $y_0 = -0.6$.

4. Solve the nonlinear system of equations

$$10 - x + \sin(x + y) - 1 = 0$$
$$8y - \cos^2(z - y) - 1 = 0$$
$$12z + \sin z - 1 = 0$$

using Newton's method with $x_0 = 0.1$, $y_0 = 0.25$, and $z_0 = 0.08$.

5. Solve the following nonlinear systems using Newton's method

(a) $f_1(x, y) = 3x - y - 3 = 0$,
   $f_2(x, y) = x - y + 2 = 0$.

(b) $f_1(x, y) = x^2 + 4y^2 - 16 = 0$,
   $f_2(x, y) = x^2 - 2x - y + 1 = 0$.

(c) $f_1(x, y, z) = x^2 + y^2 + z^2 - 9 = 0$,
   $f_2(x, y, z) = xyz - 1 = 0$,
   $f_3(x, y, z) = x + y - z^2 = 0$.

6. Consider the system of two equations

$$f_1(x, y) = x^2 + ay^2 - 1 = 0$$
$$f_2(x, y) = (x - 1)^2 + y^2 - 1 = 0$$

where a is a parameter.

(a) Write down two explicit formulas for Newton's iteration for this system. First, write it in the form

$$\begin{bmatrix} x_{n+1} \\ y_{n+1} \end{bmatrix} = \begin{bmatrix} x_n \\ y_n \end{bmatrix} - J^{-1} \begin{bmatrix} x_n^2 + ay_n^2 - 1 \\ (x_n - 1)^2 + y_n^2 - 1 \end{bmatrix}$$

where you explicitly exhibit the Jacobian J and its inverse J^{-1}. Second, evaluate this expression explicitly, i.e., multiply it out and simplify.

(b) Using these formulas, write a (very short) MATLAB program to implement Newton iteration just for this example. Try it for $a = 1$.

7. Use one iteration of Newton's method for systems with initial guess $(1/2, 1/2)^T$ on

$$x^2 + y^3 = 1$$
$$x^3 - y^2 = -1/4.$$

8. Use Newton's method for systems to find two solutions near the origin of

$$x^2 + x - y^2 = 1$$
$$y - \sin x^2 = 0.$$

APPLIED PROBLEMS FOR CHAPTER 3

1. The temperature in the interior of a material with imbedded heat sources is obtained from the solution of the equation

$$e^{-(1/2)t} \cosh^{-1}(e^{(1/2)t}) = \sqrt{k/2}.$$

Given that $k = 0.67$, find the temperature t.

2. A relationship for the compressibility factor c of real gases has the form

$$c = \frac{1 + x + x^2 - x^3}{(1 - x)^3}.$$

If $c = 0.9$, find the value of x in the interval $[-0.1, 0]$ using the bisection method.

3. L.L. Vant-Hull (*Solar Energy*, **18**,33 (1976)) derived the following equation for the geometrical concentration factor, C, in the study of solar energy

$$C = \frac{\pi(h/\cos A)^2 F}{0.5\pi D^2(1 + \sin A - 0.5\cos A)}.$$

Given $h = 290$, $C = 1100$, $F = 0.7$, and $D = 13$, what is the value of A?

4. The Legendre polynomial of the sixth order is given by

$$P_6(x) = \frac{1}{48}(693x^6 - 945x^4 + 315x^2 - 15).$$

Find the zeros of $P_6(x)$ using Newton's method. (*Note*: All the zeros of the Legendre polynomial are less than one in magnitude and, for polynomials of an even order, are symmetric about the origin.)

5. The Chebyshev polynomial of degree 6 is given by

$$T_6(x) = 32x^6 - 48^4 + 18x^2 - 1.$$

Find the zeros of $T_6(x)$ using Newton's method.
(*Note*: All the zeros of the Chebyshev polynomial are less than one in magnitude.)

6. In the study of a natural gas reservoir, Collier et al. (*Soc. Petr. Engrs.*, **21**, 5 (1981)) derived the following equation that governs the relationship of the gas pressure and reservoir volume:

$$x = y^{-1.5} \left[\frac{y - 1/(1 + k)}{1 - 1/(1 + k)} \right]^b \qquad \text{where}$$

$b = 1.5k/(1 + k)$
$x = $ normalized volume, dimensionless
$y = $ normalized pressure, dimensionless
$k = 1.06315$, dimensionless

A reservoir will be useful until the pressure drops to a value that makes production uneconomic. Determine the value of y where the solution curve intersects the cutoff line $x = 0.15$.

Note: Try the secant method with $y_0 = 1$ and $y_1 = 2$.

7. The acidity of a saturated solution of magnesium hydroxide in hydrochloric acid is given by the equation

$$\frac{3.64 \times 10^{-11}}{[H_3O^+]^2} = [H_3O^+] + 3.6 \times 10^{-4}$$

for the hydronium ion concentration $[H_3O^+]$. If we set $x = 10^4[H_3O^+]$, the equation becomes

$$x^3 + 3.6x^2 - 36.4 = 0$$

Determine the value of x and $[H_3O^+]$.

8. Planck's law for the energy density E of blackbody radiation at 1000 K states that

$$E = E(\lambda) = \frac{k\lambda^{-5}}{e^{c/\lambda} - 1}$$

where $k > 0$ is a constant and $c = 0.00014386$. The value of λ at which E is maximum must satisfy the equation

$$(5\lambda - c)e^{c/\lambda} - 5\lambda = 0.$$

Determine the value of λ correct to eight decimal places.

9. Fixed point algorithm can be used as a possible model for turbulence. Each problem deals with the equation

$$x = \lambda x(1 - x) \qquad (E1)$$

as we gradually increase λ from 2 to 5.

(a) $\lambda = 2.6$. Sketch the graphs of $y = x$ and $y = 2.6x(1 - x)$ on the same graph and solve equation (E1) by iteration. Next, solve equation (1) by simple algebra, confirming your answer.

(b) $\lambda = 3.1$. Sketch $y = x$ and $y = f(x) = 3.1x(1-x)$ on the same graph and attempt to solve (E1) by fixed point iteration. (Notice that $|f'(x)| > 1$ at the root.) You will find that x_n bounces back and forth, but gets closer to values r_1 and r_2 called attractors. Find r_1 and r_2 to seven decimal places. Let $f(x) = 3.1x(1-x)$. Superimpose the graph of

$$y = g(x) = f(f(x))$$
$$= 9.61x - 39.401x^2 + 59.582x^3 - 29.79x^4$$

on your earlier graph, and observe that r_1 and r_2 appear to be the two roots of $x = g(x)$, where $|g'(x)| < 1$.

(c) $\lambda = 3.1$ continued. Note that $f(r_1) = r_2$ and $f(r_2) = r_1$. Use this to show that $g'(r_1) = g'(r_2)$.

(d) $\lambda = 3.5$. In this case, use the iteration to find four attractors: s_1, s_2, s_3, and s_4. Guess what equation they are the solution of.

(e) $\lambda = 3.56$. Use the iteration to find eight attractors.

(f) $\lambda = 3.57$. As you keep increasing λ by smaller and smaller amounts, you will double the number of attractors at each stage, until at approximately 3.57 you should get chaos. Beyond $\lambda = 3.57$ other strange things happen.

10. K. Wark and D. E. Richards (*Thermodynamics*, 6th ed., 1999, McGraw-Hill. Boston, Example 14-2, pp. 768-769) compute the equilibrium composition of a mixture of carbon monoxide and oxygen gas at one atmosphere. Determining the final composition requires solving

$$3.06 = \frac{(1-x)(3+x)^{1/2}}{x(1+x)^{1/2}}$$

for x. Obtain a fixed point iteration formula for finding the roots of this equation. If your formula does not converge, develop one that does.

11. A uniform hydro cable $C = 80$ m long with mass per unit length $\rho = 0.5$ kg/m is hung from two supports at the same level $L = 70$ m apart. The tension T in the cable at its lowest point must satisfy the equation

$$\frac{\rho g C}{T} = e^{\rho g L/(2T)} - e^{-\rho g L/(2T)}$$

where $g = 9.81$. If we set $x = \rho g/(2T)$, then x must satisfy the equation

$$2Cx = e^{Lx} - e^{-Lx}.$$

Solve this equation for x and find T correct to seven decimal places.

12. One of the forms of the Colebrook Equation for calculating the Friction Factor f is given by

$$1 + 2\log_{10}\left(\frac{\epsilon/D}{3.7} + \frac{2.51}{Re\sqrt{f}}\right)\sqrt{f} = 0$$

where:

f is the Friction Factor and is dimensionless.

ϵ is the Absolute Roughness and is in units of length.

D is the Inside Diameter and, as these formulas are written, is in the same units as e.

Re is the Reynolds Number and is dimensionless.

ϵ/D is the Relative Roughness and is dimensionless.

This equation can be solved for f given the relative Roughness and the Reynolds number. Find f for the following values of ϵ/D and Re.

  (a) $\epsilon/D = 0.0001$, $Re = 3 \times 10^5$.

  (b) $\epsilon/D = 0.03$, $Re = 1 \times 10^4$.

  (c) $\epsilon/D = 0.01$, $Re = 3 \times 10^5$.

13. A certain nonlinear spring develops a restoring force given by

$$F = c_1 x + c_3 x^3 + c_5 x^5$$

where F is in Kg force and x is in cm. The values of the constants are

$$c_1 = 5.25 \text{ kg/cm},$$
$$c_3 = 0.60 \text{ kg/cm},$$
$$c_5 = 0.0118 \text{ kg/cm}.$$

If a 12.5 kg weight is placed on the spring, how far will it compress?

14. A relationship between V, P, and T of gases is given by the Beattle-Bridgeman equation of state

$$P = \frac{RT}{V} + \frac{a_1}{V^2} + \frac{a_3}{V^3} + \frac{a_4}{V^4}$$

where P is the pressure, V is the volume, T is the temperature and $a_1 = -1.06$, $a_2 = 0.057$, $a_3 = -0.0001$, $R = 0.082$ liter-atm/K-g mole. If $T = 293$ K and $P = 25$ atm, find V using Newton's method with $V_0 = 0.96$ (see Ayyub and McCuen, 1996 p. 91).

15. A tunnel diode is supplied by the manufacturer who provides the following voltage V-current $I = I(V)$ characteristic:

$$I = I(V) = V^3 - 1.5V^2 + 0.6V.$$

The tunnel diode is connected with a resistor R and a voltage source E. Applying Kirchhoff's voltage law, one can find that steady current through the diode must satisfy the equation:

$$I(V) = (E - V)/R.$$

For given E and R, this equation is of the form $f(V) = 0$. Find V for the following values of E and R:

(a) $E = 1.4$, $R = 19$,

(b) $E = 1.8$, $R = 19$.

16. A van der Waals fluid is one which satisfies the equation of state

$$p = \frac{R\theta}{v - b} - \frac{a}{v^2}.$$

Here R, a, b are positive constants, p is the pressure, θ the absolute temperature, and v the volume. Show that if one sets

$$P = \frac{27b^2}{a}p, \quad T = \frac{27Rb}{8a}\theta, \quad V = \frac{v}{3b}$$

the equation of state takes the dimensionless form

$$(P + \frac{3}{V^2})(3V - 1) = 8T.$$

Observe that $V > 1/3$ since $T > 0$.

Find V to five decimal places using a numerical method when

(a) $P = 6$, $T = 2$,

(b) $P = 4$, $T = 1$,

(c) $P = 5$, $T = 5$.

17. A body of mass m projected straight upward with initial velocity v_0 satisfies
the differential equation

$$m\frac{dv}{dt} = -kv - mg, \quad v(0) = v_0$$

where $g = 9.8$ m/sec^2 is the acceleration due to gravity, $k > 0$ is a constant
due to the air resistance, and $v = v(t)$ is the velocity. The solution is

$$v(t) = -\frac{mg}{k} + (v_0 + \frac{mg}{k})e^{-kt/m}.$$

If $y = y(t)$ is the height of the mass above the point from which the mass is
projected that we take to be zero, then the position at time t is

$$y(t) = -\frac{mgt}{k} + \frac{m}{k}\left(v_0 + \frac{mg}{k}\right)\left(1 - e^{-kt/m}\right).$$

Let $v_0 = 25$ m/sec.

(a) Suppose $m = 6$ kg, $k = 3$ kg/sec. How high will the projectile go? When
will it return to earth, i.e., to $y = 0$?

(b) Suppose $m = 12$ kg, $k = 3$ kg/sec. How high will the projectile go?
When will it return to earth?

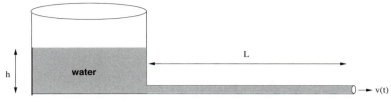

FIGURE 3.13
Water discharge.

(c) When $k = 0$, i.e., there is no air resistance, the equation governing the motion yield

$$\bar{v} = -gt + v_0, \quad \bar{y} = -\frac{gt^2}{2} + v_0 t$$

where the $\bar{v}$ and $\bar{y}$ are the values of the velocity and position when $k = 0$. Let $v_0 = 25$ m/sec. and $m = 6$ kg. Now let successively $k = 1, 0.1, 0.01, 0.001, 0.0001$ and calculate the return times and compare them with the return time for $k = 0$. The numerical evidence should suggest that as $k \to 0$, the return times converge to the value for $k = 0$.

18. Water is discharged from a reservoir through a long pipe (see Figure 3.13). By neglecting the change in the level of the reservoir, the transient velocity $v(t)$ of the water flowing from the pipe at time t is given by

$$v(t) = \sqrt{2gh} \times \tanh\left(\frac{t}{2L}\sqrt{2gh}\right)$$

where h is the height of the fluid in the reservoir, L is the length of the pipe, $g = 9.81$ m/sec^2 is gravity. Find the value of h necessary to achieve a velocity of $v = 4$ m/sec at time $t = 4$ sec, when $L = 5$ m. Use Newton's method for the calculation with the starting value $h_0 = 0.5$. The method should be stopped when the relative change in the solution is below 10^{-6}.

Chapter 4

System of Linear Equations

Linear systems of equations arise in many problems in engineering and science, as well as in mathematics, such as the study of numerical solutions of boundary-value problems and partial differential equations.

Solution algorithms for these types of problems may be either direct or iterative. In direct methods, the solution is obtained in a fixed number of steps subject only to rounding errors, while iterative methods are based on successive improvement of initial guesses for the solution.

In this chapter, both techniques will be considered to solve the linear system

$$
\begin{aligned}
a_{11}x_1 + a_{12}x_2 + \cdots + a_{1n}x_n &= b_1 \\
a_{21}x_1 + a_{22}x_2 + \cdots + a_{2n}x_n &= b_2 \\
\vdots \qquad \vdots \qquad \vdots \qquad \vdots \quad &= \vdots \\
a_{n1}x_1 + a_{n2}x_2 + \cdots + a_{nn}x_n &= b_n
\end{aligned}
\tag{4.1}
$$

for the unknowns $x_1, x_2, \ldots, x_n$, given the coefficients a_{ij}, $i, j = 1, 2, \ldots, n$ and the constants b_i, $i = 1, 2, \ldots, n$.

EXAMPLE 4.1 : Electrical Circuit Analysis Application

Consider the problem of finding the currents in different parts of an electrical circuit, with resistors as shown in Figure 4.1.

Current flows in circuits are governed by Kirchhoff's laws. The first law states that the sum of the voltage drops around a closed loop is zero. The second law states that the voltage drop across a resistor is the product of the current and the resistance. For our network in Figure 4.1 there are three separate loops. Applying Kirchhoff's second law yields the linear system of equations:

$$
\begin{aligned}
R_6 i_1 + R_1(i_1 - i_2) + R_2(i_1 - i_3) &= V_1 : \quad \text{Flow around loop I} \\
R_3 i_2 + R_4(i_2 - i_3) + R_1(i_2 - i_1) &= V_2 : \quad \text{Flow around loop II} \\
R_5 i_3 + R_4(i_3 - i_2) + R_2(i_3 - i_1) &= V_3 : \quad \text{Flow around loop III}
\end{aligned}
$$

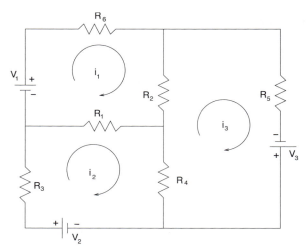

FIGURE 4.1
Electrical circuit.

which simplify to

$$\begin{bmatrix} R_1 + R_2 + R_6 & -R_1 & -R_2 \\ -R_1 & R_1 + R_3 + R_4 & -R_4 \\ -R_2 & -R_4 & R_2 + R_4 + R_5 \end{bmatrix} \begin{bmatrix} i_1 \\ i_2 \\ i_3 \end{bmatrix} = \begin{bmatrix} V_1 \\ V_2 \\ V_3 \end{bmatrix}. \qquad (4.2)$$

4.1   MATRICES AND MATRIX OPERATIONS

Before studying linear systems of equations, it is useful to consider some algebra associated with matrices.

DEFINITION 4.1   *An n by m matrix is a rectangular array of real or complex numbers that can be written as*

$$\mathbf{A} = (a_{ij}) = \begin{bmatrix} a_{11} & a_{12} & \ldots & a_{1m} \\ a_{21} & a_{22} & \ldots & a_{2m} \\ \vdots & \vdots & & \vdots \\ a_{n1} & a_{n2} & \ldots & a_{nm} \end{bmatrix}$$

a_{ij} denotes the *element* or *entry* that occurs in row i and column j of A, and the *size* of a matrix is described by specifying the number of rows and columns that occur in the matrix. If a matrix has only one row, we call it a **row vector**, and a matrix having only one column is called a **column vector**. Capital letters will be used to denote matrices and lowercase letters to denote their entries.

A matrix A with n rows and n columns is called a **square matrix** of order n, and the elements $a_{11}, a_{22}, \ldots, a_{nn}$ are called the **main diagonal** of A.

EXAMPLE 4.2

Let

$$A = \begin{bmatrix} -1 & 3 & 4 & 0 \\ 5 & -6 & 2 & 7 \\ 3 & 0 & 1 & 2 \end{bmatrix}, \quad B = \begin{bmatrix} 1 & -3 & 4 & 2 \end{bmatrix}, \quad C = \begin{bmatrix} \pi \\ 3 \\ 0 \end{bmatrix}.$$

In these examples, A is a 3 by 4 (written 3×4) matrix, B has only one row and is a row vector, and C has only one column and is a column vector.

The following definition gives the basic operations on matrices.

DEFINITION 4.2

i) *If A and B are two matrices of order $n \times m$, then the sum of A and B is the $n \times m$ matrix $C = A + B$ whose entries are*

$$c_{ij} = a_{ij} + b_{ij}$$

ii) *If A is a matrix of order $n \times m$ and λ a real number, then the product of λ and A is the $n \times m$ matrix $C = \lambda A$ whose entries are*

$$c_{ij} = \lambda a_{ij}$$

iii) *If A is a matrix of order $n \times m$ and B is a matrix of order $m \times p$, then the matrix product of A and B is the $n \times p$ matrix $C = AB$ whose entries are*

$$c_{ij} = \sum_{k=1}^{m} a_{ik} b_{kj}$$

iv) *If A is a matrix of order $n \times m$, then the transpose of A is the $m \times n$ matrix $C = A^T$ whose entries are*

$$c_{ij} = a_{ji}$$

EXAMPLE 4.3

If

$$A = \begin{bmatrix} 1 & -2 & 4 \\ 3 & 5 & -6 \end{bmatrix}, \quad B = \begin{bmatrix} 0 & 3 & -7 \\ 1 & -3 & 8 \end{bmatrix}$$

then

$$A + B = \begin{bmatrix} 1 & 1 & -3 \\ 4 & 2 & 2 \end{bmatrix}, \quad -3B = \begin{bmatrix} 0 & -9 & 21 \\ -3 & 9 & -24 \end{bmatrix},$$

$$A - B = \begin{bmatrix} 1 & -5 & 11 \\ 2 & 8 & -14 \end{bmatrix}, \quad A^T = \begin{bmatrix} 1 & 3 \\ -2 & 5 \\ 4 & -6 \end{bmatrix}$$

if

$$A = \begin{bmatrix} 1 & 3 & -1 \\ 2 & 0 & -1 \\ 4 & 2 & -2 \end{bmatrix}, \quad B = \begin{bmatrix} 2 & 1 \\ -3 & 2 \\ 3 & 0 \end{bmatrix}$$

then

$$AB = \begin{bmatrix} -10 & 7 \\ 1 & 2 \\ -4 & 8 \end{bmatrix}, \quad AA = A^2 = \begin{bmatrix} 3 & 1 & -2 \\ -2 & 4 & 0 \\ 0 & 8 & -2 \end{bmatrix}.$$

Certain square matrices have special properties. For example, if the elements below the main diagonal are zero, the matrix is called an **upper triangular** matrix. Thus,

$$U = \begin{bmatrix} 1 & -2 & -1 \\ 0 & 3 & 6 \\ 0 & 0 & 2 \end{bmatrix}$$

is an upper triangular. A square matrix, in which all elements above the main diagonal are zero, is called **lower triangular**. If only the elements in the main diagonal are nonzero, the matrix is called a **diagonal matrix**. An example of a diagonal matrix is

$$I_n = \begin{bmatrix} 1 & 0 & \cdots & 0 \\ 0 & 1 & \ddots & \vdots \\ \vdots & \ddots & \ddots & 0 \\ 0 & \cdots & 0 & 1 \end{bmatrix}$$

which is called the **identity matrix**.

EXERCISE SET 4.1

1. Given the matrices

$$A = \begin{bmatrix} 3 & 8 & 1 & -6 \\ 1 & 1 & 4 & 0 \\ 3 & 2 & 1 & -4 \end{bmatrix}, \quad B = \begin{bmatrix} -2 & 6 & 2 & -3 \\ 3 & 0 & 0 & -4 \\ -5 & 9 & 4 & -4 \end{bmatrix}, \quad C = \begin{bmatrix} -5 \\ -2 \\ 2 \\ 1 \end{bmatrix}.$$

(a) Compute $A + B$, $-4A - 3B$,

(b) Find AC, BC,

(c) Find B^T, $(AB)^T$.

2. Given the matrices

$$A = \begin{bmatrix} -1 & 4 & 6 \\ 4 & -5 & 0 \\ 1 & 1 & -2 \end{bmatrix}, \quad B = \begin{bmatrix} -1 & 3 & 3 & 4 \\ -2 & 0 & -4 & 1 \\ 6 & 2 & 2 & -5 \end{bmatrix}$$

(a) Compute AB, BA,

(b) Find A^2, B^2,

(c) $B^T A$.

3. Find vectors u and v such that

$$L = \begin{bmatrix} 1 & 0 \\ 300 & 1 \end{bmatrix} = I + uv^T.$$

Express L^{-1} in terms of u and v.

4. Given the matrices

$$A = \begin{bmatrix} 1 & -2 & 2 \\ 3 & 1 & 1 \\ 2 & 0 & 1 \end{bmatrix}, \quad B = \begin{bmatrix} -1 & -2 & 4 \\ 1 & 3 & -5 \\ 2 & 4 & -7 \end{bmatrix}, \quad C = \begin{bmatrix} 1 & 0 & 2 \\ 4 & 2 & -1 \\ 2 & 3 & 1 \end{bmatrix}.$$

(a) Show that $AB = BA = I$,

(b) Show that $AI = IA = A$,

(c) Show that $AC \neq CA$ and $BC \neq CB$.

4.2   NAIVE GAUSSIAN ELIMINATION

Consider the system (4.1) in matrix form

$$A\mathbf{x} = \mathbf{b}.$$

Let us denote the original system by $A^{(1)}\mathbf{x} = \mathbf{b}^{(1)}$. That is,

$$\begin{bmatrix} a_{11}^{(1)} & a_{12}^{(1)} & \cdots & a_{1n}^{(1)} \\ a_{21}^{(1)} & a_{22}^{(1)} & \cdots & a_{2n}^{(1)} \\ \vdots & \vdots & \vdots & \vdots \\ a_{n1}^{(1)} & a_{n2}^{(1)} & \cdots & a_{nn}^{(1)} \end{bmatrix} \begin{bmatrix} x_1 \\ x_2 \\ \vdots \\ x_n \end{bmatrix} = \begin{bmatrix} b_1^{(1)} \\ b_2^{(1)} \\ \vdots \\ b_n^{(1)} \end{bmatrix}. \tag{4.3}$$

The Gaussian elimination consists of reducing the system (4.3) to an equivalent system $U\mathbf{x} = \mathbf{d}$, in which U is an upper triangular matrix. This new system can be easily solved by back substitution.

Algorithm:

Step 1: Assume $a_{11}^{(1)} \neq 0$. Define the row multipliers by

$$m_{i1} = \frac{a_{i1}^{(1)}}{a_{11}^{(1)}}.$$

Multiply the first row by m_{i1} and subtract from the ith row $(i = 2, \ldots, n)$ to get

$$a_{ij}^{(2)} = a_{ij}^{(1)} - m_{i1}a_{1j}^{(1)}, \quad j = 2, 3, \cdots, n$$
$$b_i^{(2)} = b_i^{(1)} - m_{i1}b_1^{(1)}.$$

Here, the first rows of A and $\mathbf{b}$ are left unchanged, and the entries of the first column of A below $a_{11}^{(1)}$ are set to zeros.

The result of the transformed system is

$$\begin{bmatrix} a_{11}^{(1)} & a_{12}^{(1)} & \cdots & a_{1n}^{(1)} \\ 0 & a_{22}^{(2)} & \cdots & a_{2n}^{(2)} \\ \vdots & \vdots & \vdots & \vdots \\ 0 & a_{n2}^{(2)} & \cdots & a_{nn}^{(2)} \end{bmatrix} \begin{bmatrix} x_1 \\ x_2 \\ \vdots \\ x_n \end{bmatrix} = \begin{bmatrix} b_1^{(1)} \\ b_2^{(2)} \\ \vdots \\ b_n^{(2)} \end{bmatrix}.$$

We continue in this way. At the kth step we have

Step k: Assume $a_{kk}^{(k)} \neq 0$. Define the row multipliers by

$$m_{ik} = \frac{a_{ik}^{(k)}}{a_{kk}^{(k)}}.$$

Multiply the kth row by m_{ik} and subtract from the ith row $(i = k+1, \ldots, n)$ to get

$$a_{ij}^{(k+1)} = a_{ij}^{(k)} - m_{ik}a_{kj}^{(k)}, \quad j = k+1, \ldots, n$$
$$b_i^{(k+1)} = b_i^{(k)} - m_{ik}b_k^{(k)}.$$

At this step, the entries of column k below the diagonal element are set to zeros, and the rows 1 through k are left undisturbed. The result of the transformed system is

$$\begin{bmatrix} a_{11}^{(1)} & a_{12}^{(1)} & \cdots & a_{1k}^{(1)} & a_{1,k+1}^{(1)} & \cdots & a_{1n}^{(1)} \\ 0 & a_{22}^{(2)} & \cdots & a_{2k}^{(2)} & a_{2,k+1}^{(2)} & \cdots & a_{2n}^{(2)} \\ \vdots & \vdots & & \vdots & \vdots & & \vdots \\ 0 & 0 & \cdots & a_{kk}^{(k)} & a_{k,k+1}^{(k)} & \cdots & a_{kn}^{(k)} \\ 0 & 0 & \cdots & 0 & a_{k+1,k+1}^{(k+1)} & \cdots & a_{k+1,n}^{(k+1)} \\ \vdots & \vdots & & \vdots & \vdots & & \vdots \\ 0 & 0 & \cdots & 0 & a_{n,k+1}^{(k+1)} & \cdots & a_{nn}^{(k+1)} \end{bmatrix} \begin{bmatrix} x_1 \\ x_2 \\ \vdots \\ x_k \\ x_{k+1} \\ \vdots \\ x_n \end{bmatrix} = \begin{bmatrix} b_1^{(1)} \\ b_2^{(2)} \\ \vdots \\ b_k^{(k)} \\ b_{k+1}^{(k+1)} \\ \vdots \\ b_n^{(k+1)} \end{bmatrix}.$$

At $k = n - 1$, we obtain the final triangular system

$$
\begin{array}{rcl}
a_{11}^{(1)}x_1 + a_{12}^{(1)}x_2 + \quad \cdots \quad + a_{1n}^{(1)}x_n &=& b_1^{(1)} \\
a_{22}^{(2)}x_2 + \quad \cdots \quad + a_{2n}^{(2)}x_n &=& b_2^{(2)} \\
\cdots &=& \cdots \\
a_{n-1,n-1}^{(n-1)}x_{n-1} + a_{n-1,n}^{(n-1)}x_n &=& b_{n-1}^{(n-1)} \\
a_{nn}^{(n)}x_n &=& b_n^{(n)}.
\end{array}
$$

Using back substitution, we obtain the following solution of the system

$$
x_n = \frac{b_n^{(n)}}{a_{nn}^{(n)}}
$$

$$
x_{n-1} = \frac{b_{n-1}^{(n-1)} - a_{n-1,n}^{(n-1)}x_n}{a_{n-1,n-1}^{(n-1)}}
$$

$$
x_i = \frac{b_i^{(i)} - (a_{i,i+1}^{(i)}x_{i+1} + \cdots + a_{in}^{(i)}x_n)}{a_{ii}^{(i)}}
$$

$$
= \frac{b_i^{(i)} - \sum_{j=i+1}^{n} a_{ij}^{(i)}x_j}{a_{ii}^{(i)}}, \quad i = n - 2, n - 3, \ldots, 1.
$$

Remarks: In the Gaussian elimination algorithm described above, we used the equations in their natural order and we assumed at each step that the pivot element $a_{kk}^{(k)} \neq 0$. So the algorithm fails if the pivot element becomes zero during the elimination process. In order to avoid an accidental zero pivot, we use what is called **Gaussian elimination** with scaled partial pivoting. This method will be described in the next section.

Operation count. One way of measuring the efficiency of the algorithm of naive Gaussian elimination is to count the number of arithmetic operations required to obtain the solution of the system $A\mathbf{x} = \mathbf{b}$. The current convention is to count the number of multiplications and divisions only because most computers perform additions and subtractions much faster. Furthermore, the number of multiplications and divisions is counted together. Consider the naive Gaussian elimination algorithm described above. Step k requires

$$
(n - k)^2 + (n - k) \quad \text{operations to find the new } a_{ij}
$$

and

$$
(n - k) \quad \text{operations to find the new } b_i.
$$

The operations must be done for $k = 1, 2, \cdots, n - 1$. Hence, using the formulas

$$
\sum_{k=1}^{n} k = \frac{n(n+1)}{2}, \quad \sum_{k=1}^{n} k^2 = \frac{n(n+1)(2n+1)}{6},
$$

the total number of operations is found to be

$$
\frac{n^3 - n}{3} \quad \text{operations applied on the matrix } A
$$

and

$$\frac{n^2 - n}{2} \quad \text{operations applied on the matrix } \mathbf{b}.$$

To solve the resulting triangular system, one needs $(n - i)$ multiplications and one division to get x_i. By summing this over $i = 1, 2, \ldots, n$, we get

$$\frac{n(n + 1)}{2} \quad \text{operations to solve the triangular system.}$$

Thus, we have the following result:

LEMMA 4.1

  The total number of multiplications and divisions required to obtain the solution of an $n \times n$ linear system using naive Gaussian elimination is

$$\frac{n^3}{3} + n^2 - \frac{n}{3}.$$

Hence, for n large the total number of operations is approximately $n^3/3$.

EXAMPLE 4.4

Solve the system of equations

$$\begin{bmatrix} 1 & 1 & 1 & 1 \\ 2 & 3 & 1 & 5 \\ -1 & 1 & -5 & 3 \\ 3 & 1 & 7 & -2 \end{bmatrix} \begin{bmatrix} x_1 \\ x_2 \\ x_3 \\ x_4 \end{bmatrix} = \begin{bmatrix} 10 \\ 31 \\ -2 \\ 18 \end{bmatrix}.$$

The augmented matrix along with the row multipliers m_{i1} are

$$\begin{matrix} \text{pivotal element} \rightarrow \\ m_{21} = 2 \\ m_{31} = -1 \\ m_{41} = 3 \end{matrix} \begin{bmatrix} \mathbf{1} & 1 & 1 & 1 & | & 10 \\ 2 & 3 & 1 & 5 & | & 31 \\ -1 & 1 & -5 & 3 & | & -2 \\ 3 & 1 & 7 & -2 & | & 18 \end{bmatrix}.$$

Subtracting multiples of the first equation from the three others gives

$$\begin{matrix} \\ \text{pivotal element} \rightarrow \\ m_{32} = 2 \\ m_{42} = -2 \end{matrix} \begin{bmatrix} 1 & 1 & 1 & 1 & | & 10 \\ 0 & \mathbf{1} & -1 & 3 & | & 11 \\ 0 & 2 & -4 & 4 & | & 8 \\ 0 & -2 & 4 & -5 & | & -12 \end{bmatrix}.$$

Subtracting multiples, of the second equation from the last two, gives

$$\begin{matrix} \\ \\ \text{pivotal element} \rightarrow \\ m_{43} = -1 \end{matrix} \begin{bmatrix} 1 & 1 & 1 & 1 & | & 10 \\ 0 & 1 & -1 & 3 & | & 11 \\ 0 & 0 & -\mathbf{2} & -2 & | & -14 \\ 0 & 0 & 2 & 1 & | & 10 \end{bmatrix}.$$

```
» A=[1 1 1 1;2 3 1 5;-1 1 -5 3;3 1 7 -2];
» b=[10 31 -2 18]';
» ngaussel(A,b)

   The augmented matrix is

augm =

   1   1   1   1   10
   2   3   1   5   31
  -1   1  -5   3   -2
   3   1   7  -2   18

The transformed upper triangular augmented matrix C is =

C =
   1   1   1   1   10
   0   1  -1   3   11
   0   0  -2  -2  -14
   0   0   0  -1   -4

   The vector solution is =

x =
            1
            2
            3
            4
```

Table 4.1 Naive Gaussian elimination for Example 4.4.

Subtracting multiples, of the third equation from the last one, gives the upper triangular system

$$\left[\begin{array}{cccc|c} 1 & 1 & 1 & 1 & 10 \\ 0 & 1 & -1 & 3 & 11 \\ 0 & 0 & -2 & -2 & -14 \\ 0 & 0 & 0 & -1 & -4 \end{array}\right].$$

The process of the back substitution algorithm applied to the triangular system produces the solution

$$x_4 = \frac{-4}{-1} = 4$$
$$x_3 = \frac{-14 + 2x_4}{-2} = \frac{-6}{-2} = 3$$
$$x_2 = 11 + x_3 - 3x_4 = 11 + 3 - 12 = 2$$
$$x_1 = 10 - x_2 - x_3 - x_4 = 10 - 2 - 3 - 4 = 1$$

given in Table 4.1.

M-function 4.2

The following MATLAB function **ngaussel.m** finds the solution of a linear system using naive Gaussian elimination. INPUTS are an *nxn* matrix *A* and an *n x 1* coefficient vector *b*.

```
function ngaussel(A,b)
% Solve the system Ax=b using naive gaussian elimination
n=length(b);
x=zeros(n,1);
fprintf('\n');
disp('     The augmented matrix is')
augm =[A b]
for k=1:n-1
  for i=k+1:n
    m=A(i,k)/A(k,k);
    for j=k+1:n
      A(i,j)=A(i,j)-m*A(k,j);
    end
    A(i,k)=m;
    b(i)=b(i)-m*b(k);
  end
end
x(n)=b(n)/A(n,n);
for i=n-1:-1:1
  S=b(i);
  for j=i+1:n
    S=S-A(i,j)*x(j);
  end
  x(i)=S/A(i,i);
end
% Print the results
fprintf('\n');
disp(' The transformed upper triangular augmented matrix C is =')
fprintf('\n');
for i=1:n
  for j=1:n
    if (j<i) A(i,j)=0; end
  end
end
C=[A b]
fprintf('\n');
disp('     The vector solution is =')
x
```

EXERCISE SET 4.2

1. Consider the system

$$
\begin{bmatrix}
1 & 1 & 7 & 8 \\
-3 & -3 & -1 & 6 \\
0 & 2 & 2 & 7 \\
5 & 1 & 1 & 0
\end{bmatrix}
\begin{bmatrix}
x_1 \\
x_2 \\
x_3 \\
x_4
\end{bmatrix}
=
\begin{bmatrix}
4 \\
1 \\
2 \\
1
\end{bmatrix}.
$$

Attempt to solve the system by Gaussian elimination. Explain what happens.

2. Solve the following tridiagonal system of 10 equations by "hand" or with the computer.

$$
\begin{bmatrix}
-4 & 1 & & & \\
1 & -4 & 1 & & 0 \\
& & \ddots & & \\
0 & & 1 & -4 & 1 \\
& & & 1 & -4
\end{bmatrix}
\begin{bmatrix}
x_1 \\
x_2 \\
\vdots \\
x_9 \\
x_{10}
\end{bmatrix}
=
\begin{bmatrix}
0.5 \\
2 \\
\vdots \\
2 \\
0.5
\end{bmatrix}.
$$

3. Consider the system

$$
\begin{bmatrix}
3 \cdot 10^{-n} & 3.0 \\
1.0 & 1.0
\end{bmatrix}
\begin{bmatrix}
x_1 \\
x_2
\end{bmatrix}
=
\begin{bmatrix}
2.0 + 10^{-n} \\
1.0
\end{bmatrix}.
$$

Its exact solution is $x_1 = 1/3$, $x_2 = 2/3$.

(a) Compute the numerical solution for $n = 4, 8, 12$.

(b) Compute the residual vector $\bar{\mathbf{r}} = A\bar{\mathbf{x}} - \mathbf{b}$, and the error vector $\bar{\mathbf{e}} = \mathbf{x} - \bar{\mathbf{x}}$, for the different values of n given in (a). $\bar{\mathbf{x}}$ denotes the computed solution and $\mathbf{x}$ the exact solution.

4. The Hilbert matrix A of order n is defined by

$$
a_{ij} = \frac{1}{i + j - 1}, \quad i, j = 1, 2, \ldots, n.
$$

(a) Write down the matrix A for $n = 5$.

(b) If $\mathbf{b} = [1, 0, 0, 0, 0]^T$, solve the system $A\mathbf{x} = \mathbf{b}$ using Gaussian elimination.

5. Write the following systems of equations in matrix form and then solve them:

(a)
$$
\begin{aligned}
x + y + 7z + v + 2w &= 1 \\
2x + 1y + 6z + 6v + 3w &= -2 \\
- y + 2z + 3v + 2w &= -3 \\
5x - y + 4z + 7v + w &= 3 \\
2x + 3y + 2z + w &= 4.
\end{aligned}
$$

(b) $3x + 5y - 2z - w = -12$
         $-2x + y - 9z - 2w = 1$
         $x + 7y - 6z - 3w = 10$
         $-3x + y - z - 4w = 7.$

(c) $2x + y + z = 7$
         $2x + 2y + 3z = 10$
         $-4x + 4y + 5z = 14.$

COMPUTER PROBLEM SET 4.2

1. Write a computer program in a language of your choice to solve a system of
   n linear equations and n unknowns using naive Gaussian elimination.

   Input data to the program should be:

   (a) The number of equations n.
   (b) The augmented matrix.

   Output should consist of:

   (a) The augmented matrix.
   (b) The vector solution.

   Test your program to solve the linear system of equations

   $$\begin{bmatrix} 1 & -1 & 0 & 5 \\ 3 & -2 & 1 & -1 \\ 1 & 1 & 9 & 4 \\ 1 & -7 & 2 & 3 \end{bmatrix} \begin{bmatrix} x_1 \\ x_2 \\ x_3 \\ x_4 \end{bmatrix} = \begin{bmatrix} 18 \\ 8 \\ 47 \\ 32 \end{bmatrix}.$$

2. Use the MATLAB function ngaussel.m to solve the system of equations

   $$\begin{bmatrix} 1 & -1 & 2 & 5 & -7 & -8 \\ 3 & -9 & 1 & -1 & 8 & 1 \\ -1 & 1 & 9 & -9 & 2 & 3 \\ 1 & 7 & 2 & -3 & -1 & 4 \\ 7 & 1 & -2 & 4 & 1 & -1 \\ 2 & 3 & -9 & 12 & -2 & 7 \end{bmatrix} \begin{bmatrix} x_1 \\ x_2 \\ x_3 \\ x_4 \\ x_5 \\ x_6 \end{bmatrix} = \begin{bmatrix} -12 \\ 8 \\ 20 \\ 39 \\ 16 \\ 45 \end{bmatrix}.$$

3. Write a MATLAB function to solve the system of equations that finds the
   coefficients a_1, a_2, a_3 of a parabola defined by $y = a_1 x^2 + a_2 x + a_3$ given three
   points through which the parabola passes. Find the equation of the parabolas
   passing through the following set of points.

   (a) $(1, 0)$, $(-2, 21)$, $(0, 1)$,
   (b) $(1, 5.0)$, $(-2, 7.1)$, $(3, 6.6)$,
   (c) $(0, \sqrt{2})$, $(-1, \pi + 1 + \sqrt{2})$, $(3, 9\pi - 3 + \sqrt{2})$.

4.3 GAUSSIAN ELIMINATION WITH SCALED PARTIAL PIVOTING

Having a zero pivot element is not the only source of trouble that can arise when we apply naive Gaussian elimination. Under certain circumstances, the pivot element can become very small in magnitude compared to the rest of the elements in the pivot row. This can dramatically increase the round-off error, which can result in an inaccurate solution vector. To illustrate some of the effects of round-off error in the elimination process, we apply naive Gaussian elimination to the system

$$0.0002x_1 + 1.471x_2 = 1.473$$
$$0.2346x_1 - 1.317x_2 = 1.029$$

using four-digit floating-point arithmetic with rounding. The exact solution of this system is $x_1 = 10.00$ and $x_2 = 1.000$.

The multiplier for this system is

$$m_{21} = \frac{0.2346}{0.0002} = 1173.$$

Applying naive Gaussian elimination and performing the appropriate rounding gives

$$0.0002x_1 + 1.471x_2 = 1.473$$
$$- 1726.x_2 = -1727.$$

Hence,

$$x_2 = \frac{-1727.}{-1726.}$$
$$= 1.001$$

$$x_1 = \frac{1.473 - (1.471)(1.001)}{0.0002}$$
$$= \frac{1.473 - 1.472}{0.0002}$$
$$= 5.000.$$

As one can see, x_2 is a close approximation of the actual value. However, the relative error in the computed solution for x_1 is very large: 50%. The failure of naive Gaussian elimination in this example results from the fact that $|a_{11}| = 0.0002$ is small compared to $|a_{12}|$. Hence, a relatively small error due to round-off in the computed value, x_2, led to a relatively large error in the computed solution, x_1.

A useful strategy to avoid the problem of having a zero or very small pivot element is to use Gaussian elimination with scaled partial pivoting. In this method, the equations of the system (4.3) are used in a different order, and the pivot equation is selected by looking for the absolute largest coefficient of x_k relative to the size of the equation.

The basic idea in elimination with partial pivoting is to avoid small pivots and control the size of the multipliers. The order in which the equations would be used as pivot equations is determined by the index vector that we call $\mathbf{d} = [d_1, d_2, \ldots, d_n]$. At the beginning we set $\mathbf{d} = [1, 2, \ldots, n]$. We then define the scale vector

$$\mathbf{c} = [c_1, c_2, \ldots, c_n]$$

where

$$c_i = \max_{1 \le j \le n} |a_{ij}|, \quad i = 1, 2, \ldots, n.$$

The elimination with scaled partial pivoting consists of choosing the pivot equation such that the ratio $|a_{i,1}|/c_i$ is greatest. To do that we define the ratio vector

$$\mathbf{r} = [r_1, r_2, \ldots, r_n]$$

where

$$r_i = \frac{|a_{i1}|}{c_i}, \quad i = 1, 2, \ldots, n.$$

If r_j is the largest element in $\mathbf{r}$, we interchange d_1 and d_j in the index vector $\mathbf{d}$ to get the starting index vector

$$\mathbf{d} = [d_j, d_2, \ldots, d_1, \ldots, d_n].$$

This means that row j is the pivot equation in step 1. The Gaussian elimination is then used to get an equivalent system of equation with zeros below and above the pivot element. Note that during the elimination process, only elements in the index vector $\mathbf{d}$ have been interchanged and not the equations.

The process continues in this way until the end of step $(n - 1)$ where a final index vector is obtained containing the order in which the pivot equations were selected. The solution of the system of equation is then obtained by performing a back substitution, reading the entries of the index vector from the last to the first.

We shall now illustrate the method by solving the system

$$\begin{bmatrix} 1 & 3 & -2 & 4 \\ 2 & -3 & 3 & -1 \\ -1 & 7 & -4 & 2 \\ 3 & -1 & 6 & 2 \end{bmatrix} \begin{bmatrix} x_1 \\ x_2 \\ x_3 \\ x_4 \end{bmatrix} = \begin{bmatrix} -11 \\ 6 \\ -9 \\ 15 \end{bmatrix}. \tag{4.4}$$

At the beginning, the index vector is $\mathbf{d} = [1, 2, 3, 4]$. We first define the scaled vector by

$$\mathbf{c} = [c_1, c_2, c_3, c_4]$$

where

$$c_i = \max_{1 \le j \le 4} |a_{ij}|, \quad i = 1, 2, 3, 4.$$

For our example (4.4) the scaled vector is

$$\mathbf{c} = [4, 3, 7, 6].$$

We emphasize that the scaled vector will remain unchanged through the entire elimination process.

In order to determine the index vector, we now define the ratio vector by

$$\mathbf{r} = [r_1, r_2, r_3, r_4]$$

where

$$r_i = \frac{|a_{i1}|}{c_i}, \quad i = 1, 2, 3, 4.$$

In our example, it is

$$\mathbf{r} = \left[\frac{1}{4}, \frac{2}{3}, \frac{1}{7}, \frac{1}{2}\right];$$

since $2/3$ is the largest element in $\mathbf{r}$, we interchange elements 1 and 2 in the index vector to get

Step 1: $\mathbf{d} = [2, 1, 3, 4]$. This means that row 2 is the pivot equation.
We now use Gaussian elimination to get

$$\begin{bmatrix} 0 & 9/2 & -7/2 & 9/2 \\ 2 & -3 & 3 & -1 \\ 0 & 11/2 & -5/2 & 3/2 \\ 0 & 7/2 & 3/2 & 7/2 \end{bmatrix} \begin{bmatrix} x_1 \\ x_2 \\ x_3 \\ x_4 \end{bmatrix} = \begin{bmatrix} -14 \\ 6 \\ -6 \\ 6 \end{bmatrix}.$$

We continue in this way.

Step 2: $\mathbf{r} = \left[\frac{9}{8}, \frac{11}{14}, \frac{7}{12}\right]$.

As $9/8$ is the largest element in $\mathbf{r}$, nothing is to be interchanged in the index vector, thus

$$\mathbf{d} = [2, 1, 3, 4].$$

This means that row 1 becomes the new pivot equation.
We now apply Gaussian elimination again to get

$$\begin{bmatrix} 0 & 9/2 & -7/2 & 9/2 \\ 2 & -3 & 3 & -1 \\ 0 & 0 & 16/9 & -4 \\ 0 & 0 & 38/9 & 0 \end{bmatrix} \begin{bmatrix} x_1 \\ x_2 \\ x_3 \\ x_4 \end{bmatrix} = \begin{bmatrix} -14 \\ 6 \\ 100/9 \\ 152/9 \end{bmatrix}.$$

Step 3: $\mathbf{r} = \left[\frac{16}{63}, \frac{38}{54}\right]$.

As $38/9 > 16/9$ in $\mathbf{r}$ we interchange row 3 and 4 to get the new index vector

$$\mathbf{d} = [2, 1, 4, 3].$$

This means that row 4 becomes the new pivot row. We now apply Gaussian elimination to get

$$\begin{bmatrix} 0 & 9/2 & -7/2 & 9/2 \\ 2 & -3 & 3 & -1 \\ 0 & 0 & 0 & -4 \\ 0 & 0 & 38/9 & 0 \end{bmatrix} \begin{bmatrix} x_1 \\ x_2 \\ x_3 \\ x_4 \end{bmatrix} = \begin{bmatrix} -14 \\ 6 \\ 4 \\ 152/9 \end{bmatrix}.$$

Finally, back substitution produces the solution

$$x_4 = \frac{4}{-4} = -1$$

$$x_3 = \frac{1}{38/9}(152/9) = 4$$

$$x_2 = \frac{1}{9/2}[-14 + 4(7/2) - 1(-9/2)] = 1$$

$$x_1 = \frac{1}{1/2}[6 + 3 - 12 - 1] = -2.$$

Note that during the elimination process, only elements in the index vector **d** have been interchanged and not the equations.

EXAMPLE 4.5

Solve the system of equation using Gaussian elimination with scaled partial pivoting

$$\begin{bmatrix} 1 & 3 & -2 & 4 \\ 2 & -3 & 3 & -1 \\ -1 & 7 & -4 & 2 \\ 3 & -1 & 6 & 2 \end{bmatrix} \begin{bmatrix} x_1 \\ x_2 \\ x_3 \\ x_4 \end{bmatrix} = \begin{bmatrix} -11 \\ 6 \\ -9 \\ 15 \end{bmatrix}.$$

The computed results are shown in Table 4.2.

EXAMPLE 4.6

Solve the system of equation using Gaussian elimination with scaled partial pivoting

$$\begin{bmatrix} 0.0002 & 1.471 \\ 0.2346 & -1.317 \end{bmatrix} \begin{bmatrix} x_1 \\ x_2 \end{bmatrix} = \begin{bmatrix} 1.473 \\ 1.029 \end{bmatrix}.$$

The MATLAB function gaussel.m gives the solution

$$x_1 = 10 \quad \text{and} \quad x_2 = 1.$$

Ill-conditioning

A linear system is said to be *ill-conditioned* if the coefficient matrix tends to be singular, that is, small perturbations in the coefficient matrix will produce large

```
» A=[1 3 -2 4;2 -3 3 -1;-1 7 -4 2;3 -1 6 2];
» b=[-11 6 -9 15]';
» gaussel(A,b)

   The augmented matrix is =

augm =

   1    3   -2    4   -11
   2   -3    3   -1    6
  -1    7   -4    2   -9
   3   -1    6    2   15

The scale vector =

c =
       4    3    7    6

The index vector =

d =
       2    1    4    3

   The transformed upper triangular augmented matrix C is =

C =
         0        4.5       -3.5       4.5       -14
         2         -3          3        -1         6
         0          0          0        -4         4
         0          0       4.22         0     16.89

   The vector solution is =

x =
        -2
         1
         4
        -1
```

Table 4.2 Gaussian elimination for Example 4.5.

changes in the solution of the system. For example, consider the following system in two equations:

$$\begin{bmatrix} 1.00 & 2.00 \\ 0.49 & 0.99 \end{bmatrix} \begin{bmatrix} x_1 \\ x_2 \end{bmatrix} = \begin{bmatrix} 3.00 \\ 1.47 \end{bmatrix}.$$

The exact solution of this system is $\mathbf{x} = \begin{bmatrix} 3 & 0 \end{bmatrix}'$. Now consider the system

$$\begin{bmatrix} 1.00 & 2.00 \\ 0.48 & 0.99 \end{bmatrix} \begin{bmatrix} x_1 \\ x_2 \end{bmatrix} = \begin{bmatrix} 3.00 \\ 1.47 \end{bmatrix}$$

obtained by changing the entry $a_{21} = 0.49$ of the previous system to 0.48.

The exact solution of this new system is $\mathbf{x} = \begin{bmatrix} 1 & 1 \end{bmatrix}'$ which differs significantly from the first one. Then it is realized that ill-conditioning is present. The difficulty arising from ill-conditioning cannot be solved by simple refinements in the Gaussian elimination procedure. To find out if a system of equations $\mathbf{Ax=b}$ is ill-conditioned, one has to compute the so-called *condition number* of the matrix $\mathbf{A}$. A detailed analysis of this topic is normally introduced in advanced numerical analysis texts.

MATLAB's Methods

In MATLAB the solution of the linear systems $\mathbf{Ax=b}$ is obtained by using the backslash operator "\" ("left division"). The following is a simple example:

```
>> A=[4 -1 1; 2 5 2; 1 2 4];
>> b=[8 3 11]';
>> x=A\b
x =
      1.0000
     -1.0000
      3.0000
```

The \ operator does not solve the system $\mathbf{Ax=b}$ by evaluating $A^{-1}b$. It chooses an appropriate and efficient algorithm that depends on the properties of the matrix $\mathbf{A}$. For example, if $\mathbf{A}$ is symmetric and has positive diagonals, it attempts to use the Cholesky factorization.

EXAMPLE 4.7

Solve the Electrical Circuit problem (see Example 4.1) using the \ operator for $R_1 = 15$, $R_2 = 20$, $R_3 = 10$, $R_4 = 15$, $R_5 = 30$, $R_6 = 10$, $V_1 = 100$, $V_2 = 0$, and $V_3 = 200$.

The system of equation to be solved is

$$\begin{bmatrix} 45 & -15 & -20 \\ -15 & 40 & -15 \\ -20 & -15 & 65 \end{bmatrix} \begin{bmatrix} i_1 \\ i_2 \\ i_3 \end{bmatrix} = \begin{bmatrix} 100 \\ 0 \\ 200 \end{bmatrix}.$$

The MATLAB solution is

```
>> A=[45 -15 -20; -15 40 -15; -20 -15 65];
>> b=[100 0 200]';
>> i=A\b
i =
      6.5799
      4.7955
      6.2082
```

M-function 4.3

The following MATLAB function **gaussel.m** finds the solution of a linear system using Gaussian elimination with scaled partial pivoting. INPUTS are an $n \times n$ matrix A and an $n \times 1$ coefficient vector b.

```
function gaussel(A,b)
% Solve the system Ax=b using Gaussian elimination with scaled
% partial pivoting.
n=length(b);
x=zeros(n,1);
fprintf('\n');
disp('    The augmented matrix is =')
augm =[A b]
for i=1:n
  d(i)=i;
  smax=0;
  for j=1:n
    smax=max(smax,abs(A(i,j)));
  end
  c(i)=smax;
end
for k=1:n-1
  rmax=0;
  for i=k:n
    R=abs(A(d(i),k))/c(d(i));
    if (R>rmax)
      j=i;
      rmax=R;
    end
  end
  dk=d(j);
  d(j)=d(k);
  d(k)=dk;
  for i=k+1:n
    m=A(d(i),k)/A(dk,k);
    for j=k+1:n
      A(d(i),j)=A(d(i),j)-m*A(dk,j);
    end
    A(d(i),k)=m;
  end
end
% Perform the back substitution.
```

```
for k=1:n-1
  for i=k+1:n
    b(d(i))=b(d(i))-b(d(k))*A(d(i),k);
  end
end
x(n)=b(d(n))/A(d(n),n);
for i=n-1:-1:1
  S=b(d(i));
  for j=i+1:n
    S=S-A(d(i),j)*x(j);
  end
  x(i)=S/A(d(i),i);
end
disp('The scale vector =')
c
disp('The index vector =')
d
fprintf('\n');
disp(' The transformed upper triangular augmented matrix C is =')
fprintf('\n');
for i=1:n
  M(i,:)=A(d(i),:);
end
for i=1:n
  for j=1:n
    if (j<i) M(i,j)=0; end
  end
end
C=[M b]
fprintf('\n');
disp('   The vector solution is =')
x
```

EXERCISE SET 4.3

1. Solve the following systems using Gaussian elimination with partial scaled pivoting:

   (a) $6x + 4y + 13z = -23$
          $2x + \ y - \ \ z = \ \ \ 4$
          $-3x + 6y - \ \ z = \ \ \ 8$

   (b) $\begin{bmatrix} -2 & -3 & 1 & 2 \\ 7 & 6 & 0 & -3 \\ 0 & 3 & 1 & 5 \\ 2 & -2 & 6 & 6 \end{bmatrix} \begin{bmatrix} x_1 \\ x_2 \\ x_3 \\ x_4 \end{bmatrix} = \begin{bmatrix} 2 \\ -4 \\ 1 \\ 8 \end{bmatrix}.$

2. If the Gaussian algorithm with scaled partial pivoting is used on the following

system, what is the scaled vector? What is the second pivot row?

$$\begin{bmatrix} 2 & 4 & 29 \\ -1 & 5 & 4 \\ 1 & -1 & 2 \end{bmatrix} \begin{bmatrix} x_1 \\ x_2 \\ x_3 \end{bmatrix} = \begin{bmatrix} 15 \\ -3 \\ 1 \end{bmatrix}.$$

3. Consider the system

$$\begin{bmatrix} 3 & 3 & 7 \\ 1 & 1 & -1 \\ -2 & -2 & 3 \end{bmatrix} \begin{bmatrix} x_1 \\ x_2 \\ x_3 \end{bmatrix} = \begin{bmatrix} 1 \\ 4 \\ 3 \end{bmatrix}.$$

Attempt to solve the system by Gaussian elimination with scaled partial pivoting. Explain what happens.

4. Compute the inverse A^{-1}, where

$$A = \begin{bmatrix} 2 & 1 & 2 \\ 1 & 2 & 3 \\ 4 & 1 & 2 \end{bmatrix}$$

by solving the system $AX = I$, using scaled partial pivoting.

5. Solve the system

$$\begin{bmatrix} -1 & 2 & -1 & 0 & 4 \\ 1 & 2 & 0 & 3 & 0 \\ 0 & -3 & 1 & 1 & 2 \\ 1 & 0 & 2 & -1 & 3 \\ 2 & -2 & 2 & -2 & 1 \end{bmatrix} \begin{bmatrix} x_1 \\ x_2 \\ x_3 \\ x_4 \\ x_5 \end{bmatrix} = \begin{bmatrix} 1 \\ 2 \\ 4 \\ 0 \\ 3 \end{bmatrix}$$

using Gaussian elimination with scaled partial pivoting. Find the scaled and index vectors at each step.

6. Find the scale vector and the third pivot row if Gaussian elimination with scaled partial pivoting is used on the matrix

$$\begin{bmatrix} -2 & 4 & 5 & -1 \\ -1 & 2 & 9 & 4 \\ 6 & 2 & -9 & -5 \\ 3 & 0 & 3 & -8 \end{bmatrix}.$$

7. Use both naive Gaussian elimination and Gaussian elimination with scaled partial pivoting to solve the following linear system using four-decimal floating point arithmetic

$$0.0003x_1 + 1.354x_2 = 1.357.$$
$$0.2322x_1 - 1.544x_2 = 0.7780.$$

8. Use Gaussian elimination with scaled partial pivoting and two-digit rounding arithmetic to solve the linear system

$$0.0001x_1 + 66.x_2 = 65.$$
$$2.0x_1 - 28.x_2 = 33.$$

9. Consider the system

$$\begin{bmatrix} 2.1 & 1.8 \\ 6.2 & 5.3 \end{bmatrix} \begin{bmatrix} x_1 \\ x_2 \end{bmatrix} = \begin{bmatrix} 2.1 \\ 6.2 \end{bmatrix}.$$

Solve the system using Gaussian elimination. Change the entry $a_{21} = 6.2$ to 6.1 and solve the system again. From the solutions obtained, what conclusion can be made about the system of equations?

10. The Hilbert matrix A of order n is defined by

$$a_{ij} = \frac{1}{i + j - 1}, \quad i, j = 1, 2, \dots, n.$$

It is a classical example of an ill-conditioned matrix; small changes in its entries will produce a large change in the solution to the system $\mathbf{Ax=b}$.

(a) Solve the system $\mathbf{Ax} = [1\ 0\ 0]'$ for $n = 3$ using Gaussian elimination.

(b) Approximate the coefficient of Hilbert matrix to three decimal digits, that is take

$$\tilde{A} = \begin{bmatrix} 1.0 & 0.5 & 0.333 \\ 0.5 & 0.333 & 0.25 \\ 0.333 & 0.25 & 0.2 \end{bmatrix}$$

and solve the system $\tilde{A}\mathbf{x} = [1\ 0\ 0]'$. Compare with the solution obtained in (a).

11. Consider the linear system $Ax = b$, where

$$A = \begin{bmatrix} 1 & 1+\varepsilon \\ 1-\varepsilon & 1 \end{bmatrix}, \quad b = \begin{bmatrix} 2-\varepsilon^2 \\ 2-2\varepsilon \end{bmatrix}, \quad x = \begin{bmatrix} 1 \\ 1-\varepsilon \end{bmatrix}.$$

The inverse A^{-1} of A is given by

$$A^{-1} = \varepsilon^{-2} \begin{bmatrix} 1 & -1-\varepsilon \\ -1+\varepsilon & 1 \end{bmatrix}.$$

Compute for $\varepsilon = 10^{-1}, \dots, 10^{-6}$ the corresponding solution x of the above system by

(a) using the Gaussian elimination,

(b) direct multiplication $A^{-1}b$.

12. Consider the *sparse* matrix $A = (a_{ij})_{1 \le i,j \le 10}$ with $a_{ij} \ne 0$ for $i = 1$, $j = 1, ..., 10$, as well as $j = 1$, $i = 1, ..., 10$ and $i = j$, $j =, ..., 10$; otherwise $a_{ij} = 0$.

(a) Sketch the *sparsity pattern* of the such a matrix, i.e., where are the nonzero entries.

(b) Verify that the first step of the Gaussian elimination without pivoting transform zero entries into nonzero entries.

(c) Show that we can avoid transforming zero entries into nonzero entries, by simply interchanging rows and/or columns of the matrix A before applying the Gaussian elimination.

13. The linear system

$$\begin{bmatrix} 10^{-6} & 10^{-6} & 1 \\ 10^{-6} & -10^{-6} & 1 \\ 1 & 1 & 2 \end{bmatrix} \begin{bmatrix} x_1 \\ x_2 \\ x_3 \end{bmatrix} = \begin{bmatrix} 2 \cdot 10^{-6} \\ -2 \cdot 10^{-6} \\ 1 \end{bmatrix}$$

has the exact solution

$$x_1 = \frac{-1}{1 - 2 \cdot 10^{-6}}, \quad x_2 = 2, \quad x_3 = \frac{10^{-6}}{1 - 2 \cdot 10^{-6}}.$$

Solve the system using four-digits floating-point arithmetic without pivoting.

COMPUTER PROBLEM SET 4.3

1. Write a computer program in a language of your choice to solve a system of n linear equations and n unknowns using Gaussian elimination with scaled partial pivoting.
   Input data to the program should be

(a) The number of equations n.

(b) The augmented matrix.

Output should consist of:

(a) The augmented matrix.

(b) The vector solution.

Test your program to solve the linear system

$$\begin{bmatrix} -2 & -3 & 1 & 2 \\ 7 & 6 & 0 & -3 \\ 0 & 3 & 1 & 5 \\ 2 & -2 & 6 & 6 \end{bmatrix} \begin{bmatrix} x_1 \\ x_2 \\ x_3 \\ x_4 \end{bmatrix} = \begin{bmatrix} 2 \\ -4 \\ 1 \\ 8 \end{bmatrix}.$$

2. Use the MATLAB function gaussel.m to solve the following linear system of equations

$$\begin{bmatrix} 0.824 & -0.065 & -0.814 & -0.741 \\ -0.979 & -0.764 & 0.216 & 0.663 \\ 0.880 & 0.916 & 0.617 & -0.535 \\ 0.597 & -0.245 & 0.079 & 0.747 \end{bmatrix} \begin{bmatrix} x_1 \\ x_2 \\ x_3 \\ x_4 \end{bmatrix} = \begin{bmatrix} 0.477 \\ -0.535 \\ -0.906 \\ -0.905 \end{bmatrix}.$$

3. The linear system

$$\begin{bmatrix} 10 & 7 & 8 & 7 \\ 7 & 5 & 6 & 5 \\ 8 & 6 & 10 & 9 \\ 7 & 5 & 9 & 10 \end{bmatrix} \begin{bmatrix} x_1 \\ x_2 \\ x_3 \\ x_4 \end{bmatrix} = \begin{bmatrix} 32 \\ 23 \\ 33 \\ 31 \end{bmatrix}$$

in which the coefficient matrix is known as Wilson's matrix is badly unstable. Use the MATLAB function gaussel.m to solve the system and compare your solution with the exact one $[1\ 1\ 1\ 1]'$. Find the solution of the system if the coefficient vector is changed to

(a) $[32.1\ 22.9\ 32.9\ 31.1]'$,

(b) $[32.01\ 22.99\ 32.99\ 31.01]'$.

What conclusion can you make regarding the solutions obtained?

4. Solve the following system using the MATLAB functions gaussel.m.

$$\begin{bmatrix} -3.481 & 4.701 & -6.085 & -6.867 \\ 7.024 & -1.611 & -2.494 & -4.337 \\ 9.661 & 9.023 & -9.422 & 7.129 \\ 7.614 & -8.928 & -5.643 & -8.261 \end{bmatrix} \begin{bmatrix} x_1 \\ x_2 \\ x_3 \\ x_4 \end{bmatrix} = \begin{bmatrix} 2.511 \\ -4.144 \\ -7.448 \\ -2.143 \end{bmatrix}.$$

5. Use the MATLAB function gaussel.m to solve the linear system

$$\begin{bmatrix} 1 & -2 & 3 & 1 & 2 \\ 1 & 2 & -1 & -1 & 0 \\ 2 & 1 & -3 & -2 & -1 \\ 1 & -2 & 2 & 1 & 0 \\ 1 & -1 & -3 & -2 & 0 \end{bmatrix} \begin{bmatrix} w_1 \\ w_2 \\ w_3 \\ w_4 \\ w_5 \end{bmatrix} = \begin{bmatrix} 4 \\ 0 \\ -5 \\ 2 \\ -6 \end{bmatrix}.$$

6. Use MATLAB function gaussel.m to solve the system of equations that finds the coefficients a_1, a_2, a_3 of a parabola defined by $y = a_1 x^2 + a_2 x + a_3$ given three points through which the parabola passes. Find the equation of the parabolas passing through the following set of points.

(a) $(1, 0)$, $(-2, 21)$, $(0, 1)$,

(b) $(1, 5.0)$, $(-2, 7.1)$, $(3, 6.6)$,

(c) $(0, \sqrt{2})$, $(-1, \pi + 1 + \sqrt{2})$, $(3, 9\pi - 3 + \sqrt{2})$.

7. An electrical network has two voltage sources and six resistors. By applying both Ohm's law and Kirchhoff's Current law, we get the following linear system of equations:

$$\begin{bmatrix} R_1 + R_3 + R_4 & R_3 & R_4 \\ R_3 & R_2 + R_3 + R_5 & -R_5 \\ R_4 & -R_5 & R_4 + R_5 + R_6 \end{bmatrix} \begin{bmatrix} i_1 \\ i_2 \\ i_3 \end{bmatrix} = \begin{bmatrix} V_1 \\ V_2 \\ 0 \end{bmatrix}.$$

Use the MATLAB function gaussel.m to solve the linear system for the current i_1, i_2, and i_3 if

(a) $R_1 = 1$, $R_2 = 2$, $R_3 = 1$, $R_4 = 2$, $R_5 = 1$, $R_6 = 6$, and $V_1 = 20$, $V_2 = 30$,

(b) $R_1 = 1$, $R_2 = 1$, $R_3 = 1$, $R_4 = 2$, $R_5 = 2$, $R_6 = 4$, and $V_1 = 12.5$, $V_2 = 22.5$,

(c) $R_1 = 2$, $R_2 = 2$, $R_3 = 4$, $R_4 = 1$, $R_5 = 4$, $R_6 = 3$, and $V_1 = 40$, $V_2 = 36$.

8. To decompose the expression $\frac{4x^3+4x^2+x-1}{x^2(x+1)^2}$ into a sum of partial fractions, we write

$$\frac{4x^3 + 4x^2 + x - 1}{x^2(x+1)^2} = \frac{A_1}{x} + \frac{A_2}{x^2} + \frac{A_3}{x+1} + \frac{A_4}{(x+1)^2}.$$

The coefficients A_i, for $i = 1, \ldots, 4$, are given by the solution of a system of equations. Write the system of equations in a matrix form and then use the MATLAB function gaussel.m to find the coefficients A_i, for $i = 1, \ldots, 4$.

9. Use the MATLAB function gaussel.m to solve the following linear system:

$$\begin{bmatrix} 0.378 & 0.6321 & 0.0662 & 0.2457 & 0.1281 & 0.4414 \\ 0.6924 & 0.7461 & 0.2286 & 0.8044 & 0.3281 & 0.5646 \\ 0.7920 & 0.2994 & 0.5044 & 0.5787 & 0.0826 & 0.2399 \\ 0.7758 & 0.3962 & 0.4622 & 0.4058 & 0.1713 & 0.3482 \\ 0.9570 & 0.6141 & 0.5165 & 0.4767 & 0.3346 & 0.8807 \\ 0.0432 & 0.8976 & 0.5470 & 0.2036 & 0.8721 & 0.2812 \end{bmatrix} \begin{bmatrix} x_1 \\ x_2 \\ x_3 \\ x_4 \\ x_5 \\ x_6 \end{bmatrix} = \begin{bmatrix} 12.3428 \\ -10.6701 \\ 6.9823 \\ 8.4511 \\ 3.0012 \\ -13.1223 \end{bmatrix}.$$

4.4 LU DECOMPOSITION

Consider the system of equations

$$A\mathbf{x} = \mathbf{b}.$$

The *LU* decomposition consists of transforming the coefficient matrix A into the product of two matrices, L and U, where L is a lower triangular matrix and U is an upper triangular matrix having 1's on its diagonal.

Once L and U are found, the solution of the system $A\mathbf{x} = \mathbf{b}$ can be carried out by writing

$$LU\mathbf{x} = \mathbf{b}$$

and setting

$$U\mathbf{x} = \mathbf{y} \qquad (4.5)$$

so that

$$L\mathbf{y} = \mathbf{b}. \qquad (4.6)$$

Eqns. (4.5) and (4.6) are two triangular systems which can be easily solved by first using the forward substitution in (4.6) to get $\mathbf{y}$, and then with $\mathbf{y}$ known, we use the back substitution in (4.5) to get $\mathbf{x}$. Two types of factorizations will now be presented, the first one uses Crout's and Cholesky's methods and the second one uses the Gaussian elimination method.

4.4.1   Crout's and Cholesky's methods

We shall illustrate the method of finding L and U in the case of a 4×4 matrix: We wish to find L, having nonzero diagonal entries, and U such that

$$\begin{bmatrix} l_{11} & 0 & 0 & 0 \\ l_{21} & l_{22} & 0 & 0 \\ l_{31} & l_{32} & l_{33} & 0 \\ l_{41} & {}_{42} & l_{43} & l_{44} \end{bmatrix} \begin{bmatrix} 1 & u_{12} & u_{13} & u_{14} \\ 0 & 1 & u_{23} & u_{24} \\ 0 & 0 & 1 & u_{34} \\ 0 & 0 & 0 & 1 \end{bmatrix} = \begin{bmatrix} a_{11} & a_{12} & a_{13} & a_{14} \\ a_{21} & a_{22} & a_{23} & a_{24} \\ a_{31} & a_{32} & a_{33} & a_{34} \\ a_{41} & a_{42} & a_{43} & a_{44} \end{bmatrix}.$$

Multiplying the rows of L by the first column of U, one gets

$$l_{i1} = a_{i1}, \quad i = 1, 2, 3, 4.$$

Hence, the first column of L is given by the first column of A. Next, multiply the columns of U by the first row of L to get

$$l_{11}u_{1i} = a_{1i}, \quad i = 2, 3, 4.$$

Thus,

$$u_{1i} = \frac{a_{1i}}{l_{11}}, \quad i = 2, 3, 4,$$

which give the first row of U.

We continue in this way by getting alternatively a column of L and a row of U. The result is

$$l_{i2} = a_{i2} - l_{i1}u_{12}, \qquad\qquad i = 2, 3, 4.$$

$$u_{2i} = \frac{a_{2i} - l_{21}u_{1i}}{l_{22}}, \qquad\qquad i = 3, 4.$$

$$l_{i3} = a_{i3} - l_{i1}u_{13} - l_{i2}u_{23}, \qquad i = 3, 4.$$

$$u_{34} = \frac{a_{34} - l_{31}u_{14} - l_{32}u_{24}}{l_{33}},$$

$$l_{44} = a_{44} - l_{41}u_{14} - l_{42}u_{24} - l_{43}u_{34}.$$

In algorithmic form, the factorization may be presented as follows for an $n \times n$ matrix:

$$l_{ij} = a_{ij} - \sum_{k=1}^{j-1} l_{ik} u_{kj}, \quad j \le i, \; i = 1, 2, \ldots, n. \tag{4.7}$$

$$u_{ij} = \frac{a_{ij} - \sum_{k=1}^{i-1} l_{ik} u_{kj}}{l_{ii}}, \quad i \le j, \; j = 2, 3, \ldots, n. \tag{4.8}$$

Note that this algorithm can be applied if the diagonal elements l_{ii}, for each $i = 1, \ldots, n$, of L, are nonzero.

The LU factorization that we have just described, requiring the diagonal elements of U to be one, is known as **Crout's method**. If instead the diagonal of L is required to be one, the factorization is called **Doolittle's method**.

In the case when the $n \times n$ matrix A is symmetric ($A = A^T$), and positive definite, that is

$$\mathbf{x}^T A \mathbf{x} > 0, \quad \text{for all nonzero } n\text{-vectors } \mathbf{x}$$

then it is possible to carry out a factorization without any need for pivoting or scaling. This factorization is known as **Cholesky's method**, and A can be factored in the form

$$A = LL^T$$

where L is a lower triangular matrix. The construction of L is similar to the one used for Crout's method. Multiplying L by L^T and setting the result equal to A gives

$$l_{ii} = \left[a_{ii} - \sum_{k=1}^{i-1} (l_{ik})^2 \right]^{1/2}, \quad i = 1, 2, \ldots, n$$

$$\tag{4.9}$$

$$l_{ij} = \frac{a_{ij} - \sum_{k=1}^{j-1} l_{ik} l_{jk}}{l_{jj}}, \quad i = j+1, j+2, \ldots, n; \; j = 1, 2, \ldots, n.$$

EXAMPLE 4.8

Factor the following matrix using Cholesky's method.

$$A = \begin{bmatrix} 16 & 4 & 4 \\ 4 & 26 & 6 \\ 4 & 6 & 11 \end{bmatrix}.$$

From (4.9) we have

$l_{11} = \sqrt{16} = 4$

$l_{21} = 4/4 = 1, \quad l_{22} = \sqrt{26 - 1} = 5$

$l_{31} = 4/4 = 1, \quad l_{32} = (6 - 1)/5 = 1, \quad l_{33} = \sqrt{11 - 1 - 1} = 3.$

Thus,

$$L = \begin{bmatrix} 4 & 0 & 0 \\ 1 & 5 & 0 \\ 1 & 1 & 3 \end{bmatrix} \quad \text{and} \quad U = L^T = \begin{bmatrix} 4 & 1 & 1 \\ 0 & 5 & 1 \\ 0 & 0 & 3 \end{bmatrix}.$$

The LU method is particularly useful when it is necessary to solve a whole series of systems

$$A\mathbf{x} = \mathbf{b}_1, \ A\mathbf{x} = \mathbf{b}_2, \ \ldots, \ A\mathbf{x} = \mathbf{b}_n$$

each of which has the same square coefficient matrix A.

Many problems in science and engineering involve systems of this type, and in this case it is more efficient to use the LU method than separately applying the Gaussian elimination method to each of the k systems. Storage space may be economized since there is no need to store the zeros in either L or U, and the ones on the diagonal of U. Eqn. (4.9) shows that once any element of A, a_{ij}, is used, it never again appears in the equations and its place can be used to store elements, of either L or U. In other words, the L and U matrices are constructed by storing their elements in the space of A.

EXAMPLE 4.9

Use Crout's method to solve the system

$$\begin{bmatrix} 1 & 1 & 1 & 1 \\ 2 & 3 & 1 & 5 \\ -1 & 1 & -5 & 3 \\ 3 & 1 & 7 & -2 \end{bmatrix} \begin{bmatrix} x_1 \\ x_2 \\ x_3 \\ x_4 \end{bmatrix} = \begin{bmatrix} 10 \\ 31 \\ -2 \\ 18. \end{bmatrix}.$$

If A has a direct factorization LU, then

$$A = \begin{bmatrix} 1 & 1 & 1 & 1 \\ 2 & 3 & 1 & 5 \\ -1 & 1 & -5 & 3 \\ 3 & 1 & 7 & -2 \end{bmatrix} = \begin{bmatrix} l_{11} & 0 & 0 & 0 \\ l_{21} & l_{22} & 0 & 0 \\ l_{31} & l_{32} & l_{33} & 0 \\ l_{41} & l_{42} & l_{43} & l_{44} \end{bmatrix} \begin{bmatrix} 1 & u_{12} & u_{13} & u_{14} \\ 0 & 1 & u_{23} & u_{24} \\ 0 & 0 & 1 & u_{34} \\ 0 & 0 & 0 & 1 \end{bmatrix}.$$

By multiplying L with U and comparing the elements of the product matrix with those of A, we obtain:

(i) Multiplication of the first row of L with the columns of U gives

$$l_{11} = 1,$$
$$l_{11}u_{12} = 1 \Longrightarrow u_{12} = 1,$$
$$l_{11}u_{13} = 1 \Longrightarrow u_{13} = 1,$$
$$l_{11}u_{14} = 1 \Longrightarrow u_{14} = 1.$$

(ii) Multiplication of the second row of L with the columns of U gives

$$l_{21} = 2,$$
$$l_{21}u_{12} + l_{22} = 3 \Longrightarrow l_{22} = 3 - l_{21}u_{12} = 1,$$
$$l_{21}u_{13} + l_{22}u_{23} = 1 \Longrightarrow u_{23} = (1 - l_{21}u_{13})/l_{22} = -1,$$
$$l_{21}u_{14} + l_{22}u_{24} = 5 \Longrightarrow u_{24} = (5 - l_{21}u_{14})/l_{22} = 3.$$

(iii) Multiplication of the third row of L with the columns of U gives

$$l_{31} = -1,$$
$$l_{31}u_{12} + l_{32} = 1 \implies l_{32} = 1 - l_{31}u_{12} = 2,$$
$$l_{31}u_{13} + l_{32}u_{23} + l_{33} = -5 \implies l_{33} = -5 - l_{31}u_{13} - l_{32}u_{23} = -2,$$
$$l_{31}u_{14} + l_{32}u_{24} + l_{33}u_{34} = 3 \implies u_{34} = (3 - l_{31}u_{14} - l_{32}u_{24})/l_{33} = 1.$$

(iv) Multiplication of the fourth row of L with the columns of U gives

$$l_{41} = 3,$$
$$l_{41}u_{12} + l_{42} = 1 \implies l_{42} = 1 - l_{41}u_{12} = -2,$$
$$l_{41}u_{13} + l_{42}u_{23} + l_{43} = 7 \implies l_{43} = 7 - l_{41}u_{13} - l_{42}u_{23} = 2,$$
$$l_{41}u_{14} + l_{42}u_{24} + l_{43}u_{34} + l_{44} = -2 \implies$$
$$l_{44} = -2 - l_{41}u_{14} - l_{42}u_{24} - l_{43}u_{34} = -1.$$

Hence,

$$L = \begin{bmatrix} 1 & 0 & 0 & 0 \\ 2 & 1 & 0 & 0 \\ -1 & 2 & -2 & 0 \\ 3 & -2 & 2 & -1 \end{bmatrix} \quad \text{and} \quad U = \begin{bmatrix} 1 & 1 & 1 & 1 \\ 0 & 1 & -1 & 3 \\ 0 & 0 & 1 & 1 \\ 0 & 0 & 0 & 1 \end{bmatrix}.$$

By applying the forward substitution to the lower triangular system $L\mathbf{y} = \mathbf{b}$, we get

$$y_1 = 10$$
$$y_2 = 31 - 2(10) = 11$$
$$y_3 = [-2 + 10 - 2(11)]/(-2) = 7$$
$$y_4 = -[18 - 3(10) + 2(11) - 2(7)] = 4.$$

Finally, by applying the back substitution to the upper triangular system $U\mathbf{x} = \mathbf{y}$, we get

$$x_1 = 10 - 4 - 3 - 2 = 1$$
$$x_2 = -[11 - 4 - 3(3)] = 2$$
$$x_3 = 7 - 4 = 3$$
$$x_4 = 4.$$

The computed results are shown in Table 4.3.

4.4.2 Gaussian elimination method

We shall now illustrate a method of constructing L and U using Gaussian elimination. We process exactly as Gaussian elimination except that we keep a record of the elementary row operation performed at each step.

```
» A=[1 1 1 1;2 3 1 5;-1 1 -5 3;3 1 7 -2];
» b=[10 31 -2 18]';
» lu(A,b)


L =

            1          0          0          0
            2          1          0          0
           -1          2         -2          0
            3         -2          2         -1

        The forward substitution gives
y =
           10
           11
            7
            4


U =

            1          1          1          1
            0          1         -1          3
            0          0          1          1
            0          0          0          1

        The vector solution is =
x =
            1
            2
            3
            4
```

Table 4.3   Crout's method for Example 4.9.

We concentrate here on the factorization without pivoting. So, we assume that the naive Gaussian elimination can be successfully performed to solve the linear system $A\mathbf{x} = \mathbf{b}$. Also, we are not concerned with the coefficient vector $\mathbf{b}$. That is, we do not need to form an augmented matrix. Looking back at the elimination process described in Section 4.2, we see that the row multipliers for naive Gaussian elimination at kth step are defined by

$$m_{ik} = \frac{a_{ik}^{(k)}}{a_{kk}^{(k)}}$$

provided that $a_{kk}^{(k)} \neq 0$. It is easily verified that when the matrix

$$M_1 = \begin{bmatrix} 1 & & & & \\ -m_{21} & 1 & & 0 & \\ -m_{31} & 0 & 1 & & \\ \vdots & \vdots & & \ddots & \\ -m_{n1} & 0 & \cdots & 0 & 1 \end{bmatrix}$$

is multiplied by the matrix A on the left, the result is

$$M_1 A = \begin{bmatrix} a_{11}^{(1)} & a_{12}^{(1)} & \cdots & a_{1n}^{(1)} \\ 0 & a_{22}^{(2)} & \cdots & a_{2n}^{(2)} \\ \vdots & \vdots & \ddots & \vdots \\ 0 & a_{n2}^{(2)} & \cdots & a_{nn}^{(2)} \end{bmatrix}$$

which has the same effect of multiplying row 1 of A by the row multiplier m_{i1} and subtracting the result from row i of A for $i = 2, ..., n$. If $a_{kk}^{(k-1)} \neq 0$, the $k - 1$ step is formed by

$$M_{k-1} M_{k-2} \cdots M_1 A = \begin{bmatrix} a_{11}^{(1)} & a_{12}^{(1)} & \cdots & a_{1k}^{(1)} & \cdots & a_{1n}^{(1)} \\ 0 & a_{22}^{(2)} & \cdots & a_{2k}^{(2)} & \cdots & a_{2n}^{(2)} \\ \vdots & 0 & \cdots & \vdots & & \vdots \\ 0 & 0 & \cdots & a_{kk}^{(k)} & \cdots & a_{kn}^{(k)} \\ \vdots & \vdots & & \vdots & & \vdots \\ 0 & 0 & \cdots & a_{nk}^{(k)} & \cdots & a_{nn}^{(k)} \end{bmatrix}$$

where

$$M_{k-1} = \begin{bmatrix} 1 & & & & & \\ & \ddots & & & & \\ & & 1 & & 0 & \\ & & -m_{k,k-1} & \ddots & & \\ & & \vdots & & & \\ & & -m_{n,k-1} & & & 1 \end{bmatrix}.$$

After $(n - 1)$ steps, the elimination process without pivoting results in

$$M_{n-1} M_{n-2} \cdots M_1 A = \begin{bmatrix} a_{11}^{(1)} & a_{12}^{(1)} & \cdots & a_{1n}^{(1)} \\ 0 & a_{22}^{(2)} & \cdots & a_{2n}^{(2)} \\ \vdots & & \ddots & \vdots \\ 0 & \cdots & 0 & a_{nn}^{(n)} \end{bmatrix} = U \qquad (4.10)$$

which is an upper triangular matrix. It is easy to verify directly that $M_i^{-1} M_i = I$, $i = 1, ..., n - 1$. Using this result, we multiply Eqn. (4.10) on the left by M_{n-1}^{-1}, then M_{n-2}^{-1}, ..., to get

$$A = M_1^{-1} M_2^{-1} \cdots M_{n-1}^{-1} U.$$

It is also easy to verify directly that

$$M_{k-1}^{-1} = \begin{bmatrix} 1 & & & & & \\ & \ddots & & & & \\ & & 1 & & 0 & \\ & & m_{k,k-1} & \ddots & & \\ & & \vdots & & & \\ & & m_{n,k-1} & & & 1 \end{bmatrix}.$$

Therefore,

$$M_1^{-1} M_2^{-1} \cdots M_{n-1}^{-1} = \begin{bmatrix} 1 & & & & \\ m_{21} & 1 & & 0 & \\ m_{31} & m_{32} & \ddots & & \\ \vdots & \vdots & \ddots & 1 & \\ m_{n1} & m_{n2} & \cdots & m_{n,n-1} & 1 \end{bmatrix} = L \qquad (4.11)$$

which is a lower triangular matrix with all ones in its diagonal. Finally, we see that

$$LU = M_1^{-1} M_2^{-1} \cdots M_{n-1}^{-1} M_{n-1} M_{n-2} \cdots M_1 A = A$$

where U is given by Eqn. (4.10) and L given by Eqn. (4.11).

EXAMPLE 4.10
Use Gaussian elimination to construct the LU decomposition of the matrix

$$A = \begin{bmatrix} 1 & 1 & 6 \\ -1 & 2 & 9 \\ 1 & -2 & 3 \end{bmatrix}.$$

The matrix L will be constructed from the identity matrix placed at the left of A. We start with

$$A = IA = \begin{bmatrix} 1 & 0 & 0 \\ 0 & 1 & 0 \\ 0 & 0 & 1 \end{bmatrix} \begin{bmatrix} 1 & 1 & 6 \\ -1 & 2 & 9 \\ 1 & -2 & 3 \end{bmatrix}.$$

The row multipliers are $m_{21} = -1$ and $m_{31} = 1$. Subtract from row 2 the multiple m_{21} of row 1 and from row 3 the multiples m_{31} of row 1, to get

$$A = \begin{bmatrix} 1 & 0 & 0 \\ -1 & 1 & 0 \\ 1 & 0 & 1 \end{bmatrix} \begin{bmatrix} 1 & 1 & 6 \\ 0 & 3 & 15 \\ 0 & -3 & -3 \end{bmatrix}.$$

Note that the multipliers m_{21} and m_{31} are put below the diagonal element in the first column of the left matrix. The row multiplier of the second row is $m_{32} = -1$. Subtract from row 3 the multiple m_{32} of row 2 to get the desired LU factorization of A:

$$A = \begin{bmatrix} 1 & 0 & 0 \\ m_{21} \leftarrow -\mathbf{1} & 1 & 0 \\ m_{31} \leftarrow \mathbf{1} \; m_{32} \leftarrow -\mathbf{1} & 1 \end{bmatrix} \begin{bmatrix} 1 & 1 & 6 \\ 0 & 3 & 15 \\ 0 & 0 & 12 \end{bmatrix} = LU.$$

The operational count for triangularizing a matrix using Gaussian elimination is obtained in a way similar to the one we used for the naive Gaussian elimination. It can be shown that the number of multiplications and divisions needed for triangularizing an $n \times n$ matrix using Gaussian elimination is

$$\frac{n^3 - n}{3}$$

and the number of subtractions is

$$\frac{2n^3 - 3n^2 + n}{6}.$$

MATLAB's Methods

In MATLAB the LU factorization is performed by the built in function lu. The factorization is performed with pivoting. The MATLAB command [L,U,P] = lu(A) creates an upper-triangular matrix **U**, a lower-triangular matrix **L**, and a full permutation matrix **P** so that **PA=LU**. A **permutation matrix** is an $n \times n$ matrix that arises from the identity matrix by permuting its rows. The matrix **PA** is **A** with its rows rearranged.

```
>> A = [1 2 3; 2 8 11; 3 22 35]; b = [1 1 1]';
>> [L,U,P] = lu(A)
            1.0000         0         0
L =         0.6667    1.0000         0
            0.3333    0.8000    1.0000
U =
        3.0000   22.0000   35.0000
             0   -6.6667   12.3333
             0         0    1.2000
P =
        0    0    1
        0    1    0
        1    0    0
```

One can proceed by using the \ operator to find the solution of the system **Ax=b** as follows:

```
>> y = L\(P*b)
y =
```

```
    1.0000
    0.3333
    0.4000
>> x = U\y
x =
    1.3333
   -0.6667
    0.3333
```

We can show that **PA=LU** by verifying that $\mathbf{A}=\mathbf{P}^{-1}\mathbf{LU}$

```
>> inv(P)*L*U
ans =
    1    2    3
    2    8   11
    3   22   35
```

EXERCISE SET 4.4

1. The matrix
$$A = \begin{bmatrix} 2 & -1 & 2 \\ 2 & -3 & 3 \\ 6 & -1 & 8 \end{bmatrix}$$

   has the following LU decomposition

$$L = \begin{bmatrix} 2 & 0 & 0 \\ 2 & -2 & 0 \\ 6 & 2 & 3 \end{bmatrix}, U = \begin{bmatrix} 1 & -\frac{1}{2} & 1 \\ 0 & 1 & -\frac{1}{2} \\ 0 & 0 & 1 \end{bmatrix}.$$

   Use this decomposition of A to solve $A\mathbf{x} = \mathbf{b}$, where $\mathbf{b} = [-2, -5, 0]^T$.

2. Factor the following matrices into LU decomposition.

   (a) $\begin{bmatrix} 4 & -1 & 1 \\ 3 & 3 & 8 \\ 3 & 3 & 2 \end{bmatrix}$ (b) $\begin{bmatrix} 3 & -1.5 & 2 \\ -1 & 0 & 2 \\ 4 & -3.5 & 5 \end{bmatrix}$

   (c) $\begin{bmatrix} 2 & 0 & 0 & 0 \\ 1 & 2 & 0 & 0 \\ 0 & -3 & 1 & 0 \\ 2 & -2 & 1 & 1 \end{bmatrix}$ (d) $\begin{bmatrix} 2.75 & 4.23 & -2.88 & 5.45 \\ -4.45 & 6.00 & 0.00 & 1.19 \\ -1.00 & -5.33 & 1.11 & 0.00 \\ 6.34 & 7.00 & 0.00 & -4.54 \end{bmatrix}.$

3. The matrix A in the system of the linear equation
$$A\mathbf{x} = \mathbf{b}$$

   has the LU decomposition
$$A = LU$$

M-function 4.4a

The following MATLAB function **lufact.m** finds the solution of a linear system using Crout's LU decomposition. INPUTS are an $n \times n$ matrix A and an $n \times 1$ coefficient vector b.

```
function lufact(A,b)
% Solve the system Ax=b using the LU decomposition.
n=length(b);
y=zeros(n,1);
x=zeros(n,1);
fprintf('\n');
for i=1:n
  U(i,i)=1;
end
L(1,1)=A(1,1)/U(1,1);
for j=2:n
  L(j,1)=A(j,1)/U(1,1);
  U(1,j)=A(1,j)/L(1,1);
end
for i=2:n-1
  S=0;
  for k=1:i-1
    S=S+U(k,i)*L(i,k);
  end
  L(i,i)=(A(i,i)-S)/U(i,i);
  for j=i+1:n
    S=0;
    for k=1:i-1
      S=S+U(k,i)*L(j,k);
    end
    L(j,i)=(A(j,i)-S)/U(i,i);
    S=0;
    for k=1:i-1
      S=S+U(k,j)*L(i,k);
    end
    U(i,j)=(A(i,j)-S)/L(i,i);
  end
end
S=0;
for k=1:n-1
  S=S+U(k,n)*L(n,k);
end
L(n,n)=(A(n,n)-S)/U(n,n);
```

```
% Perform the forward substitution.
y(1)=b(1)/L(1,1);
for i=2:n
  S=b(i);
  for j=1:i-1
    S=S-L(i,j)*y(j);
  end
  y(i)=S/L(i,i);
end
% Perform the back substitution.
x(n)=y(n)/U(n,n);
for i=n-1:-1:1
  S=y(i);
  for j=i+1:n
    S=S-U(i,j)*x(j);
  end
  x(i)=S/U(i,i);
end
L
disp('        The forward substitution gives')
y
U
disp('        The vector solution is =')
x
```

with

$$
L = \begin{bmatrix} 1 & 0 & 0 & 0 & 0 \\ -1 & 1 & 0 & 0 & 0 \\ 0 & -1 & 1 & 0 & 0 \\ 0 & 0 & -1 & 1 & 0 \\ 0 & 0 & 0 & -1 & 1 \end{bmatrix}, \quad U = L^T, \quad \mathbf{b} = \begin{bmatrix} 1 \\ 1 \\ 1 \\ 1 \\ 1 \end{bmatrix}.
$$

Determine $\mathbf{x}$.

4. The matrix A in the system of the linear equations

$$A\mathbf{x} = \mathbf{b}$$

has the LU decomposition

$$A = LU$$

with

$$
L = \begin{bmatrix} 1 & 0 & 0 & 0 & 0 \\ -1 & 1 & 0 & 0 & 0 \\ 0 & -1 & 1 & 0 & 0 \\ 0 & 0 & -1 & 1 & 0 \\ -1 & -1 & 0 & -1 & 1 \end{bmatrix}, U = \begin{bmatrix} 1 & -1 & 0 & 1 & 0 \\ 0 & 1 & -1 & 0 & 1 \\ 0 & 0 & 1 & -1 & 0 \\ 0 & 0 & 0 & 1 & -1 \\ 0 & 0 & 0 & 0 & 1 \end{bmatrix}
$$

and $\mathbf{b} = [1, 1, 1, 1, 1]^T$. Determine $\mathbf{x}$.

5. Solve the following system using the LU decomposition.

(a)
$$\begin{bmatrix} 3 & -2 & 1 & 0 \\ 1 & -4 & 2 & 2 \\ 1 & 0 & 3 & 0 \\ -3 & 7 & 9 & 1 \end{bmatrix} \begin{bmatrix} x_1 \\ x_2 \\ x_3 \\ x_4 \end{bmatrix} = \begin{bmatrix} -3 \\ 6 \\ 5 \\ 5 \end{bmatrix}$$

(b)
$$\begin{bmatrix} 1 & 1 & -2 \\ 4 & -2 & 1 \\ 3 & -1 & 3 \end{bmatrix} \begin{bmatrix} x_1 \\ x_2 \\ x_3 \end{bmatrix} = \begin{bmatrix} 3 \\ 5 \\ 8 \end{bmatrix}.$$

6. Suppose Cholesky's method is applied to factor a matrix A. Determine A, if L has been found to be equal to

$$L = \begin{bmatrix} 1 & 0 & 0 & 0 \\ -3 & 2 & 0 & 0 \\ -3 & -1 & 3 & 0 \\ 4 & 2 & 6 & 4 \end{bmatrix}.$$

7. Using Cholesky's method, calculate the decomposition $A = LL^T$, given

$$U = \begin{bmatrix} 4 & -1 & 0 & 0 \\ -1 & 4 & -1 & 0 \\ 0 & -1 & 4 & -1 \\ 0 & 0 & -1 & 4 \end{bmatrix}.$$

8. Consider the matrix

$$A = \begin{bmatrix} 2 & -3 & 2 & 5 \\ 1 & -1 & 1 & 2 \\ 3 & 2 & 2 & 1 \\ 1 & 1 & -3 & -1 \end{bmatrix}.$$

Use Gaussian elimination to determine a unit lower triangular matrix M and an upper triangular matrix U such that $MA = U$.

9. Find the LU factorization of the matrix A in which the diagonal elements of L are 1 for

$$A = \begin{bmatrix} 4 & -2 & 0 & 0 \\ -2 & 2 & 2 & 0 \\ 0 & 2 & 8 & -6 \\ 0 & 0 & -6 & 10 \end{bmatrix}.$$

Use the LU factorization to find x such that $Ax = [2 \ 0 \ -8 \ 16]^T$.

COMPUTER PROBLEM SET 4.4

1. Write a computer program in a language of your choice to solve a system of n linear equations and n unknowns using the LU decomposition.

Input data to the program should be

(a) The number of equations n.

(b) The augmented matrix.

Output should consist of:

(a) The augmented matrix.

(b) The L and U matrices.

(c) The vector solution.

Test your program to solve Exercise 1(b).

2. The following 8 by 8 tridiagonal system of equations arise in the solution of partial differential equations using the finite difference method

$$\begin{bmatrix} -2 & 1 & & & \\ 1 & -2 & 1 & & \\ & \ddots & \ddots & \ddots & \\ & & 1 & -2 & 1 \\ & & & 1 & -2 \end{bmatrix} \begin{bmatrix} x_1 \\ x_2 \\ \vdots \\ x_7 \\ x_8 \end{bmatrix} = \mathbf{b}.$$

Use the MATLAB function lufact to solve the system for the following **b** vectors

(a) $[1 \quad 1 \quad \ldots \quad 1]'$

(b) $[10 \quad 0 \quad \ldots \quad 0 \quad 10]'$

(c) $[0.1 \quad 0.2 \quad \ldots \quad 0.7 \quad 0.8]'$.

3. Generate a 15×15 matrix **A** using the MATLAB command rand and then use the MATLAB function lufact to solve the linear system $\mathbf{Ax} = \mathbf{b}$ for the following **b**'s:

(a) $\mathbf{b} = [1 \quad 1/2 \quad 1/3 \ldots \quad 1/15]'$,

(b) $\mathbf{b} = [1 \quad 2 \quad 3 \quad \ldots \quad 15]'$,

(c) $\mathbf{b} = [0.1 \quad 0.2 \quad 0.3 \quad \ldots \quad 0.15]'$.

Check the accuracy of your solutions by computing the matrix difference **Ax-b**. The accuracy depends on how close the elements are to zero.

4. Use the MATLAB function lufact to factor the following matrix

$$\begin{bmatrix} 6 & 2 & 1 & -1 \\ 2 & 4 & 1 & 0 \\ 1 & 1 & 4 & -1 \\ -1 & 0 & -1 & 3 \end{bmatrix}.$$

5. Create a 10×10 matrix A with random integer entries $a_{ij} \in [-6, 6]$, $1 \leq i, j \leq 10$. To this end, make use of the MATLAB function randint. Use the MATLAB function lufact to find the solution of the linear system of equations $\mathbf{Ax=b}$ for the following coefficient vectors:

(a) $\mathbf{b} = [1, 1, ..., 1]'$,

(b) $\mathbf{b} = [1, 2, 3, 4, 5, 0.1, 0.2, 0.3, 0.4, 0.5]'$,

(c) $\mathbf{b} = [2, 4, 6, 8, 10, 1, 3, 5, 7, 9]'$.

4.5   ITERATIVE METHODS

Because of round-off errors, direct methods become less efficient than iterative methods when they are applied to large systems, sometimes with as many as 100,000 variables. Examples of these large systems arise in the solution of partial differential equations. In these cases, an iterative method is preferable. In addition to round-off errors, the amount of storage space required for iterative solutions on a computer is far less than the one required for direct methods when the coefficient matrix of the system is sparse, that is, matrices that contain a high proportion of zeros. Thus, for sparse matrices, iterative methods are more attractive than direct methods.

An iterative scheme for linear systems consists of converting the system (4.1) to the form

$$\mathbf{x} = \mathbf{b}' - B\mathbf{x}.$$

After an initial guess, $\mathbf{x}^{(0)}$ is selected, the sequence of approximation solution vectors is generated by computing

$$\mathbf{x}^{(k)} = \mathbf{b}' - B\mathbf{x}^{(k-1)}$$

for each $k = 1, 2, 3, \ldots$.

4.5.1   Jacobi iterative method

We shall illustrate the method for a 3×3 linear system.

$$a_{11}x_1 + a_{12}x_2 + a_{13}x_3 = b_1$$
$$a_{21}x_1 + a_{22}x_2 + a_{23}x_3 = b_2$$
$$a_{31}x_1 + a_{32}x_2 + a_{33}x_3 = b_3$$

where we assume that the diagonal terms a_{11}, a_{22}, and a_{33} are all nonzero.

We begin our iterative scheme by solving each equation for one of the variables, choosing, when possible, to solve for the variable with the largest coefficient

$$
\begin{aligned}
x_1 &= & u_{12}x_2 + u_{13}x_3 + c_1 \\
x_2 &= u_{21}x_1 & + u_{23}x_3 + c_2 \\
x_3 &= u_{31}x_1 + u_{32}x_2 & + c_3
\end{aligned}
$$

where $u_{ij} = -\frac{a_{ij}}{a_{ii}}$, $c_i = \frac{b_i}{a_{ii}}$, $i = 1, 2, 3$.

Let $x_1^{(0)}$, $x_2^{(0)}$, $x_3^{(0)}$ be an initial approximation of the solution. The $(n+1)$st approximation is obtained from the approximation by writing

$$
\begin{aligned}
x_1^{(n+1)} &= & u_{12}x_2^{(n)} + u_{13}x_3^{(n)} + c_1 \\
x_2^{(n+1)} &= u_{21}x_1^{(n)} & + u_{23}x_3^{(n)} + c_2 \\
x_3^{(n+1)} &= u_{31}x_1^{(n)} + u_{32}x_2^{(n)} & + c_3
\end{aligned}
$$

for $n = 0, 1, 2, \ldots$.

In algorithmic form, the Jacobi iterative method may be presented as follows for an $n \times n$ linear system:

Consider (4.1) and solve for x_i in the ith equation to obtain, provided that $a_{ii} \neq 0$,

$$
x_i = \sum_{\substack{j=1 \\ j \neq i}}^{n} \left(-\frac{a_{ij}x_j}{a_{ii}} \right) + \frac{b_i}{a_{ii}}, \quad i = 1, 2, \ldots, n
$$

and generate $x_i^{(k)}$ for $k \geq 1$ by

$$
x_i^{(k)} = \frac{-\sum_{\substack{j=1 \\ j \neq i}}^{n} \left(a_{ij}x_j^{(k-1)} \right) + b_i}{a_{ii}}, \quad i = 1, 2, \ldots, n. \tag{4.12}
$$

The iterative process is terminated when a convergence criterion is satisfied. One commonly used stopping criterion, known as the **relative change criteria**, is to iterate until

$$
\frac{|\mathbf{x}^{(k)} - \mathbf{x}^{(k-1)}|}{|\mathbf{x}^{(k)}|}, \quad \mathbf{x}^{(k)} = \left(x_1^{(k)}, \ldots, x_n^{(k)} \right)^T
$$

is less than a prescribed tolerance $\epsilon > 0$. Contrary to Newton's method for finding the roots of an equation, the convergence or divergence of the iterative process in the Jacobi method does not depend on the initial guess, but depends only on the character of the matrices themselves. However, a good first guess in case of convergence will make for a relatively small number of iterations. This is also true for the Gauss-Seidel method that will be presented in the next section.

EXAMPLE 4.11

Solve the following system using the Jacobi iterative method. Use $EPS = 10^{-3}$, $ITMAX = 30$, and $\mathbf{x}^{(0)} = \mathbf{0}$ as the starting vector.

$$
\begin{aligned}
7x_1 - 2x_2 + x_3 &= 17 \\
x_1 - 9x_2 + 3x_3 - x_4 &= 13 \\
2x_1 + 10x_3 + x_4 &= 15 \\
x_1 - x_2 + x_3 + 6x_4 &= 10.
\end{aligned}
$$

These equations can be rearranged to give

$$x_1 = (17 + 2x_2 - x_3)/7$$
$$x_2 = (-13 + x_1 + 3x_3 - x_4)/9$$
$$x_3 = (15 - 2x_1 - x_4)/10$$
$$x_4 = (10 - x_1 + x_2 - x_3)/6.$$

which provide the following Jacobi iterative process:

$$x_1^{(k+1)} = (17 + 2x_2^{(k)} - x_3^{(k)})/7$$
$$x_2^{(k+1)} = (-13 + x_1^{(k)} + 3x_3^{(k)} - x_4^{(k)})/9$$
$$x_3^{(k+1)} = (15 - 2x_1^{(k)} - x_4^{(k)})/10$$
$$x_4^{(k+1)} = (10 - x_1^{(k)} + x_2^{(k)} - x_3^{(k)})/6.$$

Substitute $\mathbf{x}^{(0)} = (0, 0, 0, 0)$ into the right-hand side of each of these equations to get

$$x_1^{(1)} = (17 + 2(0) - 0)/7 = 2.\,428\,571\,429$$
$$x_2^{(1)} = (-13 + 0 + 3(0) - 0)/9 = -1.\,444\,444\,444$$
$$x_3^{(1)} = (15 - 2(0) - 0)/10 = 1.5$$
$$x_4^{(1)} = (10 - 0 + 0 - 0)/6 = 1.\,666\,666\,667$$

and so $\mathbf{x}^{(1)} = (2.\,428\,571\,429, -1.\,444\,444\,444, 1.5, 1.\,666\,666\,667)^T$.

Similar procedure generates a sequence that converges to (see Table 4.4)

$$\mathbf{x}^{(9)} = (2.000127203, -1.000100162, 1.000118096, 1.000162172)^T.$$

4.5.2 Gauss-Seidel iterative method

The algorithm for Gauss-Seidel is almost the same as for Jacobi, except that each x-value is improved using the most recent approximations to the values of the other variables. In this case, the $(n + 1)$st approximation is obtained from the nth approximation for a 3×3 system by writing

$$x_1^{(n+1)} = \qquad\qquad u_{12}x_2^{(n)} + u_{13}x_3^{(n)} + c_1$$
$$x_2^{(n+1)} = u_{21}x_1^{(n+1)} \qquad\qquad + u_{23}x_3^{(n)} + c_2$$
$$x_3^{(n+1)} = u_{31}x_1^{(n+1)} + u_{32}x_2^{(n+1)} \qquad\qquad + c_3.$$

In algorithmic form, Gauss-Seidel may be presented as follows: $x_i^{(k)}$ is generated for $k \geq 1$ by

$$x_i^{(k)} = \frac{-\sum_{j=1}^{i-1}(a_{ij}x_j^{(k)}) - \sum_{j=i+1}^{n}(a_{ij}x_j^{(k-1)}) + b_i}{a_{ii}} \qquad (4.13)$$

```
» A=[7 -2 1 0;1 -9 3 -1;2 0 10 1;1 -1 1 6];
» b=[17 13 15 10]';
» x0=[0 0 0 0]';
» jacobi(A,b,x0,10^(-3),30)

    The augmented matrix is =

Augm =

    7   -2    1    0   17
    1   -9    3   -1   13
    2    0   10    1   15
    1   -1    1    6   10

The solution vectors are:

iter #      0         1          2          3          4         ...

x1 =    0.000000   2.428571   1.801587   2.061829   1.977515   2.008259   1.997187   2.001020   1.999639   2.000127
x2 =    0.000000  -1.444444  -0.859788  -1.047413  -0.981367  -1.006166  -0.997771  -1.000802  -0.999721  -1.000100
x3 =    0.000000   1.500000   0.847619   1.062566   0.979451   1.007360   0.997320   1.000926   0.999667   1.000118
x4 =    0.000000   1.666667   0.771164   1.081834   0.971365   1.010278   0.996369   1.001287   0.999542   1.000162

The method converges after 9 iterations to

x =
              2.0001
             -1.0001
              1.0001
              1.0002
```

Table 4.4   The Jacobi iterative method for Example 4.11.

for each $i = 1, 2, \ldots, n$.

The comments following the Jacobi algorithm regarding stopping criteria and starting vectors also apply to the Gauss-Seidel algorithm. Because the new values can be immediately stored in the location that held the old values, the storage requirements for $\mathbf{x}$ with the Gauss-Seidel method is half what it would be with the Jacobi method and the rate of convergence is more rapid.

EXAMPLE 4.12

Solve the following system using the Gauss-Seidel iterative method. Use $EPS = 10^{-3}$, $ITMAX = 30$, and $\mathbf{x}^{(0)} = \mathbf{0}$ as the starting vector.

$$
\begin{aligned}
7x_1 - 2x_2 + x_3 &= 17 \\
x_1 - 9x_2 + 3x_3 - x_4 &= 13 \\
2x_1 + 10x_3 + x_4 &= 15 \\
x_1 - x_2 + x_3 + 6x_4 &= 10
\end{aligned}
$$

From Example 4.11, we have

$$
\begin{aligned}
x_1 &= (17 + 2x_2 - x_3)/7 \\
x_2 &= (-13 + x_1 + 3x_3 - x_4)/9 \\
x_3 &= (15 - 2x_1 - x_4)/10
\end{aligned}
$$

$$x_4 = (10 - x_1 + x_2 - x_3)/6$$

which provide the following Gauss-Seidel iterative process:

$$x_1^{(k+1)} = (17 + 2x_2^{(k)} - x_3^{(k)})/7$$
$$x_2^{(k+1)} = (-13 + x_1^{(k+1)} + 3x_3^{(k)} - x_4^{(k)})/9$$
$$x_3^{(k+1)} = (15 - 2x_1^{(k+1)} - x_4^{(k)})/10$$
$$x_4^{(k+1)} = (10 - x_1^{(k+1)} + x_2^{(k+1)} - x_3^{(k+1)})/6.$$

Substitute $\mathbf{x}^{(0)} = (0, 0, 0, 0)$ into the right-hand side of each of these equations to get

$$x_1^{(1)} = (17 + 2(0) - 0)/7 = 2.\,428\,571\,429$$
$$x_2^{(1)} = (-13 + 2.\,428\,571\,429 + 3(0) - 0)/9 = -1.1746031746$$
$$x_3^{(1)} = (15 - 2(2.\,428\,571\,429) - 0)/10 = 1.0142857143$$
$$x_4^{(1)} = (10 - 2.\,428\,571\,429 - 1.1746031746 - 1.0142857143)/6$$
$$= 0.8970899472$$

and so

$$\mathbf{x}^{(1)} = (2.428571429, -1.1746031746, 1.0142857143, 0.8970899472)^T.$$

similar procedure generates a sequence that converges to (see Table 4.5)

$$\mathbf{x}^{(5)} = (2.000025, -1.000130, 1.000020, 0.999971)^T.$$

4.5.3 Convergence

The matrix formulation of the Jacobi and Gauss-Seidel iterative methods can be obtained by splitting the matrix A into the sum

$$A = D + L + U$$

where D is the diagonal of A, L the lower triangular part of A, and U the upper triangular part of A. That is,

$$D = \begin{bmatrix} a_{11} & 0 & \cdots & 0 \\ 0 & \ddots & \ddots & \vdots \\ \vdots & \ddots & \ddots & 0 \\ 0 & \cdots & 0 & a_{nn} \end{bmatrix}, \quad L = \begin{bmatrix} 0 & & \cdots & & 0 \\ a_{21} & \ddots & & & \vdots \\ \vdots & \ddots & \ddots & & \\ a_{n1} & \cdots & a_{n,n-1} & 0 \end{bmatrix}, \quad U = \begin{bmatrix} 0 & a_{12} & \cdots & a_{1n} \\ \vdots & \ddots & \ddots & \vdots \\ & & \ddots & a_{n-1,n} \\ 0 & \cdots & & 0 \end{bmatrix}.$$

Thus, the system (4.1) can be written as

$$(D + L + U)\,\mathbf{x} = \mathbf{b}.$$

```
» A=[7 -2 1 0;1 -9 3 -1;2 0 10 1;1 -1 1 6];
» b=[17 13 15 10]';
» x0=[0 0 0 0]';
» seidel(A,b,x0,10^(-3),30)

   The augmented matrix is

Augm =

   7   -2   1    0    17
   1   -9   3   -1    13
   2    0   10   1    15
   1   -1   1    6    10

The solution vectors are:

iter #    0            1           2          3          4          ...

x1 =   0.000000     2.428571    1.948073   2.000025   2.001849   2.000025
x2 =   0.000000    -1.174603   -0.989573  -0.993877  -1.000154  -1.000130
x3 =   0.000000     1.014286    1.020676   0.999300   0.999517   1.000020
x4 =   0.000000     0.897090    1.006946   1.001133   0.999747   0.999971

The method converges after 5 iterations to

x =
              2.0000
             -1.0001
              1.0000
              1.0000
```

Table 4.5   The Gauss-Seidel iterative method for Example 4.12.

The Jacobi method in matrix form is

$$D\,\mathbf{x}^{(k)} = -(L+U)\,\mathbf{x}^{(k-1)} + \mathbf{b}$$

and the Gauss-Seidel method in matrix form is

$$(D+L)\,\mathbf{x}^{(k)} = -U\,\mathbf{x}^{(k-1)} + \mathbf{b}.$$

Before stating the theorem on the convergence of the Jacobi and Gauss-Seidel methods, we make the following definition:

DEFINITION 4.3   *An $n \times n$ matrix A is **strictly diagonally dominant** if*

$$|a_{ii}| > \sum_{\substack{j=1 \\ j \neq i}}^{n} |a_{ij}|, \quad for\ i = 1, 2, ..., n.$$

We now give a sufficient condition for Jacobi and Gauss-Seidel to convergence.

THEOREM 4.1 (Jacobi and Gauss-Seidel convergence theorem)
If A is strictly diagonally dominant, then the Jacobi and Gauss-Seidel methods converge for any choice of the starting vector $\mathbf{x}^{(0)}$.

Proof: The proof can be found in advanced texts on numerical analysis.

EXAMPLE 4.13
Consider the system of equation

$$\begin{bmatrix} 3 & 1 & 1 \\ -2 & 4 & 0 \\ -1 & 2 & -6 \end{bmatrix} \begin{bmatrix} x_1 \\ x_2 \\ x_3 \end{bmatrix} = \begin{bmatrix} 4 \\ 1 \\ 2 \end{bmatrix}.$$

The coefficient matrix of the system is strictly diagonally dominant since

$$|a_{11}| = |3| = 3 > |1| + |1| = 2$$
$$|a_{22}| = |4| = 4 > |-2| + |0| = 2$$
$$|a_{33}| = |-6| = 6 > |-1| + |2| = 3.$$

Hence, if the Jacobi or Gauss-Seidel method is used to solve the system of equations, then it will converge for any choice of the starting vector $\mathbf{x}^{(0)}$.

EXERCISE SET 4.5

1. Solve the following system using the Jacobi method

$$\begin{bmatrix} 17 & 1 & 4 & 3 & -1 & 2 & 3 & -7 \\ 2 & 10 & -1 & 7 & -2 & 1 & 1 & -4 \\ -1 & 1 & -8 & 2 & -5 & 2 & -1 & 1 \\ 2 & 4 & 1 & -9 & 1 & 3 & 4 & -1 \\ 1 & 3 & 1 & 7 & 15 & 1 & -2 & 4 \\ -2 & 1 & 7 & -1 & 2 & 12 & -1 & 8 \\ 3 & 4 & 5 & 1 & 2 & 8 & -8 & 2 \\ 5 & 1 & 1 & 1 & -1 & 1 & -7 & 10 \end{bmatrix} \begin{bmatrix} x_1 \\ x_2 \\ x_3 \\ x_4 \\ x_5 \\ x_6 \\ x_7 \\ x_8 \end{bmatrix} = \begin{bmatrix} 71 \\ 43 \\ -11 \\ -33 \\ -65 \\ 52 \\ -73 \\ 20 \end{bmatrix}.$$

2. The solution of the system

$$x_1 + 3x_2 = 3$$
$$4x_1 + x_2 = 1$$

is $x_1 = 0$ and $x_2 = 1$. Apply the Jacobi and Gauss-Seidel methods to this rearrangement, starting with a vector close to the solution. Which method diverges most rapidly?

M-function 4.5a

The following MATLAB function **jacobi.m** finds the solution of a linear system using the Jacobi iteration method. INPUTS are an $n \times n$ matrix A; an $n \times 1$ coefficient vector b; an initial vector x_0; a tolerance *tol*; the maximum number of iterations *itmax*.

```
function jacobi(A,b,x0,tol,itmax)
% Solve the system Ax=b using the Jacobi iteration method.
n=length(b);
x=zeros(n,1);
fprintf('\n');
disp('      The augmented matrix is =')
Augm=[A b]
Y=zeros(n,1);
Y=x0;
for k=1:itmax+1
  for i=1:n
    S=0;
    for j=1:n
      if (j~=i)
        S=S+A(i,j)*x0(j);
      end
    end
    if(A(i,i)==0)
      break
    end
    x(i)=(-S+b(i))/A(i,i);
  end
  err=abs(norm(x-x0));
  rerr=err/(norm(x)+eps);
  x0=x;
  Y=[Y x];
  if(rerr<tol)
    break
  end
end
% Print the results
if(A(i,i)==0)
 disp('       division by zero')
elseif (k==itmax+1)
 disp('       No convergence')
else
 fprintf('\n');
```

```
disp(' The solution vectors are:')
fprintf('\n');
disp('iter #    0    1    2    3    4     ...')
fprintf('\n');
for i=1:n
  fprintf('x%1.0f =   ',i)
  fprintf('%10.6f',Y(i,[1:k+1]))
  fprintf('\n');
end
fprintf('\n');
disp(['The method converges after ',num2str(k),' iterations to']);
x
end
```

3. Show that the matrix

$$A = \begin{bmatrix} 16 & -1 & 4 & 0 \\ 2 & 9 & 1 & 1 \\ 3 & 0 & 11 & 2 \\ 1 & 1 & 0 & 13 \end{bmatrix}$$

is positive definite, then solve the linear system

$$A\mathbf{x} = \mathbf{b} \quad \text{with} \quad \mathbf{b} = [1, -1, 2, 0]^T$$

using the Gauss-Seidel iterative method.

4. Consider the linear system $\begin{bmatrix} 4 & 0 & 1 \\ 1 & 4 & 1 \\ 1 & 0 & 4 \end{bmatrix} \begin{bmatrix} x \\ y \\ z \end{bmatrix} = \begin{bmatrix} 5 \\ 3 \\ -10 \end{bmatrix}$.

   (a) Use naive Gaussian elimination to solve the system.

   (b) Find the matrix that needs to be analyzed to determine whether Jacobi's iteration method will converge for this problem.

   (c) Perform one iteration of Gauss-Seidel, with starting guess $\begin{bmatrix} 1 & 2 & -3 \end{bmatrix}'$.

5. Consider the linear system

$$\begin{aligned} x_1 + 4x_2 &= -15 \\ 5x_1 + x_2 &= 1. \end{aligned}$$

Apply the Jacobi method to this arrangement, beginning with a vector close to the solution $\mathbf{x} = [1.01, -4.01]$ and observe divergence. Now interchange the equations to solve the system again and observe convergence.

6. Consider the linear system

$$\begin{bmatrix} 4 & 1 \\ 2 & 5 \end{bmatrix} \begin{bmatrix} x \\ y \end{bmatrix} = \begin{bmatrix} 3 \\ 1 \end{bmatrix}.$$

M-function 4.5b

The following MATLAB function **seidel.m** finds the solution of a linear system using Gauss-Seidel iteration method. INPUTS are an $n \times n$ matrix A; an $n \times 1$ coefficient vector b; an initial vector x_0; a tolerance tol; the maximum number of iterations $itmax$.

```
function seidel(A,b,x0,tol,itmax)
% Solve the system Ax=b using Gauss-Seidel iteration method.
n=length(b);
x=zeros(n,1);
fprintf('\n');
disp('     The augmented matrix is =')
Augm=[A b]
Y=zeros(n,1);
Y=x0;
for k=1:itmax+1
  for i=1:n
    S=0;
    for j=1:i-1
      S=S+A(i,j)*x(j);
    end
    for j=i+1:n
      S=S+A(i,j)*x0(j);
    end
    if(A(i,i)==0)
      break
    end
    x(i)=(-S+b(i))/A(i,i);
  end
  err=abs(norm(x-x0));
  rerr=err/(norm(x)+eps);
  x0=x;
  Y=[Y x];
  if(rerr<tol)
    break
  end
end
% Print the results
if(A(i,i)==0)
 disp('      division by zero')
elseif (k==itmax+1)
 disp('      No convergence')
else
```

```
fprintf('\n');
disp(' The solution vectors are:')
fprintf('\n');
disp('iter #    0    1    2    3    4    ...')
fprintf('\n');
for i=1:n
  fprintf('x%1.0f =    ',i)
  fprintf('%10.6f',Y(i,[1:k+1]))
  fprintf('\n');
end
fprintf('\n');
disp(['The method converges after ',num2str(k),' iterations to']);
x
end
```

(a) Set up the Jacobi iteration with initial guess $x = 3$, $y = 11$ and perform two iteration of Jacobi's method.

(b) Set up the Gauss-Seidel iteration with initial guess $x = 3$, $y = 11$ and perform two iteration of the Gauss-Seidel method.

(c) Explain why both methods should converge for this case.

7. Solve the following systems using the Gauss-Seidel iterative method

(a)
$$
\begin{aligned}
x - y + 2z - w &= -1 \\
2x + y - 2z - 2w &= -2 \\
-x + 2y - 4z + w &= 1 \\
3x \qquad\qquad - 3w &= -3.
\end{aligned}
$$

(b)
$$
\begin{aligned}
5x - y + 3z &= 3 \\
4x + 7y - 2z &= 2 \\
6x - 3y + 9z &= 9
\end{aligned}
$$

8. Given the linear system

$$
\begin{bmatrix}
3 & -5 & 47 & 20 \\
11 & 16 & 17 & 10 \\
56 & 22 & 11 & -18 \\
17 & 66 & -12 & 7
\end{bmatrix}
\begin{bmatrix}
x_1 \\
x_2 \\
x_3 \\
x_4
\end{bmatrix}
=
\begin{bmatrix}
18 \\
26 \\
34 \\
82
\end{bmatrix}.
$$

Reorder the equation of the system and use the Jacobi iteration to solve it.

9. Use the Gauss-Seidel iteration to solve the system

$$
\begin{bmatrix}
18 & 1 & 4 & 3 & -1 & 2 & 3 & -7 \\
2 & 12 & -1 & 7 & -2 & 1 & 1 & -4 \\
-1 & 1 & -9 & 2 & -5 & 2 & -1 & 1 \\
2 & 4 & 1 & -12 & 1 & 3 & 4 & -1 \\
1 & 3 & 1 & 7 & -16 & 1 & -2 & 4 \\
-2 & 1 & 7 & -1 & 2 & 13 & -1 & 8 \\
3 & 4 & 5 & 1 & 2 & 8 & -20 & 2 \\
5 & 1 & 1 & 1 & -1 & 1 & -7 & 12
\end{bmatrix}
\begin{bmatrix}
x_1 \\ x_2 \\ \vdots \\ \\ \\ \\ \\ x_8
\end{bmatrix}
=
\begin{bmatrix}
71 \\ 42 \\ -11 \\ -37 \\ -61 \\ 52 \\ -73 \\ 22
\end{bmatrix}.
$$

10. The following system has an approximate solution $x_1 = 3.072$, $x_2 = -5.497$, $x_3 = -2.211$, and $x_4 = 4.579$.

$$
\begin{bmatrix}
1.20 & 0.45 & 0.35 & 0.45 \\
0.89 & 2.59 & -0.33 & -0.22 \\
0.71 & 0.78 & 4.01 & -0.88 \\
0.11 & 0.55 & 0.66 & 3.39
\end{bmatrix}
\begin{bmatrix}
x_1 \\ x_2 \\ x_3 \\ x_4
\end{bmatrix}
=
\begin{bmatrix}
2.500 \\ -11.781 \\ -15.002 \\ 11.378
\end{bmatrix}.
$$

Use both the Gauss-seidel and Jacobi methods to approximate the solution of the system.

11. Use Jacobi's method and Gauss-Seidel to solve

$$
\begin{bmatrix}
-4 & 2 & 0 & . & . & . & 0 \\
2 & -4 & 2 & 0 & . & . & 0 \\
0 & 2 & -4 & 2 & 0 & . & 0 \\
0 & 0 & . & . & . & 0 & 0 \\
0 & . & 0 & . & . & . & 0 \\
0 & . & . & 0 & 2 & -4 & 2 \\
0 & . & . & . & 0 & 2 & -4
\end{bmatrix}
\mathbf{x} =
\begin{bmatrix}
2 \\ 3 \\ . \\ . \\ . \\ . \\ 11
\end{bmatrix}.
$$

12. Consider the system of equations

$$
\begin{bmatrix}
4 & 0 & 1 \\
1 & 4 & 1 \\
1 & 0 & 4
\end{bmatrix}
\begin{bmatrix}
x \\ y \\ z
\end{bmatrix}
=
\begin{bmatrix}
5 \\ 3 \\ -10
\end{bmatrix}.
$$

(a) Find the matrix that needs to be analyzed to determine whether Jacobi's iteration method will converge for this problem.

(b) Perform one iteration of Gauss-Seidel, with starting guess $[1 \ 2 \ -3]'$.

COMPUTER PROBLEM SET 4.5

1. Write a computer program in a language of your choice to solve a system of n linear equations and n unknowns using:

(a) The Jacobi iterative method.

(b) The Gauss-Seidel iterative method.

Input data to the program should be

(a) The number of equations n.

(b) The augmented matrix.

(c) The starting vector.

Output should consist of:

(a) The augmented matrix.

(b) The vector solution.

Test your program to solve the system

$$\begin{aligned}
x - y + 2z - w &= -1 \\
2x + y - 2z - 2w &= -2 \\
-x + 2y - 4z + w &= 1 \\
3x \qquad\qquad - 3w &= -3.
\end{aligned}$$

2. Given the system

$$\begin{aligned}
x \qquad + z &= 2 \\
x - y \qquad &= 0 \\
x + 2y - 3z &= 0.
\end{aligned}$$

Use the MATLAB functions seidel and jacobi to check that the Jacobi iterative method converges whereas the Gauss-Seidel iterative method diverges. Use $TOL = 10^{-3}$ and start with $\mathbf{x}^{(0)} = \mathbf{0}$.

3. Use both the MATLAB functions seidel and jacobi iterations to approximate the solution of the system

$$\begin{bmatrix}
9.235 & 0.781 & 0.421 & 0.341 & 0.881 & 0.444 & 0.365 \\
0.701 & 8.725 & 0.751 & 0.637 & 0.915 & 0.192 & 0.617 \\
0.509 & 0.167 & 7.927 & 0.628 & 0.051 & 0.837 & 0.028 \\
0.013 & 0.436 & 0.721 & 9.222 & 0.519 & 0.915 & 0.835 \\
0.858 & 0.022 & 0.636 & 0.943 & 8.909 & 0.878 & 0.974 \\
0.965 & 0.384 & 0.154 & 0.184 & 0.109 & 9.334 & 0.403 \\
0.239 & 0.516 & 0.817 & 0.844 & 0.385 & 0.476 & 7.861
\end{bmatrix} \quad \mathbf{x} = \begin{bmatrix}
29.433 \\
76.413 \\
-18.564 \\
-44.103 \\
34.227 \\
-38.622 \\
-80.017
\end{bmatrix}.$$

4. Consider the system of equation $Ax = b$ where

$$A = \begin{bmatrix} 1 & k \\ 2k & 1 \end{bmatrix}, \quad k \text{ real.}$$

(a) Find the values of k for which the matrix A is strictly diagonally dominant.

(b) For $k = 0.25$, solve the system using the Jacobi method.

5. Use the MATLAB function seidel.m with $\mathbf{x}^{(0)} = \mathbf{0}$ to solve the linear system

$$\begin{bmatrix} 2 & -5 & 2 & 5 & 30 & -8 \\ -8 & 36 & -1 & 0 & 1 & 8 \\ 11 & -4 & 25 & 1 & -4 & 4 \\ 1 & 0 & -3 & 19 & 2 & 1 \\ 42 & -2 & -9 & 0 & 3 & 0 \\ -3 & 7 & -7 & 4 & 5 & -32 \end{bmatrix} \begin{bmatrix} x_1 \\ x_2 \\ x_3 \\ x_4 \\ x_5 \\ x_6 \end{bmatrix} = \begin{bmatrix} 27 \\ 36 \\ 42 \\ 18 \\ 54 \\ 60 \end{bmatrix}.$$

At first glance the set does not seem to be suitable for an iterative solution, since the coefficient matrix is not diagonally dominant. However, by simply reordering the equations the matrix can be made diagonally dominant. Use the MATLAB function seidel.m $\mathbf{x}^{(0)} = \mathbf{0}$ to solve the new reordered system and compare the number of iterations needed for convergence for each case.

6. Repeat the preceding computer problem using the MATLAB function jacobi.m.

APPLIED PROBLEMS FOR CHAPTER 4

1. Consider the following electrical network:

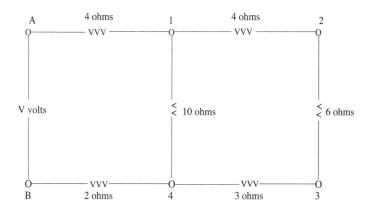

By applying both Ohm's law and Kirchhoff's Current law, we get the following system of linear equations

$$\begin{aligned} 3v_1 - 5v_2 + 2v_3 &= 0 \\ -2v_1 + 5v_2 + 2v_4 &= -5V \\ v_2 - 3v_3 + 2v_4 &= 0 \\ 3v_1 + 10v_3 - 28v_4 &= 0 \end{aligned}$$

where $v_1, \ldots, v_4$ denote the potentials and V the potential applied between A and B. Solve the system for $V = 40, 80$.

2. A resistor network with two voltage sources is shown below. By applying both

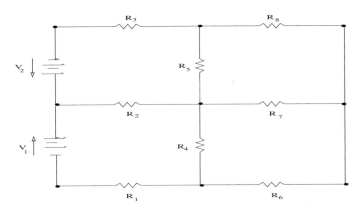

Ohm's law and Kirchhoff's Current law, we get the following system of linear equations

$$\begin{bmatrix} R_1 + R_2 + R_4 & -R_2 & 0 & -R_4 \\ -R_2 & R_2 + R_3 + R_5 & -R_5 & 0 \\ 0 & -R_5 & R_5 + R_7 + R_8 & -R_7 \\ -R_4 & 0 & -R_7 & R_4 + R_6 + R_7 \end{bmatrix} \begin{bmatrix} i_1 \\ i_2 \\ i_3 \\ i_4 \end{bmatrix} = \begin{bmatrix} -V_1 \\ V_2 \\ 0 \\ 0 \end{bmatrix}.$$

Solve the system of equation if $R_1 = R_2 = R_3 = 100$, $R_4 = R_5 = 150$, $R_6 = R_7 = R_8 = 200$, $V_1 = 10$, and $V_2 = 12$.
Note: The coefficient matrix is diagonally dominant.

3. A least-square polynomial consists of finding a polynomial that best represents a set of experimental data. A polynomial of degree three $y = a_0 + a_1 x + a_2 x^2 + a_3^3$ that best fits n data points is obtained from the solution of the system of equations (see Chapter 7):

$$\begin{aligned}
a_0 n + a_1 \sum x_i + a_2 \sum x_i^2 + a_3 \sum x_i^3 &= \sum y_i \\
a_0 \sum x_i + a_1 \sum x_i^2 + a_2 \sum x_i^3 + a_3 \sum x_i^4 &= \sum y_i x_i \\
a_0 \sum x_i^2 + a_1 \sum x_i^3 + a_2 \sum x_i^4 + a_3 \sum x_i^5 &= \sum y_i x_i^2 \\
a_0 \sum x_i^3 + a_1 \sum x_i^4 + a_2 \sum x_i^5 + a_3 \sum x_i^6 &= \sum y_i x_i^3.
\end{aligned}$$

Find the least-square polynomial of degree three that best fits to the following data:

x	1.0	1.5	2.0	2.5	3.0	3.5	4.0	4.5
y	4.3	6.2	10.1	13.5	19.8	22.6	24.7	29.2

4. Consider the loading of a statically determinate pin-jointed truss shown in Figure 4.2. The truss has seven members and five nodes, and is under the action of the forces R_1, R_2, and R_3 parallel to the y-axis.

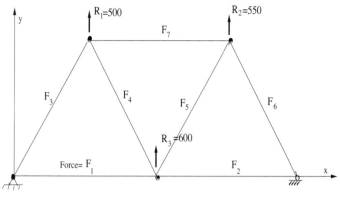

FIGURE 4.2
Determinate pin-jointed truss.

Using the result that at each pin the sum of all forces Fi acting horizontally of vertically is equal to zero, find the member forces (Fi) obtained from the following system of equations.

$$\begin{bmatrix} 1.0 & -1.0 & 0 & 0.5 & -0.5 & 0 & 0 \\ 0 & 0 & 0 & -8.66 & -8.66 & 0 & 0 \\ 0 & 1.0 & 0 & 0 & 0 & 0.5 & 0 \\ 0 & 0 & 0.5 & -0.5 & 0 & 0 & -1.0 \\ 0 & 0 & 8.66 & 8.66 & 0 & 0 & 0 \\ 0 & 0 & 0 & 0 & 0.5 & -0.5 & 1.0 \\ 0 & 0 & 0 & 0 & 8.66 & 8.66 & 0 \end{bmatrix} \begin{bmatrix} F_1 \\ F_2 \\ F_3 \\ F_4 \\ F_5 \\ F_6 \\ F_7 \end{bmatrix} = \begin{bmatrix} 0 \\ 500 \\ 0 \\ 0 \\ 550 \\ 0 \\ 600 \end{bmatrix}.$$

5. Consider the square shown below. The left face is maintained at 100^0C and the top face at 500^0C, while the other faces are exposed to an environment at 100^0C. By applying a numerical method known as the finite difference, the temperature at the various nodes is given from the solution of the system of equations:

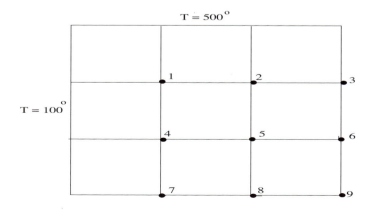

$$
\begin{bmatrix}
-4 & 1 & 0 & 1 & 0 & 0 & 0 & 0 & 0 \\
1 & -4 & 1 & 0 & 1 & 0 & 0 & 0 & 0 \\
0 & 2 & -4.67 & 0 & 0 & 1 & 0 & 0 & 0 \\
1 & 0 & 0 & -4 & 1 & 0 & 1 & 0 & 0 \\
0 & 1 & 0 & 1 & -4 & 1 & 0 & 1 & 0 \\
0 & 0 & 1 & 0 & 2 & -4.67 & 0 & 0 & 1 \\
0 & 0 & 0 & 2 & 0 & 0 & -4.67 & 1 & 0 \\
0 & 0 & 0 & 0 & 2 & 0 & 1 & -4.67 & 1 \\
0 & 0 & 0 & 0 & 0 & 1 & 0 & 1 & -2.67
\end{bmatrix}
\begin{bmatrix}
T_1 \\ T_2 \\ T_3 \\ T_4 \\ T_5 \\ T_6 \\ T_7 \\ T_8 \\ T_9
\end{bmatrix}
=
\begin{bmatrix}
-600 \\ -500 \\ -567 \\ -100 \\ 0 \\ -67 \\ -167 \\ -67 \\ -67
\end{bmatrix}.
$$

Find the temperature T_i, $i = 1, \ldots, 9$.

Note: The coefficient matrix is sparse (contains a large number of zero elements) and diagonally dominant. For this reason, iterative methods of solution may be very efficient.

6. Consider heat conduction in a small wire carrying electrical current that is producing heat at a constant rate. The equation describing the temperature $y(x)$ along the wire ($0 \leq x \leq 1$ cm) is

$$
D \frac{\partial^2 y}{\partial x^2} = -S
$$

with boundary conditions $y(0) = y(1) = 0°C$, thermodiffusion coefficient $D = 0.01 \text{cm}^2/\sec$, and normalized source term $S = 1°C/\sec$.

If we discretize the domain into 20 equal subintervals, using $x_j = j/20$ for $j = 0$ to 20, we can approximate the equation at x_j to obtain

$$
D \frac{y_{j-1} - 2y_j + y_{j+1}}{h^2} = -S
$$

where y_j is the temperature at $x = x_j$ and $h = 0.05$ is the step size. If we apply the boundary conditions at x_0 and x_{20} we are left with 19 equations for 19 unknown temperatures, y_1 to y_{19}. We can put these equations into the matrix form $Ay = b$ where

$$
A = \begin{bmatrix}
-2 & 0 & 0 & 0 & \ldots & 0 & 0 \\
1 & -2 & 1 & 0 & \ldots & 0 & 0 \\
0 & 1 & -2 & 0 & \ldots & 0 & 0 \\
 & & \vdots & & & & \\
 & & \vdots & & & & \\
0 & 0 & \ldots & 0 & 1 & -2 & 1 \\
0 & 0 & \ldots & 0 & 0 & 1 & -2
\end{bmatrix}, \quad
y = \begin{bmatrix}
y_1 \\ y_2 \\ y_3 \\ \vdots \\ \vdots \\ y_{18} \\ y_{19}
\end{bmatrix}, \quad
b = \begin{bmatrix}
-0.25 \\ -0.25 \\ -0.25 \\ \vdots \\ \vdots \\ -0.25 \\ -0.25
\end{bmatrix}.
$$

Solve the above steady-state system problem by using the Jacobi iteration and Gauss-Seidel iteration. Start with $y = 0$.

7. Suppose a university is comprised of three colleges: Sciences, Engineering, and Computer Science. The annual budgets for these colleges are $16 million,

$5 million, and $8 million, respectively. The full-time enrollments are 4000 in Sciences, 1000 in Engineering, and 2000 in Computer Science. The enrollment consists of 70% from Sciences, 20% from Engineering, and 10% from Computer Science. For example, Engineering and Computer Science courses contain some students from other colleges. The distribution is as follows

	Courses taught by		
Students from	Sciences	Engineering	Computer Science
Sciences	70%	10%	15%
Engineering	20%	90%	10%
Computer Science	10%	0%	75%

Determine to the nearest cent the annual cost of educating a student in each of the three colleges.

8. (*Solution of a mass balance problem*) For the following separation system (see figure below), we know the inlet mass flowrate (in Kg/hr) and the mass fractions of each species in the inlet (stream # 1) and each outlet (streams # 2, 4, and 5). Here we use the notation that F is the mass flow rate of stream #

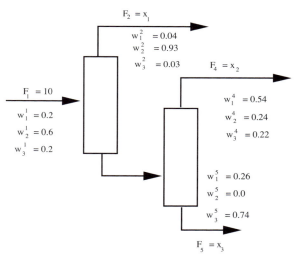

i, and w_j^i is the mass fraction of species # j in stream # i.

We want to calculate the unknown mass flow rates of each outlet stream. If we define the unknowns as

$$x_1 = F_2, \ x_2 = F_4, \ x_3 = F_5$$

and set up the mass balances for

(a) the total mass flow rate,

(b) the mass flow rate of species 1,

(c) the mass flow rate of species 2.

we obtain a set of three linear algebraic equations for the three unknown outlet flow rates x_1, x_2, and x_3

$$10w_1^1 = x_1 w_1^2 + x_2 w_1^4 + x_3 w_1^5$$
$$10w_2^1 = x_1 w_2^2 + x_2 w_2^4 + x_3 w_2^5$$
$$10w_3^1 = x_1 w_3^2 + x_2 w_3^4 + x_3 w_3^5.$$

Convert this set of equations to matrix form, and solve the system by hand using Gaussian elimination with partial pivoting.

9. A coffee shop specializes in blending gourmet coffees. From type A, type B, and type C coffees, the owner wants to prepare a blend that will sell for $8.50 for a 1-pound bag. The cost per pound of these coffees is $12, $9, and $7, respectively. The amount of type B is to be twice the amount of type A. Write the system of equation that needs to be solved to find the amount of each type of coffee that will be in the final blend and solve the system using naive Gaussian elimination.

10. A company makes three types of patio furniture: chairs, rockers, and chaise lounges. Each requires wood, plastic, and aluminum, in the amounts shown in the following table. The company has in stock 400 units of wood, 600 units of plastic, and 1500 units of aluminum. For its end-of-the-season production run, the company wants to use up all the stock. To do this how many chairs, rockers, and chaise lounges should it make? (*Hint:* Let c be the number of chairs, r the number of rockers, and l the number of chaise lounges.)

	Wood	Plastic	Aluminum
Chair	1 unit	1 unit	2 units
Rocker	1 unit	1 unit	3 units
Chaise lounge	1 unit	2 units	5 units

11. In a structure, six wires support three beams. Three weights are attached at the points shown. Assume that the structure is stationary and that the weights of the wires and beams are very small compared to the applied weights. The principles of statics state that the sum of the forces is zero and that the sum of the moments about any point is zero. Applying these principles to each beam, we obtain the following equations:

Beam 1
$$T_1 + T_2 = T_3 + T_4 + W_1 + T_6,$$
$$-T_3 - 4T_4 - 5W_1 - 6T_6 + 7T_2 = 0.$$

Beam 2
$$T_3 + T_4 = W_2 + T_5,$$
$$-W_2 - 2T_5 + 3T_4 = 0.$$

Beam 3
$$T_5 + T_6 = W_3,$$
$$-W_3 + 3T_6 = 0.$$

Write the set of equations in a matrix form and determine the forces T_i for the following cases.

(a) $W_1 = W_2 = W_3 = 0$ N,

(b) $W_1 = W_2 = W_3 = 300$ N,

(c) $W_1 = 400$ $N, W_2 = 300$ $N, W_3 = 200$ N.

Chapter 5

Interpolation

Consider the following problem:

x	x_0	x_1	$\cdots$	x_n
y	y_0	y_1	$\cdots$	y_n

Table 5.1   Table of numerical values.

Suppose that we are given a table of $(n+1)$ points where the $x's$ are distinct and satisfy

$$x_0 < x_1 < x_2 < \ldots < x_n.$$

Our objective is to find a polynomial curve that passes through the given points (x_i, y_i), $i = 0, 1, \ldots, n$. Hence, we need to find a polynomial $p(x)$ such that

$$p(x_i) = y_i \quad \text{for} \ \ i = 0, 1, \ldots, n. \tag{5.1}$$

This is a topic called **interpolation**. The polynomial $p(x)$ is said to interpolate Table 5.1 and the points x_i are called the **nodes**.

Two methods for doing this will be developed in this chapter. The first is **Newton's method**, based on using the *divided differences* and the second is the **Lagrange method**, based on using the so-called *cardinal functions*.

EXAMPLE 5.1 : Relative Viscosity of Ethanol

Viscosity can be thought of as the stickiness of a liquid. Table 5.2 give the relative viscosity V of ethanol as a function of the percent of anhydrous solute weight w (see *Handbook of Chemistry and Physics*, CRC Press, 1982-83). Estimating the values of the V at points other than the nodes is an essential feature of interpolation.

w	5	10	15	20	30	40
V	1.226	1.498	1.822	2.138	2.662	2.840

Table 5.2   Relative viscosity of ethanol.

5.1   POLYNOMIAL INTERPOLATION THEORY

Given the data in Table 5.1 where the $x_i's$ are assumed to be distinct, we want to study the problem of finding a polynomial

$$p(x) = a_0 + a_1x + a_2x^2 + \cdots + a_nx^n \tag{5.2}$$

that interpolates the given data.

Applying condition (5.1) to the given tabulated points x_i gives the system

$$\begin{aligned}
a_0 + a_1x_0 + a_2x_0^2 + \cdots + a_nx_0^n &= y_0 \\
a_0 + a_1x_1 + a_2x_1^2 + \cdots + a_nx_1^n &= y_1 \\
\vdots \quad &= \vdots \\
a_0 + a_1x_n + a_2x_n^2 + \cdots + a_nx_n^n &= y_n.
\end{aligned} \tag{5.3}$$

This is a system of $(n+1)$ linear equations in $(n+1)$ unknowns $a_0, a_1, \ldots, a_n$. In matrix form, the system is

$$\mathbf{X}a = \mathbf{Y}$$

where

$$\mathbf{X} = \left[x_i^j \right] \quad i, j = 0, 1, \ldots, n$$

$$a = [a_0, \ldots, a_n]^T, \quad \mathbf{Y} = [y_0, \ldots, y_n]^T. \tag{5.4}$$

The matrix $\mathbf{X}$ is known as the **Vandermonde** matrix.

Thus, solving the system (5.3) is equivalent to solving the polynomial interpolation problem.

THEOREM 5.1

Given $n + 1$ distinct points $x_0, x_1, \ldots, x_n$ and $n + 1$ arbitrary real values $y_0, y_1, \ldots, y_n$, there is a unique polynomial p of degree $\leq n$ that interpolates Table 5.1.

Proof: It can be easily shown that the determinant of the matrix $\mathbf{X}$ in (5.4) is (see Exercise 2)

$$\det(\mathbf{X}) = \prod_{0 \leq j < i \leq n} (x_i - x_j) \tag{5.5}$$

and is nonzero since the points x_i are assumed to be distinct. Thus, there is a unique solution for the a_i's, that is, there is a unique interpolating polynomial of degree $\leq n$.

EXAMPLE 5.2

Find the polynomial that interpolates the table

x	4.5	6.1
y	7.1	2.3

The polynomial sought has the form

$$p(x) = a_0 + a_1 x.$$

By applying condition (5.1) to the given table, we obtain the system

$$a_0 + 4.5a_1 = 7.1$$
$$a_0 + 6.1a_1 = 2.3.$$

Solve to get

$$a_0 = 20.6, \quad a_1 = -3.0.$$

Therefore,

$$p(x) = 20.6 - 3.0x.$$

EXERCISE SET 5.1

1. Determine a polynomial of degree ≤ 3 that interpolates the data

x	1.2	2.1	3.0	3.6
y	0.7	8.1	27.7	45.1

2. Find a polynomial of order 2 that interpolates the table

x	0.2	0.4	0.6
y	-.95	-.82	-.65

3. Show that the determinant of the Vandermonde matrix given in (5.4) is equal to

$$V_n(x_0, x_1, \ldots, x_n) = \det \begin{bmatrix} 1 & x_0 & x_0^2 & \cdots & x_0^n \\ 1 & x_1 & x_1^2 & \cdots & x_2^n \\ \vdots & \vdots & \vdots & & \vdots \\ 1 & x_n & x_n^2 & \cdots & x_n^n \end{bmatrix} = \prod_{0 \leq j < i \leq n} (x_i - x_j).$$

4. Use the table

x	1.00	1.20	1.40	1.60	1.80	2.00
$f(x)$	1.000000	0.918168	0.887263	0.893515	0.931338	1.000000

together with linear interpolation to construct an approximation to $f(1.51)$.

5. Use the table

x	1.00	1.20	1.40	1.60	1.80	2.00
$f(x)$	1.0000	0.9181	0.8872	0.8935	0.9313	1.0000

together with linear interpolation to construct an approximation to $f(1.44)$.

6. You are given the polynomial $p(x) = (x - 350)^4$. Suppose we want to interpolate this polynomial by another polynomial at n points equally distributed on the interval $[350, 351]$.

   (a) Under what condition on n do we expect the polynomial interpolant to coincide with the given polynomial p?

   (b) Expand p and find its five coefficients. Now, interpolate this polynomial at five and then at ten uniformly spaced points in $[350, 351]$ using the *Vandermonde* approach, and find the coefficient of the interpolating polynomial. What do you observe? What do you conclude from this observation?

5.2   NEWTON'S DIVIDED-DIFFERENCE INTERPOLATING POLYNOMIAL

In this section we want to develop an interpolating formula that will be more efficient and convenient to use than the one shown in the previous section. This will avoid the problem of finding the solution to a system of simultaneous equations.

Let us first derive the first-degree Newton interpolating polynomial ($n = 1$) that takes, in this case, the form

$$p_1(x) = a_0 + a_1(x - x_0),$$

Since a straight line can be passed through two points, then

$$p_1(x) = y_0 + \left(\frac{y_1 - y_0}{x_1 - x_0} \right)(x - x_0). \tag{5.6}$$

At this stage, we introduce the following notation

$$f[x_0, x_1, \ldots, x_n] = \frac{f[x_1, \ldots, x_n] - f[x_0, \ldots, x_{n-1}]}{x_n - x_0} \tag{5.7}$$

where

$$y_i = f(x_i) = f[x_i], \quad i = 0, \ldots, n.$$

This coefficient is called the nth **order divided difference** of f.

In view of (5.7), $p_1(x)$ can be written in the form

$$p_1(x) = f[x_0] + f[x_0, x_1](x - x_0)$$

or

$$p_1(x) = p_0(x) + f[x_0, x_1](x - x_0). \tag{5.8}$$

Note that $p_1(x)$ is obtained by adding to $p_0(x)$ the correction term

$$f[x_0, x_1](x - x_0).$$

We now proceed to the second-degree interpolating polynomial. From Eqn. (5.8) we would expect $p_2(x)$ to be in the form

$$p_2(x) = p_1(x) + a_2(x - x_0)(x - x_1). \tag{5.9}$$

Note that, for each of the points x_0 and x_1, condition (5.1) is satisfied, since $p_1(x)$ itself satisfies (5.1) and $(x - x_0)(x - x_1)$ takes the value zero.

Evaluating $p_2(x)$ at $x = x_2$ and using the fact that $p_2(x_2) = f(x_2)$, we get

$$
\begin{aligned}
a_2 &= \frac{f(x_2) - p_1(x_2)}{(x_2 - x_0)(x_2 - x_1)} \\
&= \left[\frac{f(x_2) - f(x_0) - f[x_0, x_1](x_2 - x_0)}{(x_2 - x_1)} \right] / (x_2 - x_0) \\
&= \left[\frac{f(x_2) - f(x_1) + f(x_1) - f(x_0) - f[x_0, x_1](x_2 - x_0)}{(x_2 - x_1)} \right] / (x_2 - x_0) \\
&= \left[\frac{f(x_2) - f(x_1)}{(x_2 - x_1)} + \frac{f[x_0, x_1](x_1 - x_0) - f[x_0, x_1](x_2 - x_0)}{(x_2 - x_1)} \right] / (x_2 - x_0).
\end{aligned}
$$

Simplify the second fraction inside the brackets to get

$$
\begin{aligned}
a_2 &= \frac{f[x_1, x_2] - f[x_0, x_1]}{x_2 - x_0} \\
&= f[x_0, x_1, x_2].
\end{aligned}
$$

Thus, in view of (5.9), we have

$$p_2(x) = p_1(x) + f[x_0, x_1, x_2](x - x_0)(x - x_1). \tag{5.10}$$

Repeating this entire process again, p_3, p_4, and higher interpolating polynomials can be consecutively obtained in the same way. In general, we would obtain for $p_n(x)$ the interpolating formula

$$p_n(x) = p_{n-1}(x) + f[x_0, \ldots, x_n](x - x_0)(x - x_1) \cdots (x - x_{n-1}). \tag{5.11}$$

x	x_0	x_1	$\ldots$	x_n
$f(x)$	$f(x_0)$	$f(x_1)$	$\ldots$	$f(x_n)$

Table 5.3 Values of a function f.

We now turn to the problem of calculating the divided differences. We start with Table 5.3, which gives the values of a function f where the points $x_0, \ldots, x_n$ are assumed to be distinct.

Using Eqn. (5.7), the first four divided differences are

$$f[x_0] = f(x_0)$$
$$f[x_0, x_1] = \frac{f[x_1] - f[x_0]}{x_1 - x_0}$$
$$f[x_0, x_1, x_2] = \frac{f[x_1, x_2] - f[x_0, x_1]}{x_2 - x_0}$$
$$f[x_0, x_1, x_2, x_3] = \frac{f[x_1, x_2, x_3] - f[x_0, x_1, x_2]}{x_3 - x_0}.$$

One should note that $f[x_0, x_1, \ldots, x_n]$ is invariant under all permutations of the arguments $x_0, x_1, \ldots, x_n$. More general properties of divided differences may be found in [7].

This procedure is best illustrated by arranging the divided differences in the format shown in Table 5.4. The coefficient in the top diagonal of the table is the

x_i	$f[x_i]$	$f[x_i, x_{i+1}]$	$f[x_i, x_{i+1}, x_{i+2}]$ $\ldots$
x_0	$f[x_0]$		
		$f[x_0, x_1]$	
x_1	$f[x_1]$		$f[x_0, x_1, x_2]$
		$f[x_1, x_2]$	
x_2	$f[x_2]$		$f[x_1, x_2, x_3]$
		$f[x_2, x_3]$	
x_3	$f[x_3]$		$f[x_2, x_3, x_4]$
		$f[x_3, x_4]$	
x_4	$f[x_4]$		$f[x_3, x_4, x_5]$
		$f[x_4, x_5]$	
x_5	$f[x_5]$		
$\vdots$	$\vdots$	$\vdots$	$\vdots$

Table 5.4 Format for constructing divided differences of $f(x)$.

ones needed to write the Newton interpolating polynomial (5.11), that is

$$p_n(x) = f[x_0] + f[x_0, x_1](x - x_0) + f[x_0, x_1, x_2](x - x_0)(x - x_1) + \cdots$$
$$\cdots + f[x_0, x_1, \ldots, x_n](x - x_0)(x - x_1) \cdots (x - x_{n-1}),$$

which can be expressed in the more compact form

$$p_n(x) = \sum_{i=0}^{n} \left\{ f[x_0, x_1, \ldots, x_i] \prod_{j=1}^{i} (x - x_{j-1}) \right\}. \qquad (5.12)$$

Formula (5.12) is known as the **Newton forward divided-difference** formula. If the nodes are recorded as $x_n, x_{n-1}, \ldots, x_0$, a similar formula known as the **Newton backward divided-difference** formula is obtained

$$p_n(x) = \sum_{i=0}^{n} \left\{ f[x_{n-i}, x_{n-i+1}, \ldots, x_n] \prod_{j=1}^{i} (x - x_{n-j+1}) \right\}. \qquad (5.13)$$

EXAMPLE 5.3

Form a divided-difference table for the following data and obtain Newton's interpolating polynomial.

x	2	4	6	8
y	4	8	14	16

From the top diagonal of Table 5.5 Newton's interpolating polynomial is

x_i	$f[x_i]$	f[,]	f[,,]	f[,,,]
2	4			
		$\frac{8-4}{4-2} = 2$		
4	8		$\frac{3-2}{6-2} = \frac{1}{4}$	
		$\frac{14-8}{6-4} = 3$		$\frac{-\frac{1}{2}-\frac{1}{4}}{8-2} = -\frac{1}{8}$
6	14		$\frac{1-3}{8-4} = -\frac{1}{2}$	
		$\frac{16-14}{8-6} = 1$		
8	16			

Table 5.5 Divided differences for Example 5.3.

$$p_3(x) = 4 + 2(x - 2) + \frac{1}{4}(x - 2)(x - 4) - \frac{1}{8}(x - 2)(x - 4)(x - 6)$$

$$= -\frac{1}{8}x^3 + \frac{7}{4}x^2 - 5x + 8.$$

When the nodes $x_0, x_1, \ldots, x_n$ are evenly spaced, that is,

$$x_k = x_0 + kh, \quad k = 0, 1, \ldots, n$$

with

$$h = \frac{x_n - x_0}{n}$$

Eqns. (5.12) and (5.13) can be expressed in a simplified form as follows:

Let us assume that the nodes are arranged in ascending order. We define the **forward difference** Δf_i by

$$\Delta f_i = f_{i+1} - f_i, \quad \text{with } f(x_i) = f_i, \quad i = 0, 1, \ldots.$$

Higher powers are defined recursively by

$$\Delta^n f_i = \Delta \left(\Delta^{n-1} f_i \right), \quad n = 2, 3, \ldots.$$

Using the binomial-coefficient notation

$$\binom{s}{k} = \frac{s(s-1)\cdots(s-k+1)}{k!}$$

Eqn. (5.12) can be expressed as

$$p_n(x) = \sum_{k=0}^{n} \binom{s}{k} \Delta^k f_0, \quad s = \frac{x - x_0}{h}. \tag{5.14}$$

Eqn. (5.14) is called the **Newton forward difference**. Similarly Eqn. (5.13) can be expressed as

$$p_n(x) = \sum_{k=0}^{n} (-1)^k \binom{-s}{k} \nabla^k f_n, \tag{5.15}$$

with $\nabla f_i = f_i - f_{i-1}, i \geq 1$. Eqn. (5.15) is called the **Newton backward difference**. For more details regarding the derivation of these formulas see [19].

EXAMPLE 5.4

Form a divided-difference table for the following data and obtain Newton's interpolating polynomial.

x	1	2	3	5	7
y	3	5	9	11	15

The MATLAB function newtondd produces the coefficients for the polynomial that appears in Table 5.6. The data and polynomial are shown in Figure 5.1. The equation of Newton's interpolating polynomial is

$$p_4(x) = \mathbf{3} + \mathbf{2}(x - 1) + (x - 1)(x - 2) - \frac{1}{\mathbf{2}}(x - 1)(x - 2)(x - 3)$$

$$+ \frac{1}{\mathbf{8}}(x - 1)(x - 2)(x - 3)(x - 5)$$

$$= \frac{1}{8}x^4 - \frac{15}{8}x^3 + \frac{73}{8}x^2 - \frac{113}{8}x + \frac{39}{4}.$$

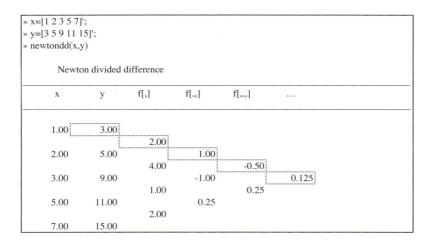

Table 5.6   Newton's divided-difference for Example 5.4.

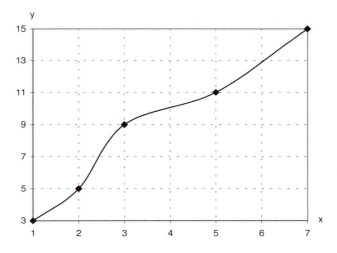

FIGURE 5.1
Interpolating polynomial for Example 5.4.

EXAMPLE 5.5

The following table lists values of the sine function at various points. The divided-difference table corresponding to these data is shown in Table 5.7.

x	0.0	0.1	0.3	0.7	0.8	0.9
$\sin x$	0.0	0.099833	0.29552	0.644218	0.717356	0.783327
x	1.3	1.9	2.2	2.7	3.1	
$\sin x$	0.963558	0.94630	0.808496	0.427380	0.041581	

M-function 5.2

The following MATLAB function **newtondd.m** forms a divided-difference table for the Newton interpolating polynomial. INPUTS are the vectors *x* and *y* .

```
function newtondd(x,y)
% Newton divided difference
disp('          Newton divided difference')
disp('_____')
disp('    x          y          f[,]        f[,,]        f[,,,]        ... ')
disp('_____')
n=length(x);
for k=1:n-1
 % Compute first divided difference
   d(k,1)=(y(k+1)-y(k))/(x(k+1)-x(k));
end
for i=2:n-1
   for k=1:n-i
   % Compute ith divided difference
      d(k,i)=(d(k+1,i-1)-d(k,i-1))/(x(k+i)-x(k));
   end
end
% print results
if (rem(n,2)==0)
   p=n/2;
   m=n/2;
else
   p=fix(n/2);
   m=fix(n/2)+1;
end
for i=1:p
  fprintf('%8.2f %8.2f',x(i),y(i));
  for k=1:i-1
    fprintf('          %8.5f',d(i-k,2*k));
  end
  fprintf('\n     ')
  for k=1:i
    fprintf('          %8.5f',d(i-k+1,2*k-1));
  end
  fprintf('\n')
end
j=p;
for i=m:-1:1
```

```
    j=j+1;
    fprintf('%8.2f %8.2f',x(j),y(j));
    for k=1:i-1
        fprintf('          %8.5f',d(j-k,2*k));
    end
    fprintf('\n     ')
    for k=1:i-1
        fprintf('          %8.5f',d(j-k+1,2*k-1));
    end
    fprintf('\n')
end
```

```
» x=[0 0.1 0.3 0.7 0.8 0.9 1.3 1.9 2.2 2.7 3.1]';
» y=sin(x);
» newtondd(x,y)

   Newton divided difference
```

x	y	f[.]	f[..]	f[...]	...					
0	0									
		0.99833								
0.1	0.099833		-0.0663167							
		0.978435		-0.1592857						
0.3	0.295520		-0.1778167		0.01533333					
		0.871745		-0.147019		0.00748016				
0.7	0.644218		-0.28073		0.02206548		-0.0009829			
		0.73138		-0.1293667		0.00620238		0.00006		
0.8	0.717356		-0.35835		0.02950833		-0.0008633		-0.00013	
		0.65971		-0.0998583		0.00464836		-0.00022		0.00008
0.9	0.783327		-0.418265		0.03694571		-0.0013179		0.00008	-0.00003
		0.4505775		-0.0555235		0.00214441		0.00000		-0.00003
1.3	0.963558		-0.4793408		0.04016233		-0.0013202		0.00001	
		-0.0287633		0.00070377		-0.0004961		0.00001	
1.9	0.946300		-0.4784259		0.03921983		-0.0012855			
		-0.4593467		0.07129947		-0.0034526				
2.2	0.808496		-0.3786067		0.03162401					
		-0.762232		0.12822269						
2.7	0.427380		-0.2247394							
		-0.9644975								
3.1	0.041581									

Table 5.7 Newton's divided differences for Example 5.5.

EXERCISE SET 5.2

1. The following tabulated function is a polynomial. Find the degree of the polynomial and the coefficient of the highest power of x.

x	0	1	2	3	4	5
$f(x)$	-7	-4	5	26	65	128

2. Determine the polynomial of degree ≤ 3 that interpolates the function $f(x) = 2x^2 - x + 2$ at $x = 0, 1, 2, 3$. What is the degree of the polynomial?

3. Show that Newton's interpolating polynomial of degree one passing through the points (x_0, y_0) and (x_1, y_1) may be written in the form

$$p_1(x) = \frac{1}{x_1 - x_0} \begin{vmatrix} y_0 & x_0 - x \\ y_1 & x_1 - x \end{vmatrix}.$$

4. Determine the polynomial of degree ≤ 5, using Newton's divided-differences, that interpolates the table

x	1.0	2.0	3.0	4.0	5.0	6.0
$f(x)$	14.5	19.5	30.5	53.5	94.5	159.5

Use the resulting polynomial to estimate the value of $f(4.5)$. Compare with the exact value, $f(4.5) = 71.375$.

5. Construct Newton's interpolating polynomial for the data given by the table

x	-3	-2	-1	0	1
y	-41	-17	-5	1	3

6. Fill in the divided-difference table:

x_i	$f[.]$	$f[.,.]$	$f[.,.,.]$	$f[.,.,.,.]$
1.1	2.45			
		0.609		
2.2	3.12		—	
		—		—
3.3	3.98		—	
		—		
4.4	5.22			

7. Given $f(x) = 2x^3 + 4x^2 + 3x + 5$. Find Newton's interpolating polynomial that interpolates the values of f at $x = -1, 0, 1, 2, 3$.

8. To investigate the relationship between yield of potatoes, y, and level of fertilizer application, x, an experimenter divided a field into 5 plots of equal size and applied differing amounts of fertilizer to each. The data recorded for each plot are given by the table (in pounds).

x	1	2	3	4	5
y	22	23	25	30	28

(a) Find an interpolating polynomial for this table.

(b) According to the interpolating polynomial, approximately how many pounds would you expect from a plot to which 2.5 pounds of fertilizer had been applied?

9. Estimate $\log(2)$ using linear interpolation. Interpolate between $\log(1)$ and $\log(6) = 1.791759$. The true value of $\log(2)$ is 0.6931472.

10. Let $f(x) = \ln(1 + x)$, $x_0 = 1$, and $x_1 = 1.1$. Use linear interpolation to calculate an approximation value for $f(1.04)$.

11. Construct a quadratic polynomial that interpolates $f(x) = \log_2 x$ at the nodes $x_0 = 1$, $x_1 = 2$, $x_2 = 4$.

12. Consider the problem of constructing the cubic polynomial $p_3(x)$ that interpolates as follows:

$$p_3(a) = f(a), \quad p_3(b) = f(b), \quad p_3(c) = f(c),$$

and

$$p_3'(c) = f'(c).$$

(a) Let $p_2(x)$ be the quadratic that interpolates f in the usual way at the nodes $x = a$, $x = b$, and $x = c$. Find a cubic polynomial $q(x)$ such that

$$q_3(x) = p_2(x) + q(x)$$

also interpolates f in the usual way at the same nodes.

(b) Now find the particular choice of q that implies that

$$q_3'(c) = f'(c).$$

13. Construct a divided-difference table for the interpolation problem $f(1) = 2$, $f(2) = 4$, $f'(2) = 0$. Write down Newton's interpolating polynomial.

14. Given the polynomial $p(x) = x^9$, try to interpolate p at 21 equidistant points on the interval $[0, 2]$. Find that interpolant, and explain how you find it.

15. Use the Newton divided-difference polynomial to approximate $f(3)$ from the following table:

x	0	1	2	4	5	6
$f(x)$	1	14	15	5	6	19

16. Given the following tabulated function

x	0	1	2	3	4	5
$f(x)$	−7	−4	5	26	65	128

This function is a polynomial. Find the degree of the polynomial and the coefficient of the highest power of x.

17. Determine the values of the missing entries of the following-divided difference table:

x_i	$f[.]$	$f[.,.]$	$f[.,.,.]$	$f[.,..,.]$
-3	y_0			
		Δy_0		
-2	y_1		-22	
		Δy_1		$\Delta^3 y_0$
-1	y_2		17	
		1		$\Delta^3 y_1$
0	y_3		$\Delta^2 y_2$	
		49		$\Delta^3 y_2$
3	y_4		67	
		Δy_4		
4	463			

18. Find the interpolation polynomial for the table

x	0.0	0.50	1.00	1.50	2.00	2.50	3.0
$f(x)$	-0.333	-0.270	-0.200	-0.133	-0.077	-0.033	0.0

Compare the interpolated values with the corresponding values of $f(x) = (x-3)/(x^2-9)$.

COMPUTER PROBLEM SET 5.2

1. Write a computer program in a language of your choice that performs Newton interpolation. Input data to the program should include a set of values of x, the number of such values, the corresponding values y, and the value α of x at which Newton's interpolating polynomial, $p(x)$, is to be evaluated. The output should be the divided-difference table and the value of $p(\alpha)$. Test your program for the following tabulation of an actual thermodynamic quantity.

x	0	0.2	0.4	0.8	1.0
$f(x)$	1.000	0.916	0.836	0.0741	0.624
x	1.4	1.6	1.8	2.0	1.2
$f(x)$	0.224	0.265	0.291	0.316	0.429

Approximate $f(0.23)$ using Newton's interpolating polynomial.

2. Use the MATLAB function newtondd.m to find the polynomial that interpolates the function e^x at 11 equally spaced points in the interval $[0, 1]$. Compare the values of polynomial to e^x at the midpoints.

3. Use the MATLAB function newtondd.m to find the polynomial that interpolates the function $\tan x$ at 21 equally spaced points in the interval $[0, \pi/4]$.

5.3   THE ERROR OF THE INTERPOLATING POLYNOMIAL

Let us now derive an expression for the error term $f(x) - p_n(x)$. In general, an error exists at the nontabular points, known as the truncation error. To determine the form of this error term, we consider a function $f(x)$, which has $n+1$ continuous derivatives in $[a, b]$. Let the error term be represented by

$$E_n(x) = f(x) - p_n(x). \tag{5.16}$$

We will now prove that for any t in $[a, b]$ containing the distinct nodes $x_0, x_1, \ldots, x_n$, there is a ζ in (a, b) for which

$$E_n(t) = \frac{1}{(n+1)!} f^{(n+1)}(\zeta) \prod_{i=0}^{n} (t - x_i). \tag{5.17}$$

Let t be fixed and distinct from the nodal points x_i. To simplify the notation, we define

$$\Psi(x) = \prod_{i=0}^{n} (x - x_i) \tag{5.18}$$

so that

$$E_n(t) = f^{(n+1)}(\zeta) \frac{\Psi(t)}{(n+1)!}.$$

We now define the function

$$e(x) = f(x) - p_n(x) - \Psi(x) \frac{f(t) - p_n(t)}{\Psi(t)}.$$

Observe that $e(x) = 0$ at the tabulated points x_i and t. Hence, $e(x)$ has $n+2$ zeros in the interval containing the tabulated points. Now apply Rolle's Theorem, which states that between any two zeros of $e(x)$ there must be at least one zero of $e'(x)$. Thus, the first derivative $e'(x)$ has at least $n+1$ zeros. Similarly $e''(x)$ has at least n zeros. Continuing in this way, we finally find that $e^{(n+1)}(x)$ has at least one zero.

Let such a zero be z. Then

$$e^{(n+1)}(z) = f^{(n+1)}(z) - p_n^{(n+1)}(z) - \Psi^{(n+1)}(z) \frac{f(t) - p_n(t)}{\Psi(t)} = 0.$$

But $p_n^{(n+1)}(z) = 0$ since the degree of $p_n(x) \leq n$. Also $\Psi^{(n+1)}(z) = (n+1)!$ (see Exercise 1 below). Hence, we have

$$f^{(n+1)}(z) - (n+1)! \frac{f(t) - p_n(t)}{\Psi(t)} = 0.$$

Finally, solve for $f(t) - p_n(t)$ to get (5.17).

$$E_n(t) = f(t) - p_n(t) = f^{(n+1)}(z)\frac{\Psi(t)}{(n+1)!}.$$

The next result addresses the special case when $f(x)$ is defined on $[a, b]$, which contains equally spaced nodes.

Assume $f(x)$ is defined on $[a, b]$, which contains equally spaced nodes $x_k = x_0 + hk$, $k = 0, 1, \ldots, n$. In addition, assume that $f^{(n+1)}$ is continuous on $[a, b]$ and satisfies $f^{(n+1)}(x) \le M$. Then on $[a, b]$

$$|f(x) - p_n(x)| \le \frac{1}{4(n+1)} M \left(\frac{b-a}{n}\right)^{n+1}. \tag{5.19}$$

The proof can be obtained from the previous result. It can be shown that (see Exercise 5 below)

$$|\Psi(x)| \le \frac{1}{4} n! \, h^{n+1}, \quad h = \frac{b-a}{n} \quad \forall x \in [a, b].$$

Therefore, from Eqn. (5.17) we get

$$|f(x) - p_n(x)| \le \frac{1}{(n+1)!} |f^{(n+1)}(\zeta)| \frac{1}{4} h^{(n+1)} n! \le \frac{1}{4(n+1)} M \left(\frac{b-a}{n}\right)^{n+1}.$$

EXAMPLE 5.6
An interpolating polynomial is to be used to approximate $f(x) = \cos x$ with 11 equally spaced nodes in $[1, 2]$. What bound can be placed on the error?

We have $n = 10$, $a = 1$, $b = 2$, $f^{(11)}(x) = \sin x$ and $|f^{(11)}(x)| \le 1 = M$. Thus,

$$|p(x) - \cos x| \le \frac{1}{4(11)} \left(\frac{2-1}{10}\right)^{11} \approx 2.27 \times 10^{-13}.$$

Formula (5.19) requires an upper bound on the second derivative in the case when a linear function is being interpolated, and this is not always available as a practical matter. However, if a table of values of f is available, we can use a difference approximation to the second derivative (see Chapter 9) to estimate the upper bound on the derivative and hence the error. Recall from Section 9.1

$$f''(x) \approx \frac{f(x-h) - 2f(x) + f(x+h)}{h^2}.$$

Assume the function values are given at the equally spaced grid points $x_k = a + kh$ for some grid spacing h. An approximation to the derivative upper bound is given by

$$\max_{x_k \leq x \leq x_{k+1}} |f''(x)| \approx \max \left\{ \left| \frac{f(x_{k-1}) - 2f(x_k) + f(x_{k+1})}{h^2} \right|, \right.$$
$$\left. \left| \frac{f(x_k) - 2f(x_{k+1}) + f(x_{k+2})}{h^2} \right| \right\}.$$

EXAMPLE 5.7

Given the table of values

x	1.5	1.6	1.7
$\Gamma(x)$	0.88622693	0.89351535	0.90863873
x	1.8	1.9	2.0
$\Gamma(x)$	0.93138377	0.96176583	1.00000000

of the gamma function $\Gamma(x)$ in $[1.5, 2]$, use linear interpolation to approximate $\Gamma(1.625) = 0.896\,574\,280\,056\,6$ and estimate the error made.

We have

$$\Gamma(1.625) \approx p_1(1.625) = \Gamma(1.7) + \frac{\Gamma(1.7) - \Gamma(1.6)}{0.1}(1.625 - 1.7)$$
$$\approx 0.90863873 + \frac{0.90863873 - 0.89351535}{0.1}(-0.075)$$
$$\approx 0.897296195.$$

We estimate the derivative as the larger of

$$|D_1| = \left| \frac{\Gamma(1.5) - 2\Gamma(1.6) + \Gamma(1.7)}{0.01} \right| = 0.783496$$

and

$$|D_2| = \left| \frac{\Gamma(1.6) - 2\Gamma(1.7) + \Gamma(1.8)}{0.01} \right| = 0.762166.$$

Thus, the error can be estimated as

$$|\Gamma(1.625) - p_1(1.625)| \approx \left(\frac{1}{8}\right)(0.1)^2(0.783\,496) = 0.000\,979\,37.$$

Another formula for the error $f(t) - p_n(t)$ that is useful in many situations can be derived as follows:

Let t be any point in the domain of f, distinct from the node points $x_0, x_1, \ldots, x_n$. Consider the polynomial p_{n+1} interpolating f at $x_0, x_1, \ldots, x_n, t$. By Newton's forward divided-difference formula, we have

$$p_{n+1}(x) = p_n(x) + f[x_0, x_1, \ldots, x_n, t] \prod_{i=0}^{n}(x - x_i).$$

Since $p_{n+1}(t) = f(t)$, let $x = t$ to obtain

$$f(t) - p_n(t) = f[x_0, x_1, \ldots, x_n, t] \prod_{i=0}^{n}(t - x_i). \tag{5.20}$$

By combining Eqns. (5.17) and (5.20), one gets the relation

$$f[x_0, x_1, \ldots, x_n] = \frac{1}{n!} f^{(n)}(\zeta). \tag{5.21}$$

EXERCISE SET 5.3

1. Show that the $(n+1)$th derivative of $\Psi(x)$ defined by Eqn. (5.18) is

$$\Psi^{(n+1)}(x) = (n+1)!$$

2. Show that the third-order divided difference $f[x_0, x_1, x_2]$ of f is invariant under all permutations of x_0, x_1, x_2.

3. Show that

$$f[x_0, x_1, x_2] = \frac{\begin{vmatrix} 1 & x_0 & f_0 \\ 1 & x_1 & f_1 \\ 1 & x_2 & f_2 \end{vmatrix}}{\begin{vmatrix} 1 & x_0 & x_0^2 \\ 1 & x_1 & x_1^2 \\ 1 & x_2 & x_2^2 \end{vmatrix}}.$$

4. In interpolating with $(n+1)$ equally spaced nodes on an interval $[a, b]$ show that for any $x \in [a, b]$

$$\prod_{i=0}^{n} |x - x_i| \leq \frac{1}{4} h^{(n+1)} n! \quad \text{with} \quad h = \frac{b - a}{n}.$$

5. Let p_n be a polynomial of degree n. Show that the divided difference of order k of p_n is zero if $k > n$.

6. Determine the maximum step size that can be used in the interpolation of $f(x) = e^x$ in $[0, 1]$, so that the error in the linear interpolation will be less than 5×10^{-4}. Find also the step size if quadratic interpolation is used.

7. Consider a table of natural logarithm values, for the interval $[1/2, 1]$. How many entries do we have to have for linear interpolation between entries to be accurate to within 10^{-3}?

8. In this exercise we investigate the error formula of polynomial interpolation. We will try to predict, based on this formula, whether polynomial interpolation provides a good fit for a given function f on $[0, 2]$. We are given the following functions defined on the interval $[0, 2]$:

   (a) $f(x) = e^{-2x}$,

   (b) $h(x) = \sin x$,

   (c) $g(x) = 1/(x+1)$,

   (d) $k(x) = \sqrt{|x-1|}$.

     (i) Find formulas that express the nth order derivatives of each function. Check the smoothness of each of these functions on $[0, 2]$ whether a derivative of some order does or does not exist at some point(s).

     (ii) Try to bound the nth order derivatives of each function on the interval $[0, 2]$ and divide the bound by $n!$ to get the error formula.

     (iii) Suppose we want to interpolate these functions at some equally spaced points on $[0, 2]$; say we start with 5 points, then 10 points, etc. .... What is your prediction concerning the convergence?

9. Write down a general error formula for polynomial interpolation of a function f at the three points 1, 2, 3. Based on this error formula, estimate the error at the point $x = 1.5$ for the function e^{2x}.

10. If we want to construct a table of logarithmic values so that cubic interpolation would be accurate to within 10^{-6}, how close together would the points in the table have to be?

5.4 LAGRANGE INTERPOLATING POLYNOMIAL

There are many forms of $p_n(x)$ that are different from the Newton form. Among them is the form known by the name of **Lagrange**, which we will derive in this section. As before, we seek a polynomial of degree n which passes through the $n+1$ points given by Table 5.3. To proceed, let us first define the functions which have the property

$$L_j(x_i) = \begin{cases} 1 \text{ if } i = j \\ 0 \text{ if } i \neq j \end{cases}. \tag{5.22}$$

These functions are known as **cardinal functions**.

We now write the nth Lagrange interpolating polynomial in the form

$$p_n(x) = \sum_{i=0}^{n} f(x_i) L_i(x). \tag{5.23}$$

Observe that $p_n(x)$ easily satisfies property (5.1). So, if we define

$$L_i(x) = \frac{(x - x_0)(x - x_1) \cdots (x - x_{i-1})(x - x_{i+1}) \cdots (x - x_n)}{(x_i - x_0)(x_i - x_1) \cdots (x_i - x_{i-1})(x_i - x_{i+1}) \cdots (x_i - x_n)}$$

$$= \prod_{\substack{j=0 \\ j \neq i}}^{n} \left[\frac{x - x_j}{x_i - x_j} \right], \quad i = 0, 1, \ldots, n \tag{5.24}$$

then the degree of $p_n(x) \leq n$ since all $L_i(x)$ have degree n.

Notice that $L_i(x)$ satisfies property (5.22) since at $x = x_i$ the numerator and denominator of Eqn. (5.24) will be equal. So $L_i(x_i) = 1$. Further, if $x = x_j$ and $i \neq j$, then the numerator of the (5.24) will be zero, and $L_i(x_j) = 0$.

The expression (5.23) is called the **Lagrange form** of $p_n(x)$.

EXAMPLE 5.8

Derive the Lagrange polynomial that interpolates the data in the following table.

x	-1	0	3
$f(x)$	8	-2	4

From Eqn. (5.24) it follows

$$L_0(x) = \frac{(x - 0)(x - 3)}{(-1 - 0)(-1 - 3)} = \frac{1}{4} x(x - 3)$$

$$L_1(x) = \frac{(x + 1)(x - 3)}{(0 + 1)(0 - 3)} = -\frac{1}{3}(x + 1)(x - 3)$$

$$L_2(x) = \frac{(x + 1)(x - 0)}{(3 + 1)(3 - 0)} = \frac{1}{12} x(x + 1).$$

Thus, from Eqn. (5.23) the Lagrange interpolating polynomial is

$$p_2(x) = 2x(x - 3) + \frac{2}{3}(x + 1)(x - 3) + \frac{1}{3} x(x + 1).$$

As one observes from this example, one of the disadvantages of the Lagrange polynomial method is that it requires many computations. Another disadvantage of the Lagrange polynomial method is that if one wants to add or subtract a point from the table used to construct the polynomial, one has to start with a complete new set of calculations.

EXAMPLE 5.9

Determine the coefficients of the Lagrange interpolating polynomial that interpolates the data in Example 5.8 and evaluates it at $x = -0.5$.

Using the MATLAB function lagrange, we get the output shown in Table 5.8.

```
» x=[-1 0 3]';
» y=[8 -2 4]';
» lagrange(x,y,-0.5)

c =

    2.0000   0.6667   0.3333

    p(a)=  2.250000
```

Table 5.8   Coefficients of the Lagrange interpolating polynomial for Example 5.9.

MATLAB's methods

In MATLAB the interpolating polynomial that is defined from a set of points can be obtained by using the built-in function polyfit, which can be called in the following ways:

p = polyfit(x,y,n)

x and y are vectors defined from a set of n+1 points, and n is the degree of the interpolating polynomial. The coefficients of the polynomial are returned in descending powers of x in vector p. For example, the cubic polynomial that passes through the four points of the table given in Example 5.3 is

```
>> x = [2   4   6   8]
x =
     2   4   6   8
>> y = [4   8   14   16]
y =
     4   8   14   16
>> p = polyfit(x,y,3)
p =
     -0.1250   1.7500   -5.0000   8.0000
```
which are the coefficients of the interpolating polynomial. One can find the y-value that corresponds to $x = 7$ with

```
>> z = polyval(p,7)
z =
     15.8750
```

EXAMPLE 5.10

Use MATLAB to estimate the relative viscosity of ethanol at $w = 27$ in Example 5.1.

The estimate may be found as follows:

```
>> w = [5   10   15   20   30   40];
>> V = [1.2260   1.4980   1.8220   2.1380   2.6620   2.8400];
>> format short e
>> p = polyfit(w,V,5)
p =
   -7.1238e-008 7.7790e-006 -3.5319e-004 7.5752e-003 -1.0626e-002 1.1293e+000
>> format long
>> V = polyval(p,27)
V =
      2.47460518400000
```

Thus, $V(27) \approx 2.47460518400000$.

M-function 5.4

The following MATLAB function **lagrange.m** determines the coefficients of the Lagrange interpolating polynomial *p(x)* and computes *p(a)*. INPUTS are the vectors *x* and *y* and the value of *a*.

```
function lagrange(x,y,a)
% Coefficients of the Lagrange interpolating polynomial.
n=length(x);
p=0;
for k=1:n
  b(k)=1;
  d(k)=1;
  for j=1:n
    if j~= k
      b(k)=b(k)*(x(k)-x(j));
      d(k)=d(k)*(a-x(j));
    end
  end
  c(k)=y(k)/b(k);
  p=p+c(k)*d(k);
end
c
fprintf('\n   p(a)= %10.6f',p)
fprintf('\n')
```

EXERCISE SET 5.4

1. Derive the Lagrange polynomial that interpolates the data in the following table.

x	0	2	4	6
y	1	-1	3	4

2. Derive the cardinal functions that can be used to interpolate the following table.

x	-1	0.5	1
y	3	1	4

3. Let

$$p(x) = 12\frac{(x-1)x(x-2)}{6} - 4\frac{(x-3)x(x-2)}{2} +$$
$$0\frac{(x-3)(x-1)(x-2)}{-6} - 2\frac{(x-3)(x-1)x}{-2},$$
$$q(x) = 12 + 8(x-3) + 4(x-3)(x-1) + 1(x-3)(x-1)x,$$
$$r(x) = -2 - 1(x-2) + 3(x-2)x + 1(x-2)x(x-1).$$

Using some simple arithmetic (which must be shown) and without any algebra, explain why p, q, and r are all the same polynomial.

4. Let $p(x)$ be the quadratic polynomial that interpolates $f(x) = x^3$ at $x = 1, 2$, and 3.

   (a) Write down the Lagrange and Newton formulas for p.
   (b) Bound the relative error $e(x) = |f(x) - p(x)|/|f(x)|$ on the interval $1 < x < 3$.
   (c) Bound the relative error $e(x) = |f(x) - p(x)|/|f(x)|$ on the interval $1.9 < x < 2.1$. Why is your bound so much smaller or larger than the bound in (b)?
   (d) State the general formula that gives the error on an interval $[a, b]$, in the degree-n polynomial p, which interpolates a smooth function f at $n + 1$ points t, in $[a, b]$. Separate the error into a product of three factors and explain why two of them are inevitable.

5. Let $f(x) = 2x^2 e^x + 1$. Construct a Lagrange polynomial of degree two or less using $x_0 = 0, x_1 = 0.5$, and $x_2 = 1$. Approximate $f(0.8)$.

6. Given that $\ln(2) = 0.6932$, $\ln(3) = 1.0986$ and $\ln(6) = 1.7918$ interpolate using Lagrange polynomial to approximate the natural logarithm at each integer from one to ten. Tabulate your results with the absolute and relative errors.

7. Consider a function $f(x)$ with the following known values:

x	0	1	3	4
$f(x)$	2	2	2	14

(a) Find the Lagrange polynomial through all the points.

(b) Find the Lagrange polynomial through $x = 0$, 1, and 3.

(c) Find the Lagrange polynomial through $x = 1$, 3, and 4.

(d) Use the results of parts (a), (b), and (c) to approximate $f(2)$ and write an expression for the error terms. (Your answer may include $f^{(n)}(\xi)$.)

8. Given the three data points $(1, 3)$, $(2, 0)$, $(3, -1)$, interpolate the data by a quadratic polynomial. Write down the three cardinal functions associated with the data and the interpolant polynomial.

COMPUTER PROBLEM SET 5.4

1. Write a computer program in a language of your choice that performs Lagrange interpolation. Input data to the program should include a set of values of x, the number of such values, the corresponding values y, and the value α of x at which the Lagrange interpolation polynomial, $p(x)$, is to be evaluated. The output should be the value of $p(\alpha)$. Test your program for the following table of values

x	-2	0	-1	1	2
$f(x)$	4	1	-1	1	9

Approximate $f(0.7)$ using the Lagrange interpolation polynomial.

2. Use the MATLAB function lagrange.m to find the polynomial that interpolates the function $f(x) = \sqrt{x}$ at 11 equally spaced points in the interval $[0, 1]$. Compare the values of polynomial to $\sqrt{x}$ at the midpoints.

3. Use the MATLAB function lagrange.m to find the polynomial that interpolates the function $f(x) = \frac{1+\cos(\pi x)}{1+x}$ at 21 equally spaced points in the interval $[0, 10]$.

APPLIED PROBLEMS FOR CHAPTER 5

1. S.H.P. Chen and S.C. Saxena experimental data for the emittance e of tungsten as a function of the temperature T (*Ind. Eng. Chem. Fund.*, **12**, 220 (1973)) are given by the following table.

$T,^0 K$	300	400	500	600	700	800	900	1000	1100
e	0.024	0.035	0.046	0.056	0.067	0.083	0.097	0.111	0.125
$T,^0 K$	1200	1300	1400	1500	1600	1700	1800	1900	2000
e	0.140	0.155	0.170	0.186	0.202	0.219	0.235	0.252	0.269

They found that the equation

$$e(T) = 0.02424 \left(\frac{T}{303.16} \right)^{1.27591}$$

correlated the data accurately to three digits. Find the coefficients for Newton's interpolating polynomial $p(T)$ that interpolates these data. Compare the values of $p(T)$ with the values of $e(T)$ at the points midway between the tabulated temperatures. Plot $p(T)$ and $e(T)$ as functions of T in the interval $[300, 2000]$.

2. The following table shows the values of the thermal conductivity, k (BTU/hr ft $^o F$), of carbon dioxide gas and the values of the viscosity, μ (lb/ft hr), of liquid ethylene glycot, at various temperature $T(^o F)$.

T	31	211	391	571	
k	0.0084	0.0132	0.0180	0.0227	
T	0	50	100	150	200
μ	241	82.2	30.3	12.5	5.54

In each case, determine the simplest interpolating polynomial that is likely to predict k and μ in the specified ranges of temperatures.

3. When the initial value problem, $dy/dt = f(t, y)$, subject to an initial condition is solved numerically, $y(t)$ is calculated at the mesh points of a given interval (see Chapter 12). When a certain problem is solved, the values given in the following table were obtained.

t	0.0	0.1	0.2	0.3	0.4	0.5
y	-1.0	-0.90	-0.79	-0.67	-0.54	-0.41

The numerical method does not give the values of y at points other than the mesh points. Use an interpolating polynomial to estimate y at $t = 0.26$, 0.35, and 0.43.

4. Consider the graph of the function

$$f(x) = \frac{1}{1 + x^2}.$$

(a) Determine the polynomial $p_{10}(x)$ of 10th degree that interpolates the points with abscissa $-5, -4, \ldots, 4, 5$, all spaced one unit apart.

(b) Plot both p and f in the interval $[-5, 5]$. Show that the wiggles in $p_{10}(x)$ occur near the ends of the interval over which the interval is used by finding the error $E = |f(x) - P_{10}(x)|$ at $x = 4.2, 4.4, 4.6,$ and 4.8. This is a characteristic behavior of interpolating polynomials of high degree.

5. A chemical experiment produces the following table

T	0	5	10	15	20	25
C	14.5	12.6	11.3	10.3	9.1	8.5

(a) Plot the points and find the interpolation polynomial.

(b) Plot the interpolation polynomial.

(c) Compare the two graphs.

6. The vapor pressure P of water (in bars) as a function of temperature $T(^oC)$ is

T	0	10	20	30	40	60	80	100
P	0.0061	0.0123	0.0234	0.0424	0.0738	0.1992	0.4736	1.0133

Find the interpolating polynomial of these data and estimate $P(5)$, $P(45)$, and $P(95)$. Compare your results with the known values of the pressure: $P(5) = 0.008721$, $P(45) = 0.095848$, $P(95) = 0.84528$.

7. The following table gives values of C for the property of titanium as a function of temperature T

T	605	685	725	765	825	855	875
C	0.622	0.655	0.668	0.679	0.730	0.907	1.336

Find the interpolating polynomial of these data and estimate $C(645)$, $C(795)$, and $C(845)$. Compare your results with the known values of C: $C(645) = 0.639$, $C(795) = 0.694$, $C(845) = 0.812$.

Chapter 6

Interpolation with Spline Functions

Many scientific and engineering phenomena being measured undergo a transition from one physical domain to another. Data obtained from these measurements are better represented by a set of piecewise continuous curves rather than by a single curve. One of the difficulties with polynomial interpolation is that in some cases the oscillatory nature of high-degree polynomials can induce large fluctuations over the entire range when approximating a set of data points. One way of solving this problem is to divide the interval into a set of subintervals and construct a lower-degree approximating polynomial on each subinterval. This type of approximation is called **piecewise polynomial interpolation**.

Piecewise polynomial functions, especially spline functions, have become increasingly popular. Most of the interest has centered on cubic splines because of the ease of their applications to a variety of fields, such as the solution of boundary value problems for differential equations and the method of finite elements for the numerical solution of partial differential equations.

In this chapter we shall discuss several types of piecewise polynomials for interpolating a given set of data. The simplest of these is piecewise linear interpolation and the most popular one is cubic spline interpolation.

EXAMPLE 6.1 : Glucose Level

In performing an arginine tolerance test, a doctor measures glucose over a 80-minute time period at 10-minute interval to obtain the following data:

time (t)	0	10	20	30	40	50	60	70	80
glucose	100	118	125	136	114	105	98	104	92

Interpolating these data with a cubic spline will give the doctor better approximate values of the glucose level at different values of time other than the knots. Polynomials are not the most effective form of representation of a large set of data, especially in data sets that include local abrupt changes in the values of the quantity to be interpolated.

6.1 PIECEWISE LINEAR INTERPOLATION

We start out with the general definition of spline functions. Let f be a real-valued function defined on some interval $[a, b]$ and let the set of data points in Table 6.1 be given. For simplicity, assume that

$$a = x_1 < x_2 < \ldots < x_n = b.$$

We have the definition:

x	$a = x_1$	x_2	$\ldots$	$x_n = b$
y	$f(x_1)$	$f(x_2)$	$\ldots$	$f(x_n)$

Table 6.1   Table of values of a function f.

DEFINITION 6.1   *A function S is called a **spline of degree** k if it satisfies the following conditions:*

1. *S is defined in the interval $[a, b]$.*

2. *$S^{(r)}$ is continuous on $[a, b]$ for $0 \leq r \leq k - 1$.*

3. *S is a polynomial of degree $\leq k$ on each subinterval $[x_i, x_{i+1}]$, $i = 1, 2, \ldots, n - 1$.*

Observe that in contrast to polynomial interpolation, the degree of the spline does not increase with the number of points. Here the degree is fixed and one uses more polynomials instead. A simple and familiar example of a piecewise polynomial approximation is piecewise linear interpolation, which consists of connecting a set of data points in Table 6.1 by a series of straight lines as shown in Figure 6.1.

This procedure can be described as follows: Let $f(x)$ be a real-valued function defined on some interval $[a, b]$. We wish to construct a piecewise linear polynomial function $S(x)$, which interpolates $f(x)$ at the data points given by Table 6.1, where

$$a = x_1 < x_2 < \ldots < x_n = b.$$

Using the formula of the equation of the line, it is easy to see that the function $S(x)$ is defined by

$$S_i(x) = f(x_i) + \frac{f(x_{i+1}) - f(x_i)}{x_{i+1} - x_i}(x - x_i), \quad i = 1, \ldots, n - 1$$
$$= f(x_i) + f[x_{i+1}, x_i](x - x_i) \tag{6.1}$$

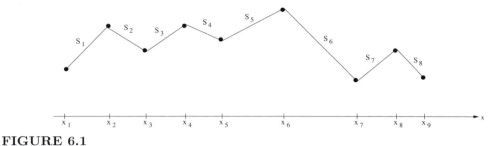

FIGURE 6.1
Piecewise linear interpolation.

on each subinterval $[x_i, x_{i+1}]$. Outside the interval $[a, b]$, $S(x)$ is usually defined by

$$S(x) = \begin{cases} S_1(x) & \text{if } x < a \\ S_{n-1}(x) & \text{if } x > b \end{cases}.$$

The points $x_2, x_3, \ldots, x_{n-1}$, where $S(x)$ changes from one polynomial to another, are called the **breakpoints** or **knots**. Because $S(x)$ is continuous on $[a, b]$, it is called a **spline of degree 1**.

EXAMPLE 6.2

Find a first degree spline interpolating the following table:

x	1	1.5	2	2.5	3
$f(x)$	1	3	7	10	15

Use the resulting spline to approximate $f(2.2)$.

From Eqn. (6.1), we have

$$S_1(x) = \frac{f(1.5) - f(1)}{1.5 - 1}(x - 1) + f(1) = \frac{3 - 1}{0.5}(x - 1) + 1 = 4x - 3$$

$$S_2(x) = \frac{f(2) - f(1.5)}{2 - 1.5}(x - 1.5) + f(1.5) = \frac{7 - 3}{0.5}(x - 1.5) + 3 = 8x - 9$$

$$S_3(x) = \frac{f(2.5) - f(2)}{2.5 - 2}(x - 2) + f(2) = \frac{10 - 7}{0.5}(x - 2) + 7 = 6x - 5$$

$$S_4(x) = \frac{f(3) - f(2.5)}{3 - 2.5}(x - 2.5) + f(2.5) = \frac{15 - 10}{0.5}(x - 2.5) + 10$$
$$= 10x - 15.$$

Hence,

$$S(x) = \begin{cases} 4x - 3 & \text{if } x \in [1, 1.5] \\ 8x - 9 & \text{if } x \in [1.5, 2] \\ 6x - 5 & \text{if } x \in [2, 2.5] \\ 10x - 15 & \text{if } x \in [2.5, 3] \end{cases}.$$

```
x=[1 1.5 2 2.5 3]';
» y=[1 3 7 10 15]';
» spl1(x,y,2.2)
```

		linear spline	
x	y		c= 2.2
1	1		
1.5	3		
2	7		
			S(c)= 8.200000
2.5	10		
3	15		

Table 6.2   First degree spline for Example 6.2.

The value $x = 2.2$ lies in $[2, 2.5]$ and so $f(2.2) \approx 6(2.2) - 5 = 8.2$. Table 6.2 shows the results obtained by using the MATLAB function spl1.

A question that one may ask is about the goodness of fit when we interpolate a function by a first-degree spline. The answer is found in the following theorem.

THEOREM 6.1   (First-Degree Spline accuracy)
Suppose f is twice differentiable and continuous on the interval $[a, b]$. If $p(x)$ is a first-degree spline interpolating f at the knots $a = x_1 < x_2 < \ldots < x_n = b$, then

$$|f(x) - p(x)| \leq \frac{1}{8} M h^2, \quad a \leq x \leq b$$

where $h = \max_i (x_{i+1} - x_i)$ and M denotes the maximum of $|f''(x)|$ on (a, b).

Proof: In Section 5.3 it has been shown that

$$f(x) - p(x) = \frac{1}{n!} f^{(n)}(\xi) \prod_{i=1}^{n} (x - x_i)$$

where n is the number nodes. On the interval $[a, b]$, we have $n = 2$. So,

$$f(x) - p(x) = \frac{1}{2} f''(\xi)(x - a)(x - b)$$

for some ξ on (a, b). Since $|f''(x)| \leq M$ on (a, b) and $\max_x |x - a| |x - b| = \frac{(b-a)^2}{4}$, it follows that

$$|f(x) - p(x)| \leq \frac{1}{2} M \frac{(b - a)^2}{4} = \frac{1}{8} M h^2.$$

From this theorem one can learn that if the only thing we know is that the second derivative of our function is bounded, then we are guaranteed that the maximum interpolation error we make decreases to zero as we increase the number of knots.

With polynomial interpolation, however, using for example 10 data points, we had an error estimate in terms of the 10-*th* derivative.

EXAMPLE 6.3

Assuming that we know M, find the smallest value of n to force the error bound for a first-degree spline to be less than a given tolerance ϵ for n equally spaced knots.

We have $|f''(x)| \leq M$, so

$$\frac{h^2 M}{8} \leq \epsilon.$$

Solve for h to get

$$h \leq \sqrt{\frac{8\epsilon}{M}}.$$

Since $h = (b-a)/(n-1)$, it follows that

$$(b-a)\sqrt{\frac{M}{8\epsilon}} \leq n - 1.$$

Now, solve for n to get

$$n = 1 + \lceil (b-a)\sqrt{\frac{M}{8\epsilon}} \rceil$$

where $\lceil x \rceil$ is the so-called **ceiling** function. That is, $\lceil x \rceil =$ the smallest integer $\geq x$.

EXERCISE SET 6.1

1. Determine whether the following functions are first-degree splines:

   (a) $f(x) = \begin{cases} 2x - 1 & \text{if } x \in [-1, 1] \\ -x + 2 & \text{if } x \in [1, 2] \\ 5x & \text{if } x \in [2, 3] \end{cases}$.

   (b) $f(x) = \begin{cases} 3x + 5 & \text{if } x \in [0, 1] \\ 2x + 6 & \text{if } x \in [1, 4] \\ x + 10 & \text{if } x \in [4, 5] \end{cases}$.

2. Given the table

x	1	2	3	4	5
$S(x)$	3	4	3	9	1

find the first-degree spline that interpolates the table. Compute $S(2.3)$.

M-function 6.1

The following MATLAB function **spl1.m** finds a first-degree spline that interpolates a table of values. INPUTS are a table of function values x and y; the value of c at which $S(x)$ is to be approximated.

```
function spl1(x,y,c)
% First-degree spline.
n=length(x);
for i=n-1:-1:2
  dis=c-x(i);
  if(dis>=0)
    break
  end
end
if(dis<0)
  i=1;
  dis=c-x(1);
end
m=(y(i+1)-y(i))/(x(i+1)-x(i));
spl1=y(i)+m*dis;
disp('        linear spline')
disp('_____')
disp(['    x            y            c = ',num2str(c),'   '])
disp('_____')
for j=1:n
  fprintf('%12.6f %12.6f ',x(j),y(j))
  if (j==i)
    fprintf('\n                    S(c)= %10.6f',spl1)
  end;
  fprintf('\n')
end
```

3. Construct a first-degree spline to approximate the function $f(x) = xe^x$ at $x = 0.4$, using the data

x	0.1	0.3	0.5	0.7
$f(x)$	.110517	.404957	.824360	1.409627

4. Determine a bound on the error in the approximation of $f(x) = \sin x$ by a first-degree spline on $[0, \pi]$ using 5 equally spaced knots.

5. Construct a first-degree spline $S(x)$ to approximate the function $f(x) = e^{\frac{2x}{3}}$ by using the values given by $f(x)$ at $x = 0, 0.02,$ and 0.04. Find an approximation for $\int_0^{0.04} e^{\frac{2x}{3}} dx$ by evaluating $\int_0^{0.04} S(x)dx$. Compare the result to the exact value 0.04053811.

6. Fit the data in the following table with a first-degree spline and evaluate the function at $x = 5$.

x	3.0	4.5	7.0	9.0
$f(x)$	2.5	1.0	2.5	0.5

7. Determine the number of knots needed to interpolate the function $f(x) = \cos x^2$ on $[0, 3]$ with a first-degree spline with an error less than 10^{-4}.

COMPUTER PROBLEM SET 6.1

1. Write a computer program in a language of your choice to find the first-degree spline $S(x)$ that interpolates a set of data points (x_i, y_i) and compute $S(x)$ at a given value α.

   Input data to the program should be

   (a) The set of data points (x_i, y_i).

   (b) The given value α.

   Output should consist of:

   (a) The value of $S(x)$ at $x = \alpha$.

   Test your program to solve Exercise 3.

2. Use the MATLAB function spl1 to find the first-degree spline that interpolates the function $f(x) = \sin x$ over the interval $[0, \pi]$ using 21 equally spaced knots.

3. Given the table of data points

x	0	$\pi/10$	$\pi/5$	$3\pi/10$	$2\pi/5$	$\pi/2$
$f(x)$	0.0	0.0985	0.3846	0.7760	1.0000	0.6243

   find the first-degree spline that interpolates the data points. Approximate the values of f at $x = \pi/20$, $\pi/4$, and $9\pi/20$. Compare with the values of $f(x) = \sin x^2$, which was used to create the data in the table.

4. Use the function $f(x) = \cos x$ to generate 11 evenly spaced data points on $[0, \pi]$. Find the first-degree spline $S(x)$ that interpolates these data points, the actual error $|S(x) - f(x)|$, and the error bound $h^2/8$.

6.2   QUADRATIC SPLINE

In many cases, linear piecewise polynomials are unsatisfactory when being used to interpolate the values of a function, which deviate considerably from a linear function. In such cases, piecewise polynomials of higher degree are more suitable to use to approximate the function. In this section, we shall discuss the simplest type of differentiable, piecewise polynomial functions, known as **quadratic splines**. As before, consider the subdivision

$$a = x_1 < x_2 < \ldots < x_n = b$$

where $x_1, \ldots, x_n$ are given in Table 6.1. For piecewise linear interpolation, we choose two points $(x_i, f(x_i))$ and $(x_{i+1}, f(x_{i+1}))$ in the subinterval $[x_i, x_{i+1}]$ and draw a line through those two points to interpolate the data. This approach is easily extended to construct the quadratic splines. Instead of choosing two points, we choose three points in the subinterval $[x_i, x_{i+1}]$ and pass a second-degree polynomial through these points as shown in Figure 6.2. We shall show that there is only one such polynomial.

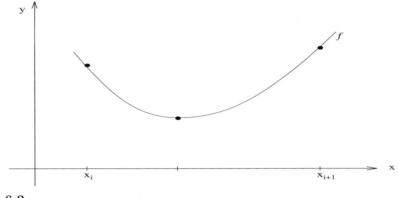

FIGURE 6.2
Quadratic spline.

To construct a quadratic spline $Q(x)$, we first define a quadratic function in each subinterval $[x_i, x_{i+1}]$ by

$$q_i(x) = a_i + b_i(x - x_i) + c_i(x - x_i)^2 \tag{6.2}$$

where a_i, b_i, and c_i are constants to be determined.

Now by Definition 6.1, $Q(x)$ must satisfy the conditions

$$Q(x) = q_i(x) \quad \text{on } [x_i, x_{i+1}] \quad \text{for } i = 1, 2, \ldots, n - 1. \tag{6.3}$$

$$q_i(x_i) = f(x_i). \tag{6.4}$$

$$q_i(x_{i+1}) = f(x_{i+1}). \tag{6.5}$$

$Q'(x)$ is continuous on $[a, b]$ if

$$q_i'(x_i) = d_i \quad \text{and} \quad q_i'(x_{i+1}) = d_{i+1}. \tag{6.6}$$

Here the values of d_i will be defined later. Using conditions (6.4) and (6.6), it is easy to see that $q_i(x)$ is uniquely defined on $[x_i, x_{i+1}]$ (see Exercise 1 below) by

$$q_i(x) = f(x_i) + d_i(x - x_i) + \frac{d_{i+1} - d_i}{2(x_{i+1} - x_i)}(x - x_i)^2. \tag{6.7}$$

We now use condition (6.5) to obtain d_i from the recursive formula

$$d_{i+1} = -d_i + 2\left[\frac{f(x_{i+1}) - f(x_i)}{x_{i+1} - x_i}\right], \quad i = 1, 2, \ldots, n-1 \tag{6.8}$$

with d_1 arbitrary.

Thus, given the data in Table 6.1 and an arbitrary value for d_1, the quadratic spline $Q(x)$ is uniquely determined by formulas (6.3), (6.7), and (6.8).

EXAMPLE 6.4

Find a quadratic spline interpolating the following table.

x	-2	-1.5	0	1.5
$f(x)$	5	3	1	2

Let $d_1 = 0$, then from (6.8) we have

$$d_2 = 2\left[\frac{3-5}{-1.5+2}\right] = -8,$$

$$d_3 = 8 + 2\left[\frac{1-3}{0+1.5}\right] = 16/3,$$

$$d_4 = -\frac{16}{3} + 2\left[\frac{2-1}{1.5-0}\right] = -4.$$

We now use (6.7) to get a quadratic spline $Q(x)$ defined by

$$q_1(x) = 5 - 8(x+2)^2 \quad \text{in} \quad [-2, -1.5],$$

$$q_2(x) = 3 - 8(x+1.5) + \frac{40}{9}(x+1.5)^2 \quad \text{in} \quad [-1.5, 0],$$

$$q_3(x) = 1 + \frac{16}{3}x - \frac{28}{9}x^2 \quad \text{in} \quad [0, 1.5].$$

EXERCISE SET 6.2

1. Show that the following function is a quadratic spline.

$$f(x) = \begin{cases} 2x^2 & \text{if } -1 \le x \le 1 \\ x^2 + 2x - 1 & \text{if } 1 \le x \le 3 \\ 8x - 10 & \text{if } 3 \le x \le 4 \end{cases}.$$

m-function 6.2

The following MATLAB function **spl2.m** finds a quadratic spline that interpolates a table of values. INPUTS are a table of function values x and y; the number of intermediate points m at which S(x) is to be approximated.

```
function spl2(x,y,m)
% Quadratic Spline
n=length(x);
% arbitary value for d1
d(1)=0;
for i=1:n-1
    d(i+1) = -d(i)+2* ((y(i+1)-y(i)) / (x(i+1)-x(i)) );
    t(i)=(x(i+1)-x(i))/(m+1);
end
disp('              Quadratic Spline        ')
disp('_____')
disp('     x          Q(x)                  ')
disp('_____')
for j=1:n-1
  r=(x(j):t(j):x(j+1)-t(j));
   for k=1:m+1
     r(k);
     spl2=y(j)+ d(j)*(r(k)-x(j))+((d(j+1)-d(j))/(2*(x(j+1)-x(j))))*(r(k)-x(j))^2;
     fprintf('%12.5f %17.5f \n',r(k),spl2);
   end
end
spl2=y(n);
fprintf('%12.5f %17.5f \n',x(n),spl2);
```

2. Determine the coefficient a and b so that the following function is a quadratic spline.

$$f(x) = \begin{cases} -x + 1 \text{ if } 0 \le x \le 1 \\ ax^2 + b \text{ if } 1 \le x \le 2 \end{cases}.$$

3. Determine the coefficient a, b, c, and d so that the following function is a quadratic spline.

$$f(x) = \begin{cases} 2x & \text{if} & -1 \le x \le 1 \\ ax^2 + bx + c & \text{if} & 1 \le x \le 2 \\ -dx & \text{if} & 2 \le x \le 4 \end{cases}.$$

4. Determine whether the following function is a quadratic spline:

$$f(x) = \begin{cases} 2x^2 - 1 & \text{if} & -2 \le x \le -1 \\ x^2 + x + 1 & \text{if} & -1 \le x \le 0 \\ 3x + 1 & \text{if} & 0 \le x \le 1 \end{cases}.$$

5. The following data are taken from a polynomial.

x	-1	0	1	2
$p(x)$	-4	-1	2	23

Find the quadratic spline $Q(x)$ that interpolates this data. Compute $Q(1.5)$.

6. Show formula (6.7) using (6.2), (6.4), and (6.6).

COMPUTER PROBLEM SET 6.2

1. Write a computer program in a language of your choice to find a quadratic spline $Q(x)$ that interpolates a set of data points (x_i, y_i) and evaluate the spline at points between the data points.
   Input data to the program should be

   (a) The set of data points (x_i, y_i).

   Output should consist of:

   (a) The value of $Q(x)$ between the data points.

   Test your program to find the quadratic spline $S(x)$ that interpolates the table

x	0.0	0.1	0.2	0.3	0.4	0.5	0.6	0.7	0.8	0.9	1.0
$f(x)$	1.0	1.25	1.58	1.99	2.51	3.16	3.93	4.44	4.87	5.12	5.5

   Evaluate $S(x)$ at the midpoints of the knots.

2. Use the MATLAB function spl2.m to find the natural cubic spline that interpolates the table

x	0	0.1	0.2	0.3	0.4	0.5	0.6	0.7	0.8	0.9	1.0
$f(x)$	0	0.01	0.04	0.10	0.19	0.32	0.52	0.8	1.21	1.82	2.72

   Plot the resulting spline. The function used to create this table is $f(x) = x^2 e^{x^2}$. Compute $|f(x) - Q(x)|$ at the midpoints of the table.

6.3   NATURAL CUBIC SPLINES

Cubic spline functions are the most popular spline functions because they are a very satisfactory device for smooth interpolation and numerical differentiation. The difficulty with quadratic splines is that their curvature is discontinuous at each breakpoint. So, if we have a smooth function to interpolate, cubic splines are more desirable to use and adequately provide accurate interpolation.

Again, we start with the Table 6.1. We now construct the cubic spline $S(x)$ subject to the conditions given by Definition 6.1 in Section 6.1.

In each subinterval $[x_i, x_{i+1}]$, we write $S(x)$ in the form

$$S_i(x) = a_i + b_i(x - x_i) + c_i(x - x_i)^2 + d_i(x - x_i)^3 \tag{6.9}$$
$$= S(x), \quad i = 1, 2, \ldots, n-1$$

where a_i, b_i, c_i, and d_i are constants to be determined. For notational convenience, let

$$h_i = x_{i+1} - x_i.$$

In order to fit the data, $S(x)$ must satisfy

$$f(x_i) = S(x_i)$$

so

$$a_i = f(x_i) \quad i = 1, 2, \ldots, n$$

and if the condition on the continuity of $S(x)$ is applied,

$$S_{i+1}(x_{i+1}) = S_i(x_{i+1})$$
$$f(x_{i+1}) = a_i + b_i(x_{i+1} - x_i) + c_i(x_{i+1} - x_i)^2 + d_i(x_{i+1} - x_i)^3$$
$$a_{i+1} = a_i + b_i h_i + c_i h_i^2 + d_i h_i^3, \quad i = 1, 2, \ldots, n-1. \tag{6.10}$$

By differentiating Eqn. (6.9), we obtain

$$S_i'(x) = b_i + 2c_i(x - x_i) + 3d_i(x - x_i)^2$$
$$S_i''(x) = 2c_i + 6d_i(x - x_i).$$

The continuity condition on the first derivative implies

$$S_i'(x_i) = S_{i-1}'(x_i)$$
$$b_i = b_{i-1} + 2c_{i-1}(x_i - x_{i-1}) + 3d_{i-1}(x_i - x_{i-1})^2$$
$$= b_{i-1} + 2c_{i-1}h_{i-1} + 3d_{i-1}h_{i-1}^2, \quad i = 2, 3, \ldots, n. \tag{6.11}$$

Similarly, imposing the continuity condition on the second derivative gives

$$S_i''(x_i) = S_{i-1}''(x_i)$$
$$2c_i = 2c_{i-1} + 6d_{i-1}(x_i - x_{i-1})$$
$$c_i = c_{i-1} + 3d_{i-1}h_{i-1}, \quad i = 2, 3, \ldots, n. \tag{6.12}$$

By increasing the index i by one, Eqn. (6.12) can be also written in the form

$$c_{i+1} = c_i + 3d_i h_i \tag{6.13}$$

for $i = 1, 2, \ldots, n-1$.

Now, by solving Eqn. (6.13) for d_i and substituting its value into Eqn. (6.10), we get, after rearranging the equation,

$$b_i = \frac{1}{h_i}(a_{i+1} - a_i) - \frac{h_i}{3}(c_{i+1} + 2c_i), \quad i = 1, 2, \ldots, n-1 \tag{6.14}$$

or

$$b_{i-1} = \frac{1}{h_{i-1}}(a_i - a_{i-1}) - \frac{h_{i-1}}{3}(c_i + 2c_{i-1}), \quad i = 2, 3, \ldots, n. \tag{6.15}$$

Similarly, solve Eqn. (6.12) for d_{i-1} and substitute its value into Eqn. (6.11) to get

$$b_i = b_{i-1} + h_{i-1}(c_i + c_{i-1}), \quad i = 2, 3, \ldots, n. \tag{6.16}$$

Finally, by combining Eqns. (6.14), (6.15), and (6.16) we get

$$h_{i-1}c_{i-1} + u_i c_i + h_i c_{i+1} = v_i, \quad i = 2, 3, \ldots, n-1 \tag{6.17}$$

where

$$u_i = 2(h_{i-1} + h_i), \quad v_i = 3w_i - 3w_{i-1} \quad \text{and} \quad w_i = \frac{1}{h_i}(a_{i+1} - a_i).$$

Eqn. (6.17) is a linear system in n unknowns and $n-2$ equations. To solve any system of equations numerically, it is necessary that the number of equations be exactly equal to the number of unknowns. This indicates that two more conditions are required.

There are several possible choices for these conditions on the second derivatives evaluated at the endpoints, which provide additional conditions to determine all the unknowns. The simplest choice is given by the natural boundary

$$S''(x_1) = S''(x_n) = 0 \tag{6.18}$$

which leads to a spline function known as a **natural cubic spline** whose graph is obtained by forcing a long flexible rod to go through each of the data points.

So, if condition (6.18) is applied to $S''(x)$, we obtain two new equations,

$$c_1 = 0 \quad \text{and} \quad c_n = 0. \tag{6.19}$$

The two Eqns. (6.19) together with the Eqn. (6.17) produce a linear, tridiagonal system of the form

$$A\mathbf{x} = \mathbf{b}$$

where

$$A = \begin{bmatrix} 1 & 0 & & & & & \\ h_1 & 2(h_1 + h_2) & h_2 & & & & \\ & \ddots & \ddots & \ddots & & & \\ & & h_{n-2} & 2(h_{n-2} + h_{n-1}) & h_{n-1} & \\ 0 & & & 0 & 1 \end{bmatrix}.$$

$$\mathbf{b} = \begin{bmatrix} 0 \\ 3(a_3 - a_2)/h_2 - 3(a_2 - a_1)/h_1 \\ \vdots \\ 3(a_n - a_{n-1})/h_{n-1} - 3(a_{n-1} - a_{n-2})/h_{n-2} \\ 0 \end{bmatrix}, \quad \mathbf{x} = \begin{bmatrix} c_1 \\ c_2 \\ \vdots \\ c_{n-1} \\ c_n \end{bmatrix}.$$

This system is strictly diagonally dominant, so an elimination method can be used to find the solution without the need for pivoting. Having obtained the values of $\{c_i\}_{i=1}^n$, the remainder of the spline coefficients for

$$S_i(x) = a_i + b_i(x - x_i) + c_i(x - x_i)^2 + d_i(x - x_i)^3$$

is obtained using the formulas

$$\begin{aligned} a_i &= f(x_i) \\ b_i &= \frac{1}{h_i}(a_{i+1} - a_i) - \frac{h_i}{3}(c_{i+1} + 2c_i) \\ d_i &= \frac{c_{i+1} - c_i}{3h_i} \end{aligned} \tag{6.20}$$

for $i = 1, 2, \ldots n - 1$.

EXAMPLE 6.5

Use the values given by $f(x) = x^3 + 2$ at $x = 0$, 0.2, 0.4, 0.6, 0.8, and 1.0 to find an approximation of $f(x)$ at $x = 0.1$, 0.3, 0.5, 0.7, and 0.9 using natural cubic spline interpolation.

We have the table

x	0.0	0.2	0.4	0.6	0.8	1.0
$f(x)$	2.0	2.008	2.064	2.216	2.512	3.0

From Eqn. (6.17), we have

$$u_1 = 0.8, \ u_2 = 0.8, \ u_3 = 0.8, \ u_4 = 0.8$$
$$v_1 = 0.72, \ v_2 = 1.44, \ v_3 = 2.16, \ v_4 = 2.88.$$

Using these values, we obtain the linear system of equations

$$\begin{bmatrix} 1 & 0 & 0 & 0 & 0 & 0 \\ 0.2 & 0.8 & 0.2 & 0 & 0 & 0 \\ 0 & 0.2 & 0.8 & 0.2 & 0 & 0 \\ 0 & 0 & 0.2 & 0.8 & 0.2 & 0 \\ 0 & 0 & 0 & 0.2 & 0.8 & 0.2 \\ 0 & 0 & 0 & 0 & 0 & 1 \end{bmatrix} \begin{bmatrix} c_1 \\ c_2 \\ c_3 \\ c_4 \\ c_5 \\ c_6 \end{bmatrix} = \begin{bmatrix} 0 \\ 0.72 \\ 1.44 \\ 2.16 \\ 2.88 \\ 0 \end{bmatrix}.$$

The solution of this system is

$$\mathbf{c} = [0.0, 0.586, 1.257, 1.585, 3.204, 0.0]^T.$$

Using Eqn. (6.20), we obtain the coefficients

$$\mathbf{a} = \begin{pmatrix} 2 \\ 2.008 \\ 2.064 \\ 2.216 \\ 2.512 \end{pmatrix}, \quad \mathbf{b} = \begin{pmatrix} 0.001 \\ 0.118 \\ 0.487 \\ 1.055 \\ 2.013 \end{pmatrix}, \quad \mathbf{d} = \begin{pmatrix} 0.976 \\ 1.120 \\ 0.545 \\ 2.699 \\ -5.340 \end{pmatrix}.$$

Hence,

$$S(x) = \begin{cases} S_1(x), \ x \in [0.0, 0.2] \\ S_2(x), \ x \in [0.2, 0.4] \\ S_3(x), \ x \in [0.4, 0.6] \\ S_4(x), \ x \in [0.6, 0.8] \\ S_5(x), \ x \in [0.8, 1.0] \end{cases}$$

with

$$S_i(x) = a_i + b_i(x - x_i) + c_i(x - x_i)^2 + d_i(x - x_i)^3, \quad i = 1, \ldots, 5.$$

That is,

$$S(x) = \begin{cases} 2 + 0.001x + 0.976x^3, & x \in [0.0, 0.2] \\ 2.008 + 0.118(x - 0.2) + 0.586(x - 0.2)^2 + 1.120(x - 0.2)^3, & x \in [0.2, 0.4] \\ 2.064 + 0.487(x - 0.4) + 1.257(x - 0.4)^2 + 0.545(x - 0.4)^3, & x \in [0.4, 0.6] \\ 2.216 + 1.055(x - 0.6) + 1.585(x - 0.6)^2 + 2.699(x - 0.6)^3, & x \in [0.6, 0.8] \\ 2.512 + 2.013(x - 0.8) + 3.204(x - 0.8)^2 - 5.340(x - 0.8)^3, & x \in [0.8, 1.0] \end{cases}.$$

For example, the value $x = 0.5$ lies in the interval $[0.4, 0.6]$, so

$$S_3(0.5) = 2.064 + 0.487(0.5 - 0.4) + 1.257(0.5 - 0.4)^2 + 0.545(0.5 - 0.4)^3$$
$$\approx 2.126.$$

A summary of the calculations is given in Table 6.3.

EXAMPLE 6.6

Given the table

```
» x=[0:0.2:1]';
» y=x.^3+2;
» spl3(x,y,2)
```
                    cubic spline

x	S(x)	Error
0	2	0
0.1	2.00107	0.000072
0.2	2.008	0
0.3	2.02678	0.000215
0.4	2.064	0
0.5	2.12579	0.000789
0.6	2.216	0
0.7	2.34006	0.002943
0.8	2.512	0
0.9	2.73998	0.010981
1	3	0

Table 6.3   Natural cubic spline for Example 6.5.

x	0	4	6	7	13	15	19	24	26
$f(x)$	0	3	8	12	7	5	2	3	5

use a cubic spline to approximate this data, and use the spline to approximate $f(x)$ at $x = 17$.

For this example, we choose $n = 3$ in order to approximate $f(17)$. A summary of the calculated values is shown in Table 6.4, and Figure 6.3 shows the graph of the corresponding cubic spline.

EXAMPLE 6.7
Determine whether the following function is a cubic spline.

$$f(x) = \begin{cases} S_1(x) = -x^3 - 3x & \text{if } 0 \le x \le 1 \\ S_2(x) = x^3 - 6x^2 + 3x - 2 & \text{if } 1 \le x \le 2 \end{cases}.$$

Since both $S_1(x)$ and $S_2(x)$ are polynomials of degree three, the only condition we need to check is the continuity of $S(x)$, $S'(x)$, and $S''(x)$ at $x = 1$. We have:

```
» x=[0 4 6 7 13 15 19 24 26]';
» y=[0 3 8 12 7 5 2 3 5]';
» spl3(x,y,3)
```

cubic spline

x	S(x)	x	S(x)
0	0	13.5	6.31667
1	0.38404	14	5.79912
2	0.91446	14.5	5.38201
3	1.73765	15	5
4	3	16	4.18356
4.5	3.85054	17	3.34531
5	4.91942	18	2.58441
5.5	6.27859	19	2
6	8	20.25	1.63444
6.25	9.00626	21.5	1.6874
6.5	10.05343	22.75	2.14665
6.75	11.07389	24	3
7	12	24.5	3.44654
8.5	14.55232	25	3.9389
10	13.25118	25.5	4.46181
11.5	10.07445	26	5
13	7		

Table 6.4   Natural cubic spline for Example 6.6.

1. $S(x)$ must satisfy the continuity condition at $x = 1$.
   $S_1(1) = -4 = S_2(1)$.

2. $S'(x)$ must satisfy the continuity condition at $x = 1$.
   $S_1'(x) = -3x^2 - 3$ and $S_1'(1) = -6$
   $S_2'(x) = 3x^2 - 12x + 3$ and $S_2'(1) = -6$.

3. $S''(x)$ must satisfy the continuity condition at $x = 1$.
   $S_1''(x) = -6x$ and $S_1''(1) = -6$
   $S_2''(x) = 6x - 12$ and $S_2''(1) = -6$.

Thus, $S(x)$ is a cubic spline on $[0, 2]$.

Smoothness Property

A practical feature of cubic splines is that they minimize the oscillations in the fit between the interpolating points. This is unlike all functions, $f(x)$, which are twice differentiable on $[a, b]$ and interpolate a given set of data points. The next theorem explains this phenomenon.

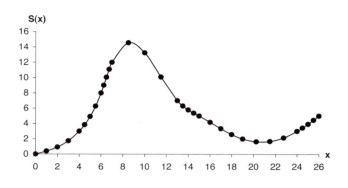

FIGURE 6.3
$S(x)$ **from Example 6.6.**

THEOREM 6.2

If f is twice-continuously differentiable on $[a,b]$ and S is the natural cubic spline that interpolates f at the knots $a = x_1 < x_2 < \cdots < x_n = b$, then

$$\int_a^b [S''(x)]^2 dx \leq \int_a^b [f''(x)]^2 dx.$$

Proof: The proof is obtained by a series of integration by parts. Let

$$e(x) = f(x) - S(x).$$

We have

$$\int_a^b [f''(x)]^2 dx = \int_a^b [S''(x) + e''(x)]^2 dx$$
$$= \int_a^b [S''(x)]^2 dx + \int_a^b [e''(x)]^2 dx + 2\int_a^b e''(x) S''(x)\, dx.$$

$$(6.21)$$

By integrating the last integral on the right-hand side, by parts, we obtain

$$\int_a^b e''(x) S''(x)\, dx = S''(x) e'(x)\big]_a^b - \int_a^b e'(x) S'''(x)\, dx. \qquad (6.22)$$

Since $S''(a) = S''(b) = 0$, we have

$$S''(x) e'(x)\big]_a^b = S''(b) e'(b) - S''(a) e'(a) = 0. \qquad (6.23)$$

Moreover, since $S'''(x) = 6 d_k = $ constant on the subinterval $[x_k, x_{k+1}]$, it follows that

$$\int_{x_k}^{x_{k+1}} e'(x) S'''(x)\, dx = 6 d_k \int_{x_k}^{x_{k+1}} e'(x)\, dx = [6 d_k e(x)]_{x_k}^{x_{k+1}} = 0.$$

Here, use has been made of the fact that $e(x_k) = f(x_k) - S(x_k) = 0$ for $k = 1, ..., n$. Hence,

$$\int_a^b e'(x) S'''(x)\, dx = \sum_{k=1}^{n-1} \int_{x_k}^{x_{k+1}} e'(x) S'''(x)\, dx = 0. \qquad (6.24)$$

Finally, using Eqns. (6.21), (6.22), (6.23), and (6.24), we get

$$\int_a^b [S''(x)]^2 dx \leq \int_a^b [S''(x)]^2 dx + \int_a^b [e''(x)]^2 dx = \int_a^b [f''(x)]^2 dx.$$

MATLAB's Methods

Cubic spline interpolation can be performed with the built-in MATLAB function spline. It can be called in the following way:

```
>> yy = spline(x,y,xx)
```

x and y are arrays of data points that define the function to be interpolated. If xx is an array, we get an array yy of interpolated values. MATLAB uses a *not-a-knot end condition*, which is different from the natural end condition. It requires continuity of $S'''(x)$ at the first internal knot, i.e., at x_2 or x_{n-1}. We now illustrate the use of the MATLAB function spline in fitting a curve to a set of data. Consider the following table:

x	0.0	0.3	0.5	0.8	1.0
y	1.0	1.09	1.25	1.64	2.0

```
>> x = [0.0   0.3   0.5   0.8   1.0];
>> y = [1.0   1.09   1.25   1.64   2.0];
>> xx = 0.0 : 0.1 : 1.0;
>> yy = spline(x,y,xx)
yy =
      1.0000 1.0100 1.0400 1.0900 1.1600 1.2500 1.3600 1.4900 1.6400 1.8100 2.0000
```

which are the interpolated values. It is possible to plot the cubic spline curve for this table using the MATLAB command plot:

```
>> plot(x,y,'o',xx,yy)
>> xlabel('x'), ylabel('y')
```

Figure 6.4 shows the resulting cubic spline curve.

EXAMPLE 6.8
Use the MATLAB function spline to solve the **Glucose level problem** at time $t = 0, 1, \ldots, 80$ (see Example 6.1).

```
>> t = [0   10   20   30   40   50   60   70   80];
>> g = [100   118   125   136   114   105   98   104   92];
>> tt = 0 : 1 : 80;
```

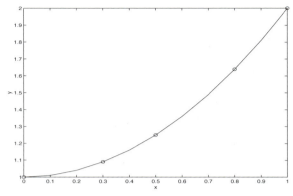

FIGURE 6.4
Cubic spline interpolation.

```
>> gg = spline(t,g,tt);
>> plot(t,g,'o',tt,gg)
```

Figure 6.5 shows the resulting cubic spline curve along with the data points.

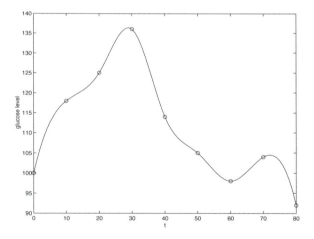

FIGURE 6.5
Glucose level.

EXERCISE SET 6.3

1. Given the table

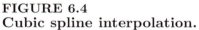

x	1.4	1.8	2.2
$f(x)$	3.12	2.81	1.7

M-function 6.3

The following MATLAB function **spl3.m** finds a natural cubic spline that interpolates a table of values. INPUTS are a table of function values x and y; the number of intermediate points m at which $S(x)$ is to be approximated.

```
function function spl3(x,y,m)
% Cubic spline
n=length(x);
for i=1:n-1
   h(i)=x(i+1)-x(i);
   b(i)=(y(i+1)-y(i))/h(i);
end
u(2)=2*(h(1)+h(2));
v(2)=6*(b(2)-b(1));
for i=3:n-1
   u(i)=2*(h(i-1)+h(i))-h(i-1)^2/u(i-1);
   v(i)=6*(b(i)-b(i-1))-h(i-1)*v(i-1)/u(i-1);
end
z(n)=0;
for i=n-1:-1:2
   z(i)=(v(i)-h(i)*z(i+1))/u(i);
end
z(1)=0;
disp('            cubic spline')
disp('_____')
disp('        x                 S(x)                     ')
disp('_____')
for i=1:n-1
   for j=1:m+1
     r=(x(i+1)-x(i))/(m+1);
     t=x(i)+(j-1)*r;
     dis=(j-1)*r;
     hh=x(i+1)-x(i);
     bb=(y(i+1)-y(i))/hh-hh*(z(i+1)+2*z(i))/6;
     q=0.5*z(i)+dis*(z(i+1)-z(i))/(6*hh);
     q=bb+dis*q;
     spl3=y(i)+dis*q;
     fprintf('%12.5f %17.5f \n',t,spl3)
   end
end
fprintf('%12.5f %17.5f \n',x(n),y(n))
fprintf('\n')
```

use natural cubic spline interpolation to approximate $f(1.6)$.

2. (Continuation). Compute the integral of the spline over $[1.4, 2.2]$.

3. Determine a, b, c, and d so that the following function is a natural cubic spline.

$$f(x) = \begin{cases} -3x^3 & \text{if } 0 \leq x \leq 2 \\ a(x-2)^3 + b(x-2)^2 + c(x-2) + d & \text{if } 2 \leq x \leq 3 \end{cases}.$$

4. Construct a natural cubic spline that interpolates the function $f(x) = xe^x$ at $x = 0, 0.25, 0.5, 0.75$, and 1.0. Compute the integral of the spline over $[0, 1]$ and compare the result to the actual value of

$$\int_0^1 xe^x dx = 1.$$

5. Show that

$$f(x) = \begin{cases} -2x^3 & \text{if } 0 \leq x \leq 1 \\ (x-1)^3 - 6(x-1)^2 - 6(x-1) - 2 & \text{if } 1 \leq x \leq 3 \end{cases}$$

is a cubic spline function.

6. Determine whether the following function is a cubic spline

$$f(x) = \begin{cases} -x^2 + 1 & \text{if } 0 \leq x \leq 1 \\ (x-1)^3 + 2(x-1) & \text{if } 1 \leq x \leq 2 \\ 2x - 1 & \text{if } 2 \leq x \leq 3 \end{cases}.$$

7. Given the table

x	2.0	2.1	2.2	2.3
$f(x)$	1.5	2.0	3.8	5.1

approximate $f(2.05), f(2.15)$, and $f(2.25)$ using natural cubic spline approximation.

8. Construct a natural cubic spline that interpolates the function $f(x) = 1/(1 + x^2)$ at $x = -2, -1, 0, 1$, and 2. Compare the interpolated values with the true values at $x = -1.5, -0.5, 0.5$, and 1.5.

9. Use the MATLAB function spl3 to interpolate the function $f(x) = \sin x$ at 10 equally spaced points in the interval $[0, \pi]$. Evaluate the spline at points between the data points (midpoints) and compare the values of the spline with the values of $\sin x$.

10. $S(x)$ is a cubic spline. Find $S(3)$

$$S(x) = \begin{cases} 1 + ax + bx^3 & \text{if} \quad 0 \le x \le 2 \\ 29 + 38(x-2) + c(x-2)^2 - 3(x-2)^2 & \text{if} \quad 2 \le x \le 3 \end{cases}.$$

11. Find the values of a, b, c, d, e, and f such that the following function defines a cubic spline and find $g(0)$, $g(2)$, and $g(3)$.

$$g(x) = \begin{cases} 2x^3 + 4x^2 - 7x + 5, & 0 \le x \le 1 \\ 3(x-1)^3 + a(x-1)^2 + b(x-1) + c, & 1 \le x \le 2 \\ (x-2)^3 + d(x-2)^2 + e(x-2) + f, & 2 \le x \le 3 \end{cases}.$$

12. Let $f(x)$ be a cubic polynomial, let $S(x)$ be the unique $\mathbb{C}^2$ cubic spline interpolating f at points $x_0 < x_1 < \ldots < x_n$ and satisfying $S'(x_0) = f'(x_0)$, $S'(x_n) = f'(x_n)$, and let $N(x)$ be the unique natural cubic spline interpolating f at points $x_0 < x_1 < \ldots < x_n$. Does $f(x) = S(x)$ for all $x_0 < x < x_n$? Does $f(x) = N(x)$ for all $x_0 < x < x_n$? Explain why or why not.

13. For what value of k is

$$f(x) = \begin{cases} (x-1)^3 + (x-1)^2 + k(x-1) + 5, & 0 \le x \le 1 \\ 3(x-1)^3 + 2(x-1)^2 + (x-1) + 5, & 1 \le x \le 2 \end{cases}$$

a cubic spline?

14. Find the natural cubic spline $S(x)$ that fits the points $(1, \ln 1)$, $(2, \ln 2)$, $(3, \ln 3)$, $(4, \ln 4)$, and $(6, \ln 6)$. Evaluate $S(x)$ at $x = 5$.

15. Find the natural cubic spline that interpolates the function $f(x) = \cos x^2$ at $x_1 = 0$, $x_2 = 0.6$, and $x_3 = 0.9$.

16. Given $f(x) = \log_2(x)$

   (a) Write the tridiagonal linear system to be solved for approximating $f(x)$ by a natural cubic spline on the nodes: 0.5, 1, 2, 4, and 8.

   (b) Use MATLAB to solve the system and find the cubic spline in the interval $[2, 4]$ only.

   (c) Evaluate the spline at $x = 3$ and compare the result with the exact value $\log_2(3)$.

17. Is this function a cubic spline?

$$g(x) = \begin{cases} -2x, & -2 \le x \le -1 \\ 1 + x^2, & -1 \le x \le 0 \\ 1 + x^2 + x^3, & 0 \le x \le 2 \end{cases}.$$

18. Consider the following piecewise polynomial function $S(x)$

$$S(x) = \begin{cases} x^3 - 1, & 0 \le x \le 1 \\ 3x^2 - 3x, & 1 \le x \le 3 \\ x^3 - 6x^2 + 24x - 27, & 3 \le x \le 4 \end{cases}.$$

Is $S(x)$ a cubic spline? Is it a spline of degree 4?

19. To approximate the top profile of the duck, we have chosen points along the curve through which we want the approximating curve to pass (see Figure 6.6). Write the (x, y) coordinates of these points and find the natural cubic spline that interpolates them. Plot the resulting spline to compare it with the given curve.

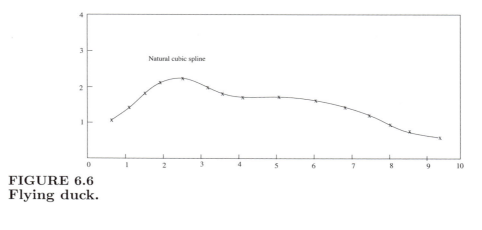

FIGURE 6.6
Flying duck.

COMPUTER PROBLEM SET 6.3

1. Write a computer program in a language of your choice to find a natural cubic spline $S(x)$ that interpolates a set of data points (x_i, y_i) and evaluate the spline at points between the data points. Input data to the program should be

   (a) The set of data points (x_i, y_i).

   Output should consist of:

   (a) The value of $S(x)$ between the data points.

   Test your program to find the natural cubic spline $S(x)$ that interpolates $f(x) = 1/(x^4 + 1)$ at 21 equally spaced knots in the interval $[-5, 5]$. Evaluate $S(x)$ at the midpoints of the knots and compare your results with the values of $f(x)$.

2. Use the MATLAB function spl3.m to find the natural cubic spline that interpolates the function $f(x) = x - \cos x^2$ at 21 equally spaced knots over the
   interval $[-3, 3]$. Evaluate the error $|f(x) - S(x)|$ at 41 equally spaced knots.

3. Use the MATLAB function spl3.m to find the natural cubic spline that interpolates the table

x	0.0	0.2	0.4	0.6	0.8	1.0
$f(x)$	0.0	0.163	0.268	0.329	0.359	0.368
x	1.2	1.4	1.6	1.8	2.0	
$f(x)$	0.361	0.345	0.323	0.2975	0.271	

Plot the resulting spline. The function used to cerate this table is $f(x) = xe^x$.
Compute $|f(x) - S(x)|$ at the midpoints of the table.

4. The data in the following table are for Bessel function J_0.

x	0	1	2	3	4	5
$J_0(x)$	1.00	0.77	0.22	-0.26	-0.40	-0.18
x	6	7	8	9	10	
$J_0(x)$	0.15	0.30	0.17	-0.09	-0.25	

Interpolate the table with a natural cubic spline. Compare your interpolated
values at $x = 0.5, 1.5, \ldots, 9.5$ with those found from the MATLAB function
besselj.

5. Given $f(x) = x^2 \cos(x + 1)$, use MATLAB to generate 41 evenly spaced data
   points in the interval $[0, 5]$. Find the natural cubic spline $S(x)$ that interpolates
   these points. Evaluate the error $|S(x) - f(x)|$ at 81 equally spaced points in
   $[0, 5]$.

6. Use the MATLAB function spline to evaluate the cubic spline at $x = 0.18$,
   0.22, 0.34 that interpolates the data points

x	0.1	0.2	0.3	0.4
y	-0.6205	-0.2840	0.0066	0.2484

APPLIED PROBLEMS FOR CHAPTER 6

1. Generate a table of values by taking

$$x_i = i - 6, \qquad i = 1, 2, \ldots, 10$$

Year	Population
1880	50,155,780
1890	62,947,710
1900	75,994,570
1910	91,972,265
1920	105,710,623
1930	122,775,625
1940	131,669,270
1950	150,697,365
1960	179,323,170
1970	203,235,290
1980	226,504,820

Table 6.5   Population model.

$$y_i = \frac{1}{1 + x_i^2}.$$

Find a natural cubic spline that interpolates this table. Compare the values of the spline with the values of $y = 1/(1 + x^2)$ at points between the data points (midpoints).

2. The U.S. population from 1880 and 1980 is given in Table 6.5.

   (a) Find a natural spline that interpolates this table.
   (b) What does this spline show for the 1975 population?

3. A biologist who is studying the growth of a bacteria culture recorded the following data

t	0	3	6	9
$p(t)$	2.0	3	4	6

where $p(t)$ denotes the number of bacteria at time t (minutes). Use a natural cubic spline to estimate the number of bacteria at $t = 5$.

4. The potential energy of two or more interacting molecules is called van der Waal's interaction energy. The following table gives the energy V of two interacting helium atoms in terms of the internuclear distance r.

r	4.6	4.8	5.0	5.2	5.4	5.6	5.8	6.0
V	32.11	9.00	-3.52	-9.23	-11.58	-12.01	-11.24	-10.12

Compute and plot the natural cubic spline interpolant for this table. Use the cubic spline to approximate the energy V at $r = 5.1$, $r = 5.5$, and $r = 5.9$.

Chapter 7

The Method of least-squares

In approximation theory, two general types of problems arise in fitting tabular data. The first one consists of finding an approximating function (perhaps a piecewise polynomial) that passes through every point in the table. The other problem consists of finding the "best" function that can be used to represent the data but does not exactly pass through every data point. This category of problems is called **curve fitting** and will be the subject of this chapter.

In Chapter 5, we constructed polynomial approximations to tabular data using interpolation methods. Interpolating polynomials are best used for approximating a function f whose values are known with high accuracy. Often, however, the tabulated data are known to be only approximate. More precisely, in most of the situations, data are given by a set of measurements having experimental errors. Hence, interpolation in this case is of little use, if not dangerous. Consider, for example, a simple physical experiment in which a spring is stretched from its equilibrium position by a known external force (see Figure 7.1).

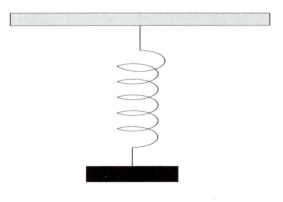

FIGURE 7.1
Spring.

EXAMPLE 7.1 : **Displacements of a Spring**

Suppose we consecutively apply forces of $1, 2, 3$, and 4 kilos to a spring and find that its displacements are $3.3, 6.1, 9.8$, and 12.3 centimeters, respectively. Hooke's law states that the force F and the displacement x are related by the linear function $F = -kx$, where $k > 0$ is a constant characteristic of the spring, called the **spring constant**. The problem we want to solve is to determine k from the experimental data. The usual approach to this problem is to seek an approximating function that best fits these data points. In the following sections, we will examine the most commonly used technique for solving the above problem, known as the method of **least-squares**.

7.1　LINEAR least-squares

Consider the experimental data in Table 7.1. The problem of data fitting consists

x	1	2	3	4	5
y	1	1	2	2	4

Table 7.1　Linear least-squares fit.

of finding a function f that "best" represents the data that are subject to errors. A reasonable way to approach the problem is to plot the data points in an xy-plane and try to recognize the shape of a *guess function* $f(x)$ (see Figure 7.2) such that

$$f(x) \approx y.$$

The best guess function would be the one that is simple in form and tends to

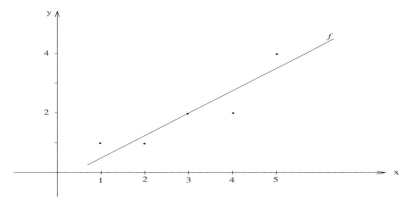

FIGURE 7.2
least-squares line that fits the data in Table 7.1.

"smooth" the data. Therefore, a reasonable guess function to the data in Figure 7.2 might be a linear one, that is

$$f(x) = ax + b.$$

Having selected a particular function $f(x)$, the problem becomes that of finding the values of the *parameters* a and b that make $f(x)$ the "best" function to fit the data.

x	x_1	x_2	$\cdots$	x_n
y	y_1	y_2	$\cdots$	y_n

Table 7.2   Table of x and y values.

Letting $e_i = f(x_i) - y_i$ with $1 \leq i \leq 5$, the least-squares criterion consists of minimizing the sum

$$E(a,b) = \sum_{i=1}^{5} (e_i)^2$$

$$= \sum_{i=1}^{5} [f(x_i) - y_i]^2$$

$$= \sum_{i=1}^{5} [(ax_i + b) - y_i]^2. \tag{7.1}$$

Here $E(a,b)$ is considered to be a function of two variables. This method of choosing a and b is commonly used in engineering, economics, and many other sciences.

We know from calculus that the minimum of (7.1) will occur when

$$\frac{\partial E(a,b)}{\partial a} = 0 \quad \text{and} \quad \frac{\partial E(a,b)}{\partial b} = 0. \tag{7.2}$$

From Table 7.1, we get, after taking the partial derivatives in (7.2) and rearranging terms:

$$15a + 5b = 10,$$
$$55a + 15b = 37.$$

The solution of this linear system of equations is $a = 0.7$ and $b = -0.1$, so the best function in the least-squares sense is

$$f(x) = 0.7x - 0.1.$$

The general problem of approximating the data in Table 7.2 with a linear function $f(x) = ax + b$ using the least-squares involves minimizing

$$E(a,b) = \sum_{i=1}^{n} [(ax_i + b) - y_i]^2 \tag{7.3}$$

with respect to the parameters a and b. In view of (7.2) we have

$$\frac{\partial}{\partial a} \sum_{i=1}^{n} [(ax_i + b) - y_i]^2 = 2 \sum_{i=1}^{n} (ax_i + b - y_i)(x_i) = 0,$$

$$\frac{\partial}{\partial b} \sum_{i=1}^{n} [(ax_i + b) - y_i]^2 = 2 \sum_{i=1}^{n} (ax_i + b - y_i) = 0.$$

Summing term by term leads to the so-called **normal equations**

$$a \sum_{i=1}^{n} x_i^2 + b \sum_{i=1}^{n} x_i = \sum_{i=1}^{n} x_i y_i,$$

(7.4)

$$a \sum_{i=1}^{n} x_i + bn = \sum_{i=1}^{n} y_i.$$

The second term, bn, in the second equation comes from the fact that $\sum_{i=1}^{n} b = nb$. The solution of this linear system is

$$a = \frac{n \sum x_i y_i - \sum x_i \sum y_i}{n \sum x_i^2 - (\sum x_i)^2},$$

(7.5)

$$b = \frac{\sum x_i^2 \sum y_i - \sum x_i y_i \sum x_i}{n \sum x_i^2 - (\sum x_i)^2}$$

where $\sum = \sum_{i=1}^{n}$.

EXAMPLE 7.2

Using the method of least-squares, find the linear function that best fits the following data (see Figure 7.3).

x	1	1.5	2	2.5	3	3.5	4
y	25	31	27	28	36	35	32

Using values in Table 7.3, we get

$$\sum_{i=1}^{7} x_i = 1 + 1.5 + 2 + 2.5 + 3 + 3.5 + 4 = 17.5,$$

$$\sum_{i=1}^{7} y_i = 25 + 31 + 27 + 28 + 36 + 35 + 32 = 214,$$

$$\sum_{i=1}^{7} x_i^2 = 1^2 + 1.5^2 + 2^2 + 2.5^2 + 3^2 + 3.5^2 + 4^2 = 50.75,$$

$$\sum_{i=1}^{7} x_i y_i = (1)(25) + (1.5)(31) + (2)(27) + (2.5)(28) +$$

$$+ (3)(36) + (3.5)(35) + (4)(32) = 554.$$

Thus, using Eqn. (7.5) we obtain

$$a = \frac{7(554) - (17.5)(214)}{7(50.75) - (17.5)^2} = 2.714\,285\,71,$$

$$b = \frac{(50.75)(214) - (17.5)(554)}{7(50.75) - (17.5)^2} = 23.78571429.$$

Therefore, the least-squares line is

$$y = 2.71428571x + 23.78571429.$$

```
» x=[1:0.5:4];
» y=[25 31 27 28 36 35 32];
» linlsqr(x,y)

            linear least squares

a =   2.714286
b =   23.785714
```

x	y	a*x+b	|y-(ax+b)|
1.0	25	26.500000	1.500000
1.5	31	27.857143	3.142857
2.0	27	29.214286	2.214286
2.5	28	30.571429	2.571429
3.0	36	31.928571	4.071429
3.5	35	33.285714	1.714286
4.0	32	34.642857	2.642857

E(a,b) = 50.142857

Table 7.3 Linear least-squares for Example 7.2.

EXERCISE SET 7.1

1. Find the linear function that best fits the data

x	1.2	2.2	3.2	4.2	5.2
y	2.7	4.2	5.8	9.1	10.1

   by using the method of least-squares and compute the error.

2. The following table lists the temperatures of a room recorded during the time interval [1:00,7:00]. Find the best linear least squares that approximate the table.

Time	1:00	2:00	3:00	4:00	5:00	6:00	7:00
Temperature	13	15	20	14	15	13	10

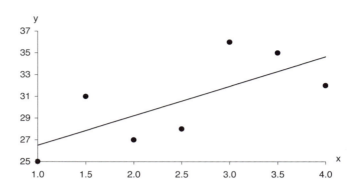

FIGURE 7.3
Linear least-squares $y = 2.7143x + 23.7857$ for Example 7.2.

Use your result to predict the temperature of the room at 8:00.

3. The following table lists the homework grades, out of 100 students. Find the linear function that best fits the data, using the method of least-squares and compute the error.

Homework #	1	2	3	4	5	6	7	8	9	10	11	12
Grades	78	65	92	57	69	60	80	91	60	70	55	45

Use this function to predict the grade of the student on the next homework, # 13.

4. Find the best least-squares line through the points $(-8, -9)$, $(-3, -4)$, $(-1, -2)$, and $(12, 11)$.

5. Given the data,

x	1	1.5	2	2.5	3	3.5	4
y	25	31	27	28	36	35	34

fit a straight line to the data using the method of least squares and compute the error.

6. Find the best least-squares line through the points $(0, 2)$, $(0, 8)$, $(1, -1)$, and $(3, 11)$.

7. Find the best least-squares line (constant) that fit the table

x	1	3	5	6
y	27/10	36/11	18/5	12/5

M-function 7.1

The following MATLAB function **linlsqr.m** constructs a least squares line $y = ax + b$ that best fits a table of x and y values. INPUTS are x and y row vectors.

```
function linlsqr(x,y)
% Construct a least square line y=ax+b
% that fits x and y row or column vectors.
n=length(x);
n=length(y);
sumx=sum(x);
sumy=sum(y);
sumx2=sum(x.*x);
sumxy=sum(x.*y);
a=(n*sumxy-sumx*sumy)/(n*sumx2-sumx^2);
b=(sumx2*sumy-sumxy*sumx)/(n*sumx2-sumx^2);
disp('            linear least squares')
fprintf('\n a =%12.6f\n',a)
fprintf(' b =%12.6f\n',b)
disp('_____')
disp('          x        y           a*x+b         |y-(ax+b)|   ')
disp('_____')
for i=1:n
   yl=a*x(i)+b;
   err(i)=abs(yl-y(i));
   fprintf('%6.2f %6.2f  %12.6f  %12.6f\n',x(i),y(i),yl,err(i))
end
err=sum(err.*err);
fprintf('\n E(a,b) =%12.6f\n',sum(err))
```

8. What line best represents the following data in the least-squares sense?

x	-1	1	2	3
y	-6.8	7.0	13.4	20.9

COMPUTER PROBLEM SET 7.1

1. Write a computer program in a language of your choice to construct the linear least-squares that best fit a given set of data points.
   Input data to the program should be

   (a) m data points (x_i, y_i).

   Output should consist of:

(a) The coefficients a and b of the linear least-squares $f(x) = ax + b$.

(b) The error.

Test your program to find the line that best fit the data

x	-3	-2	0	1	2	3	4	6	7	8
y	7.12	5.16	4.349	5.99	4.14	4.28	3.32	4.81	3.15	3.99

2. Use the MATLAB function linlsqr to find the linear function that best fit the following data

x	0	2	4	6	8	10	12	14	16	18	20
y	-3.5	1.5	6.0	13.0	19.6	26.1	32.3	39.3	42.4	50.1	55.8

Plot the resulting best line along with the data points.

7.2 LEAST-SQUARES POLYNOMIAL

The method of least-squares data fitting is not restricted to linear functions $f(x) = ax + b$ only. As a matter of fact, in many cases data from experimental results are not linear, so we need to consider some other guess functions. To illustrate this, suppose that the guess function for the data in Table 7.2 is a polynomial

$$p_m(x) = \sum_{k=0}^{m} a_k x^k \tag{7.6}$$

of degree $m \leq n - 1$. So, according to the least-squares principle, we need to find the coefficients $a_0, a_1, \ldots, a_m$ that minimize

$$E(a_0, \ldots, a_m) = \sum_{i=1}^{n} [p_m(x_i) - y_i]^2$$

$$= \sum_{i=1}^{n} \left[\sum_{k=0}^{m} (a_k x_i^k) - y_i \right]^2. \tag{7.7}$$

As before, E is minimum if

$$\frac{\partial}{\partial a_j} E(a_0, \ldots, a_m) = 0, \quad j = 0, 1, \ldots, m. \tag{7.8}$$

that is

$$\frac{\partial E}{\partial a_0} = \sum_{i=1}^{n} 2 \left[\sum_{k=0}^{m} (a_k x_i^k) - y_i \right] = 0,$$

$$\frac{\partial E}{\partial a_1} = \sum_{i=1}^{n} 2 \left[\sum_{k=0}^{m} (a_k x_i^k) - y_i \right] (x_i) = 0, \tag{7.9}$$

$$\vdots = \vdots$$

$$\frac{\partial E}{\partial a_m} = \sum_{i=1}^{n} 2 \left[\sum_{k=0}^{m} (a_k x_i^k) - y_i \right] (x_i^m) = 0.$$

Rearranging Eqn. (7.9) gives the $(m+1)$ normal equations for the $(m+1)$ unknowns $a_0, a_1, \ldots, a_m$

$$
\begin{aligned}
a_0 \, n \quad &+ a_1 \sum x_i \quad + \cdots + a_m \sum x_i^m \quad = \sum y_i \\
a_0 \sum x_i \,&+ a_1 \sum x_i^2 \quad + \cdots + a_m \sum x_i^{m+1} = \sum y_i x_i \\
a_0 \sum x_i^2 &+ a_1 \sum x_i^3 \quad + \cdots + a_m \sum x_i^{m+2} = \sum y_i x_i^2. \\
\vdots \quad &+ \quad \vdots \quad \quad + \cdots + \quad \vdots \quad \quad = \quad \vdots \\
a_0 \sum x_i^m &+ a_1 \sum x_i^{m+1} + \cdots + a_m \sum x_i^{2m} \quad = \sum y_i x_i^m
\end{aligned}
\tag{7.10}
$$

As before, $\sum$ denotes $\sum_{i=1}^{n}$. It can be shown that these normal equations have a unique solution provided that the x_i's are distinct. However, when the x-values are equally spaced, the matrix of this linear system is a Hilbert matrix (see Exercise 5, Section 4.2), which is extremely ill-conditioned. The degree of the approximating polynomial is severely limited in most cases by the round-off error. For example, using a computer with precision of about eight decimal digits will usually produce meaningless results if n is greater than six.

EXAMPLE 7.3

Find the least-squares polynomial of degree three that fit the following table of values.

x	0.0	0.5	1.0	1.5	2.0	2.5
y	0.0	0.20	0.27	0.30	0.32	0.33

From Eqn. (7.10), the normal equations are

$$
\begin{bmatrix}
6.0000 & 7.5000 & 13.750 & 28.1250 \\
7.5000 & 13.7500 & 28.1250 & 61.1875 \\
13.7500 & 28.1250 & 61.1875 & 138.2813 \\
28.1250 & 61.1875 & 138.2813 & 320.5469
\end{bmatrix}
\begin{bmatrix}
a_0 \\ a_1 \\ a_2 \\ a_3
\end{bmatrix}
=
\begin{bmatrix}
1.4200 \\ 2.2850 \\ 4.3375 \\ 9.0237
\end{bmatrix}.
$$

The solution of this linear system is

$$a_0 = 0.0033, \ a_1 = 0.4970, \ a_2 = -0.2738, \ a_3 = 0.0511$$

and the desired polynomial is

$$p_3(x) = 0.0033 + 0.4970x - 0.2738x^2 + 0.0511x^3.$$

The computed results are shown in Table 7.4.

```
» x=[0:0.5:2.5];
» y=[0 0.2 0.27 0.3 0.32 0.33];
» polylsqr(x,y,3)

The Augmented Matrix of the normal equations, [A b] =

        6.000       7.500      13.750      28.125      1.420
        7.500      13.750      28.125      61.188      2.285
       13.750      28.125      61.188     138.281      4.338
       28.125      61.188     138.281     320.547      9.024

The coefficients ao,...,an of the least-squares polynomial are

   0.0033   0.4970   -0.2738   0.0511
```

xi	yi	p(xi)	|yi-p(xi)|
0.00	0.00	0.003333	0.003333
0.50	0.20	0.189762	0.010238
1.00	0.27	0.277619	0.007619
1.50	0.30	0.305238	0.005238
2.00	0.32	0.310952	0.009048
2.50	0.33	0.333095	0.003095

$$E(ao,...,an) = 0.000292857$$

Table 7.4   Polynomial least-squares for Example 7.3.

In the following example we illustrate the use of the MATLAB function polylsqr to a higher-degree polynomial.

EXAMPLE 7.4

Find the least-squares polynomial of degree four that fit the data given by the vectors
x = -3:0.4:3
y = sin(x)

By calling the MATLAB function polysqrt(x,y,4), we obtain the results shown in Table 7.5 and represented by Figure 7.5.

The principle of least-squares, applied to polynomials given by Eqn. (7.6), can

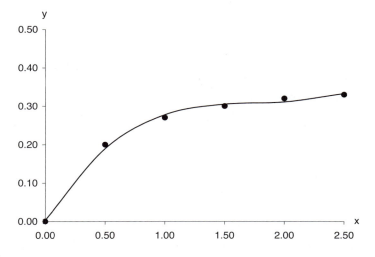

FIGURE 7.4
Polynomial least-squares for Example 7.3.

be extended to a much more general setting. Let

$$p(x) = \sum_{k=1}^{m} c_k \psi_k(x)$$

where the ψ's are polynomials of degree $\leq k$.

The polynomials ψ_k, called "basis functions," are known and held fixed. So, according to the least-squares principle, we need to find $c_1, c_2, \ldots, c_m$ that minimize

$$E(c_1, \ldots, c_m) = \sum_{k=1}^{n} \left[\sum_{i=1}^{m} c_i \psi_i(x_k) - y_k \right]^2.$$

Proceeding as before, E is minimum if

$$\frac{\partial E}{\partial c_j} = \sum_{k=1}^{n} 2 \left[\sum_{i=1}^{m} c_i \psi_i(x_k) - y_k \right] \psi_j(x_k) = 0, \quad j = 1, \ldots, m.$$

Thus,

$$\sum_{i=1}^{m} \left[\sum_{k=1}^{n} \psi_i(x_k) \psi_j(x_k) \right] c_i = \sum_{k=1}^{n} y_k \psi_j(x_k) \quad j = 1, \ldots, m.$$

Hence, $p(x)$ is found by solving these m linear normal equations for the m unknowns $c_1, c_2, \ldots, c_m$.

MATLAB's Methods

The built-in MATLAB function polyfit finds the coefficients of a polynomial that fits a set of data points in a least-squares sense; the calling syntax is

```
» x=[-3.0: 0.4: 3.3];
» y=sin(x);
» polylsqr(x,y,4)

The Augmented Matrix of the normal equations, [A b] =

1.0e+004 *

  0.0016   0.0000   0.0054  -0.0000   0.0331  -0.0000
  0.0000   0.0054  -0.0000   0.0331        0   0.0016
  0.0054  -0.0000   0.0331        0   0.2388        0
 -0.0000   0.0331        0   0.2388        0   0.0062
  0.0331        0   0.2388   0.0000   1.8648  -0.0000

The coefficients a0,...,an of the least squares polynomial are

 -0.0000   0.8552   0.0000  -0.0928  -0.0000
```

xi	yi	p(xi)	|yi-p(xi)|
-3.00	-0.14	-0.060136	0.080984
-2.60	-0.52	-0.592535	0.077033
-2.20	-0.81	-0.893326	0.084830
-1.80	-0.97	-0.998142	0.024295
-1.40	-0.99	-0.942615	0.042835
-1.00	-0.84	-0.762376	0.079095
-0.60	-0.56	-0.493058	0.071585
-0.20	-0.20	-0.170291	0.028378
0.20	0.20	0.170291	0.028378
0.60	0.56	0.493058	0.071585
1.00	0.84	0.762376	0.079095
1.40	0.99	0.942615	0.042835
1.80	0.97	0.998142	0.024295
2.20	0.81	0.893326	0.084830
2.60	0.52	0.592535	0.077033
3.00	0.14	0.060136	0.080984

```
E(a0,...,an) =  0.0685988
```

Table 7.5   Polynomial least-squares for Example 7.4.

```
>> p = polyfit(x,y,n)
```

x and y are vectors defining the data for the fit; n is the degree of the polynomial. The coefficients of the polynomial in descending powers of x are returned in vector p. Note that a warning message results if n $\geq$ length(x).

EXAMPLE 7.5

Find the linear function $F = -kx$ that best fit the data of the spring problem in Example 7.1.

```
>> x = [1   2   3   4];
>> y = [3.3    6.1   9.8    12.3];
>> p = polyfit(x,y,1)
```

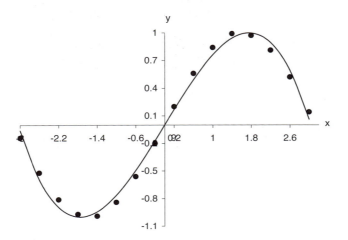

FIGURE 7.5
Polynomial of degree 4 that fits the data for Example 7.4.

p =
    3.3000

Thus

$$p(x) = 3.3x + 0 \quad \text{and} \quad F = -3.3x.$$

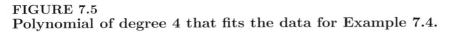

EXERCISE SET 7.2

1. Use the data of Exercise 5 to compute the least-squares polynomial of degree three that best fits the data.

2. Given the data,

x	1.0	1.2	1.5	1.8	2.1	2.4	2.8
y	5.5	7.3	14.1	26.3	48.2	80.1	111.6

   (a) Find the least-squares polynomial of degree two that best fits the data and compute the error.

   (b) Find the least-squares polynomial of degree three that best fits the data and compute the error.

3. Suppose the quadratic function

$$f(x) = x^2 + 3x + 7$$

M-function 7.2

The following MATLAB function **polylsqr.m** constructs a least squares polynomial of degree m that best fits a table of x and y values. INPUTS are x and y row vectors and the degree m of the polynomial.

```
function polylsqr(x,y,m)
% Construct the least squares polynomial of degree m
% that fits x and y row or column vectors.
n=length(x);
n=length(y);
for k=1:2*m+1
  c(k)=sum(x.^(k-1));
end
%Find the coefficient vector b of normal equations
for k=1:m+1
  b(k)=sum(y.*x.^(k-1));
end
% Find the coefficient matrix A of normal equations
for i=1:m+1
  for j=1:m+1
    A(i,j)=c(j+i-1);
  end
end
fprintf('\n')
disp(' The Augmented Matrix of the normal equations, [A b] =')
fprintf('\n')
disp([A b'])
z=A\b';
disp('The coeff.  a0,...,an of the least squares polynomial are ')
fprintf('\n')
disp(z')
% Evaluate the polynomial at xi, i=1,..,n.
disp('_____')
disp('          xi          yi            p(xi)        Iyi-p(xi)I   ')
disp('_____')
for i=1:n
  s=z(1);
  for k=2:m+1
    s=s+z(k)*x(i)^(k-1);
  end
  p(i)=s;
  err(i)=abs(y(i)-p(i));
```

```
   fprintf('%6.2f  %6.2f  %12.6f  %12.6f\n',x(i),y(i),p(i),err(i))
end
err=sum(err.*err);
fprintf('\n E(a0,...,an) =%12.6g\n',sum(err))
[x' y' p']
```

is used to generate the data,

x	0	1	2	3
$f(x)$	7	11	17	25

Find the least-squares polynomials of degree two and three that best fit these data. Compute these polynomials at the four values of x and compare your results with the exact ones.

4. Find the polynomial of degree 3 that fits the function $f(x) = \sin(\pi x)$ at the points $x_1 = -1$, $x_2 = -0.5$, $x_3 = 0$, $x_4 = 0.5$, and $x_5 = 1$ in the least-squares sense.

5. Given the following data

x	1.2	3.0	4.5	5.6	6.9	8.1
y	2.3	3.3	5.6	15.4	38.2	59.1

find the least-squares polynomial of degree two that best fits the data. Use the resulting polynomial to approximate $f(6.0)$.

6. Given the following data, fit a second degree polynomial to the data using the least-squares criterion:

x	1.0	2.3	3.8	4.9	6.1	7.2
$f(x)$	0.1	1.5	7.4	15.2	26.2	27.9

7.3   NONLINEAR least-squares

The actual selection of the **guess function** to be used for the least-squares fit depends on the nature of the experimental data. So far, we have studied only least-squares polynomials. However, many cases arise when data from experimental tests are exponentially related or tend to have vertical and horizontal asymptotes;

so we need to consider guess functions other than polynomials. In this section, we broaden our discussion of least-squares approximations to some popular forms of guess functions that arise when data are plotted.

7.3.1   Exponential form

Suppose we want to fit the data in Table 7.2 by a function of the form

$$f(x) = ae^{bx} \tag{7.11}$$

in the least-squares sense. If we develop the normal equations for (7.11) using (7.8), we will end up with nonlinear simultaneous equations, which are more difficult to solve than linear equations. So, one way to get around this difficulty is to first linearize the exponential form (7.11) by taking the logarithm of $f(x)$:

$$\ln f(x) = bx + \ln a.$$

We now set

$$F(x) = \ln f(x), \quad \alpha = \ln a \quad \text{and} \quad \beta = b$$

to get a linear function of x

$$F(x) = \beta x + \alpha. \tag{7.12}$$

Finally, transform the points (x_i, y_i) in Table 7.2 to the points $(x_i, \ln y_i)$ and use the linear least-squares described in Section 7.1 to get β and α. Having obtained β and α, we use the relations

$$b = \beta \quad \text{and} \quad a = e^{\alpha}$$

to obtain a and b.

EXAMPLE 7.6

Find the least-squares exponential that best fits the following data:

x	1	3	4	6	9	15
y	4.0	3.5	2.9	2.5	2.75	2.0

Apply the transformation (7.12) to the original points and obtain

$X = x$	1.0	3.0	4.0	6.0	9.0	15.0
$Y = \ln y$	1.39	1.25	1.06	0.92	1.01	0.69

The transformed points are shown in Table 7.6. The equation of the least-squares exponential is

$$y = 3.801404 \, e^{-0.044406x}.$$

```
» x=[1 3 4 6 9 15];
» y=[4 3.5 2.9 2.5 2.75 2];
» explsqr(x,y)

        Exponential least squares

a =   3.801404
b =  -0.044406
```

xi	yi	ln(yi)	a*exp(b*xi)	|yi-[a*exp(b*xi)]|
1	4.00	1.39	3.636293	0.363707
3	3.50	1.25	3.327274	0.172726
4	2.90	1.06	3.182757	0.282757
6	2.50	0.92	2.912280	0.412280
9	2.75	1.01	2.549046	0.200954
15	2	0.69	1.952841	0.047159

Table 7.6 Exponential least-squares for Example 7.6.

7.3.2 Hyperbolic form

Suppose we want to fit the data in Table 7.2 by a function of the form

$$f(x) = a + \frac{b}{x} \tag{7.13}$$

in the least-squares sense. This guess function is useful when the data tend to have vertical and horizontal asymptotes. Again, before finding the normal equations, we linearize $f(x)$ by setting

$$F(x) = f(x), \quad \alpha = a, \quad \beta = b, \quad \text{and} \quad X = \frac{1}{x}.$$

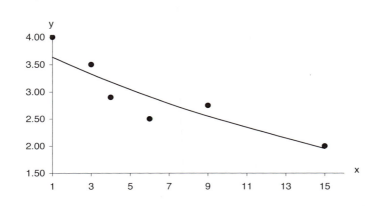

FIGURE 7.6
Exponential least-squares for Example 7.6.

Thus,

$$F(X) = \alpha + \beta X. \tag{7.14}$$

Again (7.14) is now a linear function in X, and if we transform the points (x_i, y_i) in Table 7.2 to $(1/x_i, y_i)$, we can use the linear least-squares to find $\alpha = a$ and $\beta = b$.

EXAMPLE 7.7

Find the least-squares hyperbolic that best fits the data in Example 7.6.

Apply the transformation (7.14) to the original points and obtain

$X = 1/x$	1	0.33	0.25	0.17	0.11	0.07
Y		4	3.5	2.9	2.5	2.75	2.0

The transformed points are shown in Table 7.7. The equation of the least-squares hyperbolic is

$$y = 2.364128 + \frac{1.797526}{x}.$$

We close this section by pointing out that the approximation obtained in this man-

```
x=[1 3 4 6 9 15];
» y=[4 3.5 2.9 2.5 2.75 2];
» hyplsqr(x,y)

          Hyperbolic least squares

a =   2.364128
b =   1.797526

    xi        yi        1/xi      a+b/xi    |yi-(a+b/xi)|

    1        4.00      1.000     4.161654   0.161654
    3        3.50      0.330     2.963304   0.536696
    4        2.90      0.250     2.81351    0.08649
    6        2.50      0.170     2.663716   0.163716
    9        2.75      0.110     2.563853   0.186147
    15       2.00      0.070     2.483963   0.483963
```

Table 7.7 Hyperbolic least-squares for Example 7.7.

ner for both of these examples is not the least squares approximation for the original problem, and it can, in some cases, differ from the least-squares approximation to the original problem.

Table 7.8 summarizes how to linearize a function $y = g(x)$ to the form $Y = aX + b$.

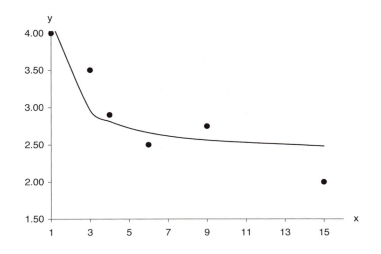

FIGURE 7.7
Hyperbolic least-squares for Example 7.7.

	Linearized form	Change of variables		
$y = f(x)$	$Y = aX + b$	$X =$	$Y =$	
$y = \alpha\,e^{\beta x}$	$\ln y = \beta x + \ln\alpha$	x	$\ln y$	$\alpha = e^b,\ \beta = a$
$y = \alpha x\,e^{-\beta x}$	$\ln(\frac{y}{x}) = -\beta x + \ln\alpha$	x	$\ln(\frac{y}{x})$	$\alpha = e^b,\ \beta = -a$
$y = \alpha\,x^{\beta}$	$\ln y = \beta\ln x + \ln\alpha$	$\ln x$	$\ln y$	$\alpha = e^b,\ \beta = a$
$y = \alpha + \beta\ln x$	$y = \beta\ln x + \alpha$	$\ln x$	y	$\beta = a,\ \alpha = b$
$y = \alpha + \frac{\beta}{x}$	$y = \beta\frac{1}{x} + \alpha$	$\frac{1}{x}$	y	$\beta = a,\ \alpha = b$
$y = \frac{\alpha}{\beta + x}$	$y = \frac{-1}{\beta}(xy) + \frac{\alpha}{\beta}$	xy	y	$\beta = -\frac{1}{a},\ \alpha = -\frac{b}{a}$
$y = \frac{\alpha x}{\beta + x}$	$y = -\beta(\frac{y}{x}) + \alpha$	$\frac{y}{x}$	y	$\beta = -a,\ \alpha = b$

Table 7.8 Summary of guess functions.

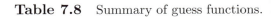

EXERCISE SET 7.3

1. Find the least-squares exponential that best fits the following data

M-function 7.3a

The following MATLAB function **explsqr.m** constructs a least squares exponential $y = ae^{bx}$ that best fits a table of *x* and *y* values. INPUTS are *x* and *y* row vectors.

```
function explsqr(x,y)
% Construct the least square exponential y=a*exp(b*x)
% that fits x and y row or column vectors.
n=length(x);
n=length(y);
z=log(y);
sumx=sum(x);
sumz=sum(z);
sumx2=sum(x.*x);
sumxz=sum(x.*z);
beta=(n*sumxz-sumx*sumz)/(n*sumx2-sumx^2);
b=beta;
alfa=(sumx2*sumz-sumxz*sumx)/(n*sumx2-sumx^2);
a=exp(alfa);
disp('            Exponential least squares')
fprintf('\n a =%12.6f\n',a)
fprintf(' b =%12.6f\n',b)
disp('_____')
disp('   xi     yi     ln(yi)    a*exp(b*xi)      |yi-[a*exp(b*xi)]|   ')
disp('_____')
for i=1:n
  ye=a*exp(b*x(i));
  err(i)=abs(ye-y(i));
   fprintf('%6.2f  %6.2f  %6.2f  %12.6f
%12.6f\n',x(i),y(i),z(i),ye,err(i))
end
```

x	0.0	1.5	2.5	3.5	4.5
y	2.0	3.6	5.4	8.1	12.0

Use your result to predict the value of y at $x = 5.0$.

2. Find the least-squares hyperbolic that best fits the data

x	1.0	1.4	1.8	2.2	2.6
y	3.7	3.5	3.4	3.1	2.9

Use your result to predict the value of y when $x = 3.0$.

M-function 7.3b

The following MATLAB function **hyplsqr.m** constructs a least squares hyperbolic $y = a + b/x$ that best fits a table of x and y values. INPUTS are x and y row vectors.

```
function hyplsqr(x,y)
% Construct the least square hyperbolic y=a+b/x
% that fits x and y row or column vectors.
n=length(x);
n=length(y);
z=ones(size(x));
z=z./x;
sumz=sum(z);
sumy=sum(y);
sumz2=sum(z.*z);
sumzy=sum(z.*y);
b=(n*sumzy-sumz*sumy)/(n*sumz2-sumz^2);
a=(sumz2*sumy-sumzy*sumz)/(n*sumz2-sumz^2);
disp('            hyperbolic least squares')
fprintf('\n a =%12.6f\n',a)
fprintf(' b =%12.6f\n',b)
disp('_____')
disp('     x      y      1/xi     a+b/x      ly-(a+b/xi)l   ')
disp('_____')
for i=1:n
  yh=a+b/x(i);
  err(i)=abs(yh-y(i));
  fprintf('%6.2f %6.2f %6.2f %12.6f  %12.6f\n',x(i),y(i),z(i),yh,err(i))
end
```

3. Given a table of data points (x_i, y_i) for $1 \le i \le n$, find the normal equations for the following guess functions

(a) $f(x) = a + b\ln x$,

(b) $f(x) = \frac{1}{a+x}$,

(c) $f(x) = ax^b$.

4. Find the normal equations for fitting a curve of the form

$$y = a\sin bx.$$

5. Find a function of the form $f(x) = a + b\sin\frac{\pi x}{10}$ that best fits the following data

x	0	1.0	2.3	4.0	5.1	6.0	6.5	7.0	8.1	9.0
$f(x)$	0.2	0.8	2.5	4.3	3.0	5.0	3.5	2.4	1.3	2.0
x	9.3	11.0	12.1	13.1	14.0	16	17.5	17.8	19.0	20.0
$f(x)$	-0.3	-1.3	-4.0	-4.9	-4.0	-3.0	-3.5	-1.6	-1.4	-0.1

6. Find the normal equations for fitting a curve of the form

$$f(x) = ax^2 + bx + c.$$

7. Determine the least-squares approximation of the type $g(x) = Ax^B$, for the data in the following table:

x	1.2	2.8	4.3	5.4	6.8	7.9
$f(x)$	2.0	11.4	28.0	41.8	72.2	91.3

8. Determine the least-squares approximation of the type $f(x) = (a - c)e^{-bx} + c$, for the data in the following table:

x	0	1	2	5	10	15	30	45	60
$f(x)$	210	198	187	155	121	103	77	71	70

9. Determine the least-squares approximation of the type $f(x) = ax^2 + bx + c$ to the function 2^x at the points $x_i = 0, 1, 2, 3, 4$.

10. We are given the following values of a function f as the variable t:

t	0.1	0.2	0.3	0.4
f	0.76	0.58	0.44	0.35

Obtain a least-squares fit of the form $g(t) = ae^{-3t} + be^{-2t}$.

11. A periodic experiment process gave the following data:

t (degrees)	0	50	100	150	200	250	300	350
y	0.75	1.76	2.04	1.41	0.30	−0.48	−0.38	0.52

Estimate the parameters a and b in the model $y = b + a \sin t$, using the least-squares approximation.

12. Consider the data in the following table:

x	1	2	3	4	5
y	2.4	6.2	10.6	16.1	21.8

Plot these data points and determine from Table 7.8 which of these functions appears better suited to fit the data.

COMPUTER PROBLEM SET 7.3

1. Use the MATLAB function explsqr to find the least squares exponential that best fit the data:

x	1	3	6	9	15
y	5.12	3.00	2.48	2.34	2.18

2. Use the MATLAB function hyplsqr to find the least squares exponential that best fit the data in the previous exercise.

7.4 TRIGONOMETRIC LEAST-SQUARES POLYNOMIAL

In this section we shall discuss how the least-squares principle may be applied to trigonometric curves of the form

$$p(x) = \frac{a_0}{2} + \sum_{k=1}^{m} a_k \cos(k\omega x) + \sum_{k=1}^{m} b_k \sin(k\omega x) \tag{7.15}$$

where ω is given. (7.15) is called a **trigonometric polynomial of order m**.

Suppose we want to approximate the data in Table 7.2 with the function (7.15) using the least-squares method. The procedure is handled in a similar manner and requires choosing the constants $a_0, \ldots, a_m, b_1, \ldots, b_m$ that minimize the least-squares error

$$E = \sum_{i=1}^{n} [p(x_i) - y_i]^2$$

$$= \sum_{i=1}^{n} y_i^2 - 2\sum_{i=1}^{n} p(x_i)y_i + \sum_{i=1}^{n} [p(x_i)]^2$$

$$= \sum_{i=1}^{n} y_i^2 - 2\sum_{i=1}^{n} \left[\frac{a_0}{2} + \sum_{k=1}^{m} a_k \cos(k\omega x_i) + \sum_{k=1}^{m} b_k \sin(k\omega x_i) \right] y_i$$

$$+ \sum_{i=1}^{n} \left[\frac{a_0}{2} + \sum_{k=1}^{m} a_k \cos(k\omega x_i) + \sum_{k=1}^{m} b_k \sin(k\omega x_i) \right]^2$$

where $E = E(a_0, \ldots, a_m, b_1, \ldots, b_m)$.

Proceeding as before, for E to be minimized, it is necessary that

$$\frac{\partial E}{\partial a_j} = 0 \quad \text{for} \quad j = 0, \ldots, m$$

and

$$\frac{\partial E}{\partial b_j} = 0 \quad \text{for} \quad j = 1, \ldots, m.$$

Taking the partial derivatives and rearranging terms, we get

$$\sum_{i=1}^{n} \left[\frac{a_0}{2} + \sum_{k=1}^{m} a_k \cos(k\omega x_i) + \sum_{k=1}^{m} b_k \sin(k\omega x_i) \right] = \sum_{i=1}^{n} y_i$$

$$\sum_{i=1}^{n} \left[\frac{a_0}{2} + \sum_{k=1}^{m} a_k \cos(k\omega x_i) + \sum_{k=1}^{m} b_k \sin(k\omega x_i) \right] \cos(j\omega x_i) = \sum_{i=1}^{n} \cos(j\omega x_i) y_i$$

(7.16)

$$\sum_{i=1}^{n} \left[\frac{a_0}{2} + \sum_{k=1}^{m} a_k \cos(k\omega x_i) + \sum_{k=1}^{m} b_k \sin(k\omega x_i) \right] \sin(j\omega x_i) = \sum_{i=1}^{n} \sin(j\omega x_i) y_i$$

for $j = 1, \ldots, m$.

These are the normal equations that form a linear system of $(2m + 1)$ equations in $(2m + 1)$ unknowns and that can be solved using Gaussian elimination. The derivation of the coefficients a_i and b_i is usually called **discrete Fourier analysis**.

EXAMPLE 7.8

Find the trigonometric least-squares polynomial that approximate the following data:

x	0.0	0.4	0.8	1.2	1.6	2.0	2.4	2.8	3.2	3.6	4.0	4.4
y	3.0	2.1	1.3	0.5	0	-0.2	0	0.3	1.1	2.0	2.9	3.5

using $m = 1$ and $\omega = 1$.

From Eqn. (7.16), we get the following normal equations

$$
\begin{aligned}
6a_0 + \left[\sum \cos x_i\right] a_1 + \left[\sum \sin x_i\right] b_1 &= \sum y_i \\
\left[\sum \cos x_i\right] a_0 + \left[\sum \cos^2 x_i\right] a_1 + \left[\sum \cos x_i \sin x_i\right] b_1 &= \sum y_i \cos x_i \\
\left[\sum \sin x_i\right] a_0 + \left[\sum \cos x_i \sin x_i\right] a_1 + \left[\sum \sin^2 x_i\right] b_1 &= \sum y_i \sin x_i
\end{aligned}
$$

with $\sum = \sum_{i=1}^{12}$.

By substituting the quantities

$$
\begin{aligned}
\sum \cos x_i &= -2.000867, & \sum \sin x_i &= 2.748837 \\
\sum \sin x_i \cos x_i &= 1.217139, & \sum y_i &= 16.5 \\
\sum \cos^2 x_i &= 6.393092, & \sum \sin^2 x_i &= 5.606908 \\
\sum y_i \sin x_i &= -4.339589, & \sum y_i \cos x_i &= -0.041184
\end{aligned}
$$

into the normal equations, we get the system

$$6.000000a_0 + -2.000867a_1 + 2.748837b_1 = 16.5$$
$$-2.000867a_0 + 6.393092a_1 + 1.217139b_1 = -0.041184$$
$$2.748837a_0 + 1.217139a_1 + 5.606908b_1 = -4.339589.$$

The solution of this system is

$$a_0 = 5.352575, \quad a_1 = 2.415546, \quad b_1 = -3.922482.$$

Thus,

$$p(x) = 2.6762875 + 2.415546\cos x - 3.922482\sin x.$$

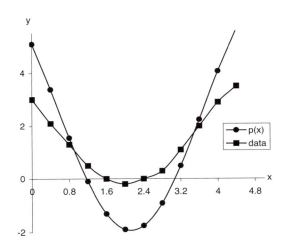

FIGURE 7.8
Plot of $f(x)$ and $p(x)$ in $[0, 4.4]$ for Example 7.8.

EXERCISE SET 7.4

1. Derive the normal equations for fitting the trigonometric curve

$$p(x) = a_0 + a_1\cos x + b_1\sin x.$$

Find the coefficients a_0, a_1, and b_1 that best fit the data

x	0	1	2	3
$f(x)$	5	54	99	181

in the least-squares sense.

Year	GNP(Current Dollars, $ millions)	GNP(Constant Dollars, $ millions)
1950	284.7	355.2
1955	398.1	438.1
1960	503.6	487.6
1965	684.8	617.7
1966	749.8	658.0
1967	763.4	674.5
1968	865.6	707.5

Table 7.9   GNP from 1950 to 1968.

APPLIED PROBLEMS FOR CHAPTER 7

1. Table 7.9 lists the Gross National Product (GNP) in current dollars and constant dollars. Current dollars are simply the dollar value with no adjustment for inflation. Constant dollars represent the GNP based on the value of the dollar in 1958.

   (a) Determine graphically which guess function seems best suited to fit these data and find the guess function using the method of least-squares.

   (b) Use the results of part (a) to predict the two GNPs in 1990 and 2000.

2. The following table lists the total water usage in the United States in billions of gallons per day.

Year	1930	1940	1950	1960	1970
Water Use	110.2	137.4	202.6	322.7	411.1

   (a) Find the least-squares exponential of the water consumption on time.

   (b) Use the results of part (a) to predict the water consumption in 1980 and 1990.

3. The following table lists the number of motor vehicle accidents in the United States for various years from 1950 to 1968.

Year	1950	1955	1960	1965	1966	1967	1968
No. Accidents (in thousands)	8,200	9,800	10,300	13,100	13,500	13,600	14,500
Accidents per 10^4 Vehicles	1,687	1,576	1,396	1,438	1,418	1,384	1,414

(a) Find the linear least-squares of the number of accidents on time. Use it to predict the number of accidents in 1990.

(b) Compute the quadratic least-squares of the number of accidents per 10,000 vehicles on time. Use it to predict the number of accidents per 10,000 vehicles in 1990.

4. A biologist is doing an experiment on the growth of a certain bacteria culture. After 4 hours the following data has been recorded:

t	0	1	2	3	4
p	1.0	1.8	3.3	6.0	11.0

where t is the number of hours and p the population in thousands. Determine the least-squares exponential that best fits these data. Use your results to predict the population of bacteria after 5 hours.

5. A small company has been in business for 3 years and has recorded annual profits (in thousands of dollars) as follows: 4 in the first year, 2 in the second year, and 1 in the third. Find a linear function that approximates the profits as a function of the years, in the least-squares sense.

6. A parachutist jumps from a plane and the distance of his drop is measured. Suppose that the distance of descent d as a function of time t can be modeled by

$$d = \alpha t + \beta t^2 e^{-0.1t}.$$

Find values of α and β that best fit the data in the table below

t	5	10	15	20	25	30
d	30	83	126	157	169	190

7. The values of the concentration C of a desired compound in terms of the time t (sec.) are given in the following table

t	3	9	12	18	24	30
C	4.1	4.3	3.9	3.4	3.1	2.7

Assuming that the guess function is $C(t) = c + ae^{-0.47t} + be^{-0.06t}$, find the values of a, b, and c that best fit this table.

8. The population p of a small city during the period $[1960, 2000]$ is given by the table

t	1960	1970	1980	1990	2000
p	12600	14000	16100	19100	23200

Use the least-squares quadratic to predict the population in the year 2010.

Chapter 8

Numerical Optimization

The single-variable optimization problem is of a central importance to optimization and practice. Not only because it is a type of problem that engineers encounter in practice, but because more general problem with n variables can often be solved by a sequence of one-variable problems. In this chapter we will introduce some basic numerical methods for locating the minimum of a function in one variable. We will only consider minimization problems with the understanding that maximizing $f(x)$ is the same as minimizing $-f(x)$. To introduce the idea consider the following example:

EXAMPLE 8.1 : **Corner Problem**
A pipe of length L and negligible diameter is to be carried horizontally around a corner from a hallway 3 ft. wide into a hallway 6 ft. wide (see Figure 8.1). What is the maximum length that the pipe can have?

L depends on the angle α. That is,

$$L(\alpha) = L_1 + L_2 = \frac{3}{\sin \alpha} + \frac{6}{\cos \alpha}.$$

So, the length of the longest pipe is given by the absolute minimum of $L(\alpha)$ on the interval $(0, \pi/2)$. The graph of $L(\alpha)$ shows that it has only one relative minimum; therefore, the absolute minimum of $L(\alpha)$ in $(0, \pi/2)$ is at its relative minimum. The approximate minimum $L(\alpha)$ of the above function can be obtained by one of the numerical methods described in this chapter.

8.1   ANALYSIS OF SINGLE-VARIABLE FUNCTIONS

DEFINITION 8.1   *(Increasing and Decreasing Function)*
Let f be a function defined on an interval I and let x_1 and x_2 be any two points on I.

(i) f is increasing on I if $x_1 < x_2$ implies that $f(x_1) < f(x_2)$.

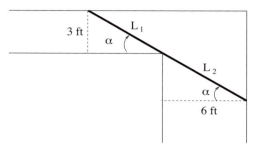

FIGURE 8.1
Corner problem.

(ii) f is decreasing on I if $x_1 < x_2$ implies that $f(x_1) > f(x_2)$.

THEOREM 8.1 (Increasing and Decreasing Function)
Let f be continuous on a closed interval $[a, b]$ and differentiable on (a, b).

(i) If $f'(x) > 0$ for all x in (a, b), then f is increasing on $[a, b]$.

(ii) If $f'(x) < 0$ for all x in (a, b), then f is decreasing on $[a, b]$.

DEFINITION 8.2   *(Local Extremum)*
Let c be in the domain of a function f.

(i) $f(c)$ is a local maximum of f if there is an open interval (a, b) containing c such that $f(x) \leq f(c)$ for any x in (a, b) that is in the domain of f.

(ii) $f(c)$ is a local minimum of f if there is an open interval (a, b) containing c such that $f(x) \geq f(c)$ for any x in (a, b) that is in the domain of f.

THEOREM 8.2 (Critical Values)
If a function f has a local extremum value at a number c in an open interval, then either $f'(c) = 0$ or $f'(c)$ does not exist. Such a number c is called a critical value of f.

THEOREM 8.3 (First Derivative Test)
Suppose f is continuous on an open interval that contains a critical value c.

(i) If f' changes sign from positive to negative at c, then f has a local maximum value at c.

(ii) If f' changes sign from negative to positive at c, then f has a local minimum value at c.

THEOREM 8.4 (Second Derivative Test)
Suppose f is twice differentiable at c.

(i) If $f'(c) = 0$ and $f''(c) > 0$, then f has a relative minimum at $x = c$.

(ii) If $f'(c) = 0$ and $f''(c) < 0$, then f has a relative maximum at $x = c$.

(iii) If $f'(c) = 0$ and $f''(c) = 0$, then the test is inconclusive.

DEFINITION 8.3   *(Unimodal Function)*
A function f is unimodal (literally "one mode") on an interval $[a, b]$, if it has a single minimum or maximum over the interval $[a, b]$.

Examples of unimodal functions that are respectively continuous, non differentiable, and discontinuous are shown in Figure 8.2.

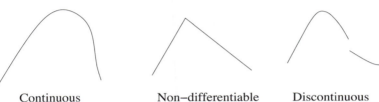

Continuous   Non–differentiable   Discontinuous

FIGURE 8.2
Unimodal functions.

EXERCISE SET 8.1

1. Determine the relative extrema of the following functions.

    (a) $f(x) = x^3 - 6x^2 + 12x + 9$,

    (b) $f(x) = x^{1/3} - x$,

    (c) $f(x) = x^2/(x^3 - 1)$.

2. Determine intervals on which the following functions are increasing and decreasing.

    (a) $f(x) = x^4 - x^2$,

    (b) $f(x) = x^2 + 1/x^2$,

    (c) $f(x) = xe^{-x}$,

    (d) $f(x) = x \ln x$.

8.2   LINE SEARCH METHODS

Line search procedures involve solving nonlinear optimization problems in a single dimension. We know from Theorem 8.2 that the critical points are given by values where either the derivative is equal to zero or does not exist. The second derivative test when applicable may then be employed to characterize the nature of these critical points. Problems can arise when either the first derivative fails to exist at some point $x = c$, or we are unable to solve the problem $f'(c) = 0$. In these cases, we can use a numerical approximation technique to approximate the optimal solution.

There are some common features shared by the algorithms of the line search methods that will be presented. Most of these algorithms require that the function be unimodal over a certain interval of interest.

8.2.1   Bracketing the minimum

The method for bracketing a minimum is similar to the bisection method for solving a nonlinear equation $f(x) = 0$. A root of a function is known to be bracketed by a pair of points when the function has opposite signs at those two points. A minimum, by contrast, is known to be bracketed only when there is a triplet of points. Suppose we have three points a, b, and c with $a < b < c$. If $f(b) < f(a)$ and $f(b) < f(c)$, then the interval (a, c) contains a minimum of $f(x)$. We say that the three points (a, b, c) *bracket the minimum.* If we now choose another point d in the interval (a, c) we can reduce the size of the interval bracketing the minimum. We have two cases to consider (see Figure 8.3).

- If $f(d) < f(b)$, then (a, d, b) brackets the minimum.

- If $f(d) > f(b)$, then (d, b, c) brackets the minimum.

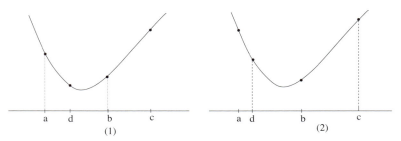

FIGURE 8.3
(1) $f(d) < f(b)$ **then** (a, d, b) **brackets the minimum.**

(2) $f(d) > f(b)$ **then** (d, b, c) **brackets the minimum.**

8.2.2   Golden section search

One of the most efficient methods for bracketing the minimum of a function in a given interval is the **golden section search**. It uses a good strategy for reducing the number of function evaluations.

Suppose we start with a triplet (a, b, c) bracketing the minimum of a function f. It remains to decide on a strategy for choosing the new point d:

Suppose b is located between a and c such that

$$\frac{b-a}{c-a} = r, \quad \frac{c-b}{c-a} = 1-r.$$

That is, b is a fraction r of the way between a and c. We now assume that the next point d is an additional fraction m beyond b,

$$\frac{d-b}{c-a} = m.$$

Therefore, the next bracketing segment will either be of length $r+m$ relative to the current one, or else of length $1-r$. If we require the size of the new interval to be the same in either case, then

$$m = 1 - 2r \tag{8.1}$$

which shows that $|b-a| = |c-d|$. When we move to the new subinterval, we require the same ratio so that

$$\frac{m}{1-r} = r. \tag{8.2}$$

By using Eqns. (8.1) and (8.2), we obtain the quadratic equation

$$r^2 - 3r + 1 = 0 \tag{8.3}$$

so that

$$r = \frac{3 - \sqrt{5}}{2} \approx 0.382.$$

Therefore, if the points b and d are chosen at relative positions r and $1 - r \approx 0.618$ in the interval (a, c), then the next subinterval will be r times the length of the original interval (as opposed to 0.5 for bisection). Hence, n reduction steps reduce the original interval by a factor

$$(1 - r)^n \approx (0.618)^n. \tag{8.4}$$

r is often referred to as the "golden ratio" and this optimal method is called the **golden section search**. Note that if the function is unimodal the method always converges to a local minimum and the convergence is linear, meaning that successive significant figures are obtained linearly with additional function evaluations.

In the preceding discussion we assumed that we have a triplet (a, b, c) bracketing the minimum of a function f. The assumption of unimodality makes golden section search a little easier to implement. An algorithm for finding the minimum of f using the golden section search method is as follows:

Suppose f is unimodal on $[a, b]$.
while $|a - b| > tol$ or $|f(b) - f(a)| > tol$
    $x_1 \leftarrow a + r(b - a)$
    $x_2 \leftarrow a + (1 - r)(b - a)$
    Compute $f(x_1)$ and $f(x_2)$
    If $f(x_1) < f(x_2)$ set $[a, b] = [x_1, b]$
    otherwise, set $[a, b] = [a, x_2]$

EXAMPLE 8.2
Use the golden section search method to find the minimum of the unimodal function $f(x) = \cos x - \sin x$ on $[1,3]$.

The graph of f shows that it has one relative minimum on $[1,3]$ and therefore f is unimodal on $[1,3]$.

Iteration 1. We start with the interval $[a,b] = [1,3]$. First, we choose x_1 and x_2 such that

$$x_1 = a + r(b-a) = 1 + r(3-1) = 1.7639320$$
$$x_2 = a + (1-r)(b-a) = 1 + (1-r)(3-1) = 2.2360680$$

and compute
$$f(x_1) = -1.1733444 \text{ and } f(x_2) = -1.4040220.$$

Since $f(x_1) > f(x_2)$ the new subinterval is

$$[a,b] = [1.7639320, 3].$$

Iteration 2. We continue in the same manner by choosing the next x_1 and x_2

$$x_1 = a + r(b-a) = 1.7639320 + r(3 - 1.7639329) = 2.2360680$$
$$x_2 = a + (1-r)(b-a) = 1.7639320 + (1-r)(3 - 1.7639329) = 2.5278640$$

and
$$f(x_1) = -1.4040220 \text{ and } f(x_2) = -1.3934259.$$

Since $f(x_1) < f(x_2)$ the new subinterval is

$$[a,b] = [1.7639320, 2.5278640].$$

Continuing in this manner leads to the values in Table 8.1 obtained by using the MATLAB function **golden**. The method converges after 39 iterations to the minimum

$$f(2.3561945) = -1.4142136$$

with tolerance values $tolx = toly = 10^{-8}$ for both the abscissas and the ordinates.

EXAMPLE 8.3
Given the unimodal function of the previous example defined in the interval $[1,3]$, find the number of iterations needed to locate the x value of the minimum to within a 0.3 range.

After n stages, the interval $[1,3]$ is reduced by $(1-r)^n = (0.61803)^n$. So, we choose n so that

$$(3-1)(0.61803)^n \leq 0.3.$$

Four stages of reduction will do, i.e., $n = 4$. Note that from Table 8.1, we have $b_4 - a_4 = 2.527864 - 2.236068 = 0.292 < 0.3$.

k	a	x1	x2	b	f(x1)	f(x2)
0	1.0000000	1.7639320	2.2360680	3.0000000	-1.1733444	-1.4040220
1	1.7639320	2.2360680	2.5278640	3.0000000	-1.4040220	-1.3934259
2	1.7639320	2.0557281	2.2360680	2.5278640	-1.3508548	-1.4040220
3	2.0557281	2.2360680	2.3475242	2.5278640	-1.4040220	-1.4141604
4	2.2360680	2.3475242	2.4164079	2.5278640	-1.4141604	-1.4116506
5	2.2360680	2.3049517	2.3475242	2.4164079	-1.4123572	-1.4141604
$\vdots$	$\vdots$	$\vdots$	$\vdots$	$\vdots$	$\vdots$	$\vdots$
37	2.3561945	2.3561945	2.3561945	2.3561945	-1.4142136	-1.4142136
38	2.3561945	2.3561945	2.3561945	2.3561945	-1.4142136	-1.4142136
39	2.3561945	2.3561945	2.3561945	2.3561945	-1.4142136	-1.4142136

Table 8.1 Minimum of $f(x) = \cos x - \sin x$ on $[1, 3]$ using the golden search method.

8.2.3 Fibonacci Search

A more sophisticated algorithm for finding the minimum of unimodal function is the Fibonacci search, where at each iteration the length of the interval is chosen according to the Fibonacci numbers defined recursively, by the equations

$$F_0 = 0$$
$$F_1 = 1$$
$$F_n = F_{n-1} + F_{n-2}, \quad n = 2, 3, \ldots.$$

n	2	3	4	5	6	7	8	...
F_n	1	2	3	5	8	13	21	...

The Fibonacci search method differs from the golden section search in that the values of r are not constant on each interval and it has the advantage that the number, n, of iterations is known a priori and based on the specified tolerance ε. The disadvantages of this method are that ε and n must be chosen a priori. An outline of the Fibonacci search algorithm is as follows:

Suppose we are given a function $f(x)$ that is unimodal on the interval $[a_0, b_0]$. The interior points c_k and d_k of the kth interval $[a_k, b_k]$ are found using the formulas

$$c_k = a_k + \left(\frac{F_{n-k-2}}{F_{n-k}}\right)(b_k - a_k)$$
$$d_k = a_k + \left(\frac{F_{n-k-1}}{F_{n-k}}\right)(b_k - a_k).$$

1. If $f(c_k) \leq f(d_k)$, then the new interval is $[a_{k+1}, b_{k+1}] = [a_k, d_k]$.

2. If $f(c_k) > f(d_k)$, then the new interval is $[a_{k+1}, b_{k+1}] = [c_k, b_k]$.

Note that this algorithm requires only one function evaluation per iteration after the initial step.

If the abscissa of the minimum is to be found with a tolerance value ε, then we want $\frac{1}{F_n}(b_0 - a_0) < \varepsilon$. That is, the number of iterations n is the smallest integer such that

$$F_n > \frac{b_0 - a_0}{\varepsilon}.$$

At the end of the required iterations, the estimated abscissa of the minimum is taken to be the midpoint of the final interval.

A useful formula for finding the sequence of Fibonacci numbers is given by

$$F_k = \frac{1}{\sqrt{5}} \left[\left(\frac{1 + \sqrt{5}}{2} \right)^k - \left(\frac{1 - \sqrt{5}}{2} \right)^k \right]. \tag{8.5}$$

EXAMPLE 8.4

Use the Fibonacci search method to approximate the minimum of the unimodal function $f(x) = \cos x - \sin x$ in $[1, 3]$ with a specified tolerance of $\varepsilon = 10^{-3}$.

The number of iterations n needed for the specified tolerance is the smallest integer such that

$$F_n > \frac{3 - 1}{10^{-3}} = 2.0 \times 10^3.$$

This implies that $n = 17$.

We have $[a_0, b_0] = [1, 3]$, the first interior points are

$$c_0 = 1 + \frac{F_{15}}{F_{17}}(3 - 1) = 1.763931888545$$

$$d_0 = 1 + \frac{F_{16}}{F_{17}}(3 - 1) = 2.236068111455$$

and

$$f(c_0) = f(1.763931888545) = -1.173344326399$$
$$f(d_0) = f(2.236068111455 = -1.404022030707.$$

Since $f(c_0) > f(d_0)$ then the new interval is

$$[a_1, b_1] = [1.763931888545, 3].$$

The next interval is obtained in the same manner that is

$$c_1 = 1.763931888545 + \frac{F_{14}}{F_{16}}(3 - 1.763\,931\,888545) = 2.236068111455$$

$$d_1 = 1.763931888545 + \frac{F_{15}}{F_{16}}(3 - 1.763931888545) = 2.527863777090.$$

Since $f(c_1) < f(d_1)$ then the new interval is

$$[a_2, b_2] = [1.763931888545, 2.527863777090].$$

We continue in this manner until we reach the last iteration, which leads to the interval

$$[a, b] = [2.35517817937519, 2.35661329451098].$$

So, the x and y coordinates of the estimated minimum are

$$x \min = \frac{2.35517817937519 + 2.35661329451098}{2} = 2.355895736943$$

$$f \min = f(2.355895736943) = -1.41421349926134.$$

A summary of the calculations obtained from the use of the MATLAB function fibonacci.m is given in Table 8.2.

n	a	c	d	b	f(c)	f(d)
17	1.0000000	1.7639319	2.2360681	3.0000000	-1.1733443	-1.4040220
16	1.7639319	2.2360681	2.5278641	3.0000000	-1.4040220	-1.3934259
15	1.7639319	2.0557282	2.2360681	2.5278641	-1.3508548	-1.4040220
14	2.0557282	2.2360681	2.3475244	2.5278641	-1.4040220	-1.4141604
⋮	⋮	⋮	⋮	⋮	⋮	⋮
5	2.3537431	2.3561349	2.3575700	2.3599617	-1.4142136	-1.4142122
4	2.3537431	2.3551782	2.3561349	2.3575700	-1.4142128	-1.4142136
3	2.3551782	2.3561349	2.3566133	2.3575700	-1.4142136	-1.4142134

Table 8.2 Minimum of $f(x) = \cos x - \sin x$ on $[1, 3]$ using the Fibonacci search method.

The Fibonacci search method has a linear convergence, since at each iteration the ratio between the interval containing the minimum and the new smaller interval is less than one. It can be shown that the asymptotic convergence rate is

$$\lim_{k \to \infty} \frac{F_{k+1}}{F_k} = \frac{\sqrt{5} - 1}{2}.$$

8.2.4 Parabolic Interpolation

Suppose we start with a triplet (a, b, c) bracketing the minimum of a function f. The x coordinate of the minimum point of the quadratic polynomial that interpolates the function f at the three points (a, b, c) is given by the formula (see Figure 8.4)

$$x = b - \frac{1}{2} \frac{(b-a)^2 [f(b) - f(c)] - (b-c)^2 [f(b) - f(a)]}{(b-a)[f(b) - f(c)] - (b-c)[f(b) - f(a)]}. \tag{8.6}$$

The value of x indicates the location where the slope of the fitted parabola is zero. We can then use this location to find the new bracket. The algorithm is as follows:

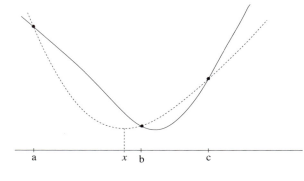

FIGURE 8.4
Parabolic interpolation.

1. If $x > b$ then

   (a) If $f(x) < f(b)$ then (b, x, c) brackets the minimum.

   (b) If $f(x) > f(b)$ then (a, b, x) brackets the minimum.

2. If $x < b$ then

   (a) If $f(x) < f(b)$ then (a, x, b) brackets the minimum.

   (b) If $f(x) > f(b)$ then (x, b, c) brackets the minimum.

Once the new bracketing interval is obtained, we continue in the same manner by finding the next quadratic polynomial that interpolates the function f at the three points (a, x, b), (x, b, c), (b, x, c), or (a, b, x). The iteration continues until a desired accuracy is obtained.

This iterating procedure is called the method of **successive parabolic interpolation**. Given three noncollinear points that bracket the minimum of $f(x)$, the method always converges and it has an order of convergence ≈ 1.324.

It is clear that the parabolic interpolation fails if the denominator of the formula is equal to zero. This happens if and only if the three given points are on the same line; that is, the three points are collinear. In this case, the information for parabolic interpolation is insufficient.

EXAMPLE 8.5

Use the successive parabolic interpolation method to find the minimum of the function $f(x) = \cos x - \sin x$ on $[1, 3]$.

The graph of f shows that $f(x)$ has one relative minimum on $[1, 3]$.

Let $(1, 2, 3)$ be the initial bracket for the minimum. Using Eqn. (8.6) with $a = 1, b = 2$, and $c = 3$, we get

$$x = 2 - \frac{1}{2} \frac{(2-1)^2 [f(2) - f(3)] - (2-3)^2 [f(2) - f(1)]}{(2-1)[f(2) - f(3)] - (2-3)[f(2) - f(1)]} = 2.3405296$$

and
$$f(x) = f(2.3405296) = -1.4140401, \quad f(b) = f(2) = -1.3254443$$

.

Since $x > b$ and $f(x) < f(b)$ the new bracket is

$$(b, x, c) = (2.0, 2.3405296, 3.0).$$

Using again Eqn. (8.6) with $a = 2, b = 2.305296$, and $c = 3$ we get the next $x = 2.3590150$ and the new bracket $(2.3405296, 2.3590150, 3.0)$. Continuing in this manner we obtain the values in Table 8.3, which converges after 5 iterations to the minimum

$$f(2.3561944902) = -1.4142135624$$

with tolerance values $tolx = toly = 10^{-8}$ for both the abscissas and the ordinates. As one can see the successive parabolic interpolation method is much faster than the golden search method since it took 39 iterations for the latter to converge to the same value.

i	a	b	c	x	f(x)	f(b)
0	1.0000000	2.0000000	3.0000000	2.3405296	-1.4140401	-1.3254443
1	2.0000000	2.3405296	3.0000000	2.3590150	-1.4142079	-1.4140401
2	2.3405296	2.3590150	3.0000000	2.3564163	-1.4142135	-1.4142079
3	2.3405296	2.3564163	2.3590150	2.3561945	-1.4142136	-1.4142135
4	2.3405296	2.3561945	2.3564163	2.3561945	-1.4142136	-1.4142136

Table 8.3 Minimum of $f(x) = \cos x - \sin x$ on $[1, 3]$ using the parabolic interpolation method.

A desirable feature of the line search methods described in this section is that they can be applied in cases when $f(x)$ is not differentiable. Here is an example.

EXAMPLE 8.6
Use the golden section search, Fibonacci search, and successive parabolic interpolation methods to approximate the minimum of the function $f(x) = |x^2 - 2| + |2x + 3|$ on $[-4, 0]$ with a tolerance value of 10^{-8} for both the abscissas and ordinates.

The graph of $f(x)$ shows that f is unimodal on $[-4, 0]$. It is easy to check that $f(x)$ is not differentiable at $x = -\sqrt{2}$ and $x = -3/2$. Table 8.4 gives the values of the approximate minimum for the three methods along with the number of iterations needed for convergence. The value of the exact minimum is $f(-\sqrt{2}) = 0.1715728753$.

Another and powerful method known as **Brent's method** is a particularly simple and robust method to find a minimum of a function $f(x)$ dependent on a single variable x. The minimum must initially be bracketed between two values $x = a$ and

Method	iterations	abscissa of min.	ordinate of min.
Golden search	41	-1.4142135593	0.1715728761
Fibonacci	42	-1.4142135610	0.1715728820
Parabolic interpolation	46	-1.4142135873	0.1715728959

Table 8.4   Use of the Golden, Fibonacci, and parabolic interpolation methods for the minimum of $f(x) = |x^2 - 2| + |2x + 3|$ on $[-4, 0]$.

$x = b$. The method uses parabolic interpolation as long as the process is convergent and does not leave the bracketing interval. The algorithm requires keeping track of six function points at all times, which are iteratively updated, reducing the minimum enclosing interval continually. The method combines parabolic interpolation through x_{min} with the golden section search (see [6] for more details).

EXERCISE SET 8.2

1. Use a graphing utility to generate the graph of the function

$$f(x) = \frac{x^4 + 1}{x^2 + 1}$$

   and use the graph to estimate the x-coordinate of the relative extrema.

2. The function

$$f(x) = cx^n e^{-x}$$

   where n is a positive integer and $c = 1/n!$ arises in the statistical study of traffic flow. Use a graphing utility to generate the graph of f for $n = 2, 3$, and 4, and make a conjecture about the number and locations of the relative extrema of f.

3. Find the point on the curve $x^2 + y^2 = 1$ that is closest to the point $(2, 0)$.

4. Find all points on the curve $x^2 - y^2 = 1$ closest to $(0, 2)$.

5. Find the point on the graph of $f(x) = x^2 + 1$ that is closest to the point $(3, 1)$.

6. Derive formula (8.6) by forcing the quadratic function $y = a_2 x^2 + a_1 x + a_0$ to pass through the points a, b, and c. (Hint: Use Lagrange interpolation)

7. Consider the function

$$f(x) = ax^2 + bx + c \quad (a > 0)$$

   and let

$$f_1 = f(\alpha - h), \ f_2 = f(\alpha), \ f_3 = f(\alpha + h)$$

M-function 8.2a

The following MATLAB function **golden.m** finds the relative minimum of a unimodal function *f* defined in an interval [a,b]. INPUTS are the end points *a* and *b* of the interval, the function *f*, a tolerance for the abscissas *tolx,* and a tolerance for the ordinates *toly. f* must be saved as an M-file.

```
function golden(f,a,b,tolx,toly)
% We assume that f is unimodal on [a,b].
% Golden Search for a Minimum of f on [a,b].
r=(3-sqrt(5))/2;
c=1-r;
x1=a+r*(b-a);
x2=a+c*(b-a);
f1=feval(f,x1);
f2=feval(f,x2);
k=0;
fprintf('\n')
disp(' Golden Section Search ')
fprintf('\n')
disp('_____')
disp(' k      a      x1      x2      b      f(x1)      f(x2) ')
disp('_____')
fprintf('\n')
while (abs(b-a) > tolx) | (abs(f2-f1) > toly)
fprintf('%2.f  %10.7f %10.7f %10.7f %10.7f %10.7f
%10.7f\n',k,a,x1,x2,b,f1,f2)
  if( f1< f2 )
    b=x2;
    x2=x1;
    x1=a+r*(b-a);
    f2=f1;
    f1=feval(f,x1);
  else
    a=x1;
    x1=x2;
    x2=a+c*(b-a);
    f1=f2;
    f2=feval(f,x2);
  end
  k=k+1;
end
fprintf( ' \n minimum = %14.10f ',f1)
fprintf(' at x = %14.10f ',b)
```

M-function 8.2b

The following MATLAB function **fibonacci.m** finds the relative minimum of a unimodal function *f* defined in an interval [a,b]. INPUTS are the function *f*, the end point *a* and *b* of the interval and a tolerance value *tol*. *f* must be saved as an M-file.

```
function fibonacci(f,a,b,tol)
% We assume that f is unimodal on [a,b]
% Fibocacci Search for the  minimum of f on [a,b]
fibonacci(1)=1;
fibonacci(2)=2;
n=2;
while ( fibonacci(n) <= (b-a)/tol )
    n=n+1;
    fibonacci(n)=fibonacci(n-1)+fibonacci(n-2);
end
c = b - (fibonacci(n-1)/fibonacci(n))*(b-a);
d = a + (fibonacci(n-1)/fibonacci(n))*(b-a);
fc = feval(f, c);
fd = feval(f, d);
fprintf('\n')
disp(' Fibonacci Search ')
fprintf('\n')
disp('_____')
disp(' n     a      c      d      b      f(c)    f(d) ')
disp('_____')
fprintf('\n')
for k = n:-1:3
    fprintf('%2.f  %10.7f  %10.7f  %10.7f  %10.7f  %10.7f
%10.7f\n',k,a,c,d,b,fc,fd)
  if (fc <= fd)
    b = d;
    d = c;
    fd = fc;
    c = a + (fibonacci(k-2)/fibonacci(k))*(b-a);
    fc = feval(f, c);
  else
    a = c;
    c = d;
    fc = fd;
    d = a + (fibonacci(k-1)/fibonacci(k))*(b-a);
    fd = feval(f, d);
   end
end
xmin=(a+b)/2;
fmin=feval(f,xmin);
fprintf('\n minimum = %14.10f',fmin)
fprintf(' at x = %14.10f',xmin)
```

M-function 8.2c

The following MATLAB function **parabint.m** finds the relative minimum of a function *f* given that *(a,b,c)* bracket a minimum of *f*. INPUTS are the points *a, b* and *c*, the function *f*, a tolerance for the abscissas *tolx,* a tolerance for the ordinates *toly,* and the maximum number of iterations itmax. *f* must be saved as an M-file.

```
function parabint(f,a,b,c,tolx,toly,itmax)
% Successive parabolic interpolation
% Given that (a,b,c) bracket a minimum of f(x).
fa = feval(f,a);
fb = feval(f,b);
fc = feval(f,c);
k=0;
fprintf('\n')
disp(' Successive Parabolic Interpolation ')
fprintf('\n')
disp('_____')
disp(' k     a       b       c       x       f(x)     f(b)  ')
disp('_____')
fprintf('\n')
deno=(fb-fc)*(b-a)-(fb-fa)*(b-c);
while ( abs(c-a)>tolx | abs(fc-fa)>toly ) & ( abs(deno)> eps ) & ( k <= itmax )
x = b - ((fb-fc)*(b-a)^2-(fb-fa)*(b-c)^2)/(2*deno);
fx = feval(f,x);
fprintf('%2.f  %10.7f %10.7f %10.7f %10.7f %10.7f %10.7f\n',k,a,b,c,x, fx,fb)
 if (x > b)
   if (fx > fb)
     c = x;
     fc = fx;
   else
     a = b;
     fa = fb;
     b = x;
     fb = fx;
   end
 else
   if (fx > fb)
     a  = x;
     fa = fx;
   else
     c = b;
     fc = fb;
     b = x;
     fb = fx;
   end
 end
 k=k+1;
 deno=(fb-fc)*(b-a)-(fb-fa)*(b-c);
end        % end while loop
fprintf(' \n minimum = %14.10f ',fx)
fprintf(' at x = %14.10f ',x)
```

where α and $h \neq 0$ are given.

Show that the minimum value $\tilde{f}$ of f is given by

$$\tilde{f} = f_2 - \frac{(f_1 - f_2)^2}{8(f_1 - 2f_2 + f_3)}.$$

8. Consider the function

$$f(x) = ax^2 + bx + c \quad (a > 0, \ b < 0).$$

Show that the minimizer x^* of f is given by

$$x^* = \frac{\bar{x}^2 f'(0)}{2(f'(0)\bar{x} + f(0) - f(\bar{x}))}$$

where $\bar{x} \in (-\infty, \infty)$ is given.

9. Evaluate formula (8.5) for $k = 0, 1, 2, 3$ and compare with the values that have been obtained from the definition and then prove it by induction.

COMPUTER PROBLEM SET 8.2

1. Use the MATLAB function golden to find the relative minimum of the function

$$f(x) = \frac{x^3 - x^2 - 8}{x - 1}, \quad \text{on } [-2, 0].$$

2. Use the MATLAB function parabint to find the relative minimum of the function

$$f(x) = \sin x + \cos x, \quad \text{on } [3, 6.5].$$

3. Use the MATLAB function golden to find the relative minimum of the function

$$f(x) = xe^x, \quad \text{on } [-2, 0].$$

4. Use the MATLAB function fibonacci to find the relative minimum of the function

$$f(x) = x \ln x, \quad \text{on } [0, 1].$$

5. Given the function

$$f(x) = e^x \cos x.$$

(a) Graph the function f and find an interval where the function has a relative minimum.

(b) Use both MATLAB functions parabint and golden to find this relative minimum.

(c) Compare both methods and see which method converges faster than the other.

8.3 MINIMIZATION USING DERIVATIVES

8.3.1 Newton's method

We know from calculus that a relative minimum of a differentiable function, $f(x)$, can be found by solving the nonlinear equation

$$f'(x) = 0.$$

This requires that we know the derivative of the function. If we use Newton's method to solve $f'(x) = 0$ then we need the second derivative as well. So, if f is twice continuously differentiable, Newton's method for approximating the minimum of a function f is given by

$$x_{n+1} = x_n - \frac{f'(x_n)}{f''(x_n)}, \quad n = 0, 1, 2, \ldots \tag{8.7}$$

where $f(x_n)$ is the approximate minimum of f, provided that the method converges. The observations made on Newton's method for solving nonlinear equations also apply to optimization. On the good side, if the initial guess x_0 is sufficiently close to the x-coordinate x_* of the relative minimum of f with $f''(x_*)$ different than zero, then the sequence x_n generated by Newton's method will converge quadratically to x_*. On the bad side, the disadvantage of Newton's method for solving $f'(x) = 0$ requires analytic derivatives, in this case f' and f''.

EXAMPLE 8.7
Use Newton's method to find the minimum of function $f(x) = \cos x - \sin x$ on $[1, 3]$.

The first and second derivatives of $f(x)$ are

$$f'(x) = -\sin x - \cos x, \quad f''(x) = -\cos x + \sin x.$$

We used the MATLAB function newton given in Section 3.5 to get after three iterations the minimum

$$f(2.35619449019234) = -1.414213562373095$$

with an accuracy of 14 decimal digits.

8.3.2 Secant method

Newton's method for minimizing f uses second derivatives of f. If the second derivative is not available, we may attempt to estimate it using first derivative information. In particular, we may approximate $f''(x_n)$ in (8.7) with

$$\frac{f'(x_n) - f'(x_{n-1})}{x_n - x_{n-1}}$$

which leads to the secant method

$$x_{n+1} = x_n - f'(x_n)\left[\frac{x_n - x_{n-1}}{f'(x_n) - f'(x_{n-1})}\right], \quad n = 1, 2, \ldots.$$

Observe that, like Newton's method, the secant method does not directly involve values of $f(x_n)$. The disadvantage of the use of the secant method to approximate the minimum of a function compared to Newton's method is that it needs two initial points to start it.

EXAMPLE 8.8
Use the secant method to find the minimum of function $f(x) = \cos x - \sin x$ on $[1, 3]$.

The derivatives of $f(x)$ is

$$f'(x) = -\sin x - \cos x.$$

We used the MATLAB function secant given in Section 3.4 to get after six iterations the minimum
$$f(2.35619449019234) = -1.414213562373095$$

with an accuracy of 14 decimal digits.

MATLAB's Methods

The built-in MATLAB function fminbnd minimizes a function of one variable on a fixed interval. The calling syntax of this function is:

```
x = fminbnd(fun,x1,x2)
```

where x is a local minimizer of the function fun, and x1, x2 are the endpoints of the interval of interest.

fminbnd(fun,x1,x2,options) minimizes with the optimization parameters specified in the structure options. We refer to the MATLAB help menu for the use of these parameters.

EXAMPLE 8.9
Use the MATLAB function fminbnd to find the minimum of $f(x) = \cos x - \sin x$ on $[1, 3]$.

First, we write an M-file that we call fun1.m containing

```
function f=fun1(x)
f = cos(x) - sin(x);
```

We now invoke fminbnd to get

```
>> x = fminbnd('fun1',1,3)
x =
        2.35619485859911
```

EXAMPLE 8.10

Solve the corner problem using the MATLAB function **parabint** (see Example 8.1).

fun2.m containing

get

08880218.

ind the minimum x^* of f over the interval

$, 0]$.

x^* with a tolerance value of $tol = 10^{-4}$.

Display all intermediate steps using a table as follows:

iteration	x_n	$f(x_n)$	$|x_n - x_{n-1}|$
1			
2			
⋮			

(c) Repeat (b) using the secant method with $tol = 10^{-5}$.

COMPUTER PROBLEM SET 8.3

1. Use secant method to find the relative minimum of the function

$$f(x) = \frac{x^3 - x^2 - 8}{x - 1}, \quad \text{on } [-2, 0].$$

2. Use Newton's method to find the relative minimum of the function

$$f(x) = \frac{x^4 + 1}{x^2 + 1}, \quad \text{on } [1/3, 3/2].$$

APPLIED PROBLEMS FOR CHAPTER 8

1. One end of a uniform beam of length L is built into a wall, and the other end is simply supported (see figure below). If the beam has a constant mass per unit length m, its deflection z from the horizontal at a distance x from the built-in end is given by

$$(47EI)z = -mg(2x^4 + 5Lx^3 - 3L^2x^2)$$

where E and I are constants depending on the material and cross section of the beam. Use a numerical method to find how far from the built-in end does maximum deflection occur. Use $L = 15$ ft, $E = 30,000$ kps/in^2, $I = 800$ in^4, and $mg = 30$ lbs/ft.

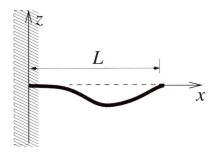

2. A closed rectangular box with a square base is to have a volume of 2250 in^3. The material for the top and bottom of the box will cost $2 per in^2, and the material for the sides will cost $3 per in^2. Find numerically the length of the side of the square base of the box with least cost.

3. A container with two square sides and an open top is to have a volume of 2000 cm^3. Use a numerical method to find the length of the square side of the box with minimum surface area.

4. A closed cylinder can is to hold 1000 cm^3 of liquid. How should we choose the radius of the cylinder to minimize the amount of material needed to manufacture the can? Use the golden section search method to find it.

5. When a shotputter projects a shot from height h above the ground, at a speed v, its range R is given by the equation

$$R = \frac{v^2 \cos \alpha}{g} \left(\sin \alpha + \sqrt{\sin^2 \alpha + \frac{2gh}{v^2}} \right),$$

where α is the angle of projection with the horizontal. Use the successive parabolic interpolation to find the angle α that maximizes R given $v = 13.7$ m/s, $h = 2.25$ m, and $g = 9.8$ m/s^2.

6. A wall 10 ft. high stands 30 ft. from a building. Find numerically the length of the shortest straight beam that will reach to the side of the building from the ground outside the wall (see figure below).

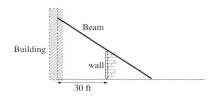

7. An open box is to be made by cutting out small squares from the corners of a 16-in.-by-30-in. piece of cardboard and bending up sides (see figure below). What size should the squares be to obtain a box with the largest volume? Use Newton's method.

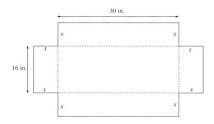

8. Two towns A and B lie on the south side of a river. A pumping station is to be located to serve the two towns. A pipeline will be constructed from the pumping station to each of the towns along the line connecting the town and the pumping station (see figure below). Use a numerical method to find the location C of the pumping station to minimize the amount of pipeline that must be constructed.

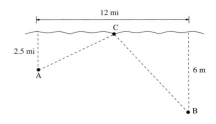

Chapter 9

Numerical Differentiation

In this chapter we investigate numerical techniques for estimating the derivatives of a function. Several formulas for approximating a first and second derivative by a difference quotient are given.

One important application of numerical differentiation is the development of algorithms for solving ordinary and partial differential equations.

EXAMPLE 9.1 : **Velocity of a Particle**

The distance s (meters) traveled by a particle moving along a coordinate line is given by the following table:

Time t (sec.)	0	2	4	6	8
s (m.)	1.00	2.72	7.38	20.08	54.59

One quantity of interest in Physics is the velocity of the particle at a given time t. It is given by $s'(t)$ and can be estimated by using numerical differentiation.

9.1 NUMERICAL DIFFERENTIATION

In many practical cases, we are faced with the problem of finding the derivative of a function whose form is either known only as a tabulation of data, or not practical to use for calculations. In this section, we will study some techniques for the numerical computation of first and second derivatives. These results will also be used to obtain the numerical solution of differential equations that will be presented in Chapters 12 and 15.

We begin with the series expansion of $f(x)$ about x. We assume that $f(x)$ has as many continuous derivatives as may be required. From Taylor's formula

$$f(x+h) = f(x) + hf'(x) + \frac{h^2}{2!}f''(x) + \frac{h^3}{3!}f'''(x) + \cdots. \tag{9.1}$$

Solving Eqn. (9.1) for $f'(x)$ yields

$$f'(x) = \frac{f(x+h) - f(x)}{h} - \frac{h}{2}f''(x) - \frac{h^2}{6}f'''(x) + \cdots \tag{9.2}$$

so that an approximation for $f'(x)$ may be written as

$$f'(x) \approx \frac{f(x+h) - f(x)}{h}, \quad \text{for a small value of } h. \tag{9.3}$$

The expression for $f'(x)$ in Eqn. (9.3) is called the **forward-divided difference** approximation. Graphically, it represents the slope of the line passing through the points $(x, f(x))$ and $(x+h, f(x+h))$. Another approximation for $f'(x)$ can be obtained by replacing h by $-h$ in Eqn. (9.1). That is,

$$f'(x) \approx \frac{f(x) - f(x-h)}{h}, \quad \text{for a small value of } h. \tag{9.4}$$

Eqn. (9.4) is called the **backward-divided difference** approximation to $f'(x)$.

Note that, from Eqn. (9.2), the **truncation error** for both the forward and backward-difference approximation to $f'(x)$ is

$$-\frac{h}{2}f''(\zeta)$$

for some ζ in the interval $(x, x+h)$, for the forward difference and for some ζ in $(x-h, x)$, for the backward difference.

One can see that for a linear function, the approximations are exact since the error term is zero in this case.

Again, if we replace h by $-h$ in Eqn. (9.1), and then subtract the resulting equation from the old one, we obtain a very popular formula for approximating $f'(x)$

$$f'(x) = \frac{f(x+h) - f(x-h)}{2h} - \frac{h^2}{3!}f'''(x) - \frac{h^4}{5!}f^{(5)} - \cdots. \tag{9.5}$$

That is,

$$f'(x) \approx \frac{f(x+h) - f(x-h)}{2h}, \quad \text{for a small value of } h. \tag{9.6}$$

Formula (9.6) is known as the **central-difference** approximation of $f'(x)$. Hence, we have an approximation to $f'(x)$ with an error of the order h^2.

EXAMPLE 9.2

Given $f(x) = e^x$, approximate $f'(1.5)$ using formulas (9.6) and (9.3) with $h = 0.1$. Compare the results with the exact value $f'(x) = e^{1.5}$.

Set $x = 1.5$ and $h = 0.1$ in (9.6) and (9.3) to get

$$f'(1.5) \approx \frac{e^{1.6} - e^{1.4}}{0.2} \approx 4.489162287752$$

$$f'(1.5) \approx \frac{e^{1.6} - e^{1.5}}{0.1} \approx 4.713433540571.$$

The absolute errors are

$$|e^{1.5} - 4.489162287752| = 0.007473$$
$$|e^{1.5} - 4.713433540571| = 0.231744.$$

Observe that the central-difference formula gives a better approximation than the forward-difference formula. We will now proceed to find approximations for the second derivative of $f(x)$. Again, consider the Taylor series expansion of $f(x)$ about x

$$f(x + h) = f(x) + hf'(x) + \frac{h^2}{2!}f''(x) + \frac{h^3}{3!}f^{(3)}(x) +$$
$$+ \frac{h^4}{4!}f^{(4)}(x) + \cdots \qquad (9.7)$$
$$f(x - h) = f(x) - hf'(x) + \frac{h^2}{2!}f''(x) - \frac{h^3}{3!}f^{(3)}(x) +$$
$$+ \frac{h^4}{4!}f^{(4)}(x) - \cdots . \qquad (9.8)$$

Add the two equations to get

$$f(x + h) + f(x - h) = 2f(x) + \frac{2h^2}{2!}f''(x) + \frac{2h^4}{4!}f^{(4)}(x) + \cdots .$$

We now solve for $f''(x)$ to obtain

$$f''(x) = \frac{f(x + h) - 2f(x) + f(x - h)}{h^2} - \frac{1}{12}h^2 f^{(4)}(x) + \cdots . \qquad (9.9)$$

Therefore, an approximation formula for $f''(x)$ is given by

$$f''(x) \approx \frac{f(x + h) - 2f(x) + f(x - h)}{h^2}, \quad \text{for a small value of } h. \qquad (9.10)$$

From (9.9) one can see that the truncation error is $-\frac{1}{12}h^2 f^{(4)}(\zeta)$ for some ζ in the interval $(x - h, x + h)$. Formula 9.10 is known as the **central-difference formula** for the second derivative.

EXAMPLE 9.3

Let $f(x) = \sin x$. Use formulas (9.3) and (9.10) with $h = 0.1$ to approximate $f'(0.5)$ and $f''(0.5)$. Compare with the true values, $f'(0.5) = 0.87758256$ and $f''(0.5) = -0.47942554$.

Using formula (9.3) we get

$$f'(0.5) \approx \frac{f(0.6) - f(0.5)}{0.1}$$

$$\approx \frac{0.56464247 - 0.47942554}{0.1} \approx 0.8521693$$

Error $= 0.02541326$.

Similarly, using formula (9.10), we get

$$f''(0.5) \approx \frac{f(0.6) - 2f(0.5) + f(0.4)}{(0.1)^2}$$

$$\approx \frac{0.56464247 - 0.95885108 + 0.38941834}{(0.1)^2} \approx -0.479027$$

Error $= 0.000399$.

MATLAB's Methods

MATLAB approximates the derivative of a function $f(x)$ with the use of the built-in function diff. We will demonstrate the uses of the diff function only for vectors. It can be used on matrices and gives the same thing for each column of the matrix that it does for a vector. For a vector with elements

$$\mathbf{x} = [x_1, x_2, ..., x_n]$$

diff computes the consecutive differences of the values of $\mathbf{x}$, that is:

$$\mathsf{diff}(\mathsf{x}) = [x_2 - x_1, x_3 - x_2, ..., x_n - x_{n-1}].$$

If we apply for example diff to the vector $\mathbf{x}$

```
>> x=[0.1    1.0    1.2    0.5    0.8    0.9];
```

```
>> diff(x)
ans =
        0.9000    0.2000    -0.7000    0.3000    0.1000
```

The derivative estimate of a function $y = f(x)$ is obtained by the quantity

$$\mathsf{diff}(\mathsf{y})./\mathsf{diff}(\mathsf{x})|_{x=x_n} \approx \frac{f(x_{n+1}) - f(x_n)}{x_{n+1} - x_n}$$

which is the forward-divided difference. As an application let us approximate the derivative of the function $y = \sin(x)$ at the entries of:

```
>> x = 0 : pi/32 : pi;
```

>> y = sin(x);

>> dy = diff(y);

>> dx = diff(x);

>> dsinx = dy./dx;

The length of each of the vectors dx, dy, dsinx is:

>> length(dsinx)

ans =

32

The entries of the vector dsinx are the estimates of the derivative of $\sin x$ at $x_1, x_2, \ldots, x_{32}$.

We will make our comparison with the exact values of the derivative $y' = \cos x$ over the same number of values of x using the command plot:

```
>> x = x(1:32);
>> plot(x,dsinx,'*',x,cos(x),'+')
>> xlabel('x')
>> text(1.5,.5,'+ Appx. with dt = pi/32')
>> text(1.5,.4,' . cos(x)')
```

The result is shown in Figure 9.1.

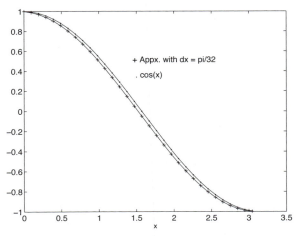

FIGURE 9.1
Numerical Approximation of the derivative of sin x.

EXAMPLE 9.4

Given $f(x) = e^x$, use the MATLAB function diff to approximate $f'(1.5)$ with $h = 0.1$.

The approximation can be obtained in the following way:

```
>> x = [1.5    1.5+0.1];
>> y = exp(x);
>> dx = diff(x);
>> dy = diff(y);
>> format long
>> dy./dx
ans =
        4.71343354057050
```

EXERCISE SET 9.1

1. Suppose a polynomial p_2 interpolates f at x_i, $i = 0, 1, 2$, where $x_i = a + ih$. By differentiating p_2, show that

$$p_2''(x) = \frac{f(x_1 + h) - f(x_1 - h)}{h^2}.$$

2. Derive the following approximate formulas:

   (a) $f'(x) \approx \frac{1}{4h} [f(x + 2h) - f(x - 2h)]$,

   (b) $f'(x) \approx \frac{1}{2h} [4f(x + h) - 3f(x) - f(x + 2h)]$.

3. Using Taylor's series, derive the following approximation for the third derivative of f

$$f'''(x) \approx \frac{1}{2h^3} [f(x + 2h) - 2f(x + h) + 2f(x - h) - f(x - 2h)].$$

4. Using Taylor's series, determine the error term for the approximate formula

$$f'(x) \approx \frac{1}{2h} [f(x + 2h) - f(x)].$$

5. Given $f(x) = e^x$, approximate $f'(1)$ using the central-difference formula with $h = 0.1$.

6. Using Taylor's series, determine the error term for the approximate formula

$$f'(x) \approx \frac{1}{5h} [f(x + 4h) - f(x - h)].$$

7. Use Taylor's series to derive the following approximation formula for the third derivative of f.

$$f'''(x) \approx \frac{1}{h^3}\left[-f(x) + 3f(x+h) - 3f(x+2h) + f(x+3h)\right].$$

8. Use Taylor's series to derive the following approximation formula for the first derivative of f.

$$f'(x) \approx \frac{2f(x+3h) - 9f(x+2h) + 18f(x+h) - 11f(x)}{6h}.$$

9. Show that the approximation formula in Exercise 8 has an error of $O(h^3)$.

10. What does the difference scheme approximates. Give its error order?

$$\frac{1}{2h}[f(x+3h) + f(x-h) - 2f(x)].$$

11. Derive a difference formula for $f''(x_0)$ through $f(x_0)$, $f(x_0-h)$, and $f(x_0+2h)$ and find the leading error term.

12. Show the approximation

$$f'(x) = \frac{8f(x+h) - 8f(x-h) - f(x+2h) + f(x-2h)}{12h}.$$

13. Compute the derivative of $f(x) = \sin x$ at $\pi/3$ using

(a) The forward-difference formula with $h = 10^{-3}$,

(b) The central-difference formula with $h = 10^{-3}$,

(c) The formula

$$f'(x_0) \approx \frac{f(x_0 + \alpha h) + (\alpha^2 - 1)f(x_0) - \alpha^2 f(x_0 - h)}{\alpha(1 + \alpha)h}$$

with $h = 10^{-3}$ and $\alpha = 0.5$.

14. Using the following data find $f'(6.0)$ with error $= O(h)$, and $f''(6.3)$, with error $= O(h^2)$

x	6.0	6.1	6.2	6.3	6.4
$f(x)$	0.1750	−0.1998	−0.2223	−0.2422	−0.2596

15. Define

$$S(h) = \frac{-f(x+2h) + 4f(x+h) - 3f(x)}{2h}.$$

(a) Show that

$$f'(x) - S(h) = c_1 h^2 + c_2 h^3 + c_3 h^4 + \dots$$

and state c_1.

(b) Compute $f'(0.398)$ using $S(h)$ and the table

x	0.398	0.399	0.400	0.401	0.402
$f(x)$	0.408591	0.409671	0.410752	0.411834	0.412915

16. The following table contains values of $f(x)$ in $[-6, -1]$

x	-6	-5	-4	-3	-2	-1
$f(x)$	932	487	225	89	28	6

Approximate $f''(-4)$ and $f''(-3)$ using the central-difference formula.

9.2   RICHARDSON'S FORMULA

In this section, we shall develop a technique for approximating derivatives of a function f that will enable us to reduce the truncation error. This technique is known as the **Richardson extrapolation** and has special utilities in computer programs for differentiation and integration of arbitrary functions.

To illustrate the method, let us consider Eqn. (9.5) in the form

$$f'(x) = \frac{f(x+h) - f(x-h)}{2h} + c_2 h^2 + c_4 h^4 + \cdots \tag{9.11}$$

where $c_2, c_4, \dots$ depend on f and x.

For x fixed, we define the function

$$a(h) = \frac{f(x+h) - f(x-h)}{2h}. \tag{9.12}$$

Now, we rewrite (9.11) as

$$f'(x) = a(h) + c_2 h^2 + c_4 h^4 + \cdots. \tag{9.13}$$

The idea for improving the truncation error is to try to eliminate the term $c_2 h^2$ in Eqn. (9.13). To do so, we replace h by $h/2$ in (9.13) to get

$$f'(x) = a\left(\frac{h}{2}\right) + c_2 \left(\frac{h}{2}\right)^2 + c_4 \left(\frac{h}{2}\right)^4 + \cdots. \tag{9.14}$$

We can now eliminate the terms in h^2 by multiplying Eqn. (9.14) by four and subtracting Eqn. (9.13) from it. The result is

$$f'(x) = a\left(\frac{h}{2}\right) + \frac{1}{3}\left[a\left(\frac{h}{2}\right) - a(h)\right] - \frac{1}{4}c_4 h^4 + \cdots. \qquad (9.15)$$

One can see that we have improved the accuracy of our estimate of the derivative by reducing the error to an order h^4. The new approximation for $f'(x)$ is given by

$$f'(x) \approx a\left(\frac{h}{2}\right) + \frac{1}{3}\left[a\left(\frac{h}{2}\right) - a(h)\right]. \qquad (9.16)$$

This extrapolation method can be extended since an estimate of order h^4 can be shown to have an error of order h^6 and so on.

EXAMPLE 9.5
Given $f(x) = e^x$, approximate $f'(1.5)$ using formula (9.16) with $h = 0.1$. Compare with the true value $e^{1.5}$.

Using formula (9.12) with $x = 1.5$ and $h = 0.1$, we get

$$a(h) = a(0.1) = \frac{f(1.6) - f(1.4)}{0.2} \approx 4.489162287752$$

$$a(h/2) = a(0.05) = \frac{f(1.55) - f(1.45)}{0.1} \approx 4.483556674219.$$

In view of (9.16), we obtain

$$f'(1.5) \approx 4.483556674219 + \frac{1}{3}(4.483556674219 - 4.489162287752)$$
$$\approx 4.481688136375.$$

The error is 9.3×10^{-7}. As we can see, formula (9.16) gives a better approximation to $f'(1.5)$ than formula (9.6) for $h = 0.1$.

In general, given an approximation $a(h)$ and having computed the values

$$D_{n,1} = a\left(\frac{h}{2^{n-1}}\right), \quad n = 1, 2, \ldots$$

for an appropriate $h > 0$, the process can be extended to m columns. Such columns and approximations are generated recursively by the formula:

$$D_{n,m+1} = \frac{4^m}{4^m - 1}D_{n,m} - \frac{1}{4^m - 1}D_{n-1,m}. \qquad (9.17)$$

The truncation error associated with the entry $D_{n,m+1}$ is of order $O(h^{2m+2})$.

The procedure is best illustrated by arranging the quantities in a table of the form shown in Table 9.1.

$$
\begin{array}{llllll}
D_{1,1} & & & & \\
D_{2,1} & D_{2,2} & & & \\
D_{3,1} & D_{3,2} & D_{3,3} & & \\
\vdots & \vdots & \vdots & \ddots & \\
D_{N,1} & D_{N,2} & D_{N,3} & \cdots & D_{N,N}
\end{array}
$$

Table 9.1 Two-dimensional triangular array $D_{N,N}$.

Observe that, since round-off error affects calculation by computers, computing $D_{N,N}$ for large values of N does not result, in general, to better accuracy of $D_{1,1}$. Therefore, one should put a limit on the number of repetitions of the process. The answer to the question, when one should stop the process in order to get the best approximation, is not known. On one hand, we want h small for accuracy, but on the other hand, we want h large for stability. One can empirically check whether the truncation error formula is being maintained from one level to the next. When this fails, the extrapolations should be broken off.

EXAMPLE 9.6
Given $f(x) = e^x$ and $h = 0.25$, compute $D_{6,6}$ to approximate $f'(1)$.

We have

$$
D_{1,1} = a(h) = \frac{e^{1.25} - e^{0.75}}{0.5} \approx 2.7466858816
$$

$$
D_{2,1} = a(h/2) = \frac{e^{1.125} - e^{0.875}}{0.25} \approx 2.7253662198.
$$

Using formula (9.17), we get

$$
D_{2,2} = \frac{4}{4-1} D_{2,1} - \frac{1}{4-1} D_{1,1} \approx 2.7182596658.
$$

Continuing in this manner leads to the values in Table 9.2.
The MATLAB function f2 used in this table is defined as follows:

```
function f=f2(x)
  f=exp(x);
```

EXERCISE SET 9.2

1. Let $f(x) = \ln x$. Approximate $f'(1)$ for $h = 0.1$ and $h = 0.01$.

```
» derive('f2',0.25,1,6)
   Derivative table
```

I	h	Di,1	Di,2	Di,3	....		
1	0.250000	2.746686					
2	0.125000	2.725366	2.718260				
3	0.062500	2.720052	2.718280	2.718282			
4	0.031250	2.718724	2.718282	2.718282	2.718282		
5	0.015625	2.718392	2.718282	2.718282	2.718282	2.718282	
6	0.007813	2.718309	2.718282	2.718282	2.718282	2.718282	2.718282

Table 9.2 Approximation of the derivative of $f(x) = e^x$ at $x = 1$ using Richardson's formula.

2. Let $a > 1$ be a given number. We consider the function

$$f(x) = a^x.$$

Using the fact that $f'(0) = \ln a$, find an approximation to $\ln 6$ by using the derivative table in Example 9.6.

3. Repeat the previous exercise to approximate $\ln 3$ to five decimal digits.

4. Given the approximation formula

$$K(x) = \frac{f(x + 3h) - f(x - h)}{4h}$$

show that

$$f'(x) - K(x) = c_1 h + c_2 h^2 + \cdots$$

and determine c_1.

5. Prove that the approximation formula

$$f'(x) \approx \frac{4f(x + h) - 3f(x) - f(x + 2h)}{2h}$$

has an error that can be written as

$$f'(x) - \frac{4f(x + h) - 3f(x) - f(x + 2h)}{2h} = c_1 h^2 + c_2 h^3 + \cdots.$$

Determine c_1 and c_2.

6. Using the following data, find $f'(6.0)$, error $= O(h)$, and $f''(6.3)$, error $= O(h^2)$.

x	6.0	6.1	6.2	6.3	6.4
$f(x)$	-0.1750	-0.1998	-0.2223	-0.2422	-0.2596

M-function 9.2

The following MATLAB function **derive.m** approximate the derivative of a function at a given point using formula 9.17 and the central difference approximation. INPUTS are a function f; a value of h; a specific point a; the number of rows n. The input function $f(x)$ should be defined as an M-file.

```
function derive(f,h,a,n)
% Approximate the derivative of a function at x = a.
disp('     Derivative table')
disp('_____')
disp('  i        h         Di,1        Di,2        Di,3    ... ')
disp('_____')
D(1,1)=(feval(f,a+h)-feval(f,a-h))/(2*h);
fprintf('%2.0f %8.4f %12.4f\n',1,h,D(1,1));
for i=1:n-1
  h=h/2;
  D(i+1,1)=(feval(f,a+h)-feval(f,a-h))/(2*h);
  fprintf('%2.0f %8.4f %12.4f',i+1,h,D(i+1,1));
  for k=1:i
    D(i+1,k+1)=D(i+1,k)+(D(i+1,k)-D(i,k))/((4^k)-1);
    fprintf('%12.4f',D(i+1,k+1));
  end
  fprintf('\n');
end
```

7. Define
$$G(h) = \frac{-f(x+2h) + 4f(x+h) - 3f(x)}{2h}$$

(a) Show that
$$f'(x) - G(h) = c_1 h^2 + c_2 h^3 + c_3 h^4 + \dots$$
and state c_1.

(b) Compute $f'(0.398)$ using the table below and $G(h)$.

x	0.398	0.399	0.400	0.401	0.402
$f(x)$	0.40859	0.40967	0.41075	0.41183	0.41292

8. Given the difference quotient known as the central-difference formula
$$D_h f = \frac{f(x+h) - f(x-h)}{2h} + O(h^2)$$

show that
$$\frac{4 D_h f - D_{2h} f}{3}$$

is a fourth-order approximation to $f'(x)$ using Taylor's series. Assume that f is six times continuously differentiable.

9. Given the continuous smooth function $g(x)$ for which $g(0) = 8$, $g(1) = 5$, $g(2) = 3$, $g(3) = 2$, and $g(4) = 3$

   (a) Use a central-difference scheme to approximate $g''(2)$.

   (b) Use the Richardson extrapolation to improve this result.

10. The following data gives approximations to the integral $I = \int_a^b f(x)dx$ for a scheme with error terms $E = K_1 h + K_2 h^3 + K_3 h^5 + \ldots$

$$I(h) = 2.3965, \quad I(h/3) = 2.9263, \quad I(h/9) = 2.9795.$$

   Construct an extrapolation table to obtain a better approximation.

11. Consider the table

x	1	2	3	4	5
$f(x)$	0.01	0.68	1.11	1.38	1.61

   (a) Approximate $f'(3)$ using the central-difference formula with $h = 2$.

   (b) Approximate $f''(3)$ using the central-difference formula with $h = 1$.

   (c) Use the Richardson extrapolation to improve the results in (a).

12. Consider the data in the table

x	1.0	1.1	1.2	1.3	1.4	1.5	1.6	1.7	1.8
$f(x)$	1.544	1.667	1.811	1.972	2.152	2.351	2.576	2.828	3.107

   (a) Approximate $f'(1.4)$ using the forward-difference formula with $h = 0.1, 0.2$.

   (b) Use the Richardson extrapolation to improve the results.

13. Compute the derivative of $f(x) = \sin x$ at $x = \pi/4$ using the Richardson extrapolation. Start with $h = 1$ and find the number of rows in the Richardson table required to estimate the derivative with six significant decimal digits.

COMPUTER PROBLEM SET 9.2

1. Write a computer program in a language of your choice that finds the first derivative of a function f at a given value α of x, using the central-difference approximation and Richardson's formula. Input data to the program should be a function f, α, the value of h, and the number of rows n.

   Test your program to find the derivative of $f(x) = e^{2x} \sin x$ at $x = 0.1, 0.5, 1, 1.5$. Compare the result with the exact answer.

2. Use the MATLAB function **derive** to approximate the derivative of the following functions:

(a) $f(x) = 3x^3 - 5x^2$ with $h = 0.2$, $N = 4$, at $x = 2.6$,

(b) $f(x) = \sqrt{x^2 + 1}$ with $h = 0.5$, $N = 4$, at $x = 1.3$,

(c) $f(x) = x \cos x - 3x$ with $h = 0.2$, $N = 4$, at $x = 1.4$,

(d) $f(x) = \sin x + x$ with $h = 0.3$, $N = 4$, at $x = \pi/4$,

(e) $f(x) = xe^x - x^2$ with $h = 0.2$, $N = 4$, at $x = 2.3$.

3. Use the MATLAB function derive to estimate the derivative of $f(x) = \arctan x$ at $x = 1$.

4. Use the MATLAB function derive to estimate the derivative of $f(x) = e^{x^2+1}$ at $x = 1$, $x = 2$, and $x = 3$. Compare with the exact values.

APPLIED PROBLEMS FOR CHAPTER 9

1. The accompanying table gives the distance $d(t)$ of a bullet at various points from the muzzle of a rifle. Use these values to approximate the speed of the bullet at $t = 60, 90$, and 120 using the central-difference formula.

$t(s)$	0	30	60	90	120	150	180
$d(m)$	0	100	220	350	490	600	750

2. A graph of the speed versus time t for a test run of an automobile is shown in the accompanying figure. Estimate the acceleration a at $t = 10, 15$, and 20 (Hint: $a(t) = v'(t)$).

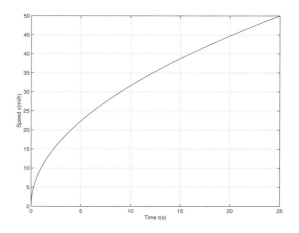

3. The graph below shows the amount of water in a city water tank during one day when no water was pumped into the tank. Approximate the rate of change of water usage at $t = 8, 12$ and 16.

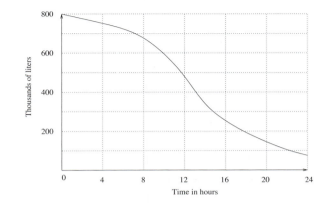

4. The following table gives the normal high temperature T for a city as a function of time (measured in days since January 1).

Days of the year	1	30	61	91	121	151	181	211	241	271
T	38	39	50	62	73	80	88	90	80	75

What is the approximate rate of change in the normal high temperature on March 2 (day number 61) and July 30 (day number 211).

5. The accompanying figure gives the speed, in miles per second, at various times of a test rocket that was fired upward from the surface of the Earth. Use the graph to approximate the acceleration of the rocket at $t = 2, 4$, and 6.

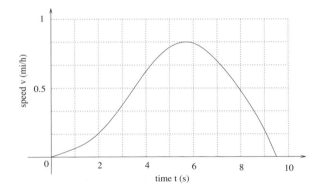

6. Water is flowing from a reservoir shaped like a hemisphere bowl of radius 12 m (see figure below). The volume of water is given by

$$V(x) = \frac{\pi}{3}x^3(3r - x).$$

The rate of change of the volume of water is

$$dV/dt = (dV/dx)(dx/dt).$$

Assume that $dx/dt = 1$ and approximate $dV/dt = dV/dx$ when $x = 3$ using the central-difference formula with $h = 0.1, 0.01$, and 0.001.

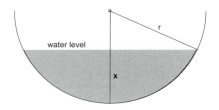

7. Let $f(x) = \sqrt{x}$ be given by the following table that has five-decimal digits of accuracy.

x	1.2	1.4	1.6	1.8	2.0
$\sqrt{x}$	1.09545	1.18322	1.26491	1.34164	1.41421
x	2.2	2.4	2.6	2.8	
$\sqrt{x}$	1.48324	1.54919	1.61245	1.673320	

Approximate $f'(x)$ at $x = 1.4, 1.8, 2.2,$ and 2.4 using the forward-difference formula with $h = 0.2$. Compare your results with the values of the exact ones.

Chapter 10

Numerical Integration

This chapter deals with the problem of estimating the definite integral

$$I = \int_a^b f(x)\,dx \qquad (10.1)$$

with $[a, b]$ finite.

In most cases, the evaluation of definite integrals is either impossible, or else very difficult, to evaluate analytically. It is known from calculus that the integral I represents the area of the region between the curve $y = f(x)$ and the lines $x = a$ and $x = b$. So the basic idea of approximating I is to replace $f(x)$ by an interpolating polynomial $p(x)$ and then integrate the polynomial by finding the area of the region bounded by $p(x)$, the lines $x = a$ and $x = b$, and the x-axis. This process is called **numerical quadrature**. The integration rules that we will be developing correspond to the varying degree of the interpolating polynomials. The most important are the ones with degree one, two, and three. However, care must be exercised when using interpolating polynomials of a higher degree; round-off errors and irregularities can cause a problem in these cases.

EXAMPLE 10.1 : The Nonlinear Pendulum

A simple pendulum is displaced initially by an angle θ_0 and then released (see Figure 10.1). We assume that the string is weightless and that no other frictional forces are present. The

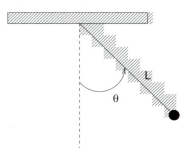

FIGURE 10.1
Pendulum.

period, T, for the general nonlinear pendulum problem is given by the integral

$$T(\theta_0) = 2\sqrt{2}\,\sqrt{\frac{L}{g}} \int_0^{\theta_0} \frac{d\theta}{\sqrt{\cos\theta - \cos\theta_0}}$$

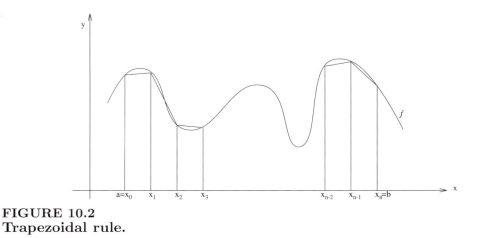

FIGURE 10.2
Trapezoidal rule.

where L is the length of the pendulum, g is acceleration due to gravity, and θ is the angle between the vertical and the string.

There is no closed-form analytical expression for the integral $T(\theta_0)$. So, one has to resort to a numerical method in order to estimate the integral for a given value of θ_0.

10.1   TRAPEZOIDAL RULE

One of the simplest methods of finding the area under a curve, known as the **trapezoidal rule**, is based on approximating $f(x)$ by a piecewise linear polynomial that interpolates $f(x)$ at the nodes $x_0, x_1, \ldots, x_n$.

In general, suppose we wish to evaluate the integral

$$I = \int_a^b f(x)\, dx.$$

We begin by dividing the interval $[a, b]$ into n subintervals, each with size h as shown in Figure 10.2, where

$$h = \frac{b-a}{n} \quad \text{and} \quad x_i = a + ih \quad \text{for} \ \ i = 0, 1, 2, \ldots, n.$$

We now consider each subinterval $[x_{i-1}, x_i]$, $i = 1, 2, \ldots, n$ and approximate the area under the curve $f(x)$ by a trapezoid with heights $f(x_{i-1})$ and $f(x_i)$. The area of the ith trapezoid is therefore

$$A_i = \frac{h}{2} \left[f(x_{i-1}) + f(x_i) \right]. \tag{10.2}$$

The total area T_n is obtained by extending (10.2) over the entire interval $[a, b]$, that is

$$T_n = A_1 + A_2 + \cdots + A_{n-1} + A_n$$

$$= \frac{h}{2}[f(x_0) + f(x_1)] + \frac{h}{2}[f(x_1) + f(x_2)] + \cdots$$
$$+ \frac{h}{2}[f(x_{n-2}) + f(x_{n-1})] + \frac{h}{2}[f(x_{n-1}) + f(x_n)].$$

By collecting common terms, we obtain the formula

$$T_n = \frac{h}{2}[f(x_0) + f(x_n)] + h\sum_{i=1}^{n-1} f(x_i). \tag{10.3}$$

Eqn. (10.3) is known as the **composite trapezoidal rule** for n subintervals, so called because it approximates the integral I by the sum of n trapezoids.

EXAMPLE 10.2

Use the composite trapezoidal rule with $n = 2$ to approximate

$$\int_0^3 x^2 e^x \, dx.$$

Using (10.3) with $n = 2$, we have

$$\int_0^3 x^2 e^x \, dx \approx \frac{3}{4}[f(0) + f(3)] + \frac{3}{2}f(1.5)$$
$$\approx 150.70307.$$

One should note that formula (10.3) gives an approximation to the integral Eqn. (10.1) with no information on the accuracy of the trapezoidal method. Because of the round-off error, taking a large number of trapezoids will not result in general to a better approximation. Since we usually do not have prior knowledge of the exact answer, it is hard to find the number of trapezoids that give the best approximation. One method often used, is by computing the integral several times with an increasing number of trapezoids, and then stoping when the difference between two successive answers is satisfactorily small.

EXAMPLE 10.3

Use the MATLAB function trapez with $n = 5$ to approximate the integral

$$\int_0^1 (7 + 14x^6) \, dx = 9.$$

```
» trapez('f1',0,1,5)
```

i	xi	f(xi)	h=0.2
0	0.0	7	
1	0.2	7.000896	
2	0.4	7.057344	
3	0.6	7.653184	
4	0.8	10.670016	
5	1.0	21	

The integral of f(x) is =   9.27628800

Table 10.1   The composite trapezoid rule for $f(x) = 7 + 14x^6$ with $n = 5$.

Before calling the function **trapez**, we first define the MATLAB function f1 as follows:

```
function f=f1(x)
f=7+14*x.^6;
```

The results are shown in Table 10.1. Therefore, the trapezoidal rule estimate with $h = \frac{1-0}{5}$ is

$$T_5 = \frac{1}{10}[f(0) + f(1)] + \frac{1}{5}[f(\frac{1}{5}) + f(\frac{2}{5}) + f(\frac{3}{5}) + f(\frac{4}{5})]$$

$$\approx \frac{1}{10}[7 + 21] + \frac{1}{5}[7.000896 + 7.057344 + 7.653184 + 10.670016]$$

$$\approx 9.276288.$$

Truncation Error

Neglecting the round-off error, let us now study the accuracy of the composite trapezoidal rule by analyzing the error in using it to estimate the integral (10.1), known as the truncation error.

By the fundamental theorem of calculus we have

$$I_i = \int_{x_{i-1}}^{x_i} f(x)\,dx = F(x_i) - F(x_{i-1}) \tag{10.4}$$

where $F'(x) = f(x)$.

Consider the Taylor series expansion of $F(x)$ about x_i. We assume that $f(x)$ has as many continuous derivatives as may be required.

$$F(x_{i-1}) = F(x_i - h)$$

$$= F(x_i) - hF'(xi) + \frac{h^2}{2!}F''(xi) - \frac{h^3}{3!}F'''(x_i) + \cdots$$

or

$$F(x_i) - F(x_{i-1}) = hF'(x_i) - \frac{h^2}{2!}F''(x_i) + \frac{h^3}{3!}F'''(x_i) - \cdots$$
$$= I_i.$$

Since $F'(x) = f(x)$, we have

$$I_i = hf(x_i) - \frac{h^2}{2!}f'(x_i) + \frac{h^3}{3!}f''(x_i) - \cdots. \tag{10.5}$$

It was found in Section 9.1 that

$$f'(x_i) = \frac{f(x_i) - f(x_{i-1})}{h} + \frac{h}{2}f''(x_i) + \cdots. \tag{10.6}$$

Substituting (10.6) into (10.5) yields

$$I_i = \frac{h}{2}\left[f(x_i) + f(x_{i-1})\right] - \frac{h^3}{12}f''(x_i) + \cdots$$
$$= A_i - \frac{h^3}{12}f''(x_i) + \cdots.$$

Hence, the error term E_i for the ith trapezoid is

$$E_i = I_i - A_i$$
$$= -\frac{h^3}{12}f''(x_i) + \text{ higher-order terms.}$$

For h small, we neglect the higher-order terms to get

$$E_i = -\frac{h^3}{12}f''(x_i). \tag{10.7}$$

If $|f''(x)| \le M$ for all x in the interval $[a, b]$, then

$$|E_i| \le \frac{h^3}{12}M. \tag{10.8}$$

Applying (10.8) over the entire interval, we obtain the total error E_T

$$|E_T| \le n\frac{h^3}{12}M = (b - a)\frac{h^2}{12}M. \tag{10.9}$$

Hence, in view of (10.7), as we expected, the error term for the composite trapezoidal rule is zero if $f(x)$ is linear; that is, for a first-degree polynomial, the trapezoidal rule gives the exact result.

EXAMPLE 10.4
Determine the number of subintervals n required to approximate

$$\int_0^2 \frac{1}{x+4}\, dx$$

with an error E_T less than 10^{-4} using the composite trapezoidal rule.

We have
$$|E_T| \leq \frac{b-a}{12} h^2 M \leq 10^{-4}.$$
In this example, the integrand is $f(x) = 1/(x + 4)$, and $f''(x) = 2(x + 4)^{-3}$. The maximum value of $|f''(x)|$ on the interval $[0, 2]$ is $1/32$, and thus, $M = 1/32$. This is used with the above formula to obtain
$$\frac{1}{192} h^2 \leq 10^{-4} \quad \text{or} \quad h \leq 0.13856.$$
Since $h = 2/n$, the number of subintervals n required is $n \geq 15$.

EXAMPLE 10.5
Use the composite trapezoidal rule with $n = 1$ to compute the integral
$$\int_1^3 (2x + 1)\, dx.$$

We have
$$T_1 = \frac{2}{2}[f(1) + f(3)] = 3 + 7 = 10.$$
As expected, the trapezoidal rule gives the exact value of the integral because $f(x)$ is linear.

EXERCISE SET 10.1

1. Approximate the integral
$$I = \int_{-2}^3 (x^4 - 3x^3 + 2x^2 - 3)\, dx$$
   using the composite trapezoidal rule with $n = 30$. Compare with the exact value $\frac{175}{12}$.

2. Eliminate the singularity at $x = 1$ of the integral
$$I = \int_0^1 \frac{2}{\sqrt{1 - x^2}}\, dx$$
   by using the transformation $x = \sin u$ and then use the composite trapezoidal rule to approximate I.

M-function 10.1

The following MATLAB function **trapez.m** computes the definite integral 10.1 using the composite trapezoidal rule. INPUTS are a function *f*; the values of *a* and *b*; the number of trapezoids *n*. The input function *f(x)* should be defined as an M-file.

```
function trapez(f,a,b,n)
% Compute the integral of f(x) from a to b using the trapezoid rule
h=(b-a)/n;
disp('_____')
disp(['        i         xi         f(xi)        h=',num2str(h) ])
disp('_____')
S=feval(f,a);
fprintf(' %2.0f %12.4f %14.6f\n',0,a,S);
for i=1:n-1
  x=a+h*i;
  g=feval(f,x);
  S=S+2*g;
  fprintf(' %2.0f %12.4f %14.6f\n',i,x,g);
end
S=S+feval(f,b);
fprintf(' %2.0f %12.4f %14.6f\n',n,b,feval(f,b));
INT=h*S/2;
fprintf('\n        The intergral of f(x) is =%16.8f\n',INT);
```

3. Use the composite trapezoidal rule to approximate the integral

$$I = \int_{0.4}^{0.6} e^{2x} \sin 3x \, dx$$

and find a bound for the error if possible.

4. Given the table

x	0	0.1	0.2	0.3
$f(x)$	2.72	3.00	3.32	3.67

use the composite trapezoidal rule to find an approximation to the integral

$$\int_0^{0.3} f(x) \, dx.$$

5. Use the composite trapezoidal rule to approximate the following integrals for the stated values of n.

(a) $\int_0^{\pi/3} \cos(3 \cos^2 x) \, dx$, $n = 7$,

(b) $\int_0^1 \ln(3 + \sin x) \, dx$, $n = 5$,

(c) $\int_0^{\pi/4} \sin x \, dx$, $n = 5$,

(d) $\int_1^2 \frac{2}{\sqrt{x}} \, dx$, $n = 4$.

6. Find an expression for the error of the following integral rule

$$\int_a^{a+h} f(x) \, dx \approx h f(a + h) - \frac{1}{2} h^2 f'(a).$$

7. Approximate the integral

$$\int_0^2 \ln \left(\frac{e^x + 2}{\cos x + 2} \right) dx$$

using the composite trapezoidal rule with $n = 5, 7$, and 9.

8. Using the smallest step size possible, approximate

$$\int_2^6 \left(\frac{1}{1 + x} \right) dx$$

using the composite trapezoidal rule using only the values of $f(x)$ at $x = 2, 3, \ldots, 6$.

9. The length of a curve $y = g(x)$ is given by

$$\int_a^b \sqrt{1 + (g'(x))^2} \, dx.$$

Use the composite trapezoidal rule to compute the length of one arch of the sine curve.

10. Apply the composite trapezoidal rule to

$$\int_0^1 \sqrt{x} \, dx$$

with $h = 1/2, 1/4, 1/8, \ldots$. Do you get the expected rate of convergence? Explain.

11. Given

x	-5	-4	-3	-2	-1	0	1	2	3	4	5	6
$f(x)$	440	0	-162	-160	-84	0	50	48	0	-64	-90	0

(a) Approximate $\int_0^6 f(x) \, dx$ using the composite trapezoidal rule with $h = 2$.

(b) Approximate $\int_0^6 f(x) \, dx$ using the composite trapezoidal rule with $h = 6$.

(c) Find the Lagrange polynomial through $f(-5)$, $f(-2)$, $f(3)$, and $f(4)$.

(d) Find the linear least squares through $f(-5)$, $f(-2)$, $f(3)$, and $f(4)$.

12. Given the continuous smooth function $g(x)$ for which $g(0) = 8$, $g(1) = 5$, $g(2) = 3$, $g(3) = 2$, and $g(4) = 3$, use the composite trapezoidal rule to approximate $\int_0^4 g(x)\,dx$.

13. Let $f(x) = e^{-x^2}$ and consider the integral

$$I = \int_0^1 f(x)\,dx.$$

(a) Use the composite trapezoidal rule with $h = 0.25$ to approximate I.

(b) Calculate the bound on the absolute error for the Trapezoidal rule.

14. Consider the data in the table

x	1.0	1.1	1.2	1.3	1.4	1.5	1.6	1.7	1.8
$f(x)$	1.544	1.667	1.811	1.972	2.152	2.351	2.576	2.828	3.107

Use the composite trapezoidal rule to approximate $\int_1^{1.8} f(x)\,dx$ with $h = 0.1, 0.2, 0.4$.

15. Consider the data in the table

x	1.6	1.8	2.0	2.23	2.4	2.6	2.8	3.0
$f(x)$	4.953	6.050	7.389	9.025	11.023	13.464	16.445	20.086

(a) Approximate $\int_{1.8}^{2.8} f(x)\,dx$ using the trapezoidal rule with $h = 0.2$.

(b) Compute the exact error knowing that $f(x) = e^x$.

(c) Compute the error bound for the Trapezoidal rule.

(d) If we did not know the true function, we would have to approximate the maximum second derivative of $f(x)$ using the data. Compute the error bound in this manner.

(e) If we want the computation to be correct to 5 decimal places, how small should the step size h be?

16. How small does h have to be for the error to be less than 10^{-4} when the trapezoidal rule is applied to

$$I = \int_1^2 \ln x\,dx?$$

17. Find an upper bound for the error incurred in estimating

$$I = \int_0^\pi x \sin x\,dx$$

with the composite trapezoidal rule with $n = 10$.

18. Consider the following rule often called the **prismoidale** rule

$$\int_a^b p(x)\,dx = \frac{b-a}{6}\left[p(a) + 4p\left(\frac{a+b}{2}\right) + p(b)\right].$$

Apply it to approximate $\int_{-1}^3 (x^3 + 2x^2 - 7)\,dx$.

19. Let $f(x) = \sqrt{1 + x^3}$.

   (a) How large must n be in the trapezoidal rule approximation of $\int_0^1 f(x)\,dx$ to insure that the absolute error is less than 10^{-3}?

   (b) Estimate the integral using the trapezoidal rule approximation with the value of n obtained in part (a).

20. Estimate the error involved in approximating

$$\int_1^3 \frac{1}{\sqrt{1+x^3}}\,dx$$

using the trapezoidal rule with $n = 100$.

COMPUTER PROBLEM SET 10.1

1. Write a computer program in a language of your choice to evaluate a definite integral using the composite trapezoidal rule. Input data to the program should be the limits of integration a and b, the integrand $f(x)$, and the number of subintervals n.

   Use your program to approximate the integral

$$\int_{-\pi}^{\pi} \frac{250x^2}{\cosh^2[500(x - t)]}\,dx$$

   for $t = 0.5, 1.0$. Tabulate the results using $n = 4, 8, \ldots, 80$ intervals.

2. Use the MATLAB function trapez to evaluate the integral

$$I = \int_{-2}^{2} e^{-2x^2}\,dx.$$

3. Use the MATLAB function trapez to evaluate the integral that arises in electrical field theory

$$H(x,r) = \frac{60r}{r^2 - x^2}\int_0^{2\pi}\left[1 - \left(\frac{x}{r}\right)^2 \sin^2\phi\right]^{1/2}\,d\phi$$

   for the following values of x and r.

(a) $r = 110$, $x = 75$,

(b) $r = 100$, $x = 70$,

(c) $r = 90$, $x = 65$.

4. Use the MATLAB function **trapez** to estimate

$$\int_0^3 \tan^{-1} x \, dx$$

with $n = 20, 40, 60$, and 100.

10.2   SIMPSON'S RULE

In this section, the partitioning of the interval $[a, b]$ is assumed to be made of an even number n of subintervals of width h. Taking them two at a time, as shown in Figure 10.3, Simpson's rule consists of approximating $f(x)$ by a quadratic polynomial interpolating at x_{i-1}, x_i, and x_{i+1}. Using the Taylor series approach,

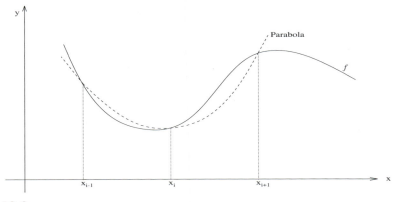

FIGURE 10.3
Simpson's rule.

we shall now try to derive a formula for Simpson's rule together with the error term.

By the fundamental theorem of calculus, we have

$$I_i = \int_{x_{i-1}}^{x_{i+1}} f(x) \, dx = F(x_{i+1}) - F(x_{i-1}). \tag{10.10}$$

The Taylor series expansion of $F(x)$ about x_i is

$$F(x_{i+1}) = F(x_i + h)$$
$$= F(x_i) + hF'(x_i) + \frac{h^2}{2!}F''(x_i) + \frac{h^3}{3!}F'''(x_i)$$
$$+ \frac{h^4}{4!}F^{(4)}(x_i) + \frac{h^5}{5!}F^{(5)}(x_i) + \frac{h^6}{6!}F^{(6)}(x_i) + \cdots$$

$$= F(x_i) + hf(x_i) + \frac{h^2}{2!} f'(x_i) + \frac{h^3}{3!} f''(x_i)$$
$$+ \frac{h^4}{4!} f'''(x_i) + \frac{h^5}{5!} f^{(4)}(x_i) + \cdots. \tag{10.11}$$

Similarly,

$$F(x_{i-1}) = F(x_i) - hf(x_i) + \frac{h^2}{2!} f'(x_i) - \frac{h^3}{3!} f''(x_i)$$
$$+ \frac{h^4}{4!} f'''(x_i) - \frac{h^5}{5!} f^{(4)}(x_i) + \cdots. \tag{10.12}$$

By substituting (10.11) and (10.12) into (10.10), we obtain

$$I_i = 2hf(x_i) + \frac{h^3}{3} f''(x_i) + \frac{h^5}{60} f^{(4)}(x_i) + \cdots. \tag{10.13}$$

It was found in Section 9.1 that

$$f''(x_i) = \frac{f(x_{i+1}) - 2f(x_i) + f(x_{i-1})}{h^2} - \frac{h^2}{12} f^{(4)}(x_i) + \cdots. \tag{10.14}$$

Finally, Eqns. (10.13) and (10.14) produce

$$I_i = \frac{h}{3} \left[f(x_{i+1}) + 4f(x_i) + f(x_{i-1}) \right] - \frac{h^5}{90} f^{(4)}(x_i) + \cdots. \tag{10.15}$$

Hence **Simpson's rule** to approximate the area over two subintervals lying between x_{i-1}, x_i and x_{i+1} is given by

$$S_i = \frac{h}{3} \left[f(x_{i+1}) + 4f(x_i) + f(x_{i-1}) \right]. \tag{10.16}$$

We can rewrite Simpson's rule over the interval $[x_0, x_2]$ as

$$\int_{x_0}^{x_2} f(x)\, dx = \frac{h}{3} \left[f(x_0) + 4f(x_1) + f(x_2) \right] - \frac{h^5}{90} f^{(4)}(\xi) \tag{10.17}$$

where $x_0 < \xi < x_2$. In view of Eqns. (10.15) and (10.17), the error term for Simpson's rule is given by

$$E_i = -\frac{h^5}{90} f^{(4)}(x_i).$$

Again, if $|f^{(4)}(x_i)| \leq M$ for all x in $[a, b]$, then

$$|E_i| \leq \frac{h^5}{90} M. \tag{10.18}$$

EXAMPLE 10.6

Use Simpson's rule to find an approximation to

$$\int_0^3 x^2 e^x\, dx.$$

Using (10.17), we have

$$\int_0^3 x^2 e^x \, dx \approx \frac{1}{2}[f(0) + 4f(1.5) + f(3)]$$
$$\approx 110.55252.$$

Simpson's composite rule

Formula (10.16) gives the approximation for the integral

$$I = \int_a^b f(x) \, dx$$

over two equal subintervals. The estimate obtained from this formula will usually not be accurate, particularly when dealing with a large interval $[a, b]$. To obtain a better accuracy, the procedure is extended to n subintervals and Simpson's rule is applied on each pair of consecutive subintervals. Since each application of Simpson's rule requires two subintervals, we assume that $n = 2m$ for some positive integer m. Therefore, by the rule of integration, we have

$$I = \sum_{i=1}^{m} \int_{x_{2i-2}}^{x_{2i}} f(x) \, dx.$$

Using (10.17), we get

$$I = \frac{h}{3} \sum_{i=1}^{m} [f(x_{2i-2}) + 4f(x_{2i-1}) + f(x_{2i})] - \frac{h^5}{90} \sum_{i=1}^{m} f^{(4)}(\xi_i). \qquad (10.19)$$

This reduces to

$$I = \frac{h}{3} \left[f(x_0) + 2\sum_{i=1}^{m-1} f(x_{2i}) + 4\sum_{i=1}^{m} f(x_{2i-1}) + f(x_{2m}) \right] - \frac{h^5}{90} \sum_{i=1}^{m} f^{(4)}(\xi_i).$$

By neglecting the error term, we get

$$S_n = \frac{h}{3} \left[f(x_0) + 2\sum_{i=1}^{m-1} f(x_{2i}) + 4\sum_{i=1}^{m} f(x_{2i-1}) + f(x_{2m}) \right] \qquad (10.20)$$

with

$$h = \frac{b-a}{n}, \quad x_i = a + ih, \quad i = 0, 1, ..., n.$$

Formula (10.20) is called **Simpson's composite rule**.

If (10.18) is applied over n subintervals of $[a, b]$, the error term E_T for the Simpson's composite rule can be expressed as

$$|E_T| \leq \frac{n\,h^5}{2\,90} M = (b-a)\frac{h^4}{180} M. \qquad (10.21)$$

Hence, the truncation error bound for Simpson's composite rule is proportional to h^4, whereas the error bound for the composite trapezoidal rule is proportional to h^2. As a result, Simpson's composite rule is more accurate than the composite trapezoidal rule, provided that the round-off error will not cause a problem. Since the error term in (10.19) involves the fourth derivative of f, Simpson's composite rule will give the exact results when applied to any polynomial of degree three or less.

EXAMPLE 10.7

Determine the number of subintervals n required to approximate

$$\int_0^2 \frac{1}{x+4}\, dx$$

with an error E_T less than 10^{-4} using Simpson's composite rule.

We have

$$|E_T| \leq \frac{b-a}{180} h^4 M \leq 10^{-4}.$$

In this example the integrand is $f(x) = 1/(x+4)$, and $f^{(4)}(x) = 24(x+4)^{-5}$. The maximum value of $|f^{(4)}(x)|$ on the interval $[0, 2]$ is $3/128$, and thus $M = 3/128$. This is used with the above formula to obtain

$$\frac{3}{90 \times 128} h^4 \leq 10^{-4} \quad \text{or} \quad h \leq \frac{2}{5}\sqrt[4]{15}.$$

Since $h = 2/(2m) = 2/n$, the number of subintervals n required is $n \geq 4$.

EXAMPLE 10.8

Use Simpson's composite rule with $n = 6$ to approximate the integral

$$\int_0^1 (7 + 14x^6)\, dx = 9.$$

From the values shown in Table 10.2, Simpson's composite rule estimate with $h = (1-0)/6$ is

$$S_6 = \frac{1}{18}\{f(0) + 2[f(\tfrac{1}{3}) + f(\tfrac{2}{3})] + 4[f(\tfrac{1}{6}) + f(\tfrac{1}{2}) + f(\tfrac{5}{6})] + f(1)\}$$

$$= \frac{1}{18}[7 + 2(7.019204 + 8.229081) + 4(7.000300 + 7.218750 +$$

$$+11.688572) + 21]$$

$$= 9.007059.$$

» simpson('f1',0,1,6)			
i	xi	f(xi)	h=0.16667
0	0	7	
1	0.1667	7.0003	
2	0.3333	7.019204	
3	0.5	7.21875	
4	0.6667	8.229081	
5	0.8333	11.688572	
6	1	21	
The integral of f(x) is =		9.00705876	

Table 10.2 Simpson's composite rule for $f(x) = 7 + 14x^6$ with $n = 6$.

EXAMPLE 10.9

Use Simpson's composite rule with $n = 2$ to approximate the integral

$$\int_1^2 (4x^3 - 2x + 3)\, dx.$$

We have

$$S_2 = \frac{1}{6}[f(1) + 4f(\frac{3}{2}) + f(2)] = 15.$$

As expected, Simpson's rule gives the exact value of the integral due to the fact that $f(x)$ has degree three.

EXERCISE SET 10.2

1. Use Simpson's composite rule to approximate the integral

$$I = \int_0^3 x^2 \cos(x^2 - 1)\, dx$$

with an accuracy of 10^{-3}.

2. How small must be h in order to compute the integral

$$I = \int_{0.2}^{1.1} e^{2x}\, dx$$

by Simpson's composite rule with an accuracy of 10^{-4}?

M-function 10.2

The following MATLAB function **simpson.m** computes the definite integral 10.1 using Simpson's composite rule. INPUTS are a function f; the values of a and b; the value of n (even). The input function $f(x)$ should be defined as an M-file.

```
function simpson(f,a,b,n)
% Compute the integral of a f from a to b using Simpson's
% composite rule. n must be even.
if n/2~=floor(n/2)
   disp(' n must be even')
   break
end;
h=(b-a)/n;
disp('_____')
disp(['    i              xi              f(xi)              h=',num2str(h) ])
disp('_____')
S=feval(f,a);
fprintf(' %2.0f %12.4f %14.6f\n',0,a,S);
for i=1:n/2
   m=2*i-1;
   x=a+h*m;
   g=feval(f,x);
   S=S+4*g;
   fprintf(' %2.0f %12.4f %14.6f\n',m,x,g);
   m=2*i;
   x=a+h*m;
   g=feval(f,x);
   if(i==n/2)
      S=S+g;
   else
      S=S+2*g;
   end;
   fprintf(' %2.0f %12.4f %14.6f\n',m,x,g);
end
INT=h*S/3;
fprintf('\n          The intergral of f(x) is =%16.8f\n',INT);
```

3. Estimate the error in the approximation of the integral

$$\int_2^3 \frac{2}{x}\,dx$$

by Simpson's composite rule with $n = 6$.

4. Use Simpson's composite rule to show that

$$\int_1^{1.6} \frac{2}{x}\,dx < 1 < \int_1^{1.7} \frac{2}{x}\,dx.$$

5. Approximate the integrals given in Exercise 5 of Section 10.1 using Simpson's composite rule.

6. Use Simpson's rule to estimate

$$\int_1^2 f(x)\,dx$$

from the following table:

x	1	1.5	2
$f(x)$	0.5	0.4	0.3

7. Let f be defined by

$$f(x) = \begin{cases} x^2 - x + 1 & \text{if } 0 \le x \le 1 \\ 2x - 1 & \text{if } 1 \le x \le 2 \end{cases}.$$

(a) Determine whether f is continuous on $[0, 2]$.

(b) Approximate the integral

$$I = \int_0^2 f(x)\,dx \quad \text{with } n = 2$$

(i) using the composite trapezoidal rule on the interval $[0, 2]$,

(ii) using Simpson's composite rule first over $[0, 1]$ and then the composite trapezoidal rule over $[1, 2]$,

(iii) using Simpson's composite rule over $[0, 2]$.

8. Compute the integral

$$\int_0^1 \frac{x^p}{x^3 + 12}\,dx$$

for $p = 0, 2$ using Simpson's composite rule with $n = 2, 8$.

9. Approximate

$$\int_{\pi/4}^{\pi/2} \frac{\cos x \log(\sin x)}{\sin^2 x + 1}\,dx$$

using Simpson's rule with $n = 8$.

10. Evaluate the following integral by Simpson's rule:

$$I = \int_1^2 (2x^5 + 5x^3 - 3x + 1)\,dx.$$

11. Using the smallest step size possible, approximate

$$\int_2^6 \left(\frac{1}{1+x}\right)\,dx$$

using the Simpson's composite rule using only the values of $f(x)$ at $x = 2, 3, \ldots, 6$.

12. The length of a curve $y = g(x)$ is given by

$$\int_a^b \sqrt{1 + (g'(x))^2}\, dx.$$

Use the Simpson's composite rule to compute the length of one arch of the sine curve.

13. Apply the Simpson's composite rule to

$$\int_0^1 \sqrt{x}\, dx$$

with $h = 1/2, 1/4, 1/8, \ldots$. Do you get the expected rate of convergence? Explain.

14. Given

x	-5	-4	-3	-2	-1	0	1	2	3	4	5	6
$f(x)$	440	0	-162	-160	-84	0	50	48	0	-64	-90	0

Approximate $\int_0^6 f(x)\ dx$ using Simpson's rule with the largest step size possible for the given data. What is the error order?

15. Given the continuous smooth function $g(x)$ for which $g(0) = 8$, $g(1) = 5$, $g(2) = 3$, $g(3) = 2$, and $g(4) = 3$, use Simpson's rule to approximate $\int_0^4 g(x)\, dx$.

16. Let $f(x) = e^{-x^2}$ and consider the integral

$$I = \int_0^1 f(x)\, dx.$$

Use Simpson's rule with $h = 0.25$ to approximate I. Use the result to estimate the error.

17. Consider the data in the table

x	1.0	1.1	1.2	1.3	1.4	1.5	1.6	1.7	1.8
$f(x)$	1.544	1.667	1.811	1.972	2.152	2.351	2.576	2.828	3.107

Use Simpson's rule to approximate $\int_1^{1.8} f(x)\, dx$ with $h = 0.1, 0.2, 0.4$.

18. Let $S(x)$ be a C^1 cubic spline with knots at $x_i = i/10$, $i = 0, 1, \ldots, 10$. Simpson's rule with $h = 0.05$ is exact for $\int_0^1 S(x)\, dx$, but it is not exact with $h = 0.04$. Explain why.

19. Given the data in the table

x	1.00	1.10	1.20	1.30	1.40	1.50
$f(x)$	1.000000	0.951351	0.918169	0.897471	0.887264	0.886227

construct an approximation to

$$\int_1^{3/2} f(x)\, dx.$$

20. Determine the number of subintervals n required to approximate

$$\int_0^2 e^{2x} \sin(3x)\, dx$$

with an error less than 10^{-4} using

(a) The composite trapezoidal rule.

(b) Simpson's composite rule.

21. Use the tabulated values of the integrand to estimate the integral

$$\int_{-\pi/2}^{\pi/2} \frac{3\cos t}{(2 + \sin t)^2}\, dt$$

using Simpson's rule with $n = 8$. Find the exact value and the approximate error.

22. Let $f(x) = \sqrt{1 + x^3}$.

(a) How large must n be in the approximation of $\int_0^1 f(x)\, dx$ by Simpson's rule to insure that the absolute error is less than 10^{-3}?

(b) Estimate the integral using Simpson's rule approximation with the value of n obtained in part (a).

23. Estimate the error involved in approximating

$$\int_0^1 e^{x^2}\, dx$$

using Simpson's rule with $n = 10$.

COMPUTER PROBLEM SET 10.2

1. Write a computer program in a language of your choice to evaluate a definite integral using Simpson's composite rule. Input data to the program should be

the limits of integration, the integrand $f(x)$, and the number of subintervals n.

Use your program to find the value of the integral

$$\int_{-\pi}^{\pi} \frac{250x^2}{\cosh^2[500(x-t)]} \, dx$$

for $t = 0.5, 1.0$. Tabulate the results using $n = 4, 8, \ldots, 40$ intervals.

2. Use the MATLAB function simpson to evaluate the integral that arises in electrical field theory

$$H(x,r) = \frac{60r}{r^2 - x^2} \int_0^{2\pi} \left[1 - \left(\frac{x}{r}\right)^2 \sin^2 \phi\right]^{1/2} d\phi$$

for the following values of x and r

(a) $r = 110$, $x = 75$,

(b) $r = 100$, $x = 70$,

(c) $r = 90$, $x = 65$.

3. Use the MATLAB function simpson to evaluate the integral

$$\int_0^1 \frac{2}{1+x^2} \, dx$$

with an accuracy of 9 decimal places. Compare your result with the exact value $\pi/2$.

4. Evaluate the integral

$$\int_0^{\pi} \frac{dx}{4 + \sin(20x)}$$

using Simpson's rule with $n = 10, 20$, and 30.

5. Compute the integral

$$\int_{-1}^1 e^{-x^2} \, dx$$

using Simpson's rules for step sizes $h = 1, 1/2, 1/4, 1/8, 1/16$, and $1/32$.

6. Evaluate

$$\int_0^3 \frac{dx}{\sqrt{1 + e^x} - x}$$

using Simpson's rule with $n = 20$.

7. Consider the integral

$$I = \int_{-1}^1 \frac{x^7 \sqrt{1 - x^2}}{(2 - x)^{13/2}} \, dx.$$

Evaluate I using Simpson's rule with $h = 0.1, 0.05, 0.02$, and 0.01. Explain the behavior of the results as h decreases.

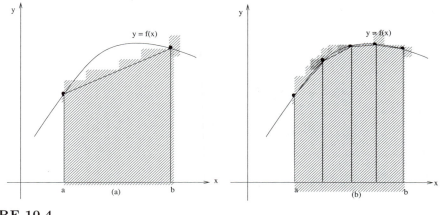

FIGURE 10.4
(a) $R_{1,1} =$ **area under** $2^0 = 1$ **trapezoid.**

(b) $R_{3,1} =$ **area under** $2^2 = 4$ **trapezoids.**

10.3   ROMBERG ALGORITHM

We now describe another powerful and popular method known as the **Romberg quadrature**, which is based on the use of the composite trapezoidal rule combined with the Richardson extrapolation.

We start out with the composite trapezoidal rule approximation

$$T_n = \frac{h}{2}\left[f(a) + f(b)\right] + h\sum_{i=1}^{n-1} f(x_i)$$

to the number I.

Here n represents the number of trapezoids related to h by

$$h = \frac{b - a}{n}$$

and

$$x_i = a + ih, \quad i = 0, 1, \dots, n.$$

For simplicity, we will consider n to be some power of 2, that is $n = 2^{k-1}$, $k = 1, 2, \dots$. Hence, n can be viewed as the number of times the interval $[a, b]$ has been halved to produce subintervals of length $h = (b - a)/2^{k-1}$ (see Figure 10.4).

To begin the presentation of the Romberg integration scheme, we adopt the following new notation for the composite trapezoidal rule

$$R_{k,1} = \frac{b - a}{2^k}\left[f(a) + f(b)\right] + \frac{b - a}{2^{k-1}}\sum_{i=1}^{2^{k-1}-1} f\left(a + \frac{b - a}{2^{k-1}}i\right), \quad k = 1, 2, \dots$$

obtained when the composite trapezoidal rule is applied to 2^{k-1} subintervals; that is, replacing h by $(b-a)/2^{k-1}$ and n by 2^{k-1} in formula (10.3).

Hence,

$$R_{1,1} = \frac{b-a}{2}[f(a) + f(b)]$$

$$R_{2,1} = \frac{b-a}{4}[f(a) + f(b)] + \frac{b-a}{2}f\left(a + \frac{b-a}{2}\right)$$

$$R_{3,1} = \frac{b-a}{8}[f(a) + f(b)] + \frac{b-a}{4}\sum_{i=1}^{3}f\left(a + \frac{b-a}{4}i\right)$$

etc....

Note that,

$$R_{2,1} = \frac{R_{1,1}}{2} + \frac{b-a}{2}f\left(a + \frac{b-a}{2}\right)$$

$$R_{3,1} = \frac{R_{2,1}}{2} + \frac{(b-a)}{4}\sum_{i=1}^{2}f\left[a + \frac{b-a}{2}\left(i - \frac{1}{2}\right)\right]$$

etc....

By induction, the general recursion relation for $R_{k,1}$ in terms of $R_{k-1,1}$ is

$$R_{k,1} = \frac{R_{k-1,1}}{2} + \frac{b-a}{2^{k-1}}\sum_{i=1}^{2^{k-2}}f\left[a + \frac{b-a}{2^{k-2}}\left(i - \frac{1}{2}\right)\right] \qquad (10.22)$$

for each $k = 2, 3, \ldots, n$.

The recursion relation (10.22) can be used to compute the sequence $R_{2,1}, R_{3,1}, \ldots, R_{n,1}$ once $R_{1,1}$ has been calculated.

Assuming that no round-off error enters into the calculation, the Richardson extrapolation technique can now be applied to Eqn. (10.22) to improve the estimate of integral I.

It can be shown that the error in the composite trapezoidal rule over 2^{n-1} subintervals can be expressed as

$$C(h^2) + D(h^4) + E(h^6) + \cdots, \quad \text{with} \quad h = \frac{b-a}{2^{n-1}}$$

where $C, D, E \ldots$ are functions of $f(x)$ and its derivatives only.

To proceed with the Richardson extrapolation, consider, for example, the approximation of I using $R_{1,1}$ and $R_{2,1}$

$$I = R_{1,1} + C(h^2) + D(h^4) + \cdots \qquad (10.23)$$

$$I = R_{2,1} + C(\frac{h^2}{4}) + D(\frac{h^4}{16}) + \cdots. \qquad (10.24)$$

Combining (10.23) and (10.24), we get another estimate for I that we will denote by $R_{2,2}$

$$R_{2,2} = \frac{4R_{2,1} - R_{1,1}}{3}.$$

This extrapolation eliminates the error term $C(h^2)$.

Similarly, if the extrapolation is carried out for $R_{2,2}$ and $R_{3,2}$, we get a new estimate for I that we will denote by $R_{3,3}$

$$R_{3,3} = \frac{16R_{3,2} - R_{2,2}}{15}.$$

This extrapolation eliminates the error term $D(h^4)$.

Following the same procedure we get the general extrapolation formula,

$$R_{i,k} = \frac{4^{k-1}R_{i,k-1} - R_{i-1,k-1}}{4^{k-1} - 1} \qquad (10.25)$$

for each $i = 2, \ldots, n$ and $k = 2, \ldots, i$. The truncation error associated with the entry $R_{i,k}$ is of order $O(h^{2k})$.

EXAMPLE 10.10

Compute $R_{2,2}$ to approximate the integral

$$\int_0^3 x^2 e^x \, dx.$$

We have

$$R_{1,1} = \frac{3}{2}[f(0) + f(3)]$$
$$= 271.154748$$
$$R_{2,1} = \frac{3}{4}[f(0) + f(3)] + \frac{3}{2}f\left(\frac{3}{2}\right)$$
$$= 150.703075$$

Using (10.25), we get

$$R_{2,2} = \frac{4R_{2,1} - R_{1,1}}{3} = 110.552517.$$

The approximation of I using Eqn. (10.25) is known as the Romberg integral and is often arranged in a form of a diagonal array, as shown in Table 10.3.

Assuming that no round-off error enters into calculations, the extrapolated values along the diagonal will converge to the correct answer more rapidly than the other values below the diagonal. It is interesting to note that the first and second columns in the array contain estimates obtained respectively by the composite trapezoidal and Simpson's rules (see Exercise 11 below).

$$R_{1,1}$$
$$R_{2,1} \ R_{2,2}$$
$$R_{3,1} \ R_{3,2} \ R_{3,3}$$
$$R_{4,1} \ R_{4,2} \ R_{4,3} \ R_{4,4}$$
$$\vdots \quad \vdots \quad \vdots \quad \vdots \quad \ddots$$
$$R_{n,1} \ R_{n,2} \ R_{n,3} \ R_{n,4} \ \ldots \ R_{n,n}$$

Table 10.3   Romberg array.

EXAMPLE 10.11

Compute the first six rows of the Romberg array to approximate

$$I = \int_0^1 (7 + 14x^6) \, dx$$

Set

$$f(x) = 7 + 14x^6, \quad a = 0, \quad b = 1, \quad \text{and} \quad n = 6.$$

We start by first computing $R_{1,1}$ and $R_{2,1}$

$$R_{1,1} = \frac{1-0}{2}[f(0) + f(1)] = \frac{1}{2}[7 + 21]$$
$$= 14$$
$$R_{2,1} = \frac{1-0}{4}[f(0) + f(1)] + \frac{1-0}{2}f(0 + \frac{1-0}{2})$$
$$= 10.609375.$$

Using (10.25), we get

$$R_{2,2} = \frac{4R_{2,1} - R_{1,1}}{3}$$
$$= \frac{4(10.609375) - 14}{3} = 9.4791666667.$$

Next, we compute

$$R_{3,1} = \frac{R_{2,1}}{2} + \frac{1}{4}[f(\frac{1}{2} \cdot \frac{1}{2}) + f(\frac{1}{2} \cdot \frac{3}{2})] = 9.4284667969$$

so,

$$R_{3,2} = \frac{4R_{3,1} - R_{2,1}}{3} = 9.0348307292$$
$$R_{3,3} = \frac{16R_{3,2} - R_{2,2}}{15} = 9.0052083333.$$

Continuing in this manner leads to the values in Table 10.4.

```
» romberg('f1',0,1,6)
                    Romberg Table

   i        h          Ri,1      Ri,2      Ri,3       ...

   1      1.0000     14.0000

   2      0.5000     10.6094    9.4792

   3      0.2500      9.4285    9.0348    9.0052

   4      0.1250      9.1088    9.0023    9.0001    9.0000

   5      0.0625      9.0273    9.0001    9.0000    9.0000    9.0000

   6      0.0313      9.0068    9.0000    9.0000    9.0000    9.0000    9.00000000
```

Table 10.4 The Romberg array for $f(x) = 7 + 14x^6$ with six rows.

M-function 10.3

The following MATLAB function **romberg.m** computes the definite integral 10.1 using the Romberg integration method. INPUTS are a function f; the values a and b; the number of rows n. The input function $f(x)$ should be defined as an M-file.

```
function romberg(f,a,b,n)
% Compute the integral of f on [a,b] using the Romberg integration.
fprintf('\n')
disp('                Romberg table')
disp('_____')
disp('   i        h        Ri,1        Ri,2        Ri,3      ... ')
disp('_____')
h=b-a;
R(1,1)=h*(feval(f,a)+feval(f,b))/2;
fprintf('%2.0f %8.4f %12.4f\n',1,h,R(1,1));
m=1;
for i=1:n-1
  h=h/2;
  S=0;
  for j=1:m
    x=a+h*(2*j-1);
    S=S+feval(f,x);
  end
  R(i+1,1)=R(i,1)/2+h*S;
  fprintf('%2.0f %8.4f %12.4f',i+1,h,R(i+1,1));
  m=2*m;
  for k=1:i
    R(i+1,k+1)=R(i+1,k)+(R(i+1,k)-R(i,k))/(4^k-1);
    fprintf('%12.4f',R(i+1,k+1));
  end
  fprintf('\n');
end
```

EXERCISE SET 10.3

1. Approximate $\int_0^2 x^2\sqrt{1+x^2}\,dx$ using the Romberg integration with $n = 4$.

2. Given $R_{4,2} = 5$ and $R_{3,2} = -2$, find $R_{4,3}$ using formula (10.25).

3. Approximate
$$\int_{0.2}^{1.1} \frac{12.1 + \ln(e^{2.4/x} - 1)}{x^4(e^{x^2} + 1)}\,dx$$
using the Romberg integration with $n = 6$.

4. Use the Romberg integration to approximate
$$\int_0^1 xe^{x^2}\,dx.$$

Complete the table until $|R_{n,n-1} - R_{n,n}| < 10^{-3}$.

5. Use Romberg integration to compute $R_{4,4}$ for the following integrals:

   (a) $\int_0^{\pi/3} x\sin x\,dx$,

   (b) $\int_{\pi/7}^{\pi/5} \sin x\,dx$.

6. Use the Romberg integration with $n = 4$ to approximate the integral
$$\int_1^3 \frac{1}{x}\,dx.$$

Compare with the exact value $\ln 3$.

7. The following data give approximations to the integral $\int_0^\pi \sin x\,dx$ for a scheme with error terms
$$E = K_1 h^2 + K_2 h^4 + K_3 h^6 + \ldots$$
$$I(h) = 1.570790, \quad I(h/2) = 1.896119, \quad I(h/4) = 1.974232,$$
$$I(h/8) = 1.993570.$$

Construct an extrapolation table to obtain better approximations.

8. Use Romberg's method to compute the integral
$$\int_0^4 f(x)\,dx$$
where $f(x)$ is defined by the following table. Do we need all the values to be used?

x	0.0	0.5	1.0	1.5	2.0	2.5	3.0	3.5	4.0
$f(x)$	-4271	-2522	-499	1795	4358	7187	10279	13633	17246

9. Compute the integral using the Romberg integration with $n = 4$

$$\frac{1}{\pi} \int_0^{2\pi} \exp(2^{-1/2} \sin x) \, dx.$$

10. Approximate

$$\int_1^4 \frac{e^x}{\sqrt{x}} \, dx$$

using the Romberg integration with $n = 5$.

11. Show that the second column in the Romberg array contains estimates obtained by Simpson's rule.

COMPUTER PROBLEM SET 10.3

1. Write a computer program in a language of your choice to evaluate a definite integral using the Romberg integral. Input data to the program should be the limits of integration a and b, the integrand $f(x)$, and the number of rows in the Romberg array.

   Use your program to find the value of the integral

   $$\int_{-\pi}^{\pi} \frac{250x^2}{\cosh^2[500(x - t)]} \, dx$$

   for $t = 0.5, 1.0$. Tabulate the results using 4, 6, and 8 rows in the Romberg array.

2. Use the MATLAB function romberg to evaluate the integral that arise in electrical field theory

   $$H(x, r) = \frac{60r}{r^2 - x^2} \int_0^{2\pi} \left[1 - \left(\frac{x}{r} \right)^2 \sin^2 \phi \right]^{1/2} d\phi$$

   for the following values of x and r.

   (a) $r = 110$, $x = 75$,
   (b) $r = 100$, $x = 70$,
   (c) $r = 90$, $x = 65$.

3. Use the MATLAB function romberg to evaluate the integral

   $$\int_0^1 \frac{2}{1 + x^2} \, dx$$

   using six rows in the Romberg array. Compare your result with the exact value $\pi/2$.

4. Use the Romberg integration to evaluate the integral

$$\int_0^2 x^4 \log(x + \sqrt{x^2 + 1}) \, dx$$

with $n = 6$.

5. Evaluate

$$\int_0^\pi \frac{\cos x}{\sqrt{x}} \, dx$$

using the Romberg integration. (*Hint:* The integral is singular. So, consider using integration by part first to remove the singularity.)

10.4   GAUSSIAN QUADRATURE

All the quadrature formulas developed so far in the preceding sections were of the form

$$\int_a^b f(x) \, dx \approx \sum_{i=1}^n w_i f(x_i) \qquad\qquad (10.26)$$

where w_i are the weights to be given to n functional values $f(x_i)$ and $x_1, x_2, \ldots, x_n$ are the nodes selected to be equally spaced. There are two groups of integration methods that are commonly used: the *Newton-Cotes formulas* that employ equally spaced nodes, and the *Gaussian quadrature formulas* that employ unequally spaced nodes. The Gaussian quadrature formula to be developed in this section has the same form as (10.26), but the nodes to be used are not equally spaced. The reason for choosing unequally spaced nodes is to minimize the error obtained in performing the approximation (10.26) by properly choosing the points x_i. Figure 10.5 shows the difference between the composite trapezoidal rule and the Gauss method with two points. In the composite trapezoidal rule, the approximation of the area under the curve between $x = a$ and $x = b$ is obtained by choosing points x_1 and x_2 at the ends of the interval $[a, b]$. Using the Gaussian quadrature, on the other hand, we choose x_1 and x_2 inside the interval (a, b) so that the area in the trapezoid equals the area under the curve. This is the basic idea of Gaussian rules.

We shall begin first by changing the interval of integration from $[a, b]$ to $[-1, 1]$ using the linear transformation

$$x = a + \frac{b - a}{2}(z + 1), \qquad -1 \le z \le 1. \qquad\qquad (10.27)$$

The corresponding function $g(z)$ is defined by

$$g(z) = f\left(\frac{(b - a)z + (a + b)}{2}\right).$$

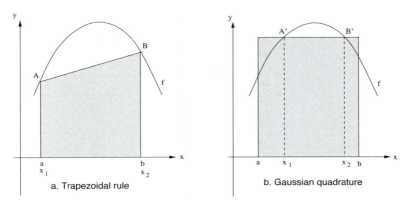

FIGURE 10.5
Comparison of Gaussian quadrature with the trapezoidal rule.

Making this substitution in (10.1) gives

$$\int_a^b f(x)\,dx = \frac{b-a}{2} \int_{-1}^1 g(z)\,dz. \qquad (10.28)$$

We will now derive the Gaussian quadrature formula for the integral

$$\int_{-1}^1 g(z)\,dz.$$

As before, we estimate the integral by approximating the function $g(z)$ with an interpolating polynomial $p_{n-1}(z)$ of degree $(n-1)$, that is

$$\int_{-1}^1 g(z)\,dz \approx \int_{-1}^1 p_{n-1}(z)\,dz = \sum_{i=1}^n w_i\,g(z_i) \qquad (10.29)$$

where the n sample points z_i and the weight's factors w_i are to be selected such that the equation

$$\int_{-1}^1 g(z)\,dz \approx \sum_{i=1}^n w_i\,g(z_i)$$

is exact.

Since the derivation of the Gaussian quadrature formula is too long (see [10]) to present for n points, we will derive the formula for three points only. In this case, $g(z)$ is approximated by a parabola that is the degree of the interpolating polynomial $p_{n-1}(z)$ is two. The points z_1, z_2, and z_3 and the weight factors w_1, w_2, and w_3 are chosen such that the areas under the curve $g(z)$ and the parabola are equal, that is

$$\text{Area} = w_1\,g(z_1) + w_2\,g(z_2) + w_3\,g(z_3). \qquad (10.30)$$

We shall adapt the method of undetermined coefficients to find z_1, z_2, z_3, w_1, w_2, and w_3. Since we have six unknowns, let us assume that the method is exact for the six sample functions

$$g(z) = 1$$
$$g(z) = z$$
$$g(z) = z^2$$
$$g(z) = z^3$$
$$g(z) = z^4$$
$$g(z) = z^5$$

so that

$$\int_{-1}^{1} 1\,dz = 2 = w_1 + w_2 + w_3$$

$$\int_{-1}^{1} z\,dz = 0 = w_1 z_1 + w_2 z_2 + w_3 z_3$$

$$\int_{-1}^{1} z^2\,dz = \frac{2}{3} = w_1 z_1^2 + w_2 z_2^2 + w_3 z_3^2$$

$$\int_{-1}^{1} z^3\,dz = 0 = w_1 z_1^3 + w_2 z_2^3 + w_3 z_3^3 \qquad (10.31)$$

$$\int_{-1}^{1} z^4\,dz = \frac{2}{5} = w_1 z_1^4 + w_2 z_2^4 + w_3 z_3^4$$

$$\int_{-1}^{1} z^5\,dz = 0 = w_1 z_1^5 + w_2 z_2^5 + w_3 z_3^5.$$

Observe that (10.31) is a nonlinear system of equations in six unknowns. It has been proved by Gauss that (10.31) has a unique solution given by

$$z_1 = -\sqrt{\frac{3}{5}}, \quad z_2 = 0, \quad z_3 = \sqrt{\frac{3}{5}}$$

$$w_1 = w_3 = \frac{5}{9}, \quad w_2 = \frac{8}{9}.$$

Substituting these constants into (10.29), we obtain the Gaussian three-points quadrature formula

$$\int_{-1}^{1} g(z)\,dz \approx \frac{5}{9}\, g\left(-\sqrt{\frac{3}{5}}\right) + \frac{8}{9}\, g(0) + \frac{5}{9}\, g\left(\sqrt{\frac{3}{5}}\right). \qquad (10.32)$$

Since the formula is exact for the polynomial in (10.31), one can show that it will be exact for all polynomials of degree ≤ 5. One should also note that Gauss's method can be extended to four or more points by using the same procedure described above. Values of the appropriate points and weight factors for $n = 2, 3, 4$, and 5 are shown in Table 10.5.

The above derivation applies only if the interval of integration is from -1 to 1. To translate any interval $[a, b]$ into $[-1, 1]$, one should use the transformation (11.13).

n	Nodes z_i	Weight Factors w_i
2	± 0.5773502692	1.0
3	± 0.7745966692	0.5555555556
	0.0	0.8888888889
4	± 0.3399810436	0.6521451549
	± 0.8611363116	0.3478548451
5	± 0.5384693101	0.4786286705
	± 0.9061798459	0.2369268851
	0.0	0.5688888889
6	± 0.2386191861	0.4679139346
	± 0.6612093865	0.3607615730
	± 0.9324695142	0.1713244924
7	± 0.9491079123	0.1294849662
	± 0.7415311856	0.2797053915
	± 0.4058451514	0.3818300505
	0.0	0.4179591837

Table 10.5   Gaussian quadrature nodes and weight factors.

EXAMPLE 10.12

Use the Gaussian quadrature formula with $n = 2$ to find an approximation to the integral

$$\int_0^3 x^2 e^x \, dx.$$

Using the transformation (11.13), we have

$$\int_0^3 x^2 e^x \, dx = \frac{3}{2} \int_{-1}^1 \left(\frac{3z + 3}{2} \right)^2 e^{\frac{3z+3}{2}} \, dz.$$

Using the values in Table 10.5, we obtain the approximation

$$\int_0^3 x^2 e^x \, dx \approx \frac{3}{2} \left[\left(\frac{3 \cdot 0.5773502692 + 3}{2} \right)^2 e^{\frac{3 \cdot 0.5773502692 + 3}{2}} \right.$$

$$\left. + \left(\frac{-3 \cdot 0.5773502692 + 3}{2} \right)^2 e^{\frac{-3 \cdot 0.5773502692 + 3}{2}} \right]$$

$$\approx 90.60740.$$

EXAMPLE 10.13

Use the Gaussian three-point quadrature formula to evaluate the integral (see Table 10.6)

$$I = \int_0^1 (7 + 14x^6)\,dx.$$

The three-point Gaussian quadrature formula uses the points 0 and $\pm\sqrt{\frac{3}{5}}$ and the coefficients $\frac{5}{9}$ and $\frac{8}{9}$. This gives, using formula (10.32).

$$I \approx \frac{1}{2} \left\{ \begin{array}{c} \frac{5}{9}\left[7 + 14\left(\frac{-\sqrt{3/5}+1}{2}\right)^6\right] + \frac{8}{9}\left(7 + \frac{14}{2^6}\right) + \\[2ex] \frac{5}{9}\left[7 + 14\left(\frac{\sqrt{3/5}+1}{2}\right)^6\right] \end{array} \right\}$$

$$\approx 8.995.$$

```
» gauss('f1',0,1,3)
```
                Gauss-Legendre quadrature three-point formula

i	zi	wi	g(zi)	wi*g(zi)
0	0.000000	0.888889	7.218750	6.416667
1	0.774597	0.555556	13.831971	7.684429
2	-0.774597	0.555556	7.000029	3.888905

                The integral of f(x) is =   8.99500000

```
» gauss('f1',0,1,4)
```
                Gauss-Legendre quadrature four-point formula

i	zi	wi	g(zi)	wi*g(zi)
0	0.339981	0.652145	8.266310	5.390834
1	-0.339981	0.652145	7.018084	4.576809
2	0.861136	0.347855	16.091115	5.597372
3	-0.861136	0.347855	7.000002	2.434984

                The integral of f(x) is =   9.00000000

Table 10.6   The Gaussian three-point and four-point quadrature formulas.

   This Gaussian three-point quadrature formula gives accuracy comparable to Simpson's rule when used with seven points. This reflects the fact that the Gaussian n-points quadrature formula is exact for polynomials of degree $\leq 2n - 1$, which is about twice that of an n-points Newton-Cotes formula.

MATLAB's methods

MATLAB has two built-in functions qad and quadl that use adaptive numerical quadrature to approximate an integral to within a specified tolerance. The syntax of these functions is:

>> quad(fun,a,b,tol,trace,arg1,arg2,...)

It approximates the integral of the function fun from a to b to within a specified tolerance tol using the recursive adaptive Simpson quadrature. Any nonzero value for the optional parameter trace causes a plot of the integrand to be created. The function fun should be a string containing the name of the M-file that evaluates $f(x)$. If the input parameters tol and trace are not specified, MATLAB uses a default error tolerance of 1.e-6 and trace=0.

The syntax for quadl is identical to quad but it approximates the integral using a high order recursive adaptive quadrature.

EXAMPLE 10.14

Use MATLAB to approximate the integral in the nonlinear pendulum problem for $\theta_0 = \pi/4$ and $c = L/g = 1$ (see Example 10.1).

Note that the integrand has a singularity at the right end point since it is infinite at $x = \theta_0$.

We first create an M-file that we name f1.m containing the definition of the integrand

```
function f=f1(x,teta0,c)
    f=2*sqrt(2)*sqrt(c). /sqrt(cos(x)-cos(teta0));
```

We now call the function quad and quadl as follows:

>> format long

>> quad('f1',0,pi/4,[ ],[ ],pi/4,1)

ans =
        6.53435329050876

>> quadl('f1',0,pi/4,[ ],[ ],pi/4,1)

ans =
        6.53434656410369

Note that quad and quadl pass the arg1,arg2,... values on to the M-file specified by the fun parameters. Supplying a null matrix [ ] for the fourth and fifth parameters instructs quad and quadl to use the defaults tolerance and zero trace.

_____

EXERCISE SET 10.4

1. Evaluate the following integral using the Gaussian four-point quadrature:

$$\int_0^{\pi/3} \ln(1 + \cos x)\, dx.$$

2. Use the Gaussian n-points quadrature formula to evaluate the integral

$$\int_{\pi/4}^{\pi/2} (\csc x + 1)(\cot x - 1)\, dx = -0.9059845977639.$$

   Can you obtain the exact result for some n?

3. Use the Gaussian quadrature with $n = 4$ to approximate

$$\int_0^1 (3x + x^3)e^x\, dx.$$

4. Evaluate

$$\int_1^{1+\sqrt{3}} (2 + 2x - x^2)^{3/2}\, dx$$

   using the Gaussian quadrature with $n = 4$.

5. Use the Gaussian three-point quadrature formula to approximate the following integrals:

   (a) $\int_0^{\pi/3} \cos(3\cos^2 x)\, dx$,

   (b) $\int_0^1 \ln(3 + \sin x)\, dx$,

   (c) $\int_0^{\pi/4} \sin x\, dx$,

   (d) $\int_1^2 \frac{2}{\sqrt{x}}\, dx$.

   Compare your answers with the results obtained in Exercise 5 of Section 10.1.

6. Compute the integral

$$\int_{2.1}^{3.1} \frac{\cos 2x}{1 + \sin x}\, dx$$

   using the Gaussian quadrature with $n = 2$ and 3.

M-function 10.4

The following MATLAB function **gauss_quad.m** computes the definite integral 10.1 using the Gaussian quadrature formula. INPUTS are a function f; the values of a and b; the value of n. The input function $f(x)$ should be defined as an M-file.

```
function gauss_quad(f,a,b,n)
% Compute the integral of f on [a,b] using Gauss-Legendre
% quadrature.
fprintf('\n')
disp('                  Gauss-Legendre quadrature')
disp('_____')
disp('      i        zi        wi        g(zi)      wi*g(zi)    ')
disp('_____')
if (n==2)
  z(1)=-sqrt(1/3);
  z(2)=-z(1);
  w(1)=1; w(2)=1;
end
if (n==3)
  z(1)=-sqrt(3/5);
  z(2)=0;
  z(3)=-z(1);
  w(1)=5/9; w(2)= 8/9; w(3)=w(1);
end
if (n==4)
  z(1)=-sqrt(1/7*(3-4*sqrt(0.3)));
  z(2)=-sqrt(1/7*(3+4*sqrt(0.3)));
  z(3)=-z(1);
  z(4)=-z(2);
  w(1)=1/2+1/12*sqrt(10/3);
  w(2)=1/2-1/12*sqrt(10/3);
  w(3)=w(1); w(4)=w(2);
end
if (n==5)
  z(1)=-sqrt(1/9*(5-2*sqrt(10/7)));
  z(2)=-sqrt(1/9*(5+2*sqrt(10/7)));
  z(3)=0;
  z(4)=-z(1);
  z(5)=-z(2);
  w(1)=0.3*((-0.7+5*sqrt(0.7))/(-2+5*sqrt(0.7)));
  w(2)=0.3*((0.7+5*sqrt(0.7))/(2+5*sqrt(0.7)));
  w(3)=128/225;
```

```
    w(4)=w(1); w(5)=w(2);
  end;
  S=0;
  for i=1:n
    x=((b-a)*z(i)+a+b)/2;
    g=feval(f,x);
    S=S+w(i)*g;
    fprintf('%2.0f %12.4f %12.4f% 12.4f %12.4f\n',i-
  1,z(i),w(i),g,g*w(i));
    end;
  INT=S*(b-a)/2;
  fprintf('\n        The intergral of f(x) is =%16.8f\n',INT);
```

7. Use the Gaussian quadrature with $n = 2, 3$, and 4 to approximate the integral

$$\int_0^1 \frac{1}{x^2 + 1} \, dx.$$

8. The quadrature formula

$$\int_{-1}^1 f(x) \, dx = c_0 f(-1) + c_1 f(0) + c_2 f(1)$$

is exact for all polynomials of degree less than or equal to 2. Determine c_0, c_1, and c_2.

9. The quadrature formula

$$\int_0^1 f(x) \, dx = \frac{1}{2} f(x_0) + c_1 f(x_1)$$

has the highest degree of precision. Determine x_0, c_1, and x_1.

10. Evaluate

$$\int_0^1 (\sin x^2) \ln x \, dx$$

using the Gaussian quadrature with $n = 4$.

11. The length of a parametric curve $\{x(t), y(t)\}$ is given by the integral

$$\int_a^b (\sqrt{[x'(t)]^2 + [y'(t)]^2} \, dx,$$

where $x(t) = \frac{3}{2} \cos t + \cos 3t$ and $y(t) = \frac{3}{2} \sin t - \sin 3t$, $0 \le t \le 2\pi$.
Estimate the length of a parametric curve to six significant decimal digits applying a numerical method of your choice.

12. Find constants w_0, w_1, w_2 such that the integration formula

$$\int_0^2 f(x) \, dx \approx w_0 f(0) + w_1 f'(0) + w_2 f(1)$$

is exact if $f(x)$ is a polynomial of degree ≤ 2.

13. Evaluate

$$\int_0^1 \frac{\sin(x^2 + 3)}{\sqrt{x}} \, dx$$

using the Gaussian quadrature with $n = 3$.

14. Prove that the formula

$$\int_{-1}^1 f(x) \, dx \approx \frac{1}{9} \left[5f(\sqrt{0.6}) + 8f(0) + 5f(-\sqrt{0.6}) \right]$$

is exact for polynomials of degree 5, and apply it to evaluate

$$\int_0^1 \frac{\sin x}{1 + x} \, dx.$$

15. Derive a two-point integration formula for integrals of the form

$$\int_{-1}^1 f(x)(1 + x^2) \, dx$$

which is exact when $f(x)$ is a polynomial of degree 3. Apply the formula for $f(x) = 2x^2$.

16. Evaluate the following integrals

(a) $I = \int_0^{100} e^{-5x} \, dx$,

(b) $I = \int_0^{100} e^{-x} \log(1 + x) \, dx$,

using the three-point Gaussian quadrature formula.

COMPUTER PROBLEM SET 10.4

1. Write a computer program in a language of your choice to evaluate a definite integral using the Gaussian quadrature formula with $n = 2, 3, 4$, or 5. Input data to the program should be the limits of integration, the integrand $f(x)$, the number of points at which the function is to be evaluated, and the nodes and weights corresponding to the different values of n.

Use your program to find the value of the integral

$$\int_{-\pi}^{\pi} \frac{250x^2}{\cosh^2[500(x - t)]} \, dx$$

for $t = 0.5, 1.0$. Tabulate the results using $n = 2, 3$, and 4.

2. Use the MATLAB function gauss_quad to evaluate the integral that arises in electrical field theory

$$H(x,r) = \frac{60r}{r^2 - x^2} \int_0^{2\pi} \left[1 - \left(\frac{x}{r} \right)^2 \sin^2 \phi \right]^{1/2} d\phi$$

for the following values of x and r

(a) $r = 110$, $x = 75$,

(b) $r = 100$, $x = 70$,

(c) $r = 90$, $x = 65$.

3. Use the MATLAB function gauss_quad to evaluate the integral

$$\int_0^1 \frac{2}{1 + x^2} dx$$

with $n = 4$. Compare your result with the exact value $\pi/2$.

4. Apply the Gaussian quadrature formula with $n = 3$ for the approximation of the integral

$$\int_0^{2\pi} e^{-x} \sin x \, dx.$$

5. Compute

$$\int_0^{0.999} \tan \frac{\pi}{2} x \, dx$$

using the Gaussian quadrature formula with $n = 4$.

6. Compute the integral

$$\int_{-1}^1 e^{-x^2} dx$$

using Gaussian quadrature with $n = 1, 2, 3, 4$, and 5. State the errors between your computed values and MATLAB's function quadl computed values for the integral.

7. Modify the MATLAB function gauss_quad given in the text so that it can be also used for $n = 6$. Use the modified function to compute

$$\int_0^{\pi/2} \frac{\cos x \log(\sin x)}{\sin^2 x + 1} dx$$

using Gaussian quadrature formula with $n = 6$.

8. The length L of the ellipse $4x^2 + 9y^2 = 36$ is given by

$$L = \frac{4}{3} \int_0^3 \sqrt{\frac{81 - 5x^2}{9 - x^2}} \, dx.$$

The integral L has a singularity at $x = 3$. Set $x = 3 \sin \theta$ and then approximate the θ-integral using the MATLAB function gauss_quad with $n = 5$.

APPLIED PROBLEMS FOR CHAPTER 10

1. The value of π can be estimated by approximating the area of the first quadrant inside the circle $x^2 + y^2 = 16$. The area is equal to 4π. Use the Gaussian quadrature with $n = 4$ to approximate π.

2. The gamma function is defined by

$$\Gamma(\alpha) = \int_0^\infty e^{-x} x^{\alpha-1}\, dx.$$

Approximate $\Gamma(\alpha)$ for values of $\alpha = 1.0,\ 1.4,$ and 2.0.

3. The error function $\operatorname{erf}(x)$ is defined by

$$\operatorname{erf}(x) = \frac{2}{\sqrt{\pi}} \int_0^x e^{-t^2}\, dt.$$

Use the Gaussian quadrature with $n = 4$ to approximate $\operatorname{erf}(0.5)$ and compare with the tabulated value, $\operatorname{erf}(0.5) = 0.5200500$.

4. A rod of length l located along the x-axis has a uniform charge per unit length and a total charge Q. The electric potential at a point P along the y-axis at a distance d from the origin is given by

$$V = k\frac{Q}{l} \int_0^l \frac{dx}{\sqrt{x^2 + d^2}}$$

where $k = 9 \times 10^9\ N \cdot m^2/C^2$. Estimate V if $Q = 10^{-6}$ C, $l = 5$ m, and $d = 4$ m.

5. Suppose we wish to calculate the temperature of the earth at certain depths over a period of time. We assume that the earth is flat and initially at zero temperature. The temperature, $T(h, t)$, at the depth h taken to be positive at the time t, is given by

$$T(h, t) = \frac{h}{2a} \int_0^t \frac{e^{-\frac{h^2}{4a(t-\tau)}}}{(t-\tau)\sqrt{4\pi a(t-\tau)}} T_s(\tau) d\tau$$

where $T_s(t)$ is the temperature at the surface of the earth. The constant a is the thermal diffusivity and is a function of the medium. If t is in hours then

$$T_s(t) = 15 + 20 \sin\left(\frac{2\pi t}{8766}\right).$$

Suppose $a = 0.009$ m^2/hr; evaluate $T(h, t)$ for the following values of t and h

(a) $t = 200, 400, 500$ and $h = 1$ m,

(b) $t = 200, 400, 500$ and $h = 5$ m,

(c) $t = 200, 400, 500$ and $h = 10$ m.

6. The steady state temperature distribution $u(r, \theta)$ of an insulated disk satisfies the partial differential equation

$$\Delta u = \frac{1}{r}\frac{\partial}{\partial r}\left(r\frac{\partial u}{\partial r}\right) + \frac{1}{r}\frac{\partial^2 u}{\partial \theta^2} = 0, \quad 0 \le \theta \le 2\pi, \ 0 \le r \le a,$$
$$u(a, \theta) = f(\theta), \quad f(0) = f(2\pi), \quad 0 \le \theta \le 2\pi.$$

Here a is the radius of the disk, the origin of the coordinate is taken to be at the center of the disk, and $f(\theta)$ is the given temperature on the periphery of the disk. The solution of this problem is

$$u(r, \theta) = \frac{1}{2\pi}\int_0^{2\pi} \frac{a^2 - r^2}{a^2 - 2ar\cos(\varphi - \theta) + r^2} f(\varphi)\, d\varphi.$$

To evaluate the integral at various points, (r, θ), use the integral formulae developed in this chapter. For this problem we use $a = 1$.

(a) Let $f(\theta) = 1$. Evaluate the integral at several random points that the temperature distribution throughout the disk is probably the constant one.

(b) Suppose $f(\theta) = \cos^2\theta - \sin^2\theta$. Evaluate the integral at several random points to conclude that $u(r, \theta)$ is probably equal to $u(r, \theta) = r^2\cos^2\theta - r^2\sin^2\theta$.

(c) Suppose $f(\theta) = \theta(2\pi - \theta)$, $0 \le \theta \le 2\pi$. Draw the graph of $u(0.2, \theta)$, $u(0.4, \theta)$, $u(0.6, \theta)$, $u(0.2, \theta)$, $u(0.8, \theta)$ for $0 \le \theta \le 2\pi$. What is $u(0, \theta)$?

(d) As in question (c) $f(\theta) = \theta(2\pi - \theta)$. Evaluate $u(r, 0)$, $u(r, \pi/4)$, $u(r, 5\pi/6)$, $u(r, 3\pi/2)$ for $r = 0.9$, 0.95, and 0.99 and observe that they satisfy the estimate
$$0 = \min_{0 \le \theta \le 2\pi} f(\theta) \le u(r, \theta) \le \max_{0 \le \theta \le 2\pi} f(\theta) = \pi^2.$$

The estimate is called the **maximum-minimum principle** for solutions to $\Delta u = 0$ in a disk.

7. In heat conduction problems, the heat flux of a pipe is given by the integral

$$C(\phi) = \frac{\sin^2\phi}{16}\int_\phi^{\pi/2} \sin x \left[\cos\phi - \frac{\pi - 2x - \sin 2x}{2\cos^2 x}\right]^3 dx$$

where ϕ is in radian.

(a) Estimate the integral for $\pi/5$, $\pi/4$, $\pi/3$ using Gaussian quadrature formula with $n = 4$. Note that the integral has a singularity at $x = \pi/2$.

(b) If we use the power series expansion to sort out the singularity we get

$$C(\phi) \approx \frac{\sin^2 \phi}{16} \int_\phi^{\pi/2} \left(x - \frac{1}{6}x^3\right) \left[\cos \phi - \left(\frac{1}{2}\pi - 2x +\right.\right.$$
$$\left.\left. + \frac{1}{2}\pi x^2 - \frac{4}{3}x^3 + \frac{1}{3}\pi x^4\right)\right]^3 dx.$$

Evaluate the integral for the same values of ϕ given in (a) and compare the results obtained in (a) and (b).

8. The velocity potential for an incompressible, irrotational fluid satisfies the Laplace equation. Let us suppose that we can think of the fluid as being two dimensional. Let us denote the potential so that the velocity is

$$\vec{v} = \nabla u.$$

Let us exit the fluid motion by imparting a normal velocity to the fluid at the boundary so that the boundary condition is given by a function g. If the container can be taken to be a disk of radius a in this two dimensional setting, then the solution u in polar coordinates is given by the integral

$$u(r, \theta) = C - \frac{a}{2\pi} \int_0^{2\pi} \ln\left(a^2 - 2ar\cos(\varphi - \theta) + r^2\right) g(\varphi) \, d\varphi$$

where C is the arbitrary constant. In this problem we take $a = 1$.

Let $g(\theta) = \sin \theta$. Take $C = 0$ and evaluate $u(r, \theta)$ at the points $(0.1, \pi/4), (0.2, \pi/4)$, and $(0.4, 3\pi/6)$.

9. To monitor the thermal pollution of a river, a biologist takes hourly temperature T reading (in $^\circ F$) from 9 AM to 4 PM. The results are shown in the following table.

Time of day	9	10	11	12	13	14	15	16
Temperature	75.3	77.0	83.2	84.8	86.5	86.4	81.1	78.6

Use Simpson's rule to estimate the average water temperature between 9 AM and 4 PM given by

$$T_{av} = \frac{1}{b - a} \int_a^b T(t) \, dt.$$

10. An observer measures the outside temperature every hour from noon until 8 PM. The recorded temperatures are given by the table:

Time of day	Noon	1	2	3	4	5	6	7	8
Temperature	62	65	69	70	68	68	61	63	62

Approximate the average temperature between 12 PM and 8 PM given by

$$av(T) = \frac{1}{b - a} \int_a^b T(x) \, dx$$

using the composite Trapezoidal rule.

11. A radar was used to record the speed of a runner during the first 4 seconds of a race (see table below). Use Simpson's rule to estimate the distance the runner covered during these 4 seconds.

Time $t(s)$	0.0	0.5	1.0	1.5	2.0	2.5	3.0	3.5	4.0
v(m/s)	0	4.66	7.72	8.76	9.72	10.10	10.53	10.77	10.91

12. A perfect radiator emits energy at a rate proportional to the fourth power of its absolute temperature, according to the Stefan-Boltzmann equation

$$E = 36.9 \cdot 10^{-12} T^4$$

where E is the emissive power in watts/cm^2, and T is the temperature in degrees Kelvin. We are interested in the fraction of this total energy contained in the visible spectrum, which is taken here to be $4 \cdot 10^{-5}$ to $7 \cdot 10^{-5}$ cm. The visible part E_v is obtained by integrating Planck's equation between these limits:

$$E_v = \int_{4 \cdot 10^{-5}}^{7 \cdot 10^{-5}} \frac{2.39 \cdot 10^{-11}}{x^5 (e^{\frac{1.432}{Tx}} - 1)} \, dx$$

where x is the wavelength in cm. Estimate E_v using Simpson's rule with $n = 20$ and $T = 3500°$.

13. A straight stretch of highway runs alongside a lake. A surveyor who wishes to know the approximate area of the lake measures the distance from various points along the road to the near and far shores of the lake according to the following table (see Figure below):

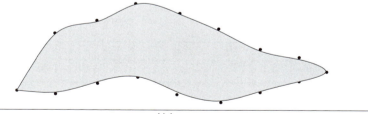

highway

Distance along highway (km)	0.0	0.5	1.0	1.5	2.0	2.5	3.0	3.5	4.0
Distance to near shore (km)	0.5	0.3	0.7	1.0	0.5	0.2	0.5	0.8	1.0
Distance to far shore (km)	0.5	2.3	2.2	3.0	2.5	2.2	1.5	1.3	1.0

Use Simpson's rule to estimate the lake's area.

14. When the truss is subject to a force F at its center, point A deflects an amount x from its equilibrium position (see the figure below). Forces in Kilonewtons required to produce deflections from 0 to 4 cm are listed in the following table.

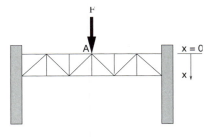

x (cm)	0.0	0.5	1.0	1.5	2.0	2.5	3.0	3.5	4.0
F	0	1.44	2.91	4.42	5.91	7.42	9.04	10.71	13.22

Use Simpson's rule to approximate the work W done by F in deflecting A given by

$$W = \int_0^4 F(x)\, dx.$$

Chapter 11

Numerical Methods for Linear Integral Equations

11.1 INTRODUCTION

Integral equations are those equations in which the unknown function appears under the integral sign. Such equations have many applications in applied mathematics, physics, and engineering and more recently in economics and biology. One of the first integral equations to be investigated is due to Abel. He considered the case of a bead sliding down a frictionless wire and asked what the shape of the wire would be for the bead to slide from a given position, say $(0, h)$, to a final position, say $(x, 0)$, in a predetermined time, $f(x)$ (see Fig. 11.1). He was led to the integral equation

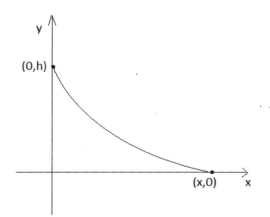

FIGURE 11.1
Sliding of a bead.

$$f(x) = \int_0^x (x - t)^{-1/2} u(t) \, dt. \tag{11.1}$$

$f(x)$ is a given function and $u(t)$, which is related to the equation for the shape of the wire, is the sought-after unknown. More generally, this is an equation of the form

$$f(x) = \int_a^x k(x,t)u(t)\, dt. \tag{11.2}$$

$f(x)$ and $k(x,t)$ are given functions and u is the unknown function. The function, $u(t)$, is raised to the first power so (11.2) is said to be linear. (11.2) is an integral equation of the Volterra type of the first kind. The Abel integral equation is a special case of (11.2). The equation

$$f(x) = u(x) + \int_a^x k(x,t)u(t)\, dt \tag{11.3}$$

is a linear Volterra integral equation of the second kind. The Volterra-type integral equation of the second kind arise in many different contexts but a type context arises in converting a differential equations' problem to an integral equations' problem.

EXAMPLE 11.1
Find an integral equation equivalent to the initial-value problem

$$y'' + p(x)y = f(x)$$

$$y(0) = a, \quad y'(0) = b.$$

Integrate the differential equation from 0 to x to find

$$y' - b + \int_0^x p(t)y(t)\, dt = \int_0^x f(t)\, dt.$$

Now integrate this result again from 0 to x. Observe that

$$\int_0^x \int_0^s p(t)y(t)\, dt ds = \int_0^x \int_t^x p(t)y(t)\, ds dt = \int_0^x (x-t)p(t)y(t)\, dt.$$

We find

$$y(x) - a - bx + \int_0^x (x-t)p(t)y(t)\, dt = \int_0^x (x-t)f(t)\, dt$$

or

$$y(x) + \int_0^x k(x,t)y(t)\, dt = F(x)$$

where $k(x,t) = (x-t)p(t)$ and $F(x) = \int_0^x (x-t)f(t)\, dt + a + bx$.

The advantage of converting a differential equation to an integral equation is that integration is a smoothing operation and often greater accuracy can be achieved.

Volterra-type integral equations have a variable upper limit. An integral equation with a fixed upper limit is called a **Fredholm type integral equation**. The integral equation

$$\int_a^b k(x,t)u(t)\, dt = f(x) \tag{11.4}$$

where $k(x,t)$ and $f(x)$ are given functions and $u(t)$ is the sought after unknown function called a **linear, Fredholm integral equation** of the first kind and the integral equation

$$u(x) + \int_a^b k(x,t)u(t)\,dt = f(x) \tag{11.5}$$

is Fredholm integral equation of the second kind.

The function, $k(x,t)$, in these integral equations is called the **kernel**. If $f(x)$ is identically zero, the equation is said to be homogeneous.

Fredholm integral equations of the first kind, especially when (a,b) is an infinite interval, arise in connection with integral transforms and special techniques have been developed to treat them. In this chapter we treat only integral equations of the second kind on the finite interval $[a,b]$.

Initial-value problems for ordinary differential equations give rise to Volterra integral equations. Boundary-value problems for ordinary differential equations give rise to Fredholm integral equations.

EXAMPLE 11.2

Find an equivalent integral equation for the boundary-value problem

$$u'' + p(x)u = f(x), \quad a < x < b$$

$$u(a) = 0, \ u(b) = 0.$$

Again integrate the differential equation from a to x to find

$$u' + \int_a^x p(s)u(s)\,ds = \int_a^x f(s)\,ds + C$$

where C is a constant. Integrate again from a to x to find

$$u(x) + \int_a^x (x-s)p(s)u(s)\,ds = \int_a^x (x-s)f(s)\,ds + C(x-a).$$

Choose C so that $u(b) = 0$, i.e., so that

$$\int_a^b (b-s)p(s)u(s)\,ds = \int_a^b (b-s)f(s)\,ds + C(b-a).$$

We find after some manipulations that

$$u(x) - \int_a^b k(x,s)p(s)u(s)\,ds = -\int_a^b k(x,s)f(s)\,ds$$

where

$$k(x,s) = \begin{cases} \frac{(s-a)(b-x)}{b-a}, & a \le s \le x \\ \frac{(x-a)(b-s)}{b-a}, & x \le s \le b. \end{cases}$$

EXERCISE SET 11.1

1. Show that $u(x) = \sin x$ is a solution to

$$u(x) = x + \int_0^x (t - x)u(t)dt.$$

2. Show that $u(x) = e^x$ is a solution to

$$u(x) = 1 + \int_0^x u(t)dt.$$

3. Show that $u(x) = 1 + 3x/4$ is a solution to

$$u(x) = 1 + x \int_0^1 tu(t)dt.$$

4. Show that $u(x) = -3$ is a solution to

$$u(x) = x + \int_0^1 (1 + xt^2)u(t)dt.$$

5. Find a Volterra integral equation for the differential equation's problem

$$u'' + q(x)u' + p(x)u = f(x), \quad u(0) = a, \ u'(0) = b.$$

6. Find a Volterra integral equation for the differential equation's problem

$$u'' - xu = e^x, \quad u(0) = 1, \ u'(0) = 2.$$

7. Find a Fredholm integral equation for the boundary-value problem

$$u'' - x^2 u = \sin x, \quad u'(0) = 0, \ u'(1) = 0.$$

11.2   QUADRATURE RULES

The first class of methods that would occur to one to try in approximating solutions to integral equations would be to try quadrature rules and so to reduce the continuous problem to a system of linear equations. In this section we will use the Trapezoidal rule and Gaussian quadrature to approximate the solution to the Fredholm integral equation

$$f(x) = u(x) + \int_a^b k(x,t)u(t)\, dt. \tag{11.6}$$

11.2.1 Trapezoidal rule

To apply the Trapezoidal rule we first divide the interval $[a, b]$ into N subintervals, each with size h, where

$$t_j = a + jh, \quad j = 0, 1, \ldots, N, \quad h = \frac{b-a}{N},$$

.

Recall from Section 10.1, the Trapezoidal rule for estimating the definite integral $\int_a^b k(x, t)u(t)\, dt$ is:

$$\int_a^b k(x, t)u(t)\, dt = \frac{h}{2}\left[k(x, t_0)u(t_0) + k(x, t_N)u(t_N)\right] + h \sum_{j=1}^{N-1} k(x, t_j)u(t_j). \quad (11.7)$$

Substitute into equation (11.4) to get

$$f(x) = u(x) + \frac{h}{2}\left[k(x, t_0)u(t_0)) + k(x, t_N)u(t_N)\right] + h \sum_{j=1}^{N-1} k(x, t_j)u(t_j). \quad (11.8)$$

A solution to the functional equation (11.8) may be obtained if we assign t_i's to x in which $i = 0, 1, \ldots, N$ and $a \leq t_i \leq b$. In this way, (11.8) is reduced to the system of equations

$$f(t_i) = u(t_i) + \frac{h}{2}\left[k(t_i, t_0)u(t_0)) + k(t_i, t_N)u(t_N)\right] + h \sum_{j=1}^{N-1} k(t_i, t_j)u(t_j). \quad (11.9)$$

Let $f(t_i) = f_i$, $k(t_i, t_j) = k_{ij}$ and $u(t_i) = u_i$, then equation (11.9) becomes

$$f_i = u_i + \frac{h}{2}\left[k_{i0}u_0 + k_{iN}u_N\right] + h \sum_{j=1}^{N-1} k_{ij}u_j \quad (11.10)$$

where $i = 0, 1, \ldots, N$. In matrix form (11.10) can be represented by

$$(\mathbf{I} + \mathbf{KD}) \cdot \mathbf{u} = \mathbf{f} \quad (11.11)$$

where

$$\mathbf{f} = [f_i]^T, \quad \mathbf{u} = [u_i]^T, \quad \mathbf{K} = [k_{ij}], \quad \text{and} \quad \mathbf{D} = \mathbf{diag}(\frac{h}{2}, h, \ldots, h, \frac{h}{2}).$$

This is a set of $N + 1$ linear algebraic equations in $N + 1$ unknowns $u_0, \ldots, u_N$ and can be solved by standard decomposition techniques.

EXAMPLE 11.3

Use the Trapezoidal rule with $N = 5$ to solve the Fredholm integral equation

$$x = u(x) - \int_0^1 x^2\, t\, u(t)\, dt.$$

Compare with the values of the exact solution $u(x) = x + \frac{4}{9}x^2$.

___

Here

$$h = \frac{1}{5} = 0.2, \quad t_j = hj = 0.2j, \quad j = 0, 1, ..., 5, \quad k(x, t) = x^2 t.$$

It follows from (11.9):

$$u(x_i) - \frac{h}{2}\left[k(x_i, t_0)u(t_0) + k(x_i, t_N)u(t_N)\right] - h\sum_{j=1}^{4} x_i^2 t_j u(t_j) = x_i$$

$$u_i - 0.1\left[k(x_i, 0)u(0) + k(x_i, 1)u(1)\right] - 0.2\sum_{j=1}^{4} x_i^2 t_j u_j = x_i$$

where $u(t_j) = u_j$ and $x = x_i = t_i$. Expand the last equation to get the system of equations

$$u_0 - 0.1[0.0\, t_0^2\, u_0 + 1.0\, t_0^2\, u_5] - 0.2\left[t_0^2 t_1\, u_1 + \ t_0^2 t_2\, u_2 + \ t_0^2 t_3\, u_3 + t_0^2 t_4\, u_4\right] = t_0$$

$$u_1 - 0.1[0.0\, t_1^2\, u_0 + 1.0\, t_1^2\, u_5] - 0.2\left[t_1^2 t_1\, u_1 + \ t_1^2 t_2\, u_2 + \ t_1^2 t_3\, u_3 + t_1^2 t_4\, u_4\right] = t_1$$

$$u_2 - 0.1[0.0\, t_2^2\, u_0 + 1.0\, t_2^2\, u_5] - 0.2\left[t_2^2 t_1\, u_1 + \ t_2^2 t_2\, u_2 + \ t_2^2 t_3\, u_3 + t_2^2 t_4\, u_4\right] = t_2$$

$$u_3 - 0.1[0.0\, t_3^2\, u_0 + 1.0\, t_3^2\, u_5] - 0.2\left[t_3^2 t_1\, u_1 + \ t_3^2 t_2\, u_2 + \ t_3^2 t_3\, u_3 + t_3^2 t_4\, u_4\right] = t_3$$

$$u_4 - 0.1[0.0\, t_4^2\, u_0 + 1.0\, t_4^2\, u_5] - 0.2\left[t_4^2 t_1\, u_1 + \ t_4^2 t_2\, u_2 + \ t_4^2 t_3\, u_3 + t_4^2 t_4\, u_4\right] = t_4$$

$$u_5 - 0.1[0.0\, t_5^2\, u_0 + 1.0\, t_5^2\, u_5] - 0.2\left[t_5^2 t_1\, u_1 + \ t_5^2 t_2\, u_2 + \ t_5^2 t_3\, u_3 + t_5^2 t_4\, u_4\right] = t_5$$

which can be written in the form

$$u_0 = 0$$

$$[1 - 0.2t_1^2 t_1]u_1 - 0.2\, t_1^2\, t_2\, u_2 - 0.2\, t_1^2\, t_3\, u_3 - 0.2t_1^2\, t_4\, u_4 - 0.1t_1^2\, u_5 = t_1$$

$$-0.2t_2^2 t_1 u_1 + [1 - 0.2t_2^2\, t_2]u_2 - 0.2t_2^2\, t_3\, u_3 - 0.2t_2^2\, t_4\, u_4 - 0.1t_2^2\, u_5 = t_2$$

$$-0.2t_3^2 t_1 u_1 - 0.2\, t_3^2\, t_2\, u_2 + [1 - 0.2t_3^2\, t_3]u_3 - 0.2t_3^2\, t_4\, u_4 - 0.1t_3^2\, u_5 = t_3$$

$$-0.2t_4^2 t_1 u_1 - 0.2\, t_4^2\, t_2\, u_2 - 0.2\, t_4^2\, t_3\, u_3 + [1 - 0.2t_4^2\, t_4]\, u_4 - 0.1t_4^2\, u_5 = t_4$$

$$-0.2t_5^2 t_1 u_1 - 0.2\, t_5^2\, t_2\, u_2 - 0.2\, t_5^2\, t_3\, u_3 - 0.2\, t_5^2\, t_4\, u_4 + [1 - 0.1t_5^2\,]u_5 = t_5.$$

Substituting the values of the t_j''s into these equations to get the system of equations in matrix form

x	u_i	Exact	Error
0.2	0.21838	0.21778	0.60×10^{-3}
0.4	0.47351	0.47111	2.40×10^{-3}
0.6	0.76541	0.76000	5.41×10^{-3}
0.8	1.09405	1.08440	9.65×10^{-3}
1.0	1.45946	1.44440	1.51×10^{-2}

Table 11.1   Trapezoidal rule for Example 11.3.

$$
\begin{bmatrix}
\frac{624}{625} & -\frac{2}{625} & -\frac{3}{625} & -\frac{4}{625} & -\frac{1}{250} \\[2mm]
-\frac{4}{625} & \frac{617}{625} & -\frac{12}{625} & -\frac{16}{625} & -\frac{2}{125} \\[2mm]
-\frac{9}{625} & -\frac{18}{625} & \frac{598}{625} & -\frac{36}{625} & -\frac{9}{250} \\[2mm]
-\frac{16}{625} & -\frac{32}{625} & -\frac{48}{625} & \frac{561}{625} & -\frac{8}{125} \\[2mm]
-\frac{1}{25} & -\frac{2}{25} & -\frac{3}{25} & -\frac{4}{25} & \frac{9}{10}
\end{bmatrix}
\begin{bmatrix}
u_1 \\[2mm] u_2 \\[2mm] u_3 \\[2mm] u_4 \\[2mm] u_5
\end{bmatrix}
=
\begin{bmatrix}
\frac{1}{5} \\[2mm] \frac{2}{5} \\[2mm] \frac{3}{5} \\[2mm] \frac{4}{5} \\[2mm] 1
\end{bmatrix}
$$

By using any of the standard methods described in Chapter 4, we get the approximate solution given in Table 11.1 along with the values of the exact solution of the integral equation $u(x) = x + \frac{4}{9}x^2$.

11.2.2   The Gauss-Nyström method

We now apply the N-point Gaussian quadrature rule to estimate the solution of the Fredholm equation

$$
f(x) = u(x) + \int_{-1}^{1} k(x,t)u(t)\,dt. \tag{11.12}
$$

Here we take the interval $[a, b]$ equal to $[-1, 1]$ knowing that to change the interval of integration from $[a, b]$ to $[-1, 1]$ one can use the linear transformation

$$
t = a + \frac{b-a}{2}(z+1), \qquad -1 \leq z \leq 1. \tag{11.13}
$$

The method is similar to the Trapezoidal except that here the integral $\int_{-1}^{1} k(x,t)u(t)\,dt$ is replaced by

$$
\int_{-1}^{1} k(x,t)u(t)\,dt = \sum_{j=1}^{N} w_j k(x, t_j)u(t_j) \tag{11.14}
$$

with weights w_j and abscissae t_j. The quadrature rule is discussed at length in Chapter 10.

By applying (11.14) to equation (11.4) we get

$$f(x) = u(x) + \sum_{j=1}^{N} w_j k(x, t_j) u(t_j). \qquad (11.15)$$

Evaluating equation (11.15) at the quadrature points, we get

$$f(x_i) = u(x_i) + \sum_{j=1}^{N} w_j k(x_i, t_j) u(t_j). \qquad (11.16)$$

Let $f(t_i) = f_i$, $k(t_i, t_j) = k_{ij}$, $u(t_i) = u_i$, and define

$$\bar{K}_{ij} = k_{ij} w_j$$

to get the matrix form of equation (11.16)

$$(\mathbf{I} + \bar{\mathbf{K}}) \cdot \mathbf{u} = \mathbf{f}. \qquad (11.17)$$

This is a set of N linear algebraic equations in N unknowns u_j. Hence, a table of values for $u(t)$ is obtained at each of the points t_j.

The above quadrature methods can be utilized for the Volterra linear equation by using a step-by-step procedure. If, for example, the Trapezoidal rule is used with equal steps, then the matrix generated in the quadrature method is a lower triangular.

EXAMPLE 11.4

Use a four-point Gaussian quadrature to solve the Fredholm integral equation

$$x = u(x) - \int_0^1 x^2 \, t \, u(t) \, dt.$$

Compare with the values of the exact solution $u(x) = x + \frac{4}{9} x^2$.

We have

$$\int_0^1 G(t) dt = \frac{1}{2} \int_{-1}^1 G\left(\frac{z+1}{2}\right) dz.$$

The weights and abscissae for the four-point Gaussian quadrature are given in Table 11.2.

It follows from (11.16):

$$u(x_i) - \frac{1}{2} \sum_{j=1}^4 w_j \, x_i^2 \, t_j \, u(t_j) = x_i$$

$$u_i - \frac{1}{2} \sum_{j=1}^4 w_j \, x_i^2 \, t_j \, u_j = x_i \qquad (11.18)$$

i	z_i	w_i	$\frac{1}{2}(z_i+1)$
1	0.33998	0.65215	0.669 99
2	0.86114	0.34785	0.930 57
3	−0.33998	0.65215	0.330 01
4	−0.86114	0.34785	0.069 43

Table 11.2   Weights and abscissae for the four-point Gaussian quadrature.

where $t_j = \frac{1}{2}(z_j+1)$, $x = x_i = t_j$, $j = 1, ..., 4$, and $u(t_j) = u_j$.

Expand equation (11.18) to get the system of equations

$$u_1 - 0.5\left[w_1\,t_1^2\,t_1\,u_1 + w_2 t_1^2\,t_2\,u_2 + w_3\,t_1^2\,t_3\,u_3 + w_4 t_1^2\,t_4\,u_4\right] = t_1$$

$$u_2 - 0.5\left[w_1\,t_2^2\,t_1\,u_1 + w_2 t_2^2\,t_2\,u_2 + w_3\,t_2^2\,t_3\,u_3 + w_4 t_2^2\,t_4\,u_4\right] = t_2$$

$$u_3 - 0.5\left[w_1\,t_3^2\,t_1\,u_1 + w_2\,t_3^2\,t_2\,u_2 + w_3\,t_3^2\,t_3\,u_3 + w_4 t_3^2\,t_4\,u_4\right] = t_3$$

$$u_4 - 0.5\left[w_1\,t_4^2\,t_1\,u_1 + w_2\,t_4^2\,t_2\,u_2 + w_3\,t_4^2\,t_3\,u_3 + w_4 t_4^2\,t_4\,u_4\right] = t_4.$$

which can be written in the form

$$u_1(1 - 0.5w_1\,t_1^2\,t_1) - 0.5w_2\,t_1^2\,t_2\,u_2 - 0.5w_3\,t_1^2\,t_3\,u_3 - 0.5w_4 t_1^2\,t_4\,u_4 = t_1$$

$$-0.5w_1\,t_2^2\,t_1\,u_1 + (1 - 0.5w_2\,t_2^2\,t_2)\,u_2 - 0.5w_3\,t_2^2\,t_3\,u_3 - 0.5w_4 t_2^2\,t_4\,u_4 = t_2$$

$$-0.5w_1\,t_3^2\,t_1\,u_1 - 0.5w_2\,t_3^2\,t_2\,u_2 + (1 - 0.5w_3\,t_3^2\,t_3)\,u_3 - 0.5w_4 t_3^2\,t_4\,u_4 = t_3$$

$$-0.5w_1\,t_4^2\,t_1\,u_1 - 0.5w_2\,t_4^2\,t_2\,u_2 - 0.5w_3\,t_4^2\,t_3\,u_3 + (1 - 0.5w_4 t_4^2\,t_4\,)u_4 = t_4$$

Substituting the values of w_j and t_j in Table 11.2 into these equations gives the system of equations in matrix form

$$\begin{bmatrix} 0.90194 & -0.07265 & -4.8304 \times 10^{-2} & -5.4205 \times 10^{-3} \\ -0.18919 & 0.85985 & -9.3185 \times 10^{-2} & -1.0457 \times 10^{-2} \\ -2.3793 \times 10^{-2} & -1.7627 \times 10^{-2} & 0.9883 & -1.3151 \times 10^{-3} \\ -1.0532 \times 10^{-3} & -7.8020 \times 10^{-4} & -5.1875 \times 10^{-4} & 0.99995 \end{bmatrix} \begin{bmatrix} u_1 \\ u_2 \\ u_3 \\ u_4 \end{bmatrix} = \begin{bmatrix} 0.66999 \\ 0.93057 \\ 0.33001 \\ 0.06943 \end{bmatrix}$$

Using any of the standard methods described in Chapter 4, we get the approximate solution given in Table 11.3 along with the values of the exact solution $u(x) = x + 4/9\,x^2$ of the integral equation.

One should know that the accuracy of this method depends entirely on the quadrature rule employed. To get the solution at some other points x, one could use equation (11.15) as an interpolating formula, in order to maintain the accuracy at the solution.

As an Example, let us estimate the solution $u(x)$ at $x = 0.2, 0.3, 05, 07$ by using (11.15) as the interpolation formula.

We have

$$u(x) = x + \frac{1}{2}\sum_{j=1}^{4} w_j\,x^2\,t_j\,u_j \tag{11.19}$$

x_i	u_i	Exact	Error
0.93057	1.3154	1.31544130	4.14×10^{-5}
0.66999	0.8695	0.86949516	4.84×10^{-6}
0.33001	0.3784	0.37841293	1.29×10^{-5}
0.06943	0.0716	0.07157246	2.75×10^{-5}

Table 11.3   Four-point Gauss-Nyström method for Example 11.4.

Gauss-Nyström method				
x_i	0.2	0.3	0.5	0.7
$u(x_i)$	0.21777	0.33999	0.61108	0.91772
Exact	0.21778	0.34000	0.61111	0.91778
Trapezoidal rule					
x_i	0.1	0.3	0.5	0.7	0.9
$u(x_i)$	0.10459	0.34135	0.61487	0.92514	1.27220
Exact	0.10444	0.34000	0.61111	0.91778	1.26000

Table 11.4   Approximation of the solution using an interpolating formula.

$$= x + x^2 \frac{1}{2} \sum_{j=1}^{4} w_j\, t_j\, u_j. \tag{11.20}$$

The summation in the right-hand side of the last equation is known and is computed by substituting the values of w_j, t_j, and $u(t_j)$ in Tables 11.2-11.3. That is

$$u(x) = x + x^2 \frac{1}{2} \left[\begin{array}{l} (0.65215)\,(0.66999)(0.8695) + (0.34785)\,(0.93057)(1.3154) \\ + (0.65215)\,(0.33001)(0.3784) + (0.34785)\,(0.06943)(0.0716) \end{array} \right].$$

Simplify, to obtain the interpolating formula

$$u(x) \simeq x + 0.4444x^2.$$

The interpolating formula using the Trapezoidal rule can be obtained in a similar way using the values in Table 11.1. That is

$$u(x) \simeq x + 0.4595x^2.$$

Some numerical values of these two formulas along with the values of exact solution are shown in Table 11.4.

EXERCISE SET 11.2

1. Approximate the solution of the following integral equations using the Trapezoidal rule with $n = 4$ and compare with the exact solution.

m-function 11.2

The MATLAB function **FredholmTrapez.m** approximates the solution of Fredholm integral equation of 2nd kind using the Trapezoidal rule. INPUTS are functions k and f; the limits of integration a and b; the number of subintervals n. The input functions $f(x)$ and $k(x,t)$ should be defined as M-files.

```
I=eye(n+1);
h=(b-a)/n;
for i=1:n+1
  v(i)=h;
% Find the vector f
  x=a+(i-1)*h;
  F(i)=feval(f,x);
  for j=1:n+1
% Find the matrix K
   t=a+(j-1)*h;
   K(i,j)=feval(k,x,t);
  end
end
v(1)=h/2;
v(n+1)=h/2;
% Find the matrix D
D=diag(v);
% Find the matrix (I+KD)
disp(' The matrix M = I+KD is  ')
M=I+K*D
disp(' The vector f is')
f=F'
% Solve the system
u=inv(M)*F';
y=[a:h:b]';
disp('Trapezoidal rule for solving Fredholm integral equation')
disp('_____')
disp('   xi          ui                           ')
disp('_____')
for i=1:n+1
  fprintf('%9.4f %12.6f \n',y(i),u(i))
end
```

m-function 11.2b

The MATLAB function **FredholmGN45.m** approximates the solution of Fredholm integral equation of 2nd kind using Gauss-Nystrom method. INPUTS are functions k and f; the limits of integration a and b; the number of points (4 or 5) n. The input functions $f(x)$ and $k(x,t)$ should be defined as M-files.

```
function FredholmGN45(k,f,a,b,n)
% Solve the Fredholm integral equation of the second kind
% using the 4 or 5 point Gauss_Nystrom method. k(x,t)=kernal, f=f(x) given.
if (n==4)
   % abscissae
   z(1)=-sqrt(1/7*(3-4*sqrt(0.3)));
   z(2)=-sqrt(1/7*(3+4*sqrt(0.3)));
   z(3)=-z(1);
   z(4)=-z(2);
   % weights
   w(1)=1/2+1/12*sqrt(10/3);
   w(2)=1/2-1/12*sqrt(10/3);
   w(3)=w(1); w(4)=w(2);
end
if (n==5)
   % abscissae
   z(1)=-sqrt(1/9*(5-2*sqrt(10/7)));
   z(2)=-sqrt(1/9*(5+2*sqrt(10/7)));
   z(3)=0;
   z(4)=-z(1);
   z(5)=-z(2);
   % weights
   w(1)=0.3*((-0.7+5*sqrt(0.7))/(-2+5*sqrt(0.7)));
   w(2)=0.3*((0.7+5*sqrt(0.7))/(2+5*sqrt(0.7)));
   w(3)=128/225;
   w(4)=w(1); w(5)=w(2);
end
I=eye(n);
fprintf('\n')
disp(' Gauss-Nystrom method for solving Fredholm integral equation')
fprintf('\n')
for i=1:n
% Compute the vetor f
   x=((b-a)*z(i)+a+b)/2;
   y(i)=x;
```

```
F(i)=feval(f,x);
 for j=1:n
% compute the matrix K
 t=((b-a)*z(j)+a+b)/2;
 K(i,j)=(b-a)/2*feval(k,x,t)*w(j);
 end
end
% Compute the matrix I+K
disp(' The matrix M = I+K is ')
M=I+K
disp(' The vector f is')
f=F'
u=inv(M)*F';
disp('_____')
disp('   xi        ui        ')
disp('_____')
for i=1:n
 fprintf('%8.4f %12.6f \n',y(i),u(i))
end
```

(a) $u(x) = \frac{5}{6}x + \frac{1}{2}\int_0^1 x\,t\,u(t)\,dt$, $u(x) = x$,

(b) $u(x) = -8x - 6x^2 + \int_0^1 (20\,x\,t^2 + 12\,x^2 t)u(t)\,dt$, $u(x) = x^2 + x$,

(c) $u(x) = \sec^2 x - 1 + \int_0^{\pi/4} u(t)\,dt$, $u(x) = \sec^2 x$,

(d) $u(x) = e^x + e^{-1}\int_0^1 u(t)\,dt$, $u(x) = e^x + 1$,

2. Approximate the solution of the following integral equations using a four-point Gauss-Nyström method

(a) $u(x) = xe^x - x + \int_0^1 x\,u(t)\,dt$,

(b) $u(x) = \sec^2 x - 1 + \int_0^{\pi/4} u(t)\,dt$, $u(x) = \sec^2 x$,

(c) $u(x) = x^2 - \frac{25}{12}x + 1 + \int_0^1 x\,t\,u(t)\,dt$,

(d) $u(x) = \cos x + \int_0^{2\pi} \sin x \, \cos t\, u(t)\,dt$, $u(x) = \cos x + \pi \sin x$,

(e) $u(x) = x\sin x - x + \int_0^{\pi/2} x\,u(t)\,dt$,

(f) $u(x) = x + \int_0^1 e^x\,e^t\,u(t)\,dt$,

(g) $u(x) = x^2 - \frac{1}{3}x + \frac{1}{4} + \int_0^1 (x+2)\,u(t)\,dt$, $u(x) = x^2 - \frac{11}{18}x - \frac{11}{36}$,

(h) $u(x) = x - \frac{1}{\pi}\int_0^{2\pi} \sin(x+t)\,u(t)\,dt$.

3. Approximate the solution of the following integral equation using both the Trapezoidal rule and the four-point Gaussian-Nyström method. Compare the results obtained by the two methods

$$u(x) = \sin x - \frac{x}{4} + \frac{1}{4}\int_0^{\pi/2} x\,t\,u(t)\,dt, \quad u(x) = \sin x.$$

11.3   THE SUCCESSIVE APPROXIMATION METHOD

Consider the Fredholm integral equation of the second kind

$$u(x) = f(x) + \int_a^b k(x,t)\, u(t)\, dt.$$

The successive approximation method consists of replacing the unknown function $u(x)$ under the integral by an initial guess function $u_0(x)$, $a \le x \le b$, to get the first approximation function $u_1(x)$. That is

$$u_1(x) = f(x) + \int_a^b k(x,t)\, u_0(t)\, dt.$$

The second approximation $u_2(x)$ is obtained in a similar manner

$$u_2(x) = f(x) + \int_a^b k(x,t) u_1(t)\, dt.$$

Similarly, we obtain the nth approximation

$$u_n(x) = f(x) + \int_a^b k(x,t) u_{n-1}(t)\, dt, \quad n \ge 1.$$

For the initial guess function, the most commonly selected function for $u_0(x)$ is $f(x)$ and $u(x)$ is obtained by taking the limit

$$u(x) = \lim_{n\to\infty} u_n(x). \tag{11.21}$$

It can be shown that the iteration converges if $k(x,t)$ is continuous for $a \le x \le b$, $a \le t \le b$ and if

$$\max_{a \le x \le b} \int_a^b |k(x,t)|\, dt < 1. \tag{11.22}$$

We now illustrate the method by the following example.

EXAMPLE 11.5

Consider the Fredholm integral equation

$$u(x) = x + \int_0^1 x^2\, t\, u(t)\, dt.$$

Use the successive approximation method to find $u(x)$. The exact solution is $u(x) = x + \frac{4}{9} x^2$.

Let $u_0(x) = x$. Substitute to get

$$u_1(x) = x + \int_0^1 \left(x^2 t\right) t \, dt = x + x^2 \int_0^1 t^2 dt$$
$$= x + \frac{1}{3} x^2 = x + \left(\frac{4}{9} - \frac{1}{9}\right)x^2.$$

The next approximation is

$$u_2(x) = x + \int_0^1 x^2 t \left(t + \frac{1}{3} t^2\right) dt = x + x^2 \int_0^1 t^2 dt + \frac{x^2}{3} \int_0^1 t^3 dt$$
$$= x + \frac{5}{12} x^2 = x + \left(\frac{4}{9} - \frac{1}{9 \cdot 4}\right)x^2.$$

The third approximation is

$$u_3(x) = x + \int_0^1 x^2 t \left(t + \frac{5}{12} t^2\right) dt = x + x^2 \int_0^1 t \left(t + \frac{5}{12} t^2\right) dt$$
$$= x + \frac{7}{16} x^2 = x + \left(\frac{4}{9} - \frac{1}{9 \cdot 4^2}\right) x^2.$$

Continuing in this manner, we get the nth approximation

$$u_n(x) = x + \left(\frac{4}{9} - \frac{1}{9 \cdot 4^{n-1}}\right) x^2.$$

Now, take the limit as $n \to \infty$ to get

$$\lim_{n \to \infty} u_n(x) = x + \frac{4}{9} x^2 = u(x).$$

EXERCISE SET 11.3

1. Find the solution of the following Fredholm integral equations using the successive approximation method.

   (a) $u(x) = \frac{5}{6} x + \frac{1}{2} \int_0^1 x \, t \, u(t) \, dt$,

   (b) $u(x) = -8x - 6x^2 + \int_0^1 (20 \, x \, t^2 + 12 \, x^2 t) u(t) \, dt$,

   (c) $u(x) = \frac{2}{e^2 - 1} \int_0^1 e^x \, e^t \, u(t) \, dt$,

   (d) $u(x) = \sec^2 x - 1 + \int_0^{\pi/4} u(t) \, dt$,

   (e) $u(x) = e^x + e^{-1} \int_0^1 u(t) \, dt$.

11.4   SCHMIDT's METHOD

We begin by considering integral equations with a degenerate kernel, i.e., kernels of the form

$$k(x,t) = \sum_{j=1}^{n} \alpha_j(x)\beta_j(t) \tag{11.23}$$

where the $\alpha_j(x)$ and $\beta_j(t)$ are continuous functions. The integral equation (11.6) takes the form

$$u(x) + \sum_{j=1}^{n} \alpha_j(x) \int_a^b \beta_j(t)u(t)\, dt = f(x). \tag{11.24}$$

Observe that the $\int_a^b \beta_j(t)u(t)dt$ are constants. They must be determined. Set

$$C_j = \int_a^b \beta_j(t)u(t)\, dt$$

so that (11.24) becomes

$$u(x) + \sum_{j=1}^{n} \alpha_j(x)C_j = f(x). \tag{11.25}$$

Multiply (11.25) by $\beta_i(x)$ and integrate the result from a to b. We find

$$C_i + \sum_{j=1}^{n} A_{ij}C_j = \varphi_i, \quad i = 1, ..., n. \tag{11.26}$$

where

$$A_{ij} = \int_a^b \beta_i(x)\alpha_j(x)dx \quad \text{and} \quad \varphi_i = \int_a^b f(x)\beta_i(x)\, dx.$$

(11.26) is a system of n equations in the n unknowns, C_i. Solve the system, insert the values obtained for the C_i into (11.25) to find the solution.

EXAMPLE 11.6
Solve

$$u(x) + \int_0^1 e^{x-t}u(t)\, dt = x.$$

The degenerate kernel is $k(x,t) = e^x e^{-t}$. Let $C = \int_0^1 e^{-t}u(t)dt$. We obtain

$$u(x) + Ce^x = x.$$

Multiply by e^{-x} and integrate from 0 to 1 to find

$$C + C = \int_0^1 xe^{-x}dx = -2e^{-1} + 1$$

$$C = -e^{-1} + \frac{1}{2}.$$

Thus, the solution is

$$u(x) = x + e^{x-1} - \frac{1}{2}e^x.$$

The method now involves approximating the given kernel, $k(x,t)$, by a degenerate kernel, solving the resulting integral equation with the degenerate kernel and using the result as our approximation.

EXAMPLE 11.7
Approximate the solution of the integral equation

$$u(x) + \int_0^1 \sin(xt)u(t)\, dt = 1.$$

Approximate the sine function by its power series representation. In this example, we take three terms

$$\sin(xt) \approx tx - \frac{1}{3!}t^3x^3 + \frac{1}{5!}t^5x^5.$$

The approximating integral equation is

$$u(x) + x\int_0^1 tu(t)\, dt - \frac{x^3}{3!}\int_0^1 t^3u(t)\, dt + \frac{x^5}{5!}\int_0^1 t^5u(t)\, dt = 1.$$

Let

$$C_1 = \int_0^1 tu(t)dt, \quad C_2 = \int_0^1 t^3u(t)dt, \quad C_3 = \int_0^1 t^5u(t)dt.$$

Multiply the last equation by x, x^3 and then x^5 and integrate the resulting equations from 0 to 1 to find three equations in three unknowns

$$\frac{4}{3}C_1 - \frac{1}{3!5}C_2 + \frac{1}{5!7}C_3 = \frac{1}{2},$$

$$\frac{1}{5}C_1 + \frac{41}{3!7}C_2 + \frac{1}{5!9}C_3 = \frac{1}{4},$$

$$\frac{1}{7}C_1 - \frac{1}{3!9}C_2 + \frac{1321}{5!11}C_3 = \frac{1}{6}.$$

The solution of the system to four decimal places is

$$C_1 = 0.3794, \quad C_2 = 0.1783, \quad C_3 = 0.1157.$$

So the approximate solution is

$$u(x) = 1 - 0.3794x + \frac{0.1783}{3!}x^3 - \frac{0.1157}{5!}x^5.$$

EXERCISE SET 11.4

Solve exactly or approximately the following four exercises using Schmidt's method.

1. $u(x) + \int_0^\pi \sin(x+t)u(t)\,dt = \cos x$.

2. $u(x) + \int_0^1 (1 + x^2 t)u(t)\,dt = x^3$.

3. $u(x) + \int_{-1}^1 e^{-xt}u(t)\,dt = 1$.

4. $u(x) + \int_0^1 \frac{1}{1+xt}u(t)\,dt = 1$.

5. Solve $u(x) + \int_0^\pi \cos(xt)u(t)\,dt = \sin x$ using the Trapezoidal rule, the Gauss-Nyström method, and Schmidt's method. Display your results graphically and compare the methods.

6. As in the above exercise, solve the integral equation

$$u(x) + \int_{-1}^1 e^{xt}u(t)\,dt = x.$$

11.5   VOLTERRA-TYPE INTEGRAL EQUATIONS

There are several approaches to solving approximately the Volterra-type integral equation

$$u(x) = f(x) + \int_a^x k(x,t)u(t)\,dt, \quad a \le x \le b. \tag{11.27}$$

The first technique is to observe that it can be reduced to a Fredholm equation if we set

$$K(x,t) = \begin{cases} k(x,t), & a \le t \le x \\ 0, & x < t \le b \end{cases}$$

and note that with this definition (11.27) becomes

$$u(x) = f(x) + \int_a^b K(x,t)u(t)\,dt. \tag{11.28}$$

This technique works but is usually inefficient.

A more direct approach is to note that (11.27) often arises from initial value problems in ordinary differential equations and try to adopt a method from ordinary differential equations.

For this purpose, choose an $h > 0$ and let $x_j = a + jh$, $j = 0, 1, 2, \ldots$ If $u(x_j)$ is the value of the solution at x_j, let u_j denote the approximation to u at x_j. Observe also that

$$u(a) = u_0 = f(a).$$

11.5.1 Euler's method

Set $x = x_1$ in (11.27) to get

$$u(x_1) = f(x_1) + \int_a^{x_1} k(x_1, s) u(s) \, ds$$

so

$$u_1 = f(x_1) + k(x_1, a) u_0 h$$
$$u(x_2) = f(x_2) + \int_a^{x_2} k(x_2, s) u(s) \, ds$$
$$= f(x_2) + \int_a^{x_1} k(x_2, s) u(s) \, ds + \int_{x_1}^{x_2} k(x_2, s) u(s) \, ds$$

so

$$u_2 = f(x_2) + k(x_2, a) u_0 h + k(x_2, x_1) u_1 h,$$
$$u_3 = f(x_3) + k(x_3, a) u_0 h + k(x_3, x_1) u_1 h + k(x_3, x_2) u_2 h.$$

More generally the approximation of $u(x_n)$ is

$$u_n = f(x_n) + \sum_{j=0}^{n-1} h k(x_n, x_j) u_j, \quad n = 0, 1, 2, ...$$

EXAMPLE 11.8

Approximate the solution of

$$u(x) = x + \int_0^x xt u(t) \, dt$$

using Euler's method with $h = 0.2$. The true solution is $u(x) = x e^{x^3/3}$.

___

We have $k(x, t) = xt$, $f(x) = x$, and $x_j = 0.2j$.

$$u_0 = f(0) = 0.0$$
$$u_1 = f(0.2) + k(0.2, 0)(0)(0.2) = 0.2$$
$$u_2 = f(0.4) + k(0.4, 0)(0)(0.2) + k(0.4, 0.2)(0.2)(0.2)$$
$$= 0.4 + 0.0032 = 0.4032$$

and so on. The results of the calculations together with the true values and the errors are given in Table 11.5.

Clearly the values of the approximation become worse as the calculations progress. Part of the problem comes from the coarse choice of h. Another source of the error comes from the rough approximation of the integral. Just as in the case of ordinary differential equations, we can improve the accuracy by using a better approximation to the integral and smaller values of h.

j	x_j	u_j	$u(x_j)$	$\|u_j - u(x_j)\|$	$\frac{\|u_j - u(x_j)\|}{\|u(x_j)\|}$
0	0.0	0.0	0.0	0.0	0.0
1	0.2	0.200000	0.200534	0.000534	0.002663
2	0.4	0.403200	0.408625	0.005425	0.013276
3	0.6	0.624154	0.644793	0.020640	0.032010
4	0.8	0.892124	0.948876	0.056753	0.059810
5	1.0	1.257894	1.395612	0.137718	0.098679

Table 11.5 Euler's method for Example 11.8.

11.5.2   Heun's method

Heun's method is based on the trapezoidal rule. The mesh spacing, h, must be chosen so that

$$\left(\max_{\substack{a \le x \le b \\ a \le t \le b}} |k(x,t)|\right)h < 1.$$

Again we note that in solving (11.27)

$$u_0 = u(x_0) = f(x_0).$$

We now note that

$$u(x_1) = f(x_1) + \int_a^{x_1} k(x_1, t)u(t)\, dt$$

and apply the trapezoidal rule to approximate $u(x_1)$:

$$u_1 = f(x_1) + \frac{h}{2}\left[k(x_1, t_0)u_0 + k(x_1, t_1)u_1\right]$$

and we now solve for u_1

$$\left[1 - \frac{k(x_1, t_1)h}{2}\right]u_1 = f(x_1) + \frac{h}{2}k(x_1, t_0)u_0$$

and now division yields u_1. To obtain u_2, we write

$$u(x_2) = f(x_2) + \int_a^{x_1} k(x_2, t)u(t)\, dt + \int_{x_1}^{x_2} k(x_2, t)u(t)\, dt.$$

The approximation is

$$u_2 = f(x_2) + \frac{h}{2}\left[k(x_2, t_0)u_0 + k(x_2, t_1)u_1\right] + \frac{h}{2}\left[k(x_2, t_1)u_1 + k(x_2, t_2)u_2\right]$$

or

$$\left[1 - \frac{k(x_2, t_2)h}{2}\right]u_2 = f(x_2) + \frac{h}{2}k(x_2, t_0)u_0 + k(x_2, t_1)u_1 h.$$

j	x_j	u_j	$u(x_j)$	$\|u_j - u(x_j)\|$	$\frac{\|u_j - u(x_j)\|}{\|u(x_j)\|}$
0	0.0	0.0	0.0	0.0	0.0
1	0.2	0.200803	0.200534	0.000269	1.341418×10^{-3}
2	0.4	0.409769	0.408625	0.001144	2.799633×10^{-3}
3	0.6	0.647809	0.644793	0.003016	4.677470×10^{-3}
4	0.8	0.956026	0.948876	0.007150	7.535231×10^{-3}
5	1.0	1.412794	1.395612	0.017182	1.231144×10^{-2}

Table 11.6   Heun's method for Example 11.9.

More generally the approximation of $u(x_{n+1})$ is

$$\left[1 - \frac{h}{2}k(x_{n+1}, t_{n+1})\right]u_{n+1} = f(x_{n+1}) + \frac{h}{2}k(x_{n+1}, t_0)u_0 + \sum_{j=1}^{n} k(x_{n+1}, t_j)u_j h.$$

EXAMPLE 11.9

Approximate the solution of

$$u(x) = x + \int_0^x xtu(t)\, dt$$

using Heun's method.

Note that $0 \le k(x,t) \le 1$ so we can choose $h = 0.2$. We have $k(x,t) = xt$, $f(x) = x$, and $x_j = 0.2j$.

$$u_0 = f(0) = 0$$
$$\left[1 - \frac{0.2}{2}k(0.2, 0.2)\right]u_1 = f(0.2) + \frac{0.2}{2}k(0.2, 0)u_0$$
$$u_1 = 0.200803$$
$$\left[1 - \frac{0.2}{2}k(0.4, 0.4)\right]u_2 = f(0.4) + \frac{0.2}{2}k(0.4, 0)u_0 + (0.2)\, k(0.4, 0.2)u_1$$
$$u_2 = 0.409769$$

The results of the calculations together with the true values and the errors are given in Table 11.6.

Clearly, Heun's method tracks the solution better and actually not that much more work.

m-function 11.5a

The MATLAB function **VolterraEuler.m** approximates the solution of Volterra integral equation of 2nd kind using Euler's method. INPUTS are functions k and f; the limits of integration a and b; the number of points n. The input functions $f(x)$ and $k(x,t)$ should be defined as M-files.

```
function VolterraEuler(k,f,a,b,n)
% Solve Volterra integral equation of the second kind
% using Euler's method. k(x,t)=kernel, f=f(x) given function.
fprintf('\n')
disp(' Euler''s method for solving Volterra integral equation')
disp('_____')
disp('   xi        ui        ')
disp('_____')
h=(b-a)/n;
u(1)=feval(f,a);
for i=2:n+1
% Compute the vector f
  x=a+(i-1)*h;
  F=feval(f,x);
  S=0;
  for j=1:i-1
    t=a+(j-1)*h;
    S=S+u(j)*h*feval(k,x,t);
  end
  u(i)=F+S;
end
y=[a:h:b]';
for i=1:n+1
  fprintf(' %9.4f %12.6f \n',y(i),u(i))
end
```

EXERCISE SET 11.5

1. The exact solution of the integral equation

$$u(x) = 1 + x^2 - \int_0^x u(t)\, dt, \quad 0 \le x \le 1$$

is $u(x) = 2x - 2 + 3e^{-x}$. Approximate the solution to this equation using both Euler's and Heun's methods with $h = 0.2$ and $h = 0.1$.

2. Given the differential equation

$$u'' - \frac{1}{x+1}u = x^2, \quad 0 \le x \le 2, \quad u(0) = 1, \quad u'(0) = 0$$

(a) Find an integral equation that is equivalent to this differential equation.

(b) Solve the integral equation using Heun's method.

m-function 11.5b

The MATLAB function **VolterraHeun.m** approximates the solution of Volterra integral equation of 2nd kind using Euler's method. INPUTS are functions k and f; the limits of integration a and b; the number of points n. The input functions $f(x)$ and $k(x,t)$ should be defined as M-files.

```
function VolterraHeun(k,f,a,b,n)
% Solve Volterra integral equation of the second kind
% using Heun's method. k(x,t)=kernel, f=f(x) given function.
fprintf('\n')
disp(' Heun''s method for solving Volterra integral equation')
disp('_____')
disp('    xi        ui                    ')
disp('_____')
h=(b-a)/n;
u(1)=feval(f,a);
for i=2:n+1
% Compute the vector f
  x=a+(i-1)*h;
  F=feval(f,x);
  S=0;
  for j=1:i-1
    t=a+(j-1)*h;
    if j==1
      S=S+u(j)*h/2*feval(k,x,t);
    else
      S=S+u(j)*h*feval(k,x,t);
    end
  end
  D=1-h/2*feval(k,x,x);
  u(i)=(F+S)/D;
end
y=[a:h:b]';
for i=1:n+1
  fprintf(' %9.4f %12.6f \n',y(i),u(i))
end
```

3. The solution of the differential equation

$$u'' + u = 0, \quad u(0) = 0, \quad u'(0) = 1$$

is $u(x) = \sin x$.

   (a) Find an equivalent integral equation for this problem.

   (b) Solve the integral equation on $0 \le x \le 2\pi$ using Heun's method.

   (c) Compare both numerical results with the true solution.

4. Heun's method can be modified so that it is a predictor-corrector method and can even be used to solve nonlinear equations. The point of this problem is to develop that algorithm. Consider

$$u(x) = f(x) + \int_a^x k(x, t, u(t)) \, dt.$$

The notation is the same as above. Again,

$$u(a) = u_0 = f(a)$$
$$u_1^* = f(x_1) + k(x_1, t_0, u_0)h$$
$$u_1 = f(x_1) + \frac{h}{2}[k(x_1, t_0, u_0) + k(x_1, t_1, u_1^*)].$$

Use Euler's method for the predictor to calculate u_2^* and then the trapezoidal rule with u_2^* in place of u_2 to calculate u_2. Continue in this way to calculate u_j for $j = 0, 1, 2, \ldots$.

Use the modified Heun's method to solve the integral equation

$$u(x) = \sin x + \int_a^x xt \cos u(t)dt, \quad 0 \le x \le 1.$$

APPLIED PROBLEMS FOR CHAPTER 11

1. The integral equation

$$p(t) = p_0 g(t) + \int_0^t g(t - \tau)p(\tau)d\tau, \quad 0 < \tau < t$$

forecast the human population $p(t)$, which depends on the initial population $p_0 = p(0)$ (number of people present at time $t = 0$) and the previous population $p(\tau)$. Here, $g(t)$ is the survival function, which gives the fraction of the number of people that survives to age t.

Solve the integral equation for $g(t) = e^{-t}$ and $p(0) = 1000$ at $t = 10$.

2. Suppose a new artificial lake has no fish and then it is stocked with a known number of a certain fish with different sizes. The fish is stocked at a known rate $s(t)$ per year. The fish multiply at an unknown rate $r(t)$. Assuming that the mortality rate is given by the function $e^{-\lambda t}$ (λ is a constant), we wish to determine $r(t)$. The solution to this model problem is given by the integral equation

$$N(t) = N(0)e^{-\lambda t} + \int_0^t [r(x) + s(x)]e^{-\lambda(t-x)}dx$$

where $N(t)$ and $s(t)$ are known to $t \ge 0$. Write the integral equation in the form

$$\phi(t) = \int_0^t r(x)e^{-\lambda(t-x)}dx$$

and then solve the corresponding integral equation.

3. One of the first integral equations to be investigated is due to Abel. It describes the case of a bead sliding down a frictionless wire at distance x in a

predetermined time $f(x)$. He was led to the integral equation in $\phi(x)$

$$-\sqrt{2g}\, f(x) = \int_0^x \frac{\phi(t)}{\sqrt{x-t}} dt$$

where g is the acceleration due to gravity. Verify that $\phi(t) = 1/2$ is a solution of Abel's problem for the case when $f(x) = x^{1/2}$.

4. In geometry, Bernouilli's problem consists of finding the shape of a curve $y = f(x)$ for which the area A under the curve on the interval $(0, x)$ is only a fraction k of the area of the rectangle circumscribing it, which is $k\,x\,f(x)$. The problem is represented by the integral equation

$$k\,x\,f(x) = \int_0^x f(\zeta)d\zeta.$$

Verify that the area under the parabola $y = x^2$ is one-third of the rectangle circumscribing it.

Chapter 12

Numerical Methods for Ordinary Differential Equations

Since the time of Newton, physical phenomena have been investigated by writing them in the form of differential equations. Because the rate of change of a quantity is more easily determined than the quantity itself from a physical experiment, most physical laws of science and engineering are expressed in terms of differential equations. A **differential equation** is an equation that involves at least one derivative of some unknown function.

In this chapter we shall be concerned with numerical methods for solving problems of the form

$$\frac{dy}{dt} = f(t, y) \quad \text{for } a \leq t \leq b \tag{12.1}$$

subject to an initial condition

$$y(a) = y_0.$$

Such problems are called **initial-value problems** (IVPs).

Fortunately, there are many good methods available for solving the initial-value problem (12.1). The methods we shall present here will serve to introduce the basic idea of most other methods and can be easily applied to systems of simultaneous first-order equations.

EXAMPLE 12.1 Newton's Law of Cooling

According to Newton's empirical law of cooling, the rate at which the temperature of a body changes is proportional to the difference between the temperature of the body and the temperature of the surrounding medium, the so-called ambient temperature. If $T(t)$ represents the temperature of a body at time t, T_m the temperature of the surrounding medium, and dT/dt the rate at which the temperature of the body changes, then Newton's law of cooling translates into the differential equation

$$\frac{dT}{dt} = k(T - T_m) \tag{12.2}$$

where k is a constant of proportionality. This equation is in the form of the prototype Eqn. (12.1).

EXAMPLE 12.2 : Chemical Reaction

The reaction between nitrous oxide and oxygen to form nitrogen dioxide is given by the balanced chemical equation $2NO + O_2 = NO_2$. At high temperatures, the dependence of the rate of this reaction on the concentrations of NO, O_2, and NO_2 is complicated. However, at 25^0C the rate at which NO_2 is formed obeys the law of mass action and is given by the differential equation

$$\frac{dx}{dt} = k(\alpha - x)^2(\beta - \frac{x}{2})$$

where $x(t)$ denotes the concentration of NO_2 at time t, k is the rate constant, α is the initial concentration of NO, and β is the initial concentration of O_2. This equation is in the form of the prototype Eqn. (12.1).

12.1   EULER'S METHOD

The simplest method of all the numerical methods for ordinary differential equations is **Euler's method**. Although this method is seldom used, it is relatively easy to analyze, and it will illustrate the central ideas involved in the construction of some more advanced methods.

Suppose we want to approximate the solution of the initial-value problem

$$\frac{dy}{dt} = f(t, y), \quad y(a) = y_0 \tag{12.3}$$

on some interval $[a, b]$.

To begin the presentation of Euler's method, let us divide the interval $[a, b]$ into N equal subintervals and define the *mesh points*

$$t_i = a + ih \quad \text{for } i = 0, 1, \ldots, N$$

with $h = (b - a)/N$. h is known as the **step size**. A numerical method for solving the IVP (12.3) will start with the initial condition y_0 and then generate values $y_1, y_2, \ldots, y_N$, which approximate the exact solution $y(t)$ at $t_1, t_2, \ldots, t_N$.

To derive Euler's method, we consider the Taylor expansion of y about t_i for each $i = 0, 1, \ldots, N - 1$. Assuming that $y(t)$ is twice continuously differentiable on $[a, b]$, we get

$$y(t_{i+1}) = y(t_i + h)$$
$$= y(t_i) + hy'(t_i) + \frac{h^2}{2}y''(\zeta) \tag{12.4}$$

for some ζ between t_i and $t_i + h$.

By dropping the error term in Eqn. (12.4) and using Eqn. (12.3), we obtain the formula

$$y(t_i + h) \approx y(t_i) + hf(t_i, y_i).$$

If we denote $y_i \approx y(t_i)$, then

$$y_{i+1} = y_i + hf(t_i, y_i), \quad i = 0, 1, \ldots, N - 1. \tag{12.5}$$

Eqn. (12.5) is known as **Euler's method**.

An algorithm for this method is shown below.

$$h \longleftarrow \frac{b-a}{N}$$
$$y(a) \longleftarrow y_0$$
$$\textbf{for} \qquad i = 0, 1, \ldots, N - 1$$
$$\lfloor \qquad y_{i+1} \longleftarrow y_i + hf(t_i, y_i)$$

A geometric interpretation of this method is given in Figure 12.1 and can be explained as follows:

Suppose we have found y_i at $t = t_i$. The equation of the tangent line to the graph

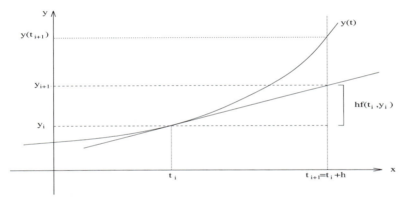

FIGURE 12.1
Geometric interpretation of Euler's method.

of $y(t)$ at $t = t_i$ is given by

$$y - y_i = m(t - t_i)$$

where $m = \frac{dy}{dt}\big|_{t=t_i} = f(t_i, y_i)$.

For $t = t_{i+1}$ and $y = y_{i+1}$, we have

$$y_{i+1} - y_i = m(t_{i+1} - t_i).$$

Thus,

$$y_{i+1} = y_i + hf(t_i, y_i).$$

This shows that the next approximation y_{i+1} is obtained at the point where the tangent to the graph of $y(t)$ at $t = t_i$ intersects with the vertical line $t = t_{i+1}$.

EXAMPLE 12.3

Solve the initial-value problem

$$\frac{dy}{dt} = 2t - y, \quad y(0) = -1$$

with $N = 10$ to get the value of y at $t = 1$. Compare with the values of the exact solution $y(t) = e^{-t} + 2t - 2$.

With $h = 1/10 = 0.1$ and $f(x, y) = 2t - y$, we have

$$y(0.1) \approx y_1 = y_0 + 0.1f(0, -1)$$
$$= -1 + (0.1)[2(0) - (-1)]$$
$$= -0.9$$

To calculate $y(0.2)$, the process is repeated starting from the point $(0.1, -0.9)$. The result is

$$y(0.2) \approx y_2 = -0.9 + (0.1)[2(0.1) - (-0.9)].$$
$$= -0.79.$$

Continuing in this manner leads to the values in Table 12.1. At $t = 1.0$ the Euler value is $y_{10} = 0.348678$ and the error is 0.0192.

Note that before calling the MATLAB function euler in Table 12.1, we must define the MATLAB function f1 as follows:

```
function f=f1(t,y)
f=2*t-y;
```

```
>> euler('f1',0,1,-1,10)

        Euler method
```

ti	f(t,y)	yi	Exact	Error
0.0	----	-1.000000	-1.000000	0
0.1	1.000000	-0.900000	-0.895163	4.84e-003
0.2	1.100000	-0.790000	-0.781269	8.73e-003
0.3	1.190000	-0.671000	-0.659182	1.18e-002
0.4	1.271000	-0.543900	-0.529680	1.42e-002
0.5	1.343900	-0.409510	-0.393469	1.60e-002
0.6	1.409510	-0.268559	-0.251188	1.74e-002
0.7	1.468559	-0.121703	-0.103415	1.83e-002
0.8	1.521703	0.030467	0.049329	1.89e-002
0.9	1.569533	0.187420	0.206570	1.91e-002
1.0	1.612580	0.348678	0.367879	1.92e-002

Table 12.1   Euler's method for Example 12.3.

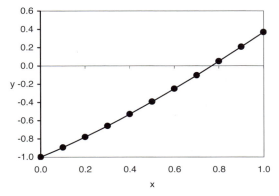

FIGURE 12.2
Euler solution and exact solution $y = e^{-t} + 2t - 2$, $n = 10$.

M-function 12.1

The following MATLAB function **euler.m** finds the solution of the initial-value problem 11.2 using Euler's method. INPUTS are a function $f(t,y)$; the initial and final values a, b of t; the initial condition y_0; the number of steps n. The input function $f(t,y)$ should be defined as an M-file.

```
function euler(f,a,b,y0,n)
% Solve the initial-value problem y'=f(x,y), y(a)=y0
% using Euler's method.
fprintf('\n')
disp('          Euler method')
disp('_____')
disp('   ti        f(ti,yi)        yi        Exact       error   ')
disp('_____')
fprintf('\n')
h=(b-a)/n;
y=y0;
fprintf('%6.2f      ----   %12.6f %12.6f  %4.2f\n',a,y,y,0)
for i=1:n
  t=a+(i-1)*h;
  m=feval(f,t,y);
  y=y+h*m;
% Write the exact solution g if known as g=g(t) otherwise set g='n'.
  t=t+h;
  g='n';
  if (g~='n')
    err=abs(g-y);
    fprintf('%6.2f %12.6f %12.6f %12.6f  %8.2e\n',t,m,y,g,err)
  else
    fprintf('%6.2f %12.6f %12.6f\n',t,m,y)
  end
end
```

EXERCISE SET 12.1

1. Given the initial-value problem

$$y' = -3y\sin(t), \quad \text{with} \quad y(0) = \frac{1}{2}$$

use Euler's method with $N = 10$ to approximate the solution. Compare with the values of the exact solution $y = \frac{1}{2}e^{3(\cos t - 1)}$.

2. Use Euler's method with $h = 0.1$ to approximate the solution of the IVP

$$y' = y - \frac{y}{t}, \quad 1 \le t \le 2, \quad y(1) = \frac{1}{2}.$$

Use the data points generated by Euler's method to find the best function that fits this data in the least squares sense. Use the resulting function to approximate the following values of y:

(a) $y(1.02)$,

(b) $y(1.67)$,

(c) $y(1.98)$.

Compare with the values of the exact solution $y = \frac{e^{t-1}}{2t}$.

3. Given the IVP

$$y' = -\frac{1}{1+t}y, \quad 0 \le t \le 1.6, \quad y(0) = 1$$

with the exact solution $y = \frac{1}{1+t}$:

(a) Use Euler's method to approximate $y(1.6)$ with $N = 8, 16$, and 32.

(b) Compare your results with the value of the exact solution and find which N gives the best approximation.

4. Given the IVP

$$y' = t - 2ty, \quad 0 \le t \le 1, \quad y(0) = 0$$

with the exact solution $y = \frac{1}{2}(1 - e^{-t^2})$, use Euler's method to approximate the solution with $N = 10$ and compare it with the actual values of y.

5. Use Euler's method to approximate the solution of the following IVPs in the interval indicated:

(a) $y' = -ty$, $[0, 1]$, $y(0) = 2$ with $h = 0.1$,

(b) $y' = -t\sin(t)y$, $[0, 1]$, $y(0) = 1$ with $h = 0.2$,

(c) $y' = \frac{1}{t}y$, $[1, 2]$, $y(1) = 1$ with $h = 0.2$,

(d) $y' = -t\sin(t)y$, $[0,1]$, $y(0) = 1$ with $h = 0.2$,

(e) $y' = \frac{1+y^2}{ty}$, $[2,3]$, $y(2) = 3$ with $h = 0.1$,

(f) $y' = \sin t + e^t$, $[0,5]$, $y(0) = 1$ with $h = 0.5$.

6. Use Euler's method to approximate the solution of the following IVPs in the interval indicated:

   (a) $y' = -e^t y$, $[0,1]$, $y(0) = 3$ with $N = 10$,

   (b) $y' = -4t^3 y$, $[1,2]$, $y(1) = 1$ with $N = 10$,

   (c) $y' = -2y$, $[0,1]$, $y(0) = 4$ with $N = 10$,

   (d) $y' = -\cos(t)y$, $[0,1.2]$, $y(0) = 2$ with $N = 12$.

COMPUTER PROBLEM SET 12.1

1. Write a computer program in a language of your choice to solve the initial-value problem (12.1) using Euler's method.

   Input data to the program should be

   (a) The function $f(t,y)$,

   (b) The interval $[a,b]$,

   (c) The initial value $y(a)$,

   (d) The step size h.

   Output should consist of the values of y at each step.

   Test your program to solve the IVP

   $$y' = (1-t)y, \qquad 0 \le t \le 3, \quad y(0) = 3$$

   with $h = 1$.

2. Use the MATLAB function euler to solve the IVP

   $$y' = ty^2, \qquad 0 \le t \le 2, \qquad y(0) = \frac{2}{5}.$$

   Approximate $y(t)$ up to $t = 2$ for step sizes $h = 1/2^n$, with $n = 1, 2, \ldots, 10$. Given the exact solution $y = \frac{2}{5-t^2}$, make log-log plots of the absolute error in $y(2)$ vs. h. What do these plots tell you?

3. Solve the IVP $y' = -t^2 + y^2$, $y(0) = 1$ using Euler's method. Use the step size $h = 0.5, 0.1$ to approximate $y(2)$.

4. Solve the IVP $y' = -t^2 y$ with $y(2) = 2$ using a Euler routine from a program library. Solve it with $h = 0.1, 0.01, 0.001$ to approximate $y(3)$.

5. Consider the IVP $y' = t - y + 1$, $y(0) = 1$. Use Euler's method with a sequence of decreasing h to approximate $y(1)$. The actual solution is $y = t + e^{-t}$. Plot the exact and numerical solutions, and report the largest error at the discrete values of t.

6. Consider the IVP $y' = 2ty$, $y(1) = 1$. Approximate $y(1.5)$ using Euler's method and compare the approximate solution with the exact one given by $y = e^{t^2-1}$. Show that with $h = 0.1$, there is a 16% relative error in the calculation of $y(1.5)$.

7. Use the MATLAB function euler to solve the IVP

$$y' = 2 + \sqrt{y - 2t + 3}, \quad y(0) = 1$$

on the interval $[0, 1.5]$. Compare with the exact solution $y(t) = 1 + 4t + \frac{1}{4}t^2$.

12.2   ERROR ANALYSIS

In solving an initial-value problem, it is useful to distinguish two types of errors: *local* and *global* truncation errors. Of course there is always the presence of the round-off error, but we assume here that there is no round-off error involved in any of the calculations.

The **local truncation error** is the error committed in the single step from t_i to t_{i+1}. For Euler's method, the local error is simply the remainder of Taylor's approximation (12.4), that is, $\frac{h^2}{2}y''(\zeta)$. Thus, we see that the local error is $O(h^2)$.

Since at each step of Euler's method an additional truncation error is introduced, the accumulation of these errors is called the **global truncation error**. We will now derive an error bound on the global error for Euler's method. Let us denote the global error after n steps by

$$e_n = y(t_n) - y_n, \quad n = 0, 1, \dots \tag{12.6}$$

where

$$y(t_n) = \text{value of the true solution at } t = t_n,$$
$$y_n = \text{value of the approximate solution at } t = t_n.$$

We have from Eqn. (12.4)

$$y(t_{n+1}) = y(t_n) + hf(t_n, y(t_n)) + \frac{h^2}{2}y''(\zeta_n), \quad t_n < \zeta_n < t_{n+1}. \tag{12.7}$$

Subtract (12.5) from (12.7) and use (12.6) to get

$$e_{n+1} = e_n + h \left[f(t_n, y(t_n)) - f(t_n, y_n) \right] + \frac{h^2}{2} y''(\zeta_n)$$

$$= e_n + h \frac{f(t_n, y(t_n)) - f(t_n, y_n)}{y(t_n) - y_n} [y(t_n) - y_n] + \frac{h^2}{2} y''(\zeta_n)$$

$$= e_n + h \frac{f(t_n, y(t_n)) - f(t_n, y_n)}{y(t_n) - y_n} e_n + \frac{h^2}{2} y''(\zeta_n). \tag{12.8}$$

By assuming continuity on $f(t, y)$ and f_y and using the mean value theorem, we get

$$e_{n+1} = e_n + h f_y(t_n, \xi_n) e_n + \frac{h^2}{2} y''(\zeta_n), \quad y_n < \xi_n < y(t_n). \tag{12.9}$$

Further, if we assume that on the region of interest in the ty-space

$$\left| \frac{\partial}{\partial y} f(t, y) \right| \leq M \quad \text{and} \quad |y''(t)| \leq K$$

where M and K are positive constants, we obtain

$$|e_{n+1}| < (1 + hM)|e_n| + \frac{h^2}{2} K. \tag{12.10}$$

We will now show by induction that

$$|e_n| \leq c[(1 + hM)^n - 1], \quad n = 0, 1, ..., \quad c = \frac{hK}{2M}. \tag{12.11}$$

Since $y(t_0) = y_0$, $e_0 = 0$, then (12.11) is true for $n = 0$. Assuming that (12.11) is true for an integer n, we need to show that it is true for $(n + 1)$. We have

$$|e_{n+1}| \leq c[(1 + hM)^{n+1} - 1]. \tag{12.12}$$

From (12.10) and (12.11) it follows that

$$|e_{n+1}| \leq c[(1 + hM)^n - 1](1 + hM) + \frac{h^2}{2} K$$

$$= c(1 + hM)^{n+1} - c(1 + hM) + \frac{h^2}{2} K.$$

Since $chM = \frac{h^2}{2} K$, simplify the last equation to get (12.12).

Finally, by using Taylor's theorem, it is easy to show that (see Exercise 1 below)

$$1 + x \leq e^x, \quad \text{for all } x \geq 0.$$

It follows that

$$(1 + hM) \leq e^{hM}.$$

Hence,

$$(1 + hM)^n \leq e^{nhM}.$$

Applying this result to Eqn. (12.11), we get

$$|e_n| \leq \frac{hK}{2M}(e^{nhM} - 1) = \frac{hK}{2M}(e^{(t_n - t_0)M} - 1). \tag{12.13}$$

This shows that the global error for Euler's method is $O(h)$. In other words, this means that the error bound is proportional to the step size h.

EXAMPLE 12.4

Consider the initial-value problem

$$y' = y - 1, \quad y(0) = 2, \quad 0 \leq t \leq 1 \quad \text{with } h = 0.1.$$

We have $f(t, y) = y - 1$ and $|\partial f / \partial y| = 1 = M$. Since the exact solution is $y = e^t + 1$, we have

$$|y''| \leq e = K, \quad \text{for all } t \text{ in } [0, 1].$$

Thus, Eqn. (12.13) gives the error bound

$$|e_n| \leq \frac{0.1e}{2}(e^{t_n} - 1).$$

Using a step size of $h = 0.1$ Table 12.2 gives the actual error, together with the error bound, obtained by using the MATLAB function euler.

In this example, the error bounds turn out to be reasonably sharp. However,

t_n	Actual error	Error bound
0.0	0.0	0.0
0.1	0.005171	0.014294
0.2	0.011403	0.030092
0.3	0.018859	0.047551
0.4	0.027725	0.066846
0.5	0.038211	0.088170
0.6	0.050558	0.111738
0.7	0.065036	0.137783
0.8	0.081952	0.166568
0.9	0.101655	0.198381
1.0	0.124539	0.233539

Table 12.2   Comparison of the actual error with the error bound for Example 12.4 using Euler's method with $h = 0.1$.

the global error bounds are usually hopelessly conservative as can be seen in the following example.

EXAMPLE 12.5

Consider the initial-value problem

$$y' = -y, \quad y(0) = 1, \quad 0 \le t \le 4 \quad \text{with } h = 0.1.$$

We have $f(t, y) = -y$ and $|\partial f / \partial y| = 1 = M$. The exact solution of this IVP is $y(t) = e^{-t}$. This implies that

$$|y''| = |e^{-t}| \le 1 = K \quad \text{for all } t \text{ in } [0, 4].$$

Thus,

$$|e_n| \le \frac{0.1}{2}(e^{t_n} - 1)$$

Table 12.3 lists the actual error together with this error bound. Note that in this

t_n	Actual error	Error bound
0.0	0.0	0.0
0.1	0.004837	0.005259
0.2	0.008731	0.011070
0.3	0.011818	0.017493
0.4	0.014220	0.024591
...	...	...
3.5	0.005166	1.605773
3.6	0.004795	1.779912
3.7	0.004448	1.972365
3.8	0.004123	2.185059
3.9	0.003819	2.420122
4.0	0.003535	2.679908

Table 12.3 Comparison of the actual error with the error bound for Example 12.5 using Euler's method with $h = 0.1$.

example, the error bound gets considerably larger than the actual error as t increases.

EXERCISE SET 12.2

1. Given the initial-value problem

$$y' = -\frac{1}{1+t}y, \quad 0 \le t \le 1, \quad y(0) = 1$$

with the exact solution $y = \frac{1}{1+t}$, use the latter along with Eqn. (12.13) to obtain an error bound for Euler's method with $h = 0.1$.

2. Show that
$$0 \le 1 + x \le e^x, \quad \text{for all } x \ge 0.$$

3. Let $f(t, y)$ be a function defined on $a \le t \le b$, $c \le y \le d$. We say that $f(t, y)$ satisfies a **Lipschitz condition** with respect to y if there exists a constant L such that
$$|f(t, y_1) - f(t, y_2)| \le L|y_2 - y_1|$$

for all $a \le t \le b$ and $c \le y_1, y_2 \le d$. If f and its partial derivative $\partial f / \partial y$ are continuous, and if
$$\left| \frac{\partial}{\partial y} f(t, y) \right| \le L,$$

show that f satisfies a Lipschitz condition on $a \le t \le b$ and $c \le y_1, y_2 \le d$ in the variable y with Lipschitz constant $L = $ constant.
Hint: Apply the mean value theorem to f for fixed t.

4. Given the initial-value problem
$$y' = -y - 2t - 1, \quad 0 \le t \le 1, \quad y(0) = 2$$

with the exact solution $y = e^{-t} - 2t + 1$, find the value of the step size h that produces at least two decimal digits of accuracy at $t = 1$, using Euler's method.

5. Show that the function
$$f(t, y) = t|y|$$

satisfies a Lipschitz condition on $[2, 3] \times [1, 2]$.

6. Find a bound for the local truncation error for Euler's method applied to $y' = 2ty$, $y(1) = 1$ given that the exact solution is $y = e^{t^2 - 1}$.

7. Consider the initial-value problem
$$y' = 2t - 3y + 1, \quad y(1) = 5.$$

The analytic solution is $y(t) = \frac{1}{9} + \frac{2}{3}t + \frac{38}{9}e^{-3(t-1)}$.

   (a) Find a formula involving c and h for the local truncation error in the nth step if Euler's method is used.

   (b) Find a bound for the local truncation error in each step $h = 0.1$ if Euler's method is used to approximate $y(1.5)$.

   (c) Approximate $y(1.5)$ using $h = 0.1$ and $h = 0.05$ with Euler's method.

   (d) Calculate the errors in part (c) and verify that the global truncation error of Euler's method is $0(h)$.

12.3   HIGHER-ORDER TAYLOR SERIES METHODS

Euler's method was derived from the Taylor Series using two terms in the series. It should be clear that we can construct an approximate solution of the initial-value problem (12.3) using a larger number of terms. In order to formalize this procedure, we will now derive an approximate solution to our initial-value problem using four terms in the Taylor series. Assuming that the solution $y(t)$ of the IVP (12.3) has four continuous derivatives, we have from the Taylor series expansion of $y(t)$ about $t = t_i$

$$y(t_{i+1}) = y(t_i) + hy'(t_i) + \cdots + \frac{h^3}{3!}y^{(3)}(t_i) + \frac{h^4}{4!}y^{(4)}(\zeta_i) \qquad (12.14)$$

with $t_i < \zeta_i < t_{i+1}$.

Neglecting the remainder term in (12.14), we get the difference equation

$$y_{i+1} = y_i + hf(t_i, y_i) + \frac{h^2}{2}f'(t_i, y_i) + \frac{h^3}{6}f''(t_i, y_i) \qquad (12.15)$$

where $i = 0, 1, \ldots, N - 1$.

By using the chain rule, we obtain

$$y' = f(t, y)$$
$$y'' = f' = f_t + f_y y' = f_t + f_y f$$
$$y''' = f'' = f_{tt} + 2f_{ty}f + f_{yy}f^2 + f_t f_y + f_y^2 f.$$

Putting the last three equations into Eqn. (12.15) gives the so-called **third-order Taylor's method** formula. It should be clear that continuing in this manner, any derivative of $y(t)$ can be expressed in terms of f and its partial derivatives. However, the practical difficulty of this method is that the various derivatives of y are complicated and, in some cases, impossible to find. Even though we have reduced the truncation error by adding more terms in the series, the method is generally impractical.

EXAMPLE 12.6

Use the third-order Taylor's method for the initial-value problem

$$\frac{dy}{dt} = 2t - y, \quad y(0) = -1$$

with $N = 10$ to approximate $y(1)$.

We have

$$f(t, y(t)) = 2t - y.$$

The first two derivatives of $f(t, y)$ are

$$f'(t, y) = 2 - y' = 2(1 - t) + y$$
$$f''(t, y) = -2 + y' = 2(t - 1) - y$$

so

$$y_{i+1} = y_i + h\psi_3(t_i, y_i) \tag{12.16}$$

with

$$\psi_3(t_i, y_i) = 2t_i - y_i + \frac{h}{2}[2(1 - t_i) + y_i] + \frac{h^2}{6}[2(t_i - 1) - y_i]$$
$$= 2t_i - y_i + \left(\frac{h}{2} - \frac{h^2}{6}\right)[2(1 - t_i) + y_i]$$

for $i = 0, 1, \ldots, N - 1$.

Since $N = 10$, $h = (1 - 0)/10 = 0.1$, then by substituting the value of h into (12.16), we get the difference equation

$$y_{i+1} = 0.9048333y_i + 0.1903333t_i + 0.0096667$$

with $t_i = 0.1i$ for $i = 0, 1, \ldots, 9$.

The values of the approximate solution y_i are shown in Table 12.4, along with the values of the exact solution $y(t) = e^{-t} + 2t - 2$ and the error.

ti	Taylor's method of order three	Exact solution	Error $|y(ti)-yi|$
0.00	-1.0	-1.0	0.0
0.10	-0.89516680	-0.89516258	4.22E-06
0.20	-0.78127689	-0.78126925	7.64E-06
0.30	-0.65919216	-0.65918178	1.04E-05
0.40	-0.52969250	-0.52967995	1.25E-05
0.50	-0.39348354	-0.39346934	1.42E-05
0.60	-0.25120380	-0.25118836	1.54E-05
0.70	-0.10343101	-0.10341470	1.63E-05
0.80	0.04931208	0.04932896	1.69E-05
0.90	0.20655245	0.20656966	1.72E-05
1.00	0.36786213	0.36787944	1.73E-05

Table 12.4 $y' = 2t - y$ using third-order Taylor's method with $h = 0.1$.

Note that one could also apply Taylor's method of order k given by the formula

$$y_{i+1} = y_i + h\psi_k(t_i, y_i) \quad \text{for } i = 0, 1, \ldots, N - 1, \tag{12.17}$$

where

$$\psi_k(t_i, y_i) = f(t_i, y_i) + \frac{h}{2!}f'(t_i, y_i) + \cdots + \frac{h^{k-1}}{k!}f^{(k-1)}(t_i, y_i)$$

to obtain a better accuracy to the solution of the initial-value problem (12.3). But as we mentioned before, the calculations of higher derivatives of $f(t, y)$ may become very complicated and time-consuming.

EXERCISE SET 12.3

1. Use one step of Taylor's method of order 2 with $h = 0.1$ to calculate an approximate value for $y(0.1)$ of the following initial-value problems:

   (a) $y' = -2ty^2$, $\quad 0 \le t \le 1$, $\quad y(0) = 1$,

   (b) $y' = 3(t - 1)^2$, $\quad 0 \le t \le 1$, $\quad y(0) = 1$.

2. Determine y'', given

   (a) $y' = 2t^2 y^2$,

   (b) $y' = ye^t$.

3. Given the IVP

$$y' = -3 - \frac{2}{t}y, \quad 1 \le t \le 1.4, \quad y(1) = 1$$

   and the exact solution $y(t) = \frac{2}{t^2} - t$, use Taylor's series of order two with $h = 0.2$ to approximate $y(1.4)$. Compare with the actual value.

4. Consider the IVP
$$y' = t + y + ty, \quad y(0) = 1.$$

   Use Taylor series method with terms through t^3 to approximate $y(0.1)$ and $y(0.5)$.

5. Derive the difference equation corresponding to Taylor's method of order two for the initial-value problems in Exercise 1:

   (a) $y' = -2ty^2$, $\quad 0 \le t \le 1$, $\quad y(0) = 1$,

   (b) $y' = 3(t - 1)^2$, $\quad 0 \le t \le 1$, $\quad y(0) = 1$.

6. Use Taylor series method of order 4 to approximate the solution of the IVP

$$y' = 4\cos t - ty, \quad y(0) = 1$$

   over the interval $[0, 3]$ with $h = 0.2$.

7. Consider the IVP
$$y' = t + y, \quad y(0) = 1$$

   which has the analytic solution $y(t) = 2e^t - t - 1$. Use Taylor's series of order 4 to estimate $y(0.5)$ with $h = 0.1$. Compute the error.

8. Solve the IVP
$$y' = \cos t - \sin y + t^2, \quad y(-1) = 3$$

   by using both the first- and second-order Taylor series methods to estimate $y(-0.8)$.

9. Use the Taylor series method with terms through t^4 to approximate the solution of the IVP

$$y' = ty^{1/3}, \quad y(1) = 1$$

in the interval $[1, 5]$ with $h = 0.5$.

COMPUTER PROBLEM SET 12.3

1. Write a computer program in a language of your choice for applying the Taylor series to the IVP

$$y' = y^2 - t^2, \quad y(0) = 1.$$

Generate a solution in the interval $[0, 0.5]$ with $h = 0.1$. Use derivatives up to $y^{(5)}$ in the Taylor series.

2. Write a computer program for applying the Taylor series of order three to the IVP

$$y' = t - 2y, \quad y(0) = 1$$

in $[0, 1]$ with $h = 0.1$. Compare with the analytic solution $y = \frac{1}{2}t - \frac{1}{4} + \frac{5}{4}e^{-2t}$.

3. Solve the IVP

$$y' = 2t - 3y + 1, \quad y(1) = 5$$

using the Taylor series of order 5 in $[1, 2]$ with $h = 0.1$.

12.4   RUNGE-KUTTA METHODS

As pointed out earlier, the Taylor methods are generally not practical because of the partial derivatives of $f(t, y)$, which need to be determined analytically. Consequently, Taylor's methods are rarely used. To avoid the differentiation of f, we will now derive methods that produce values y_i of the same accuracy as some of the Taylor methods without the necessity of differentiating $f(t, y)$. We shall now present the simplest of these methods known as the **Runge-Kutta method of order 2 (RK2)**.

The form of the formula for this method is obtained by replacing the function $h\psi_k$ in Eqn. (12.17) by the function $ak_1 + bk_2$. The result is

$$y_{i+1} = y_i + ak_1 + bk_2 \tag{12.18}$$

where

$$k_1 = hf(t_i, y_i)$$
$$k_2 = hf(t_i + \alpha h, y_i + \beta k_1),$$

and a, b, α, and β are constants to be determined so that (12.18) is as accurate as possible.

To derive these constants, we will make Eqn. (12.18) agree with the Taylor series expansion of $y(t)$ about t_i. We have

$$y(t_{i+1}) = y(t_i) + hy'(t_i) + \frac{h^2}{2}y''(t_i) + \cdots$$

$$= y(t_i) + hf(t_i, y(t_i)) + \frac{h^2}{2}f'(t_i, y(t_i)) + \cdots.$$

Since $f'(t_i, y(t_i)) = f_t + f_y f$, it follows that

$$y(t_{i+1}) = y(t_i) + hf + \frac{h^2}{2}(f_t + f_y f) + O(h^3) \tag{12.19}$$

where all the functions in Eqn. (12.19) are evaluated at the point $(t_i, y(t_i))$. We now expand $f(t_i + \alpha h, y_i + \beta k_1)$ in the Taylor's series for a function of two variables

$$f(t_i + \alpha h, y_i + \beta k_1) = f(t_i, y_i) + \alpha h f_t + \beta k_1 f_y + O(h^2). \tag{12.20}$$

So, Eqns. (12.18) and (12.20) imply that

$$y(t_{i+1}) = y(t_i) + h(a + b)f + bh^2(\alpha f_t + \beta f f_y) + O(h^3). \tag{12.21}$$

Again, all functions in (12.20) and (12.21) are evaluated at $(t_i, y(t_i))$.

Finally, by comparing (12.19) and (12.21), we obtain the system

$$\begin{cases} a + b = 1 \\ \alpha = \beta = \frac{1}{2b} \end{cases} \tag{12.22}$$

which is a system of three equations and four unknowns, and thus one variable can be chosen arbitrarily. One of the commonly used solutions is

$$a = b = \frac{1}{2} \quad \text{and} \quad \alpha = \beta = 1.$$

This leads to the Runge-Kutta method of order 2, sometimes known as the **modified Euler method**

$$y_{i+1} = y_i + \frac{h}{2}\left[f(t_i, y_i) + f(t_i + h, y_i + hf(t_i, y_i))\right], \quad i = 0, 1, \ldots, N-1$$

or

$$y_{i+1} = y_i + \frac{h}{2}(k_1 + k_2) \tag{12.23}$$

where

$$k_1 = f(t_i, y_i) \quad \text{and} \quad k_2 = f(t_i + h, y_i + hk_1).$$

Other choices for the parameters are $\alpha = \beta = 1/2$, $a = 0$, $b = 1$. This leads to a formula known as the **midpoint method** given by

$$y_{i+1} = y_i + hf\left(t_i + \frac{h}{2}, y_i + \frac{h}{2}f(t_i, y_i)\right), \quad i = 0, 1, \ldots, N-1. \tag{12.24}$$

An algorithm for the modified Euler method is given below.

$$h = \frac{a-b}{N}$$
$$y_0 \leftarrow y(a)$$
$$\textbf{for} \qquad i = 0, 1, ..., N - 1$$
$$k_1 \leftarrow f(t_i, y_i)$$
$$k_2 \leftarrow f(t_i + h, y_i + hk_1)$$
$$\lfloor \qquad y_{i+1} \leftarrow y_i + \frac{h}{2}(k_1 + k_2)$$

Similarly, an algorithm for the midpoint method is

$$h = \frac{a-b}{N}$$
$$y_0 \leftarrow y(a)$$
$$\textbf{for} \qquad i = 0, 1, \ldots, N - 1$$
$$k_1 \leftarrow f(t_i, y_i)$$
$$k_2 \leftarrow f(t_i + \frac{h}{2}, y_i + \frac{h}{2}k_1)$$
$$\lfloor \qquad y_{i+1} \leftarrow y_i + hk_2$$

Since (12.21) is $O(h^3)$, it follows that the local truncation error for both the midpoint and modified Euler methods is $O(h^3)$. Therefore, both methods are of the second order.

A very popular and often commonly used Runge-Kutta method is the **fourth-order method (RK4)** given by the formula

$$y_{i+1} = y_i + \frac{h}{6}(k_1 + 2k_2 + 2k_3 + k_4) \quad \text{for } i = 0, 1, \ldots, N - 1 \qquad (12.25)$$

with

$$k_1 = f(t_i, y_i)$$
$$k_2 = f(t_i + \frac{h}{2}, y_i + \frac{h}{2}k_1)$$
$$k_3 = f(t_i + \frac{h}{2}, y_i + \frac{h}{2}k_2)$$
$$k_4 = f(t_i + h, y_i + hk_3).$$

The local truncation error is $O(h^5)$. So, by using the Runge-Kutta method of order four, one can get approximations with accuracies equivalent to Taylor's method of order 4 and without requiring analytic differentiation of $f(t, y)$. To find a bound for the global error for the Runge-Kutta method of order four and higher is difficult (see [32]). However, if the local truncation error is of order $O(h^n)$, it can be shown that the global truncation error is of order $O(h^{n-1})$.

EXAMPLE 12.7

Use the midpoint method to obtain an approximation to the solution of the IVP

$$y' = 2t - y, \quad y(0) = -1$$

```
» midpoint('f1',0,1,-1,10)

        Midpoint method
```

ti	k1	k2	yi	Exact	Error
0.0	----	----	-1	-1	0
0.1	1.000000	1.050000	-0.895	-0.895163	1.63e-004
0.2	1.095000	1.140250	-0.780975	-0.781269	2.94e-004
0.3	1.180975	1.221926	-0.658782	-0.659182	3.99e-004
0.4	1.258782	1.295843	-0.529198	-0.52968	4.82e-004
0.5	1.329198	1.362738	-0.392924	-0.393469	5.45e-004
0.6	1.392924	1.423278	-0.250596	-0.251188	5.92e-004
0.7	1.450596	1.478067	-0.10279	-0.103415	6.25e-004
0.8	1.502790	1.527650	0.049975	0.049329	6.46e-004
0.9	1.550025	1.572524	0.207228	0.20657	6.58e-004
1.0	1.592772	1.613134	0.368541	0.367879	6.62e-004

Table 12.5 $y' = 2t - y$ using the midpoint method.

at $t = 1$ with $N = 10$.

With $h = 1/10 = 0.1$ and $f(x, y) = 2t - y$, the first step of the midpoint method for the approximation of $y(0.1)$ is

$$k_1 = f(0, -1) = 1.0$$
$$k_2 = f[0 + 0.1/2, -1 + 0.1(1/2)] = 1.05$$
$$y_1 = -1 + (0.1)(1.05) = -0.895.$$

In Table 12.5, the numerical solution is continued and compared with values obtained from the analytical solution. At $t = 1$, the midpoint value is

$$y_{10} = 0.3685409848.$$

EXAMPLE 12.8
Use the Runge-Kutta method of orders 4 and 2 to obtain an approximation to the solution of the IVP

$$y' = 2t - y, \quad y(0) = -1$$

at $t = 1$ with $N = 10$.

With $h = 1/10 = 0.1$ and $f(x, y) = 2t - y$, the first step of RK2 for the approximation of $y(0.1)$ is

$$k_1 = f(0, -1) = 1$$

$$k_2 = f(0 + 0.1, -1 + 0.1) = 1.1$$

$$y_1 = -1 + \frac{0.1}{2}(1 + 1.1) = -0.895.$$

The first step of RK4 with $h = 0.1$ and $f(t, y) = 2t - y$ is

$$k_1 = f(0, -1) = 1$$

$$k_2 = f(0 + \frac{0.1}{2}, -1 + \frac{0.1}{2}(1)) = 1.05$$

$$k_3 = f(0 + \frac{0.1}{2}, -1 + \frac{0.1}{2}(1.05)) = 1.0475$$

$$k_4 = f(0 + 0.1, -1 + (0.1)(1.0475)) = 1.09525$$

$$y_1 = -1 + \frac{0.1}{6}[1 + 2(1.05) + 2(1.0475) + 1.09525] = -0.8951625.$$

In Tables 12.6 and 12.7, the numerical solutions for RK2 and RK4, respectively, are continued and compared with values obtained from the analytical solution. At $t = 1$ the Runge-Kutta value of order 2 is

$$y_{10} = 0.3685409848$$

and the Runge-Kutta value of order 4 is

$$y_{10} = 0.3678797744.$$

Table 12.8 gives a comparison of the Euler's, midpoint, modified Euler's, and RK4

```
» rk2_4('f1',0,1,-1,10,2)

   Runge-Kutta method of order  2
```

ti	k1	k2	yi	Exact	Error
0.0	----	----	-1	-1	0
0.1	1.000000	1.100000	-0.895	-0.895163	1.63e-004
0.2	1.095000	1.185500	-0.780975	-0.781269	2.94e-004
0.3	1.180975	1.262878	-0.658782	-0.659182	3.99e-004
0.4	1.258782	1.332904	-0.529198	-0.52968	4.82e-004
0.5	1.329198	1.396278	-0.392924	-0.393469	5.45e-004
0.6	1.392924	1.453632	-0.250596	-0.251188	5.92e-004
0.7	1.450596	1.505537	-0.10279	-0.103415	6.25e-004
0.8	1.502790	1.552511	0.049975	0.049329	6.46e-004
0.9	1.550025	1.595022	0.207228	0.20657	6.58e-004
1.0	1.592772	1.633495	0.368541	0.367879	6.62e-004

Table 12.6 Runge-Kutta method of order 2 for Example 12.8.

methods for solving the IVP

$$y' = t + 3\frac{y}{t} \text{ over } [1, 2], \quad y(1) = 0 \quad \text{with } h = 0.1.$$

```
» rk2_4k('f1',0,1,-1,10,4)
```

Runge-Kutta method of order 4

ti	k1	k2	k3	k4	yi	Exact	Error
0.0	----	----	----	----	-1	-1	0
0.1	1.000000	1.050000	1.047500	1.095250	-0.895162	-0.895163	8.20e-008
0.2	1.095163	1.140404	1.138142	1.181348	-0.781269	-0.781269	1.48e-007
0.3	1.181269	1.222206	1.220159	1.259253	-0.659182	-0.659182	2.01e-007
0.4	1.259182	1.296222	1.294370	1.329745	-0.52968	-0.52968	2.43e-007
0.5	1.329680	1.363196	1.361520	1.393528	-0.393469	-0.393469	2.75e-007
0.6	1.393469	1.423796	1.422279	1.451241	-0.251188	-0.251188	2.98e-007
0.7	1.451188	1.478629	1.477257	1.503462	-0.103414	-0.103415	3.15e-007
0.8	1.503414	1.528244	1.527002	1.550714	0.049329	0.049329	3.26e-007
0.9	1.550671	1.573137	1.572014	1.593469	0.20657	0.20657	3.31e-007
1.0	1.593430	1.613759	1.612742	1.632156	0.36788	0.367879	3.33e-007

Table 12.7 Runge-Kutta method of order 4 for Example 12.8.

The exact solution to this problem is $y(t) = t^3 - t^2$.

ti	Exact solution	Euler's method	Error Euler	Midpoint method	Error Midpoint	Mod. Euler's method	Error Mod. Euler	RK4 method	Error RK4
1.0	0	0	0	0	0	0	0	0	0
1.1	0.121	0.100000	2.10E-02	0.119286	1.71E-03	0.118636	2.36E-03	0.120989	1.05E-05
1.2	0.288	0.237273	5.07E-02	0.283995	4.01E-03	0.282438	5.56E-03	0.287977	2.29E-05
1.3	0.507	0.416591	9.04E-02	0.500074	6.93E-03	0.497325	9.67E-03	0.506963	3.74E-05
1.4	0.784	0.642727	1.41E-01	0.773468	1.05E-02	0.769219	1.48E-02	0.783946	5.41E-05
1.5	1.125	0.920455	2.05E-01	1.110124	1.49E-02	1.104040	2.10E-02	1.124927	7.31E-05
1.6	1.536	1.254545	2.81E-01	1.515989	2.00E-02	1.507711	2.83E-02	1.535905	9.46E-05
1.7	2.023	1.649773	3.73E-01	1.997010	2.60E-02	1.986154	3.68E-02	2.022881	1.19E-04
1.8	2.592	2.110909	4.81E-01	2.559133	3.29E-02	2.545290	4.67E-02	2.591854	1.46E-04
1.9	3.249	2.642727	6.06E-01	3.208305	4.07E-02	3.191043	5.80E-02	3.248824	1.76E-04
2.0	4.000	3.250000	7.50E-01	3.950473	4.95E-02	3.929334	7.07E-02	3.999791	2.09E-04

Table 12.8 Comparison of the Euler's, Midpoint, Modified Euler's, and RK4 methods for solving $y' = t + 3\frac{y}{t}$, $y(1) = 0$.

MATLAB's Methods

MATLAB toolbox contains several built-in functions for solving ordinary differential equations. These are listed in Table 12.9 with their performance characteristics. For the theory and implementation of these MATLAB's routines we refer to [58]. The calling syntax is the same for all these functions. For example, the function ode23 can be called in the following ways:

```
[t,Y]=ode23(odefun,[t0 tn],y0)
[t,Y]=ode23(odefun,[t0 tn],y0,options, arg1,arg2,...)
```

odefun is the name of an M-file that defines the function $f(t, y)$. [t0 tn] defines the interval in which the solution is to be solved and y0 is the initial condition. If t0

Function	Description
ode113	Uses an explicit predictor-corrector method.
ode15s	Uses an implicit multistep method.
ode23	Uses Runge-Kutta schemes of order 2 and 3.
ede45	Uses Runge-Kutta schemes of order 4 and 5.

Table 12.9   MATLAB's Built-in function for solving ODE.

is not given, it is assumed to be zero. The **options** structure parameter allows you to control the details of computations. Use the null matrix [ ] as a place holder if no options are set. The **arg1,arg2,...** parameters are passed through the user-defined function **odefun**.

EXAMPLE 12.9 Chemical Reaction

Let $\alpha = 0.0010$ mole/L, $\beta = 0.0041$ mole/L, $k = 7.13 \times 10^3$ (L)^2/(mole)^2(sec.), and $x(0) = 0$ mole/L. Use the MATLAB function **ode45** to approximate the concentration of NO in the time interval $[0, 10]$ (see Example 12.2).

First, we create an M-file that we name **odefun1.m**, which defines the ODE:

```
function f=odefun1(t,x,flag,k,alpha,beta)
   f = k*(alpha-x).^2*(beta-x);
```

The **flag** parameter is required for compatibility with **ode45**.

Using MATLAB, we get

```
>> tn = 10; x0 = 0;
>> k = 7130; alpha = 0.0010; beta = 0.0041;
>> [t,x] = ode45('odefun1',tn,x0,[ ],k,alpha,beta);
>> plot(t,x,'*')
```

The result is shown in Fig. 12.3

EXAMPLE 12.10 Newton's Law of Cooling

Use the MATLAB function **ode23** to approximate the temperature T at $t = 4$ min. of a ceramic insulator baked at 400^0C and cooled in a room in which the temperature is 25^0C. Use $k = -0.213$ (see Example 12.1).

First, we create an M-file that we name **odefun2.m**, which defines the ODE:

```
function f=odefun2(t,y,flag,k,Tm)
```

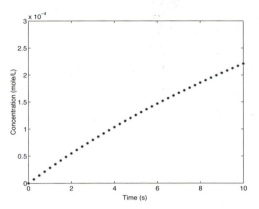

FIGURE 12.3
ode45 solution of the chemical reaction problem.

```
    f = k*(y-Tm);
```

Using `ode23` with $T_0 = T(0) = 400$ and $T_m = 25$, we get

```
>> tn = 4; y0 = 400;
>> k = -0.213; Tm = 25;
>> [t,y] = ode23('odefun2',tn,y0,[ ],k,Tm);
```

MATLAB gives the approximate value $T(4) = 184.96$.

EXERCISE SET 12.4

1. Use the Runge-Kutta method of order 4 with $h = 0.1$ to approximate the solution of IVP
$$y' = y - \frac{y}{t}, \quad 1 \le t \le 2, \quad y(1) = \frac{1}{2}.$$
Use the data points generated by the Runge-Kutta method to find the best function that fits this data in the least squares sense. Use the resulting function to approximate the following values of y:

   (a) $y(1.02)$,

   (b) $y(1.67)$,

   (c) $y(1.98)$.

   Compare with the values of the exact solution $y = \frac{e^{t-1}}{2t}$.

2. Given the IVP
$$y' = e^{t^2} - \frac{y}{t}, \quad 1 \le t \le 2, \quad y(1) = \frac{e}{2}$$

M-function 12.4a

The following MATLAB function **midpoint.m** finds the solution of the initial-value problem 11.2 using the midpoint method. INPUTS are a function $f(t,y)$; the initial and final values a, b of t; the initial condition y_0; the number of steps n; The input function $f(t,y)$ should be defined as an M-file.

```
function midpoint(f,a,b,y0,n)
% Solve the initial-value problem y'=f(t,y), y(a)=y0
% using the midpoint method.
fprintf('\n')
disp('             Midpoint method')
disp('_____')
disp('   t         k1        k2        y       Exact      error   ')
disp('_____')
fprintf('\n')
h=(b-a)/n;
y=y0;
m=feval(f,a,y0);
fprintf('%6.2f      ----        ----  %12.6f\n',a,y)
for i=1:n
  t=a+(i-1)*h;
  k1=feval(f,t,y);
  k2=feval(f,t+h/2,y+h*k1/2);
  y=y+h*k2;
  t=t+h;
% Enter the exact solution if known as g=g(t) otherwise set g='n'.
  g='n';
  if (g~='n')
    err=abs(g-y);
    fprintf('%6.2f %12.6f %12.6f %12.6f %12.6f  %8.2e\n', t, k1, k2, y, g, err)
  else
    fprintf('%6.2f %12.6f %12.6f %12.6f\n',t,k1,k2,y)
  end
end
```

use the Runge-Kutta method of order 2 with $N = 10$ to approximate the solution and compare with the values of the exact solution $y = \frac{1}{2t}e^{t^2}$.

3. Use the answers generated in Exercise 4 below and the linear-least squares method to approximate y at

   (a) $t = 1.25$,
   (b) $t = 1.65$.

   Compare your results with the actual values of y.

4. Use the Runge-Kutta method of order 2 to approximate the solution of the following IVPs in the interval indicated:

M-function 12.4b

The following MATLAB function **rk2_4.m** finds the solution of the initial-value problem 11.2 using the Runge-Kutta methods. INPUTS are a function *f(t,y)*; the initial and final values *a*, *b* of *t*; the initial condition y_0; the number of steps *n*; the RK order 2 or 4. The input function *f(t,y)* should be defined as an m-file.

```
function rk2_4(f,a,b,y0,n,order)
% solve the initial-value problem y'=f(t,y), y(a)=y0
% using the Runge-Kutta methods.
fprintf('\n')
disp(['      Runge-Kutta method of order = ',num2str(order)])
h=(b-a)/n;
y=y0;
if (order==2)
disp('_____')
disp('   t       k1       k2       y       Exact     error    ')
disp('_____')
fprintf('\n')
fprintf('%6.2f      ----       ----   %12.6f %12.6f  %4.2f\n',a,y,y,0)
   for i=1:n
     t=a+(i-1)*h;
     k1=feval(f,t,y);
     k2=feval(f,t+h,y+h*k1);
     y=y+h*(k1+k2)/2;
     t=t+h;
% Enter the exact solution if known as g=g(t) otherwise set g='n'.
     g='n';
     if (g~='n')
       err=abs(g-y);
       fprintf('%6.2f %12.6f %12.6f %12.6f %12.6f  %8.2e\n',t,k1,k2,
y, g, err)
     else
       fprintf('%6.2f %12.6f %12.6f %12.6f\n',t,k1,k2,y)
     end
   end
end
if (order==4)
disp('_____')
disp('   t    k1     k2     k3     k4      y        Exact    error ')
disp('_____')
fprintf('\n')
```

```
fprintf('%6.2f     ----      ----      ----     ---- %12.6f %12.6f
%4.2f\n',a,y,y,0)
  for i=1:n
    t=a+(i-1)*h;
    k1=feval(f,t,y);
    k2=feval(f,t+h/2,y+h*k1/2);
    k3=feval(f,t+h/2,y+h*k2/2);
    k4=feval(f,t+h,y+h*k3);
    y=y+h*(k1+2*k2+2*k3+k4)/6;
    t=t+h;
% Enter the exact solution if known as g=g(t) otherwise set g='n'.
    g=exp(-t)+2*t-2;
    if (g~='n')
      err=abs(g-y);
      fprintf('%6.2f %12.6f %12.6f %12.6f %12.6f %12.6f %12.6f
%8.2e\n', t, k1, k2, k3, k4, y, g, err)
    else
        fprintf('%6.2f %12.6f %12.6f %12.6f %12.6f %12.6f\n',t,k1,k2,
k3, k4, y)
    end
  end
end
```

(a) $y' = -e^t y$, $[0, 1]$, $y(0) = 3$ with $N = 10$,

(b) $y' = -4t^3 y$, $[1, 2]$, $y(1) = 1$ with $N = 10$,

(c) $y' = -2y$, $[0, 1]$, $y(0) = 4$ with $N = 10$,

(d) $y' = -\cos(t)y$, $[0, 1.2]$, $y(0) = 2$ with $N = 12$.

5. Repeat Exercise 4 using the Runge-Kutta method of order 4.

6. Use the Runge-Kutta method of order 4 to approximate the solution of the following IVPs in the interval indicated:

(a) $y' = -ty$, $[0, 1]$, $y(0) = 2$ with $h = 0.1$,

(b) $y' = -t \sin(t)y$, $[0, 1]$, $y(0) = 1$ with $h = 0.2$,

(c) $y' = \frac{1}{t}y$, $[1, 3]$, $y(1) = 1$ with $h = 0.2$,

(d) $y' = \frac{1+y^2}{ty}$, $[2, 3]$, $y(2) = 3$ with $h = 0.1$,

(e) $y' = \sin t + e^t$, $[0, 5]$, $y(0) = 1$ with $h = 0.5$.

7. When $f(t, y)$ depends only on t, i.e., $f(t, y) = f(t)$, show that the Runge-Kutta method of order 4 reduces to Simpson's rule

$$\int_{t_n}^{t_n+h} f(t)dt = y_{n+1} - y_n$$

$$\approx \frac{h}{6}\left[f(t_n) + 4f(t_n + \frac{h}{2}) + f(t_n + h)\right].$$

8. The exact solution to the IVP

$$y' = y \quad \text{with} \quad y(0) = 1$$

is $y = e^t$. Apply the Runge-Kutta method of order 4 to approximate e^t for $t = 0.5$ and $t = 1.0$, and show that the first five decimal places of the result are correct.

9. Show that the solution of the IVP

$$y' = e^{-x^2}, \quad y(0) = 0$$

is

$$y(x) = \int_0^x e^{-t^2} dt.$$

Use the Runge-Kutta method of order 4 with $h = 0.05$ to approximate the value of

$$y(1) = \int_0^1 e^{-t^2} dt.$$

COMPUTER PROBLEM SET 12.4

1. Write a computer program in a language of your choice to solve the IVP (12.1) using the Runge-Kutta method of order 4.

   Input data to the program should be

   (a) The function $f(t, y)$,
   (b) The interval $[a, b]$,
   (c) The initial value $y(a)$,
   (d) The step size h.

   Output should consist of the values of y at each time step.

   Test your program to solve the IVP

   $$y' = (1 - t)y, \qquad 0 \le t \le 3, \quad y(0) = 3$$

   with $h = 1$.

2. Solve the IVP $y' = -ty^2$, $y(2) = 1$ using Runge-Kutta method of order 4. Use the step size $h = 1, 0.1$ to approximate $y(3)$.

3. Solve the IVP $y' = -t^2 y$ with $y(2) = 2$ using a Runge-Kutta routine from a program library. Solve it with $h = 0.1, 0.01, 0.001$ to approximate $y(3)$.

4. Consider the IVP $y' = \sqrt{t+y}$, $y(0.4) = 0.41$. Use Runge-Kutta method of order 2 to approximate $y(0.8)$ with $h = 0.01$.

5. Consider the IVP $y' = -y + 10\sin 3t$ with $y(0) = 0$.

   (a) Use the Runge-Kutta method of order 4 with $h = 0.1$ to approximate the solution in the interval $[0, 2]$.

   (b) Using the result in (a) obtain an interpolating function and graph it. Find the positive roots of the interpolating function on the interval $[0, 2]$.

6. To show an interesting fact about Runge-Kutta methods, we consider the following initial-value problems

$$y' = 2(t+1), \quad y(0) = 1$$
$$y' = \frac{2y}{t+1}, \quad y(0) = 1.$$

   They both have the same solution $y = (t+1)^2$.

   (a) Use the Runge-Kutta method of order 2 with $h = 0.1$ to approximate the solution in $[0, 1]$ of the two initial-value problems.

   (b) Compare the approximate solutions with the actual values of y.

   (c) Show that for the first equation RK2 gives the exact results but not for the second equation, although the exact solution is the same for both equations. The interesting fact is that the error for Runge-Kutta methods depends on the form of the equation as well as on the solution itself.

7. Consider the IVP $y' = 2ty$, $y(1) = 1$. Approximate $y(1.5)$ using Runge-Kutta method of order 4 and compare the approximate solution with the exact one given by $y = e^{t^2-1}$.

8. Use the MATLAB function rk2_4 to solve the IVP

$$y' = 2 + \sqrt{y - 2t + 3}, \quad y(0) = 1$$

   on the interval $[0, 1.5]$. Compare with the exact solution $y(x) = 1 + 4t + \frac{1}{4}t^2$.

9. Use the MATLAB function rk2_4 to solve the IVP

$$y' = ty + \sqrt{y}, \quad y(0) = 1$$

   in $[0, 1]$ with $h = 0.1$.

10. A definite integral $\int_a^b f(t)dt$ can be evaluated by solving the initial-value problem for $y(b)$

$$y' = f(t), \quad a \le t \le b, \quad y(a) = y_0$$

    since

$$I = \int_a^b f(t)dt = \int_a^b y'(t)dt = y(b) - y(a) = y(b) - y_0.$$

Thus, $y(b) = I + y_0$. Use Runge-Kutta method of order 4 with $h = 0.1$ to approximate the value of the error function given by

$$erf(t) = \frac{2}{\sqrt{\pi}} \int_0^t e^{-x^2} dx$$

at $t = 1, 1.5, 2$.

11. Use the MATLAB function rk2_4 to solve the IVP

$$y' = ty^2, \qquad 0 \le t \le 2, \qquad y(0) = \frac{2}{5}.$$

Approximate $y(t)$ up to $t = 2$ for step sizes $h = 1/2^n$, with $n = 1, 2, \ldots, 10$. Given the exact solution $y = \frac{2}{5-t^2}$, make log-log plots of the absolute error in $y(2)$ vs. h. What do these plots tell you?

12. The solution of the IVP

$$y' = \frac{2}{t^4} - y^2, \quad y(1) = -0.414$$

crosses the t-axis at a point in the interval $[1, 2]$. By experimenting with the MATLAB function rk2_4 determine this point.

13. The solution to the IVP

$$y' = y^2 - 2e^t y + e^{2t} + e^t, \quad y(0) = 3$$

has a vertical asymptote at some point in the interval $[0, 2]$. By experimenting with the MATLAB function rk2_4 determine this point.

14. Use the Runge-Kutta method of order 4 with $h = 0.05$ to approximate the solution in $[1, 1.5]$ of the IVP

$$y' = \frac{y^2 + ty - t^2}{t^2}, \quad y(1) = 2.$$

Plot the error of these approximate values given that the exact solution is $y(t) = \frac{t(1+t^2/3)}{1-t^2/3}$.

15. The implicit solution of the IVP

$$y' = -\frac{4t^3 y^3 + 2ty^5 + 2y}{3t^4 y^2 + 5t^2 y^4 + 2t}, \quad y(1) = 1$$

is $t^4 y^3 + t^2 y^5 + 2ty = 4$. Use the Runge-Kutta method of order 4 with $h = 0.1, 0.05$ to approximate the values of $y(t)$ at $t = 1.1, 1.2, \ldots, 2.0$. To check the error in these approximate values, construct a table of the approximate values and the residual

$$R(t, y) = t^4 y^3 + t^2 y^5 + 2ty - 4.$$

12.5 MULTISTEP METHODS

All the methods discussed so far in the previous sections of this chapter have one thing in common, that is, y_{i+1} was computed knowing only y_i and values of f and its derivatives. So, these are called **single-step methods**. On the other hand, if the knowledge of $y_i, y_{i-1}, \ldots, y_{i-k+1}$ is required for the computation of y_{i+1}, the method is referred to as a **multistep method**. One approach to the derivation of these methods is based on the idea of numerical integration and the use of finite-difference methods.

By integrating the solution of the IVP (12.3) between t_i and t_{i+1}, we obtain

$$y(t_{i+1}) - y(t_i) = \int_{t_i}^{t_{i+1}} f(t, y(t))dt.$$

Thus,

$$y(t_{i+1}) = y(t_i) + \int_{t_i}^{t_{i+1}} f(t, y(t))dt. \tag{12.26}$$

To carry out the integration in (12.26) we can use a finite-difference method to approximate $f(t, y)$ at some of the data points $t_0, t_1, \ldots, t_i$. This will lead to the formula

$$y_{i+1} = y_i + \int_{t_i}^{t_{i+1}} p(t)dt \tag{12.27}$$

where $y_i \approx y(t_i)$ and $p(t) \approx f(t, y)$.

12.6 ADAMS-BASHFORTH METHODS

In the Adams-Bashforth (AB) method, we suppose that $p(t)$ is given by the Newton backward-difference polynomial $p_{m-1}(t)$, derived in Section 5.2, through m points $(t_i, y(t_i)), \ldots, (t_{i-m+1}, y(t_{i-m+1}))$. Substituting $p_{m-1}(t)$ into Eqn. (12.27), we obtain

$$y_{i+1} = y_i + \int_{t_i}^{t_{i+1}} \sum_{k=0}^{m-1} (-1)^k \binom{-s}{k} \nabla^k f(t_i, y_i)dt.$$

By making the change of variable $s = \frac{t - t_i}{h}$, we obtain

$$y_{i+1} = y_i + h \sum_{k=0}^{m-1} \nabla^k f(t_i, y_i)(-1)^k \int_0^1 \binom{-s}{k} ds \tag{12.28}$$

where

$$\frac{1}{k!h^k} \nabla^k f(t_n, y_n) = f[(t_{n-k}, y_{n-k}), \ldots, (t_n, y_n)]$$

and

$$\binom{-s}{k} = (-1)^k \frac{s(s+1)\cdots(s+k-1)}{k!}.$$

The coefficients

$$\gamma_k = (-1)^k \int_0^1 \binom{-s}{k} ds$$

can be easily computed from the definition of the binomial coefficient taken as a function of s. As an example, let us calculate γ_k for $k = 0, 1, 2, 3$. We have

$$\gamma_k = \int_0^1 \frac{s(s+1)\cdots(s+k-1)}{k!} ds.$$

Thus,

$$\gamma_0 = \int_0^1 ds = 1$$

$$\gamma_1 = \int_0^1 s\, ds = \frac{1}{2}$$

$$\gamma_2 = \int_0^1 \frac{s(s+1)}{2} ds = \frac{1}{2} \int_0^1 (s^2 + s) ds = \frac{5}{12}$$

$$\gamma_3 = \int_0^1 \frac{s(s+1)(s+2)}{6} ds = \frac{1}{6} \int_0^1 (s^3 + 3s^2 + 2s) ds = \frac{3}{8}.$$

Formula (12.28) is known as **Adams-Bashforth m-step method**. Observe that if $m = 1$ in (12.28), the Adams-Bashforth method leads to Euler's method. A list for the cases $m = 2, 3, 4$, along with the local errors follows:

Adams-Bashforth two-step method (AB2) ($m = 2$):

$$y_{i+1} = y_i + \frac{h}{2} \left[3f(t_i, y_i) - f(t_{i-1}, y_{i-1}) \right], \qquad (12.29)$$

$$y_0 = a_0, \quad y_1 = a_1,$$

where $i = 1, 2, \ldots, N - 1$. The local truncation error is $O(h^3)$.

Adams-Bashforth three-step method (AB3) ($m = 3$):

$$y_{i+1} = y_i + \frac{h}{12} \left[23f(t_i, y_i) - 16f(t_{i-1}, y_{i-1}) + 5f(t_{i-2}, y_{i-2}) \right], \qquad (12.30)$$

$$y_0 = a_0, \quad y_1 = a_1, \quad y_2 = a_2,$$

where $i = 2, 3, \ldots, N - 1$. The local truncation error is $O(h^4)$.

Adams-Bashforth four-step method (AB4) $(m = 4)$:

$$y_{i+1} = y_i + \frac{h}{24} \left[55f(t_i, y_i) - 59f(t_{i-1}, y_{i-1}) + 37f(t_{i-2}, y_{i-2}) \right.$$
$$\left. -9f(t_{i-3}, y_{i-3}) \right], \tag{12.31}$$

$$y_0 = a_0, \quad y_1 = a_1, \quad y_2 = a_2, \quad y_3 = a_3,$$

where $i = 3, 4, \ldots, N - 1$. The local truncation error is $O(h^5)$.

EXAMPLE 12.11

Use Eqn. (12.28) to derive the Adams-Bashforth three-step method.

We have for $m = 3$,

$$y_{i+1} = y_i + h \left[f(t_i, y_i) + \frac{1}{2} \nabla f(t_i, y_i) + \frac{5}{12} \nabla^2 f(t_i, y_i) \right].$$

Since

$$\nabla f(t_i, y_i) = hf[(t_{i-1}, y_{i-1}), (t_i, y_i)]$$
$$= f(t_i, y_i) - f(t_{i-1}, y_{i-1})$$
$$\nabla^2 f(t_i, y_i) = 2h^2 f[(t_{i-2}, y_{i-2}), (t_{i-1}, y_{i-1}), (t_i, y_i)]$$
$$= f(t_i, y_i) - 2f(t_{i-1}, y_{i-1}) + f(t_{i-2}, y_{i-2})$$

thus,

$$y_{i+1} = y_i + h \left\{ f(t_i, y_i) + \frac{1}{2}[f(t_i, y_i) - f(t_{i-1}, y_{i-1})] + \frac{5}{12}[f(t_i, y_i) \right.$$
$$\left. - 2f(t_{i-1}, y_{i-1}) + f(t_{i-2}, y_{i-2})] \right\}.$$

Simplify the last equation to get (12.30).

The local truncation errors given in (12.29) through (12.31) have been derived using the Newton backward divided-difference formula. For an example, let us derive the local error of (12.30):

The remainder term R_3 for Newton's backward formula with $m = 3$ is

$$-h^3 f^{(3)}(\zeta) \frac{s(s+1)(s+2)}{3!}.$$

Thus, the local error E_3 for the formula (12.30) is given by

$$E_3 = h \int_0^1 R_3 \, ds = h^4 f^{(3)}(\zeta) \frac{1}{6} \int_0^1 [s(s+1)(s+2)] ds.$$

Since the sign of $s(s+1)(s+2)$ does not change on $[0,1]$, it follows that there exists a μ_i between t_{i-2} and t_{i+1} such that

$$E_3 = h^4 f^{(3)}(\mu_i) \frac{3}{8} = h^4 y^{(4)}(\mu_i) \frac{3}{8}. \tag{12.32}$$

Therefore, E_3 is $O(h^4)$. The local error for Adams-Bashforth m-step method takes the form

$$E_m(h) = \gamma_m h^{m+1} y^{m+1}(\mu_i) \quad \text{for some } \mu_i \in (t_{i-m+1}, t_{i+1}).$$

The local errors for (12.29) and (12.31) can be derived similarly.

The major disadvantage common to all multistep methods, in general, and Adams-Bashforth methods, in particular, is that they are not self-starting. Take, for instance, the Adams-Bashforth four-step method. Four starting values are needed before the formula can be used. In practice, a single-step method with the same order of accuracy is used to determine the starting values. The Adams-Bashforth of order 4 method is widely used in combination with the Runge-Kutta method of order 4, because both kinds have local errors of $O(h^5)$. The advantage that Adams-Bashforth methods have over single-step methods is that the determination of y_{i+1} requires only one evaluation of $f(t, y)$ per step, whereas Runge-Kutta methods for $n \geq 3$ require four or more function evaluations. For this reason, the multistep methods can be twice as fast as the Runge-Kutta methods of comparable accuracy.

EXAMPLE 12.12
Use the Adams-Bashforth method of orders 2, 3, and 4 to obtain an approximation to the solution of the IVP

$$y' = 2t - y, \quad y(0) = -1$$

with $N = 10$, at $t = 1$. Obtain the starting values from the exact solution $y = e^{-t} + 2t - 2$.

We illustrate the use of Adams-Bashforth of order four to estimate $y(0.4)$. The starting values can be obtained using Runge-Kutta method of order four, but since the exact solution is known, we have

$$y_1 = -0.8951625820, \quad y_2 = -0.7812692469, \quad y_3 = -0.6591817793$$

and

$$f(0.1, -0.8951625820) = 2(0.1) - (-0.8951625820) = 1.095\,162\,582$$
$$f(0.2, -0.7812692469) = 2(0.2) - (-0.7812692469) = 1.181\,269\,247$$
$$f(0.3, -0.6591817793) = 2(0.3) - (-0.6591817793) = 1.259\,181\,779.$$

Then, from Eqn. (12.31):

$$y_4 = -0.6591817793 + \frac{0.1}{24} [55(1.259\,181\,779) - 59(1.181\,269\,247)$$
$$+ 37(1.095\,162\,582) - 9(1)]$$
$$y_4 = -0.5296770801.$$

```
» abash('f1',0,1,-1,10,2)

Runge-kutta method gives

y(0.0) =          -1
y(0.1) =      -0.8951625

        Adams-Bashforth method of order 2
```

ti	fi-3	fi-2	fi-1	fi	yi	Exact	error
0.2			1	1.095163	-0.780888	-0.781269	3.81e-004
0.3			1.095163	1.180888	-0.658513	-0.659182	6.69e-004
0.4			1.180888	1.258513	-0.52878	-0.52968	8.99e-004
0.5			1.258513	1.32878	-0.392389	-0.393469	1.08e-003
0.6			1.32878	1.392389	-0.24997	-0.251188	1.22e-003
0.7			1.392389	1.44997	-0.102094	-0.103415	1.32e-003
0.8			1.44997	1.502094	0.050722	0.049329	1.39e-003
0.9			1.502094	1.549278	0.208009	0.20657	1.44e-003
1.0			1.549278	1.591991	0.369344	0.367879	1.46e-003

Table 12.10   The Adams-Bashforth method of order 2 for Example 12.12.

A summary of the calculations using Adams-Bashforth of orders 2, 3, and 4 is given in Tables 12.10, 12.11, and 12.12, respectively.

There are many other multistep formulas available for solutions of first-order ordinary differential equations, but we will not consider them here. We finish this section by mentioning a multistep method that has the advantage of simplicity and is suitable in many applications,

$$y_{i+1} = y_{i-1} + 2hf(t_i, y_i), \quad \text{for} \quad i = 1, 2, \ldots, N - 1 \qquad (12.33)$$
$$y_0 = a_0, \quad y_1 = a_1.$$

Eqn. (12.33) is known as the **midpoint predictor** formula and is comparable in simplicity to Euler's method but has a better local truncation error of order $O(h^3)$.

EXERCISE SET 12.6

1. Use the Adams-Bashforth of order 4 method to approximate the solution of the following IVPs in the interval indicated.

   (a) $y' = -4t^3 y$, $[1, 2]$, $y(1) = 1$ with $N = 10$,

   (b) $y' = -ty$, $[0, 1]$, $y(0) = 2$ with $N = 10$,

   (c) $y' = -t\sin(t)y$, $[0, 2]$, $y(0) = 1$ with $h = 0.2$,

   (d) $y' = -2y$, $[0, 2]$, $y(0) = 4$ with $N = 10$,

   (e) $y' = -\cos(t)y$, $[0, 1.4]$, $y(0) = 2$ with $N = 14$,

M-function 12.6

The following MATLAB function **abash.m** finds the solution of the initial-value problem 11.2 using Adams-Bashforth methods. INPUTS are a function $f(t,y)$; the initial and final values a, b of t; the initial condition y_0; the number of steps n; the AB order 2, 3, or 4. The input function $f(t,y)$ should be defined as an M-file.

```
function function abash(f,a,b,y0,n,order)
% solve the initial-value problem y'=f(t,y), y(a)=y0
% using Adams-Bashforth methods.
fprintf('\n Runge-Kutta method gives \n\n')
h=(b-a)/n;
t=(a:h:b+h);
y(1)=y0;
fprintf('y(%2.1f) =%14.8f\n',a,y(1))
   % RK4 to start
for i=1:(order-1)
  k1=feval(f,t(i),y(i));
  k2=feval(f,t(i)+h/2,y(i)+h*k1/2);
  k3=feval(f,t(i)+h/2,y(i)+h*k2/2);
  k4=feval(f,t(i)+h,y(i)+h*k3);
  y(i+1)=y(i)+h*(k1+2*k2+2*k3+k4)/6;
  fprintf('y(%2.1f) =%14.8f\n',t(i)+h,y(i+1))
end
fprintf('\n')
disp(['            Adams-Bashforth method of order ',num2str(order)])
disp('_____')
disp('   t      fi-3    fi-2    fi-1    fi    y       Exact    error ')
disp('_____')
fprintf('\n')
if(order==4)
  % 4th order AB
  for i=order:n
    f1=feval(f,t(i),y(i));
    f2=feval(f,t(i-1),y(i-1));
    f3=feval(f,t(i-2),y(i-2));
    f4=feval(f,t(i-3),y(i-3));
    y(i+1)=y(i)+h*(55*f1-59*f2+37*f3-9*f4)/24;
% Enter the exact solution g if known as g=g(x) otherwise set g='n'.
    x=t(i+1);
    g='n';
    if (g~='n')
      err=abs(g-y(i+1));
```

```
      fprintf('%6.2f %12.6f %12.6f %12.6f %12.6f %12.6f %12.6f
%8.2e\n',t(i)+h,f4,f3,f2,f1,y(i+1),g,err)
     else
       fprintf('%6.2f %12.6f %12.6f %12.6f %12.6f
%12.6f\n',t(i)+h,f4,f3,f2,f1,y(i+1))
     end
  end
elseif (order==3)
  % 3rd order AB
  for i=order:n
    f1=feval(f,t(i),y(i));
    f2=feval(f,t(i-1),y(i-1));
    f3=feval(f,t(i-2),y(i-2));
    y(i+1)=y(i)+h*(23*f1-16*f2+5*f3)/12;
 % Enter the exact solution g if known as g=g(x) otherwise set g='n'.
    x=t(i+1);
    g='n';
    if (g~='n')
      err=abs(g-y(i+1));
      fprintf('%6.2f         %12.6f %12.6f %12.6f %12.6f %12.6f
%8.2e\n',t(i)+h,f3,f2,f1,y(i+1),g,err)
     else
       fprintf('%6.2f         %12.6f %12.6f %12.6f
%12.6f\n',t(i)+h,f3,f2,f1,y(i+1))
     end
  end
else
  % 2nd order AB
  for i=order:n
    f1=feval(f,t(i),y(i));
    f2=feval(f,t(i-1),y(i-1));
    y(i+1)=y(i)+h*(3*f1-f2)/2;
 % Enter the exact solution g if known as g=g(x) otherwise set g='n'.
    x=t(i+1);
    g='n';
    if (g~='n')
       err=abs(g-y(i+1));
       fprintf('%6.2f         %12.6f %12.6f %12.6f %12.6f  %8.2e\n',
t(i)+h, f2, f1, y(i+1), g, err)
     else
        fprintf('%6.2f           %12.6f %12.6f %12.6f\n',t(i)+h,f2,f1,y(i+1))
     end
  end
end
```

```
» abash('f1',0,1,-1,10,3)

Runge-kutta method gives

y(0.0) =          -1
y(0.1) =     -0.8951625
y(0.2) =     -0.7812691


        Adams-Bashforth method of order 3
```

t	fi-3	fi-2	fi-1	fi	y	Exact	error
0.3		1	1.095163	1.181269	-0.659214	-0.659182	3.24e-005
0.4		1.095163	1.181269	1.259214	-0.529736	-0.52968	5.56e-005
0.5		1.181269	1.259214	1.329736	-0.393545	-0.393469	7.59e-005
0.6		1.259214	1.329736	1.393545	-0.25128	-0.251188	9.15e-005
0.7		1.329736	1.393545	1.45128	-0.103518	-0.103415	1.04e-004
0.8		1.393545	1.45128	1.503518	0.049216	0.049329	1.13e-004
0.9		1.45128	1.503518	1.550784	0.206451	0.20657	1.19e-004
1.0		1.503518	1.550784	1.593549	0.367757	0.367879	1.23e-004

Table 12.11   The Adams-Bashforth method of order 3 for Example 12.12.

(f) $y' = -e^t y$, $[0, 2]$, $y(0) = 3$ with $N = 10$.

2. Solve the IVP
$$y' = -y + e^{-t}, \quad y(0) = 0, \quad 0 \le t \le 1$$

using the Adams-Bashforth method of order 4 with $N = 10$. Estimate the accuracy of your results. Obtain the starting values from the exact solution $y = te^{-t}$.

3. Use the Adams-Bashforth method of order 4 with $h = 0.1$ to solve the IVP
$$y' = (y + t)^2, \quad y(0) = -1.$$

Use the three starting values:

$y(0.1) = -0.917628$,
$y(0.2) = -0.862910$,
$y(0.3) = -0.827490$.

4. For the initial-value problem
$$y' = y - t^2, \quad y(0) = 1$$

with $y(0.2) = 1.2186$, $y(0.4) = 1.4682$, and $y(0.6) = 1.7379$, use the Adams-Bashforth method of order 4 to compute the solution through $t = 1.2$.

5. Derive formula (12.33) using the central-difference approximation for the first derivative.

» abash('f1',0,1,-1,10,4)

Runge-kutta method gives

y(0.0) =   -1
y(0.1) =   -0.8951625
y(0.2) =   -0.7812691
y(0.3) =   -0.65918158

Adams-Bashforth method of order 4

ti	fi-3	fi-2	fi-1	fi	yi	Exact	error
0.4	1	1.095163	1.181269	1.259182	-0.529677	-0.52968	3.05e-006
0.5	1.095163	1.181269	1.259182	1.329677	-0.393464	-0.393469	4.98e-006
0.6	1.181269	1.259182	1.329677	1.393464	-0.251181	-0.251188	6.92e-006
0.7	1.259182	1.329677	1.393464	1.451181	-0.103406	-0.103415	8.22e-006
0.8	1.329677	1.393464	1.451181	1.503406	0.049338	0.049329	9.31e-006
0.9	1.393464	1.451181	1.503406	1.550662	0.20658	0.20657	1.01e-005
1.0	1.451181	1.503406	1.550662	1.59342	0.36789	0.367879	1.06e-005

Table 12.12   The Adams-Bashforth method of order 4 for Example 12.12.

COMPUTER PROBLEM SET 12.6

1. Write a computer program in a language of your choice to solve the IVP 12.1 using the Adams-Bashforth method of order 4.

   Input data to the program should be

   (a) The function $f(t, y)$,
   (b) The interval $[a, b]$,
   (c) The initial value $y(a)$,
   (d) The step size h.

   Output should consist of the values of y at each time step.

   Test your program to solve the IVP

   $$y' = (1 - t)y, \qquad 0 \leq t \leq 3, \quad y(0) = 3$$

   with $h = 1$.

2. Use the MATLAB function abash to solve the IVP

   $$y' = ty^2, \qquad 0 \leq t \leq 2, \qquad y(0) = \frac{2}{5}.$$

   Approximate $y(t)$ up to $t = 2$ for step sizes $h = 1/2^n$, with $n = 1, 2, \ldots, 10$. Given the exact solution $y = \frac{2}{5-t^2}$, make log-log plots of the absolute error in $y(2)$ vs. h. What do these plots tell you?

3. Solve the IVP $y' = \frac{1}{4}y(1 - \frac{y}{20})$, $y(0) = 1$ using the Adams-Bashforth method of order 2. Use the step size $h = 0.1, 0.01$ to approximate $y(1)$.

4. Solve the IVP $y' = -2ty^2$ with $y(0) = 1$ using an Adams-Bashforth routine from a program library. Solve it with $h = 0.1, 0.01, 0.001$ to approximate $y(1)$.

5. Consider the IVP $y' = t - y + 1$, $y(0) = 1$. Use the Adams-Bashforth method of order 4 with a sequence of decreasing h to approximate $y(1)$. The actual solution is $y = t + e^{-t}$. Plot the exact and numerical solutions, and report the largest error at the discrete values of t.

6. Consider the IVP $y' = 100(\sin t - y)$, $y(0) = 0$. Approximate $y(1)$ using the Adams-Bashforth method of order 4.

7. Use the MATLAB function **abash** to solve the IVP

$$y' = 2 + \sqrt{y - 2t + 3}, \quad y(0) = 1$$

on the interval $[0, 1.5]$. Compare with the exact solution $y(t) = 1 + 4t + \frac{1}{4}t^2$.

8. The implicit solution of the IVP

$$y' = \frac{2t + 1}{5y^4 + 1}, \quad y(2) = 1$$

is $y^5 + y = t^2 + t - 4$. Use the Adams-Bashforth method of order 4 with $h = 0.1, 0.05$ to approximate the values of $y(t)$ at $t = 2.1, 2.2, \ldots, 3.0$. To check the error in these approximate values, construct a table of the approximate values and the residual

$$R(t, y) = y^5 + y - t^2 - t + 4.$$

12.7   PREDICTOR-CORRECTOR METHODS

In Section 12.6, Adams-Bashforth methods were derived from Eqn. (12.27) using the Newton backward-difference polynomial, which fits only the previously calculated values $f_i, f_{i-1}, f_{i-2}, \ldots$. These are called **explicit** or **open formulas**. Multistep formulas can be also derived using an interpolating polynomial that fits not only the previously calculated values $f_i, f_{i-1}, \ldots$, but the unknown f_{i+1} as well. These are called **implicit** or **closed formulas**. In order to describe the procedure, we consider the simple formula obtained by replacing the integral in Eqn. (12.27) with the trapezoidal rule (Section 10.1). The result gives the implicit formula

$$y_{i+1} = y_i + \frac{h}{2}[f(t_i, y_i) + f(t_{i+1}, y_{i+1})] \tag{12.34}$$

where $i = 0, 1, \ldots, N - 1$.

A different approach is used to solve (12.34), combining the implicit formula (12.34) and an explicit formula. Suppose, for instance, that a predicted estimate of y_{i+1} has been obtained by using an explicit method such as Euler's formula. Setting $f_{i+1}^{(0)} = f(t_{i+1}, y_{i+1}^{(0)})$, $f_{i+1}^{(0)}$ can then be used in Eqn. (12.34) to obtain the corrected estimate $y_{i+1}^{(1)}$. To improve, y_{i+1}, $y_{i+1}^{(1)}$ is used to obtain the next corrected estimate $y_{i+1}^{(2)}$. This is continued until convergence occurs to as many digits as desired. The general step of this iteration procedure is given by the formula

$$y_{i+1}^{(n+1)} = y_i + \frac{h}{2}\left[f(t_i, y_i) + f(t_{i+1}, y_{i+1}^{(n)})\right] \tag{12.35}$$

where $n = 0, 1, 2, \ldots$.

A suitable way of terminating the procedure is by using the criterion

$$\left| y_{i+1}^{(n+1)} - y_{i+1}^{(n)} \right| < \epsilon \tag{12.36}$$

where ϵ denotes a specified tolerance. Experience has shown that the first values $y_{i+1}^{(1)}$ obtained by using one correction of y_{i+1} are as accurate as $y_{i+1}^{(2)}$, $y_{i+1}^{(3)}, \ldots$.

The explicit formula used to predict the approximation $y_{i+1}^{(n)}$ is called a **predictor formula** and the implicit formula (12.35) is called a **corrector formula**. In general, if an explicit and implicit formula are used as a pair of formulas, the explicit formula is called the predictor and the implicit one is called the corrector.

12.8 ADAMS-MOULTON METHODS

If the additional interpolating node t_{i+1} is used in the approximation of the integral

$$\int_{t_i}^{t_{i+1}} f(t_i, y(t_i))dt,$$

formula (12.27) becomes

$$y_{i+1} = y_i + h \sum_{k=0}^{m} \nabla^k f(t_{i+1}, y_{i+1})(-1)^k \int_0^1 \binom{1-s}{k} ds. \tag{12.37}$$

The case when $m = 3$ leads to the most frequently used formula known as the **fourth-order Adams-Moulton (AM4) formula**,

$$y_{i+1} = y_i + \frac{h}{24}\left[9f(t_{i+1}, y_{i+1}) + 19f(t_i, y_i) - 5f(t_{i-1}, y_{i-1}) + f(t_{i-2}, y_{i-2})\right] \tag{12.38}$$

where $i = 2, 3, \ldots, N - 1$. The local truncation error is $O(h^5)$.

EXAMPLE 12.13
Derive the fourth-order Adams-Moulton formula using Eqn. (12.37).

Let

$$\gamma'_k = (-1)^k \int_0^1 \binom{1-s}{k} ds.$$

We have

$$\gamma'_0 = \int_0^1 ds = 1,$$

$$\gamma'_1 = \int_0^1 (1-s)ds = -\frac{1}{2},$$

$$\gamma'_2 = \int_0^1 (-s)(1-s)ds = -\frac{1}{12},$$

$$\gamma'_3 = \int_0^1 (-s)(1-s)(-s-1)ds = -\frac{1}{24}.$$

By substituting these values into (12.37), we obtain

$$
\begin{aligned}
y_{i+1} = y_i + h & \left[\gamma'_0 f(t_{i+1}, y_{i+1}) + \gamma'_1 \nabla f(t_{i+1}, y_{i+1}) + \gamma'_2 \nabla^2 f(t_{i+1}, y_{i+1}) \right. \\
& \left. + \gamma'_3 \nabla^3 f(t_{i+1}, y_{i+1}) \right] \\
= y_i + h & \left\{ f(t_{i+1}, y_{i+1}) - \frac{1}{2} \left[f(t_{i+1}, y_{i+1}) - f(t_i, y_i) \right] \right. \\
& - \frac{1}{12} \left[f(t_{i+1}, y_{i+1}) - 2f(t_i, y_i) + f(t_{i-1}, y_{i-1}) \right] \\
& \left. - \frac{1}{24} \left[f(t_{i+1}, y_{i+1}) - 3f(t_i, y_i) + 3f(t_{i-1}, y_{i-1}) - f(t_{i-2}, y_{i-2}) \right] \right\}.
\end{aligned}
$$

Simplify the last equation to get (12.38).

An algorithm for the fourth-order Adams-Moulton method is given below:

$$h = \frac{a-b}{N}$$
$$y_0 \leftarrow y(a)$$
$y_1, y_2 \leftarrow$ determined by using a single-step formula.
for $i = 2, 3, \ldots, N-1$
$\quad y_{i+1}^{(k-1)}$ determined by using an explicit formula.
$\quad$ Compute
$\quad y_{i+1}^{(k)} \leftarrow y_i + \frac{h}{24}[9f(t_{i+1}, y_{i+1}^{(k-1)}) + 19f(t_i, y_i)$
$\quad -5f(t_{i-1}, y_{i-1}) + f(t_{i-2}, y_{i-2})]$ for $k = 1, 2, \ldots$.
$\quad$ Iterate on k until $|y_{i+1}^{(k)} - y_{i+1}^{(k-1)}| < \epsilon$
$\quad y_{i+1} \leftarrow y_{i+1}^{(k)}$

A popular predictor-corrector method uses the fourth-order Adams-Moulton formula as the *corrector* method and the fourth-order Adams-Bashforth formula as the *predictor* method. Since four starting values are needed before the Adams-Bashforth method can be used, a single-step formula such as the Runge-Kutta method of order four is frequently employed to meet this need.

One question that can arise is, why should one bother using the predictor-corrector methods when the single-step methods are of comparable accuracy to the predictor-corrector methods of the same order? A partial answer to that relies in the actual number of functional evaluations. Take, for example, the Runge-Kutta method of order four. Each step requires four evaluations whereas the Adams-Moulton method of the same order requires only as few as two evaluations. For this reason, predictor-corrector formulas are, in general, considerably more accurate and faster than the single-step ones.

EXAMPLE 12.14

Use Adams-Moulton method of order 4 for the IVP

$$y' = 2t - y, \quad y(0) = -1$$

with $N = 10$ to approximate $y(1)$.

We illustrate the use of Adams-Moulton method of order 4 to estimate $y(0.4)$. The starting values can be obtained using Runge-Kutta method of order 4, that is

$$y_1 = -0.89516250, \quad y_2 = -0.78126921, \quad y_3 = -0.65918158$$

and

$$f(0.1, -0.89516250) = 1.095\,162\,50$$
$$f(0.2, -0.78126921) = 1.181\,269\,21$$
$$f(0.3, -0.65918158) = 1.259\,181\,58.$$

Then, the Adams-Bashforth formula of order 4 gives

$$y_4 = -0.65918158 + \frac{0.1}{24}\left[55(1.259\,18158) - 59(1.181\,269\,21)\right.$$
$$\left. +37(1.095\,162\,50) - 9(1)\right]$$
$$y_4 = -0.5296769299$$

and

$$f(0.4, y_4) = f(0.4, -0.5296769299) = 1.32967693.$$

The corrector formula (12.38) then gives

$$y_4 = -0.65918158 + \frac{0.1}{24}\left[9(1.329\,676\,93\,) + 19(1.259\,181\,58)\right.$$
$$\left. -5(1.181\,269\,21) + (1.095\,162\,50)\right]$$
$$y_4 = -0.529680082.$$

In Table 12.13, the numerical solution is continued and compared with values obtained from the analytical solution. At $t = 1$, the Adams-Moulton value of order 4 is $y_{10} = 0.367878366$.

```
» amoulton('f1',0,1,-1,10)

 Runge-kutta method of order 4 gives

y(0.0) =              -1
y(0.1) =         -0.8951625
y(0.2) =         -0.7812691
y(0.3) =         -0.65918158

        Adams-Moulton method of order 4
```

ti	fi-2	fi-1	fi	fi+1	yi	Exact	error
0.4	1.095163	1.181269	1.259182	1.329677	-0.52968	-0.52968	1.28e-007
0.5	1.181269	1.259182	1.32968	1.393467	-0.39347	-0.393469	3.91e-007
0.6	1.259182	1.32968	1.39347	1.451186	-0.251189	-0.251188	6.04e-007
0.7	1.32968	1.39347	1.451189	1.503413	-0.103415	-0.103415	7.72e-007
0.8	1.39347	1.451189	1.503415	1.55067	0.049328	0.049329	9.04e-007
0.9	1.451189	1.503415	1.550672	1.593429	0.206569	0.20657	1.00e-006
1.0	1.503415	1.550672	1.593431	1.63212	0.367878	0.367879	1.08e-006

Table 12.13 The Adams-Moulton method for Example 12.14.

EXERCISE SET 12.8

1. Use the Adams-Moulton method of order 4 to estimate the solution of the IVP

$$y' = -ty^2, \quad y(0) = 2$$

over the interval $[0, 1]$ with $h = 0.5, 0.2, 0.1, 0.01$. Do the results appear to converge towards the exact value $y(1) = 1$?

2. The IVP

$$y' = 2t(y - 1), \quad y(0) = 0$$

has values as follows:

$y(0.1) = -0.01005017,$
$y(0.2) = -0.04081077,$
$y(0.4) = -0.09417427.$

Using the Adams-Moulton method of order 4, compute $y(0.4)$.

3. For the IVP

$$y' = y - t^2, \quad y(0) = 1$$

starting values are known:

$y(0.2) = 1.218593,$
$y(0.4) = 1.468167,$

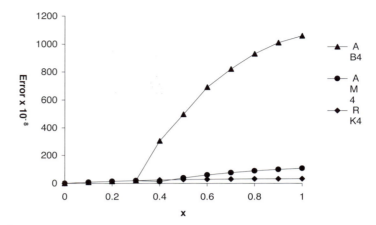

FIGURE 12.4
Errors for the RK4, AB4, and AM4 solutions to the IVP
$y' = 2t - y$, $y(0) = -1$ with N=10.

$y(0.6) = 1.737869$.

Use the Adams-Moulton method of order 4 to advance the solution to $t = 0.8$.

4. Solve the equation

$$y' = y - \frac{y}{t}, \quad y(1) = \frac{1}{2}, \quad 1 \le t \le 2$$

using the Adams-Moulton method of order 4 with $N = 10$. Compare your results with the values of the exact solution $y = \frac{e^{t-1}}{2t}$.

5. Repeat Exercise 6 of Section 12.1 using the Adams-Moulton method of order 4.

6. Applying numerical differentiation to approximate the ODE

$$y'(t) = f(t, y)$$

we can arrive at the backward-differentiation formula (BDF)

$$\sum_{i=0}^{n} A_i y_{k+1-i} \approx h f(x_{k+1}, y_{k+1}).$$

Derive a BDF method of the form

$$A y_{k+1} + b y_k + C y_{k-1} = h f(t_{k+1}, y_{k+1}),$$

find its error, and compare it with the error of the Adams-Moulton method of the same order.

M-function 12.8

The following MATLAB function **amoulton.m** finds the solution of the initial-value problem 11.2 using the Adams-Moulton method of order 4. INPUTS are a function $f(t,y)$; the initial and final values a, b of t; the initial condition y_0; the number of steps n. The input function $f(t,y)$ should be defined as an M-file.

```
function amoulton(f,a,b,y0,n)
% Solve the initial value problem y'=f(t,y), y(a)=y0
% using the Adams-Moulton method of order 4.
fprintf('\n Runge-Kutta method of order 4 gives \n\n')
h=(b-a)/n;
t=(a:h:b+h);
y(1)=y0;
fprintf('y(%2.1f) =%14.8f\n',a,y(1))
    % RK4 method to start
for i=1:3
  k1=feval(f,t(i),y(i));
  k2=feval(f,t(i)+h/2,y(i)+h*k1/2);
  k3=feval(f,t(i)+h/2,y(i)+h*k2/2);
  k4=feval(f,t(i)+h,y(i)+h*k3);
  y(i+1)=y(i)+h*(k1+2*k2+2*k3+k4)/6;
  fprintf('y(%2.1f) =%14.8f\n',t(i)+h,y(i+1))
end;
fprintf('\n')
disp('           Adams-Moulton method of order 4')
disp('_____')
disp('  t    fi-2    fi-1    fi    fi+1        y        Exact    error   ')
disp('_____')
fprintf('\n')
for i=4:n
  f1=feval(f,t(i),y(i));
  f2=feval(f,t(i-1),y(i-1));
  f3=feval(f,t(i-2),y(i-2));
  f4=feval(f,t(i-3),y(i-3));
  % AB to predict
  w=y(i)+h*(55*f1-59*f2+37*f3-9*f4)/24;
  % AM to correct
  f5=feval(f,t(i+1),w);
  y(i+1)=y(i)+h*(9*f5+19*f1-5*f2+f3)/24;
% Write the exact solution g if known as g=g(x) otherwise set g='n'.
  x=t(i+1);
  g='n';
```

```
if (g~='n')
    err=abs(g-y(i+1));
    fprintf('%6.2f %12.6f %12.6f %12.6f %12.6f %12.6f %12.6f
%8.2e\n',t(i)+h,f3,f2,f1,f5,y(i+1),g,err)
    else
        fprintf('%6.2f %12.6f %12.6f %12.6f %12.6f %12.6f\n', t(i)+h, f3,
    f2, f1, f5, y(i+1))
    end
end
```

___

COMPUTER PROBLEM SET 12.8

1. Write a computer program in a language of your choice to solve the IVP (12.3) using the Adams-Moulton method of order 4. Use the Runge-Kutta method of order 4 to determine the starting values.

   Input data to the program should be

   (a) The function $f(t, y)$,
   (b) The interval $[a, b]$,
   (c) The initial value $y(a)$,
   (d) The step size h.

   Output should consist of the values of y at each time step.

   Test your program to solve the IVP

   $$y' = (1 - t)y, \qquad 0 \le t \le 3, \quad y(0) = 3$$

   with $h = 1$.

2. Use the MATLAB function amoulton to solve the IVP

   $$y' = ty^2, \qquad 0 \le t \le 2, \qquad y(0) = \frac{2}{5}.$$

   Approximate $y(t)$ up to $t = 2$ for step sizes $h = 1/2^n$, with $n = 1, 2, \ldots, 10$. Given the exact solution $y = \frac{2}{5-x^2}$, make log-log plots of the absolute error in $y(2)$ vs. h. What do these plot tell you?

3. Use the MATLAB function amoulton to approximate the solution of the IVP

   $$y' = (1 + t^2 + y^2)^{-1}, \quad y(0) = 0$$

   at $t = 2$ with $h = 0.2$.

4. Use the MATLAB function amoulton to solve the IVP

$$y' = e^{-t} - y, \ y(0) = 1 \ \text{in} \ [0, 5]$$

with $h = 0.05$. Use the following starting values
$y(0.05) = 0.998779090,$
$y(0.10) = 0.99532116,$
$y(0.15) = 0.98981417.$

Plot the approximate solution along with the exact one given by $y(t) = e^{-t}(t + 1)$ on the same coordinate system.

5. Solve the following IVP using the Adams-Moulton method of order 4

$$y' = \frac{t^2 + y^2}{t^2 - y^2 + 2}, \quad y(0) = 0$$

to approximate $y(2)$ with $h = 0.2$. Use the starting values

$y(0.2) = 0.00131696943446691243,$
$y(0.4) = 0.0101844007651568535,$
$y(0.6) = 0.0325971175536952176.$

6. Solve the following IVP using the Adams-Moulton method of order 4

$$y' = \cos t - \sin t - y, \quad y(0) = 2 \ \text{in} \ [0, 10]$$

with $h = 0.5$. Compare with the exact solution $y = \cos t + e^{-t}$. Use the starting values from the exact solution.

7. Solve the following IVP using the Adams-Moulton method of order 4

$$y' = ty^2 + 3ty, \quad y(0) = -0.5 \ \text{in} \ [0, 3]$$

with $h = 0.2$. Use the starting values

$y(0.2) = -0.525501249395768229,$
$y(0.4) = 0.608125849624477853,$
$y(0.6) = -0.76650823677160316.$

8. Use the Adams-Moulton method of order 4 with $h = 0.01$ to approximate the integral (see Exercise 10 in Section 12.4)

$$Si(t) = \int_0^t \frac{\sinh x}{x} dx$$

at $t = 0.5, 1.0, 1.5, 2.0$ (Use $\sinh(0)/0 = 1$).

12.9   NUMERICAL STABILITY

Because of round-off error in the computer, errors are committed as the explicit calculation is carried out. Whether these errors amplify or decay characterizes the stability property of the scheme. A method is said to be **stable** if a small error at one step does not have increasingly great effect on subsequent calculations. In this section, we will show how a method that has the property of being accurate can produce results that are completely incorrect. To begin our discussion of stability, let us consider the initial-value problem known as the **test equation**

$$y' = -3y, \quad y(0) = 1. \tag{12.39}$$

The solution of this equation is easily obtained: $y(t) = e^{-3t}$. Problem (12.39) is useful as a test equation for the study of stability of IVP because the exact solution of the test equation has a constant relative error as t moves away from y_0. This means that any relative error growth of a numerical solution must be caused by the method that generates the approximate values of the $y(t_i)$'s.

Let us approximate $y(3)$ by using both the Euler and midpoint predictor methods. To use the latter, given by Eqn. (12.33), we need two starting values y_0 and y_1. For y_1 we use the exact value $y(t_1)$. Table 12.14 illustrates the results obtained with $h = 0.1$.

The error columns show that the midpoint predictor method produces a smaller error for small values of t_n than Euler's method. However, as t approaches 3, the results with the midpoint predictor method grow to have very large error, while the results by Euler's method do not show a deviation of the numerical solution from the true solution. This behavior is known as numerical **instability**.

Let us now study the nature of this instability. Applying the midpoint method to our example gives the difference equation

$$y_{i+1} - y_{i-1} + 6hy_i = 0, \quad y(0) = 1. \tag{12.40}$$

The general solution of (12.40) is

$$y_i = c_1\lambda_1^i + c_2\lambda_2^i \quad \text{where} \quad c_1 + c_2 = 1 \tag{12.41}$$

and λ_1 and λ_2 are the roots of the *characteristic equation*

$$\lambda^2 + 6h\lambda - 1 = 0, \tag{12.42}$$

given by
$$\lambda_1 = -3h + \sqrt{1 + 9h^2} \quad \text{and} \quad \lambda_2 = -3h - \sqrt{1 + 9h^2}. \tag{12.43}$$

The binomial approximation of $\sqrt{1 + 9h^2}$ for a small h gives

$$\lambda_1 = -3h + [1 + \frac{9}{2}h^2 + O(h^4)] = e^{-3h} + O(h^3)$$

ti	Exact solution	Euler's method	Error Euler	Midpoint Predictor	Error Midpt. Pred.
0	1	1	0	1	0
0.1	0.740818	0.700000	0.040818	0.740818	0
0.2	0.548812	0.490000	0.058812	0.555509	0.006697
0.3	0.406570	0.343000	0.063570	0.407513	0.000943
0.4	0.301194	0.240100	0.061094	0.311001	0.009807
0.5	0.223130	0.168070	0.055060	0.220912	0.002218
0.6	0.165299	0.117649	0.047650	0.178454	0.013155
0.7	0.122456	0.082354	0.040102	0.113839	0.008617
0.8	0.090718	0.057648	0.033070	0.110151	0.019433
0.9	0.067206	0.040354	0.026852	0.047749	0.019456
1.0	0.049787	0.028248	0.021540	0.081501	0.031714
1.1	0.036883	0.019773	0.017110	-0.001152	0.038035
1.2	0.027324	0.013841	0.013482	0.082192	0.054868
1.3	0.020242	0.009689	0.010553	-0.050467	0.070709
1.4	0.014996	0.006782	0.008213	0.112472	0.097477
1.5	0.011109	0.004748	0.006361	-0.117950	0.129059
1.6	0.008230	0.003323	0.004906	0.183243	0.175013
1.7	0.006097	0.002326	0.003770	-0.227896	0.233993
1.8	0.004517	0.001628	0.002888	0.319980	0.315464
1.9	0.003346	0.001140	0.002206	-0.419884	0.423230
2.0	0.002479	0.000798	0.001680	0.571910	0.569432
2.1	0.001836	0.000559	0.001278	-0.763030	0.764867
2.2	0.001360	0.000391	0.000969	1.029729	1.028368
2.3	0.001008	0.000274	0.000734	-1.380867	1.381875
2.4	0.000747	0.000192	0.000555	1.858249	1.857503
2.5	0.000553	0.000134	0.000419	-2.495817	2.496370
2.6	0.000410	0.000094	0.000316	3.355739	3.355330
2.7	0.000304	0.000066	0.000238	-4.509260	4.509564
2.8	0.000225	0.000046	0.000179	6.061296	6.061071
2.9	0.000167	0.000032	0.000134	-8.146038	8.146204
3.0	0.000123	0.000023	0.000101	10.948918	10.948795

Table 12.14 Comparison of the exact and approximate solution for the IVP $y' = -3y$, $y(0) = 1$ using Euler's and the midpoint predictor methods.

and

$$\lambda_2 = -3h - [1 + \frac{9}{2}h^2 + O(h^4)] = -e^{3h} + O(h^3).$$

Thus, the general solution of (12.41) is

$$y_i = c_1 \left(e^{-3h}\right)^i + c_2(-1)^i \left(e^{3h}\right)^i + O(h^3).$$

Since $hi = t_i$ and $c_1 = 1 - c_2$, we have the solution

$$y_i = (1 - c_2)e^{-3t_i} + c_2(-1)^i e^{3t_i} + O(h^3). \tag{12.44}$$

Observe that the first term of (12.44) represents the exact solution of the differential equation and the second is extraneous. It is called the **parasitic** term of the numerical solution because it is unrelated to either the exact solution term $(1 - c_2)e^{-3t_i}$ or the $O(h^3)$ per-step error term. The exact solution of the problem requires that we choose $c_2 = 0$. However, even a small round-off error may cause c_2 to be nonzero, and the parasitic term will affect the solution unless c_2 is exactly zero. As $t \to \infty$, the true solution $y(t) \to 0$ and the oscillatory term $(-1)^i e^{3t_i}$ will eventually dominate the solution y_i, and the resulting numerical solution will no longer resemble the exact one. Hence, the midpoint predictor method is an unstable method.

In general, to determine whether a given multistep method is stable, we examine the error produced when the method is applied to the *test equation*

$$y' = \lambda y, \quad y(0) = \alpha, \quad \lambda = \text{negative constant.}$$

This procedure is usually considered sufficient. A stability analysis of single-step or self-starting methods such as Euler's and all Runge-Kutta methods has shown that these methods are stable for sufficiently small h. The stability of a predictor-corrector scheme will depend on both the predictor and corrector formulas but is affected more strongly by the latter if the correction is small.

EXERCISE SET 12.9

1. Show that the general solution of the difference equation obtained by applying the test equation to Milne's corrector formula

$$y_{i+1} = y_{i-1} + \frac{h}{3}[f(t_{i+1}, y_{i+1}) + 4f(t_i, y_i) + f(t_{i-1}, y_{i-1})]$$

   is given by

$$y_i = c_1 e^{\lambda t_i} + c_2(-1)^i e^{-\frac{\lambda}{3}t_i}.$$

2. Find a and b such that the method for solving the ODEs given by

$$w_{n+1} = aw_{i-1} + bh[f(t_{n-1}, w_{n-1}) + 4f(t_n, w_n) + f(t_{n+1}, w_{n+1})]$$

   is consistent. Is it stable? (*Hint*: For consistency you must consider whether w_{n+1} differs from the true value y_{n+1} with an error $O(h^2)$ or higher.)

3. Consider the Gear solver for solving ODE

$$w_{n+1} = \frac{18}{11}w_i - \frac{9}{11}w_{i-1} + \frac{2}{11}w_{i-2} + ahf(t_{i+1}, w_{i+1})$$

   where a is a constant. Find the characteristic equation to study the stability of this method and determine whether the method is stable.

4. Use the approximation

$$y'(t_{n-1}) = \frac{-y(t_{n+1}) + 4y(t_n) - 3y(t_{n-1})}{2h}$$

   to derive a numerical method for ODEs. What is the truncation error? Is the method consistent?

5. Investigate the following multi step difference scheme for the solution to differential equations of the form $y' = f(t, y)$ for stability.

$$w_{n+1} = \frac{3}{2}w_i - \frac{1}{2}w_{i-1} + \frac{1}{2}hf(t_i, w_i).$$

6. Show that the method

$$y_{n+1} = 4y_n - 3y_{n-1} - 2hf(t_{n-1}, y_{n-1})$$

   is unstable.

7. Consider the one-step method for solving ODEs:

$$w_{i+1} = w_i + \frac{h}{4}\left[f(t_i, w_i) + 3f(t_i + \frac{2h}{3}, w_i + \frac{2h}{3}f(t_i, w_i))\right].$$

   What inequality must be satisfied by the step size h such that the method will be stable for the ODE $y' = -y$?

8. Consider the multi step method

$$w_{i+1} = \frac{1}{2}(w_i + w_{i-1}) + \frac{h}{4}\left[f(t_i, w_i) + 5f(t_{i-1}, w_{i-1})\right].$$

   Is the method stable?

12.10 HIGHER-ORDER EQUATIONS AND SYSTEMS OF DIFFERENTIAL EQUATIONS

In this section, we present a brief introduction to higher-order equations and systems. The methods developed in the previous sections can now be extended easily to solve higher-order initial-value problems.

Consider the general nth-order initial-value problem

$$y^{(n)}(t) = f(t, y(t), y'(t), \cdots, y^{(n-1)}(t)), \quad a \le t \le b \tag{12.45}$$

subject to the initial conditions

$$y(a) = \alpha_0,$$
$$y'(a) = \alpha_1,$$
$$\vdots$$
$$y^{(n-1)}(a) = \alpha_{n-1}.$$

The general approach to solving the initial-value problem (12.45) is to convert it to a system of n first-order equations as follows. Let us define the new variables

$$y_1(t) = y(t), \ y_2(t) = y'(t), \cdots, \ y_n(t) = y^{(n-1)}(t). \tag{12.46}$$

Thus, Problem (12.45) becomes

$$y_1' = y_2$$
$$y_2' = y_3$$
$$\vdots \qquad\qquad\qquad\qquad (12.47)$$
$$y_{n-1}' = y_n$$
$$y_n' = f(t, y_1, y_2, ..., y_n)$$

with initial conditions

$$y_1(a) = \alpha_0, \; y_2(a) = \alpha_1, \ldots, \; y_n(a) = \alpha_{n-1}.$$

The solution of the system (12.47) can now be carried out by using one of the methods developed in the previous sections of this chapter. As an example, we consider the Runge-Kutta method of order 4 applied to two simultaneous equations of the form

$$x' = f(t, x(t), y(t)), \quad x(a) = \alpha$$

$$(12.48)$$

$$y' = g(t, x(t), y(t)), \quad y(a) = \beta.$$

From the Runge-Kutta formula (12.25), we have

$$x_{n+1} = x_n + \frac{h}{6}(k_1 + 2k_2 + 2k_3 + k_4)$$

$$(12.49)$$

$$y_{n+1} = y_n + \frac{h}{6}(c_1 + 2c_2 + 2c_3 + c_4)$$

where

$$k_1 = f(t_n, x_n, y_n)$$
$$c_1 = g(t_n, x_n, y_n)$$
$$k_2 = f\left(t_n + \frac{h}{2}, x_n + \frac{h}{2}k_1, y_n + \frac{h}{2}c_1\right)$$
$$c_2 = g\left(t_n + \frac{h}{2}, x_n + \frac{h}{2}k_1, y_n + \frac{h}{2}c_1\right)$$
$$k_3 = f\left(t_n + \frac{h}{2}, x_n + \frac{h}{2}k_2, y_n + \frac{h}{2}c_2\right)$$
$$c_3 = g\left(t_n + \frac{h}{2}, x_n + \frac{h}{2}k_2, y_n + \frac{h}{2}c_2\right)$$
$$k_4 = f(t_n + h, x_n + hk_3, y_n + hc_3)$$
$$c_4 = g(t_n + h, x_n + hk_3, y_n + hc_3).$$

Note that the $k's$ and $c's$ should be computed following the above order. For higher-order systems, the Runge-Kutta methods can be developed in a way that is entirely similar to second-order systems.

EXAMPLE 12.15

Solve the following initial-value problem using the Runge-Kutta method of order 4 with $N = 10$.

$$y''(t) - 3y' + 2y = 6e^{3t}, \quad 0 \le t \le 1$$

with initial conditions

$$y(0) = 1, \quad y'(0) = -1.$$

Compare with the exact solution $y(t) = -8e^{2t} + 6e^{t} + 3e^{3t}$.

Using the substitution in (12.46), we get the reformulated problem:

$$\begin{aligned}
x'(t) &= 3x - 2y + 6e^{3t}, & x(0) &= -1, \\
y'(t) &= x, & y(0) &= 1, & 0 \le t \le 1.
\end{aligned}$$

Thus,

$$f(t, x, y) = 3x - 2y + 6e^{3t} \quad \text{and} \quad g(t, x, y) = x.$$

We illustrate the use of the Runge-Kutta method of order 4 by estimating $y(0.1)$. With $h = 0.1$, we have

$$\begin{aligned}
k_1 &= f(0, -1, 1) = 1 \\
c_1 &= g(0, -1, 1) = -1 \\
k_2 &= f(0 + \frac{0.1}{2}, -1 + \frac{0.1}{2}(1), 1 + \frac{0.1}{2}(-1)) = 2.221 \\
c_2 &= g(0 + \frac{0.1}{2}, -1 + \frac{0.1}{2}(1), 1 + \frac{0.1}{2}(-1)) = -0.95 \\
k_3 &= f(0 + \frac{0.1}{2}, -1 + \frac{0.1}{2}(2.221), 1 + \frac{0.1}{2}(-0.95)) = 2.399155 \\
c_3 &= g(0 + \frac{0.1}{2}, -1 + \frac{0.1}{2}(2.221), 1 + \frac{0.1}{2}(-0.95)) = -0.88895 \\
k_4 &= f(0 + 0.1, -1 + (0.1)(2.399155), 1 + (0.1)(-0.88895)) = 3.99669 \\
c_4 &= g(0 + 0.1, -1 + (0.1)(2.399155), 1 + (0.1)(-0.88895)) = -0.76008.
\end{aligned}$$

Then,

$$\begin{aligned}
x(0.1) &\approx -1 + \frac{0.1}{6}[1 + 2(2.221) + 2(2.399155) + 3.99669] \\
&\approx -0.762\,716\,4 \\
y(0.1) &\approx 1 + \frac{0.1}{6}[-1 + 2(-0.95) + 2(-0.88895) - 0.76008] \\
&\approx 0.909\,3669.
\end{aligned}$$

In Table 12.15, the numerical solution is continued and compared with values obtained from the analytical solution.

sysrk4('f1','g1',0,1,-1,1,10)

System of differential equation using Runge-Kutta method of order 4

t	x	y	x(t)	y(t)	|x-x(t)|	|y-y(t)|
0.0	-1.000	1.0000	-1.000	1.0000	0.000	0.000
0.1	-0.7627	0.9094		0.9094		1.29e-005
0.2	-0.1418	0.8601		0.8602		3.61e-005
0.3	1.0815	0.9009		0.9010		7.50e-005
0.4	3.2231	1.1068		1.1070		1.38e-004
0.5	6.7345	1.5909		1.5911		2.35e-004
0.6	12.2569	2.5203		2.5207		3.85e-004
0.7	20.6936	4.1388		4.1394		6.09e-004
0.8	33.3114	6.7976		6.7985		9.42e-004
0.9	51.8779	10.9982		10.9996		1.43e-003
1.0	78.8502	17.4517		17.4539		2.13e-003

Table 12.15   RK4 solution for Example 12.15.

EXAMPLE 12.16

Solve the following system of differential equations using the Runge-Kutta method of order 4 with $N = 10$.

$$x'(t) = -3x + 4y, \quad x(0) = 1$$
$$y'(t) = -2x + 3y, \quad y(0) = 2, \quad 0 \le t \le 1.$$

Compare the results with the values of the exact solution

$$x(t) = 3e^t - 2e^{-t} \quad \text{and} \quad y(t) = 3e^t - e^{-t}.$$

Set

$$f(t, x, y) = -3x + 4y \quad \text{and} \quad g(t, x, y) = -2x + 3y.$$

The numerical computations are given in Table 12.16 along with the true solution values for comparison.

EXERCISE SET 12.10

1. Solve the following systems of first-order differential equations

(a) $\begin{aligned} x' &= t - y + x, \ x(0) = 1, \\ y' &= x + 2y, \quad y(0) = 0, \ 0 \le t \le 1, \text{ with } h = 0.2. \end{aligned}$

(b) Solve the following IVPs

```
» sysrk4('f2','g2',0,1,1,2,10)
```

System of differential equation using Runge-Kutta method of order 4

t	x	y	x(t)	y(t)	|x-x(t)|	|y-y(t)|
0.0	1	2	1	2	0	0
0.1	1.50584	2.41067	1.50584	2.41068	4.2e-007	3.4e-007
0.2	2.02675	2.84548	2.02675	2.84548	8.6e-007	7.1e-007
0.3	2.56794	3.30876	2.56794	3.30876	1.3e-006	1.1e-006
0.4	3.13483	3.80515	3.13483	3.80515	1.9e-006	1.6e-006
0.5	3.7331	4.33963	3.7331	4.33963	2.4e-006	2.2e-006
0.6	4.36873	4.91754	4.36873	4.91754	3.1e-006	2.8e-006
0.7	5.04808	5.54467	5.04809	5.54467	3.9e-006	3.6e-006
0.8	5.77796	6.22729	5.77796	6.22729	4.7e-006	4.4e-006
0.9	6.56566	6.97223	6.56567	6.97224	5.8e-006	5.4e-006
1.0	7.41908	7.78696	7.41909	7.78697	6.9e-006	6.6e-006

Table 12.16   RK4 solution for Example 12.16.

i. $y'' = y' + ty^2$, $y(0) = 1$, $y'(0) = 0$, $0 \le t \le 1$ with $N = 10$,

ii. $y'' = 3y' - 2y$, $y(0) = 1$, $y'(0) = 1$, $0 \le t \le 1$ with $N = 10$,

iii. $y'' = ty' - 2y$, $y(0) = -1$, $y'(0) = 1$, $0 \le t \le 1$ with $N = 10$,

iv. $y'' = t^2 y - ty + t$, $y(0) = 1$, $y'(0) = 1$, $0 \le t \le 1$ with $N = 10$.

(c) Approximate the solution of the Van Pol's equation

$$y'' - (0.1)(1 - y^2)y' + y = 0, \quad y(0) = 1, \quad y'(0) = 0$$

using the Runge-Kutta method of order 4 on $[0, 6.4]$ with $h = 0.2$.

(d) Write the third-order differential equation

$$y''' = -2y'' + y' + 2y + e^t, \quad y(0) = 1, \ y'(0) = 2, \ y''(0) = 0$$

as a system of first-order differential equations.

(e) $\begin{aligned} x' &= 2 - 4y - x\sin t, \ x(0) = 0, \\ y' &= x, \qquad\qquad\quad y(0) = 0, \ 0 \le t \le 1, \text{ with } h = 0.2. \end{aligned}$

(f) $\begin{aligned} x' &= t^2 - 3x^2 + y, \ x(0) = 0, \\ y' &= 2t + x - y, \quad y(0) = 1, \ 0 \le t \le 1, \text{ with } h = 0.1. \end{aligned}$

(g) $\begin{aligned} x' &= t^3 - y^2, \qquad x(0) = 0, \\ y' &= t + y - xy, \ y(0) = 1, \ 0 \le t \le 1, \text{ with } h = 0.2. \end{aligned}$

M-function 12.10

The following MATLAB function **sys_rk4.m** solves the system 11.48 using the Runge-Kutta method of order 4. INPUTS are functions *f(t,x,y)* and *g(t,x,y)*; the initial and final values *a, b* of *t*; the initial conditions x_0, y_0 of *x* and *y*; the number of steps *n*. The input functions *f* and *g* should be defined as M-files.

```
function sys_rk4(f,g,a,b,x0,y0,n)
% Solve the system of differential equations x'=f(t,x,y), x(a)=x0 and
% y'=g(t,x,y), y(a)=y0 using the Runge-Kutta method of order 4.
fprintf('\n')
disp('        Solution of system of differential equations using RK4 ')
h=(b-a)/n;
x=x0;
y=y0;
disp('_____')
disp(' t       x        y        x(t)      y(t)   Ix-x(t)I    Iy-y(t)I ')
disp('_____')
fprintf('\n')
fprintf('%8.2f     %10.3f %10.3f %10.3f %10.3f %8.2f %8.2f\n', a, x,
y, x, y, 0, 0)
for i=1:n
  % RK of order 4
  t=a+(i-1)*h;
  k1=feval(f,t,x,y);
  c1=feval(g,t,x,y);
  k2=feval(f,t+h/2,x+h*k1/2,y+h*c1/2);
  c2=feval(g,t+h/2,x+h*k1/2,y+h*c1/2);
  k3=feval(f,t+h/2,x+h*k2/2,y+h*c2/2);
  c3=feval(g,t+h/2,x+h*k2/2,y+h*c2/2);
  k4=feval(f,t+h,x+h*k3,y+h*c3);
  c4=feval(g,t+h,x+h*k3,y+h*c3);
  x=x+h*(k1+2*k2+2*k3+k4)/6;
  y=y+h*(c1+2*c2+2*c3+c4)/6;
  %If the exact solution is not known set xs='n' and ys='n'.
  % For a system with two Eqns. write the exact solutions
  % if known as xs=xs(t) and ys=ys(t).
  % For a second order IVP write the exact solution
  % if known as ys=ys(t) and set xs='n'.
  t=t+h;
  xs='n';
  ys='n';
  if (xs=='n' & ys=='n')
```

```
    fprintf('%8.2f    %10.3f %10.3f\n',t,x,y)
  elseif (ys~='n' & xs=='n')
   err2=abs(ys-y);
   fprintf('%8.2f    %10.3f %10.3f %10.3f %10.1e\n', t,x,y,ys,err2)
  else
   err1=abs(xs-x);
   err2=abs(ys-y);
   fprintf('%8.2f    %10.3f %10.3f %10.3f %10.3f %10.1e %10.1e
  \n', t, x, y, xs, ys, err1, err2)
  end
end
```

COMPUTER PROBLEM SET 12.10

1. Solve the following system of differential equations using the MATLAB function sys_rk4

$$x' = 2x - 5y, \quad x(0) = 1$$
$$y' = 2x - 4y, \quad y(0) = 1$$

on the interval $[0, 1]$ with $h = 0.1$. Compare your results with the analytical solution $x(t) = -2e^{-t}\sin(t) + e^{-t}\cos(t)$, $y(t) = -e^{-t}\sin(t) + e^{-t}\cos(t)$.

2. Use the MATLAB function sys_rk4 to approximate the solution of the second-order differential equation

$$y''(t) = y + t, \quad y(0) = 1, \quad y'(0) = -1$$

on the interval $[0, 1]$ with $h = 0.1$.

3. Solve the second-order differential equation

$$y'' - y'e^y + ty = t^2, \quad y(1) = 0, \quad y'(1) = 1$$

on the interval $[1, 2]$.

4. Solve the second-order differential equation

$$y'', \quad y(0) = -3, \quad y'(1) = 1$$

on the interval $[1, 2]$.

5. Use a routine from a program library to solve the system of ODE

$$x' = -82x - 54y + 5, \quad x(0) = 5$$
$$y' = 108x + 71y + 4, \quad y(0) = 3.$$

The exact solution to this system is $x(t) = 2e^{-t} + 3e^{-10t} + 5t$, $y(t) = -3e^{-t} - 4e^{-10t} + 4t + 10$. Compare your results with the exact solution.

6. Consider the IVP

$$y'' = -(y')^2 t, \quad y(-1) = 0, \ y'(-1) = 1.$$

The solution of this IVP at $t = 1$ is $y(1) = \pi$. Solve this problem using the Runge-Kutta method of order 4. Output the absolute error in approximating $y(1) = \pi$ with $h = 0.1, 0.01$, and $h = 0.001$.

12.11   IMPLICIT METHODS AND STIFF SYSTEMS

Consider the system of differential equations

$$\begin{aligned}
x'(t) &= 131\,x - 931\,y, & x(0) &= 6 \\
y'(t) &= 133\,x - 933\,y, & y(0) &= -6.
\end{aligned} \tag{12.50}$$

The exact solution of this system is

$$x(t) = 14e^{-2t} - 8e^{-800t}, \qquad y(t) = 2e^{-2t} - 8e^{-800t}.$$

Suppose we approximate the values of $x(0.1)$ and $y(0.1)$ using Euler's method with $h = 0.1$. The result is

$$x(0.1) \approx 643.2 \qquad \text{and} \qquad y(0.1) \approx 633.6$$

If we compare these approximate values to the exact ones $x(0.1) = 11.462$ and $y(0.1) = 1.638$, one can clearly see that the absolute error is large. What caused the error to be large is the presence of the two terms e^{-2t} and e^{-800t} in the exact solutions. Note that the exponents of these two terms are both negative and of different magnitude. In such cases, using an explicit method such as Euler's method with a small value for the step size leads to rapidly growing round-off error. Systems such as (12.50) are called **stiff systems**.

The way to deal with stiff systems is to use an implicit method such as the Adams-Moulton method of order one known as the implicit Euler method given by

$$\begin{aligned}
x_{i+1} &= x_i + hf(x_{i+1}, y_{i+1}) \\
y_{i+1} &= y_i + hg(x_{i+1}, y_{i+1}).
\end{aligned} \tag{12.51}$$

A computer implementation of the implicit form of Euler's method produced the values in Table 12.17. The true solution values $\{x(t_k), y(t_k)\}$ are included for comparison. The results shown in Table 12.17 demonstrate the superiority of the implicit method.

To understand stiff systems better, we consider the general form of a 2×2, linear system of differential equations. The matrix form of the system is

$$\begin{bmatrix} dx/dt \\ dy/dt \end{bmatrix} = \begin{bmatrix} a_{11} & a_{12} \\ a_{21} & a_{22} \end{bmatrix} \begin{bmatrix} x \\ y \end{bmatrix}. \tag{12.52}$$

t_k	$x(t_k)$	$y(t_k)$	x_k	y_k
0.1	11.462	1.6375	11.568	1.567 9
0.01	13.72	1.9577	12.837	1.071 9
0.001	10.377	−1.598 6	9.527 6	−2.448 4
0.0001	6.612 3	−5.3853	6.589 8	−5.407 8

Table 12.17 Comparison of the exact solution and implicit Euler's method for the system (12.50).

It can be shown that the general solution of (12.52) is

$$\begin{bmatrix} x \\ y \end{bmatrix} = C_1 e^{\lambda_1 t} \mathbf{v}_1 + C_2 e^{\lambda_2 t} \mathbf{v}_2$$

where $\mathbf{v}_1$ and $\mathbf{v}_2$ are the eigenvectors of A, and λ_1 and λ_2 are the corresponding eigenvalues. C_1 and C_2 are constants, which depend on the initial conditions of the system.

A stiff system can be defined in terms of the eigenvalues of A. When the real parts of the eigenvalues of A are negative and differ largely in magnitude, we say that the system is stiff.

Here we used the implicit Euler method as an example to solve the system (12.50); however, its accuracy is limited to its order. To get a better accuracy, one can use the higher-order Adams-Moulton methods such as the the fourth-order Adams-Moulton method given by equation (12.38) in Section 12.8.

EXERCISE SET 12.11

1. Solve the following stiff IVP using Euler's method

   (a)

   $$x' = 3x - 37y, \quad x(0) = 16, \quad 0 \le t \le 1, \quad \text{with } h = 0.1$$
   $$y' = 5x - 39y, \quad y(0) = -16, \quad 0 \le t \le 1, \quad \text{with } h = 0.1.$$

   Compare the approximate results with the exact solution $x(t) = 37e^{-2t} - 21e^{-34t}$ and $y(t) = 5e^{-2t} - 21e^{-34t}$.

   (b)

   $$x' = 21x - 211y, \quad x(0) = -94, \quad 0 \le t \le 1, \quad \text{with } h = 0.1,$$
   $$y' = 23x - 213y, \quad y(0) = 94, \quad 0 \le t \le 1, \quad \text{with } h = 0.1.$$

   Compare the approximate results with the exact solution $x(t) = 117e^{-190t} - 211e^{-2t}$ and $y(t) = -23e^{-2t} + 117e^{-190t}$.

12.12   PHASE PLANE ANALYSIS: CHAOTIC DIFFERENTIAL EQUATIONS

In this section, we outline a method for understanding the behavior of nonlinear systems of ordinary differential equations, which is gaining in importance and popularity. We shall restrict ourselves to the case of two differential equations for the unknown functions $x = x(t)$, $y = y(t)$

$$\frac{dx}{dt} = f(x, y) \quad \text{and} \quad \frac{dy}{dt} = g(x, y). \tag{12.53}$$

The equations (12.53) represent a physical situation in which the time variable does not occur explicitly but only implicitly through the dependence of x and y on t and so describes a self-governing or autonomous system. We shall assume that f and g are smooth functions of x and y. Solutions to (12.53) subject to given initial conditions

$$x(t_0) = x_0 \quad \text{and} \quad y(t_0) = y_0 \tag{12.54}$$

are functions of time, but here we wish to think of them as being the parametric equations of a curve

$$(x(t), \ y(t)) \tag{12.55}$$

in the xy-plane passing through the given point (x_0, y_0). This curve is called the **orbit on trajectory of the solution** and the xy-plane is called the **phase plane**.

EXAMPLE 12.17
The solution to

$$\frac{dx}{dt} = -(x + y) \quad \text{and} \quad \frac{dy}{dt} = x - y$$

passing through $(1, 0)$ is $x = e^{-t}\cos t$ and $y = e^{-t}\sin t$. The trajectory is depicted in Figure 12.5.

Observe that in this picture, the trajectory spirals into the origin $(0, 0)$, which is an equilibrium or fixed point for the system (12.53). The equilibrium or fixed points are points in the phase plane where the system remains unchanged so that $dx/dt = 0$ and $dy/dt = 0$ at such a point.

More precisely, (x^*, y^*) is an equilibrium or a fixed point for (12.53) if

$$f(x^*, y^*) = 0 \quad \text{and} \quad g(x^*, y^*) = 0. \tag{12.56}$$

EXAMPLE 12.18
Find the fixed points for

$$\frac{dx}{dt} = y + 3x^2 \quad \text{and} \quad \frac{dy}{dt} = x - 3y^2.$$

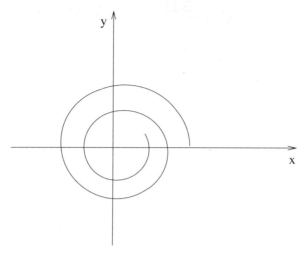

FIGURE 12.5
Spiral into a fixed point.

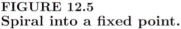

The fixed points are the solutions to the system

$$\begin{cases} y + 3x^2 = 0 \\ x - 3y^2 = 0 \end{cases}.$$

By inspection one sees that $x = 0$ if and only if $y = 0$ so $(0,0)$ is a fixed point. If $x \neq 0$, then $y \neq 0$ as well and the equations show that $x > 0$ and $y < 0$ in that case. We find that $\left(\frac{1}{3}, -\frac{1}{3}\right)$ is the only other fixed point.

Nonlinear equations are often "linearized" in the neighborhood of fixed points, and very often the behavior of the linear system is very close to that of the nonlinear system. The general theory would take us too far afield to describe here, but in the problems we shall explore the behavior of solutions to nonlinear problems on the basis of concrete examples.

To guide us in our investigation, let us make the following observations. Suppose (x^*, y^*) is a fixed point for (12.53). Then near (x^*, y^*)

$$f(x, y) = f_x(x^*, y^*)(x - x^*) + f_y(x^*, y^*)(y - y^*) + \cdots$$
$$g(x, y) = g_x(x^*, y^*)(x - x^*) + g_y(x^*, y^*)(y - y^*) + \cdots.$$

Let $u = x - x^*$ and $v = y - y^*$ so that sufficiently close to (x^*, y^*), the solutions $(u(t), v(t))$ to the linearized system

$$\frac{du}{dt} = f_x(x^*, y^*) u + f_y(x^*, y^*) v$$
$$\frac{dv}{dt} = g_x(x^*, y^*) u + g_y(x^*, y^*) v \tag{12.57}$$

should furnish a good approximation to $(u(t), v(t))$, that is, we would expect

$$x(t) \approx x^* + u(t) \quad \text{and} \quad y(t) \approx y^* + v(t).$$

EXAMPLE 12.19

Consider the linearization of the system in Example 12.18 about the fixed points $(0,0)$ and $\left(\frac{1}{3}, -\frac{1}{3}\right)$.

The linearization of the system about the fixed points $(0,0)$ is

$$\frac{du}{dt} = v \quad \text{and} \quad \frac{dv}{dt} = u.$$

The linearization of the system about the point $\left(\frac{1}{3}, -\frac{1}{3}\right)$ is

$$\frac{du}{dt} = 2u + v \quad \text{and} \quad \frac{dv}{dt} = u + 2v.$$

The behavior of the solutions to the linearized problem depends on the eigenvalues of the Jacobian matrix

$$\begin{pmatrix} f_x(x^*, y^*) & f_y(x^*, y^*) \\ g_x(x^*, y^*) & g_y(x^*, y^*) \end{pmatrix}. \tag{12.58}$$

If the real parts of the eigenvalues are nonzero, then, at least qualitatively, the solutions to the linearized problems will simulate those of the nonlinear problems quite well, whereas if the real parts are zero, that may not be the case.

It is also possible for solutions to spiral into or away from simple closed curves which are, themselves, solutions. These curves are called **limit cycles**. In general, finding limit cycles can be difficult analytically, but experimentally with the help of a computer they can readily be found.

EXAMPLE 12.20

Consider the system

$$\frac{dx}{dt} = -y + x\left(1 - x^2 - y^2\right) \quad \text{and} \quad \frac{dy}{dt} = x + y\left(1 - x^2 - y^2\right).$$

This system has one fixed point, $(0,0)$. However, if one sets $x = \gamma \cos\theta$ and $y = \gamma \sin\theta$ and rewrites the equation in terms of polar coordinates, one finds

$$\frac{d\gamma}{dt} = \gamma\left(1 - \gamma^2\right) \quad \text{and} \quad \frac{d\theta}{dt} = 1.$$

We see that the circle $\gamma = 1$ and $\theta = t + \theta_0$ is a circle that also solves the differential equations. In the problems, we will verify numerically that this circle is also a limit cycle.

We shall be content with this informal discussion of the general properties of nonlinear systems. When the systems are forced with a time-dependent forcing function, the solutions can behave erratically and lead naturally to a discussion of chaotic, dynamical systems. The fixed points themselves can be further classified and studied, but that will not be done here.

Remark: Differential equations whose independent variable is the time t are called dynamical system. The long-term behavior of solutions to **dynamical systems** can be complicated and unpredictable. The modern theory includes the topics of stability, asymptotic stability, periodicity, limit cycles, and chaos.

EXERCISE SET 12.12

Recall that every second-order differential equation

$$\frac{d^2x}{dt} = F\left(t, x, \frac{dx}{dt}\right)$$

can be reduced to a system of two first-order equations by letting $dx/dt = y$. The resulting system is

$$\frac{dx}{dt} = y \quad \text{and} \quad \frac{dy}{dt} = F(t, x, y).$$

Rewrite the differential equations in Problems 1–4 as equivalent systems.

1. $\frac{d^2x}{dt^2} = \sin\left(x + \frac{dx}{dt}\right)$.

2. $\frac{d^2x}{dt^2} = -t\frac{dx}{dt} + x^2 \cos t$.

3. $\frac{d^2x}{dt^2} = -\frac{dx}{dt} + \sin x$.

4. $\frac{d^2x}{dt^2} = (1 - x^2)\frac{dx}{dt} + x - x^3$.

In Problems 5–8, find the fixed points. The fixed points for a second-order differential equation are the points in the phase plane for the equivalent second-order system and, hence, all lie on the x-axis.

5. $\frac{dx}{dt} = \sin(x - y)$, $\quad \frac{dy}{dx} = y \cos x$.

6. $\frac{dx}{dt} = 1 - e^{-x} - y$, $\quad \frac{dy}{dx} = \sin y + x(1 - x)$.

7. $\frac{dx}{dt} = x - xy$, $\quad \frac{dy}{dt} = -y + yx^3$.

8. $\frac{d^2x}{dt^2} = (1 - x^2)\frac{dx}{dt} + (x - x^3)$.

The following problems deal with constructing phase plane plots and comparing the plots of the solutions of the nonlinear problems with those of the associated linear problems.

9. Consider the system of Example 12.20:

(a) Show analytically that $x = \cos t$, $y = \sin t$ is a solution to the differential equation that satisfies the initial condition $x(0) = 1$, $y(0) = 0$.

(b) Show that near the fixed point $(0,0)$, the linearized problem is

$$\frac{du}{dt} = u - v, \qquad \frac{dv}{dt} = u + v.$$

Solve this system and draw the trajectories for the sets of initial conditions: $(x(0) = 0.1, y(0) = 0)$; $(x(0) = 0.5, y(0) = 0)$; $(x(0) = 2, y(0) = 0)$; and $(x(0) = 5, y(0) = 0)$.

(c) Solve the nonlinear system using the same initial conditions as in (b) and compare the two sets of phase plots.

10. Consider the two differential equations:

(a) $\frac{d^2 x}{dt^2} + \left(\frac{dx}{dt}\right)^3 + x = 0$,

(b) $\left(\frac{d^2 x}{dt^2}\right) - \left(\frac{dx}{dt}\right)^3 + x = 0$.

In both cases, the origin is the unique fixed point and the linear form of both equations is $\frac{d^2 u}{dt^2} + u = 0$. Solve both the nonlinear equations subject to the initial conditions $x(0) = 0$, $x'(0) = 1$, and compare the phase plots so obtained with each other and with that obtained for the linearized problem.

11. Does the equation $\frac{d^2 x}{dt^2} + (x^2 - 1)\frac{dx}{dt} + x = 0$ have a limit cycle? Try various initial conditions and carry out some numerical experiments.

12. Consider the following differential equations:

(a) $\frac{d^2 x}{dt^2} - x + x^3 = 0$,

(b) $\frac{d^2 x}{dt^2} + x - x^3 = 0$,

(c) $\frac{d^2 x}{dt^2} + x + x^3 = 0$.

Make phase plane plots for each of them for each of the initial conditions: $(x(0) = .5, x'(0) = 0)$; $(x(0) = 1.5, x'(0) = 0)$; and $(x(0) = 4, x'(0) = 0)$.

13. Forcing introduces additional complications. Make phase plane plots for the solutions to the following differential equations:

(a) $\frac{d^2 x}{dt^2} + x + x^3 = \cos t$,

(b) $\frac{d^2 x}{dt^2} + x - x^3 = \cos t$,

subject to the initial conditions $x(0) = 0$ and $x'(0) = 0$.

14. Make phase plane plots for each of the following differential equations:

(a) $\frac{d^2 x}{dt^2} - \frac{dx}{dt} + x = \cos t$,

(b) $\frac{d^2x}{dt^2} + \frac{dx}{dt} - \left(\frac{dx}{dt}\right)^3 + x = \cos t$,

(c) $\frac{d^2x}{dt^2} + \frac{dx}{dt} + \left(\frac{dx}{dt}\right)^3 + x = \cos t$,

(d) $\frac{d^2x}{dt^2} - \frac{dx}{dt} + \left(\frac{dx}{dt}\right)^3 + x = \cos t$,

(e) $\frac{d^2x}{dt^2} + \frac{dx}{dt} + x = \cos t$,

subject to the initial conditions $x(0) = 0$ and $x'(0) = 0$.

APPLIED PROBLEMS FOR CHAPTER 12

1. The differential equation

$$R\frac{dQ}{dt} + \frac{Q}{C} = V$$

describes the charge Q on a capacitor with capacitance C during a charging process involving a resistance R and an electromotive force V. If the charge is 0 when $t = 0$, approximate Q in the time interval $[0, 4]$ with $R = 2$, $C = 3$, and $V = 4$ using the Runge-Kutta of order 4.

2. Suppose an undamped weight of mass m is attached to a spring having spring constant k, and external periodic force given by $F = \sin \alpha t$ is applied to the weight. The differential equation that describes the motion of the weight y is

$$\frac{d^2y}{dt^2} + \omega^2 y = \frac{F}{m} \sin \alpha t$$

where $\omega^2 = k/m$. If $k = 2.7 \times 10^2 N/m$, $m = 0.55kg$, $F = 2N$, and $\alpha = 10 \times s^{-1}$, approximate y at $t = 3$ sec.

3. An automobile heading light mirror is designed to reflect the light given off by the headlamp in rays parallel to the real surface. By using the principle of optics that the angle of light incidence equals the angle of light reflection, we can derive a first-order differential equation that models the desired shape of the mirror

$$\frac{dy}{dx} = \frac{-x + (x^2 + y^2)^{1/2}}{y}.$$

The mirror is designed so that the distance of the mirror directly above the lamp is 1 cm, so $y(0) = 1$. Use the Runge-Kutta method of order 4 with $h = 0.1$ to estimate y at $x = 1$ and $x = 2$.

4. Suppose a 24-lb weight stretches a spring 1 ft beyond its natural length. If the weight is pushed downward from its equilibrium position with an initial

velocity of 2 ft/sec and if the damping force is $-9(dy/dt)$, the displacement y of the weight at any time t is given by the differential equation

$$\frac{d^2y}{dt^2} + 12\frac{dy}{dt} + 32y = 0$$

with $y(0) = 0$ and $y'(0) = 2$. Compute the displacement of the spring at $t = 4$ using the Runge-Kutta method of order 4 for the system of equations.

5. The opening and closing of small, high-performance hydraulic valves used to control guidance jets in spacecrafts are accomplished by the use of a torque motor. If the mass m of the rotor is 32 slugs, the coefficient of viscous damping is 28 lb-sec/ft, the spring constant is 96 lb/ft, the radius r is 1 ft, and the motor is driven by a torque M equal to 32 $\cos t$ (in lb/ft), then the IVP describing the rotation of the motor shaft is

$$\theta'' + 4\theta' + 3\theta = \cos t, \quad \theta(0) = \theta'(0) = 0.$$

Approximate the angular displacement of the motor shaft at $t = 4$ using the Runge-Kutta method of order 4 for a system of equations.

6. An object of mass m is released from a hot-air balloon. The velocity of the object after t seconds is given by

$$\frac{dv}{dt} + cv = g \quad \text{with } v(0) = 0.$$

Given that $c = 2.2$ and $g = 32$, use Euler's method to approximate the velocity of the object after 3 sec.

7. The population in a city changes at the rate

$$\frac{dp}{dt} = (k_0 + k_1 p)p$$

where $p(t)$ denotes the number of members in the population at time t (in years) and k_0, k_1 are specified constants. Given that the initial population size is 1000 and $k_0 = 0.02$, $k_1 = 0.4$, approximate the size of the population after 10 years, using the Runge-Kutta method of order 4 with $N = 10$.

8. Consider the series circuit given in the figure below. The differential equation for the current i is

$$L\frac{di}{dt} = -Ri + E$$

where $i(0) = 0$. Suppose $E = 100V$, $R = 10\Omega$, and $L = 1h$. Use Euler's method to approximate i at $t = 0.6$ and 0.9 with $h = 0.1$.

9. If air resistance is proportional to the square of the instantaneous velocity, then the velocity v of the mass m dropped from a height h is determined from

$$m\frac{dv}{dt} = mg - kv^2, \quad k > 0.$$

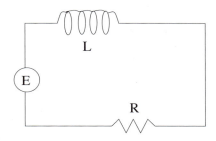

Let $v(0) = 0$, $k = 0.125$, $m = 5$ slugs, and $g = 32$ ft/s^2. Use the Runge-Kutta method of order 4 with $h = 1$ to find an approximation of the velocity of the falling mass at $t = 5$ s.

10. Suppose water is leaking from a tank through a circular hole of area A_h at its bottom. Friction and contraction of a water stream near the hole reduce the volume of water leaving the tank per second to $cA_h\sqrt{2gh}$, where $0 < c < 1$ is a constant. The differential equation for the height h of water at time t for a cubical tank with a side 10 ft and a hole of radius 2 ft is

$$\frac{dh}{dt} = -\frac{c\pi}{450}\sqrt{h}.$$

Suppose the tank is initially full and $c = 0.4$; find the height of water after 3 seconds using the Runge-Kutta method of order 2.

11. A model for population $P(t)$ of a large city is given by the initial-value problem

$$\frac{dP}{dt} = P(10^{-1} - 10^{-6}P), \quad P(0) = 10000$$

where t is in months. Use a numerical method to predict the population after 12 months.

12. In the study of dynamical systems, the phenomena of period doubling and chaos are observed. These phenomena can be seen when one uses a numerical scheme to approximate the solution to an initial-value problem for a nonlinear differential equation such as the following logistic model for population growth

$$\frac{dP}{dt} = 10P(1 - P), \quad P(0) = 0.1.$$

Use Euler's method with $h = 0.25$ to approximate $P(30)$. Note how the values of P jump from 1.22, 0.54, 1.16, and 0.70.

13. The mixing of two salt solutions of differing concentration gives rise to a first-order differential equation for the amount of salt contained in the mixture. Suppose a large tank holding 1000 liters (L) of water into which a brine solution of salt begins to flow at a constant rate of 6 L/min. The mixture is kept uniform by stirring, and the mixture flows out at a rate of 5 L/min (see figure

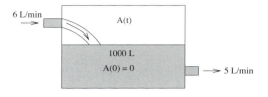

below). Let $A(t)$ denote the amount of salt in the tank at time t, then the rate at which $A(t)$ changes is a net rate

$$\frac{dA}{dt} = (\text{input rate}) - (\text{output rate}).$$

If we assume that the concentration of salt in the brine entering the tank is 1 kg/L, we obtain the initial-value problem

$$\frac{dA}{dt} = 6 - \frac{5A}{1000 + t}, \quad A(0) = 0.$$

Use a numerical method to find the amount of salt in the tank at $t = 3$.

14. In the study of an electrical field that is induced by two nearby transmission lines, an equation of the form

$$\frac{dy}{dt} = -h(t)y^2 + f(t)$$

arises. Let $f(t) = 5t + 2$ and $h(t) = t^2$. If $y(0) = 1$, use the Runge-Kutta method of order 2 to approximate $y(2)$ with $h = 0.001$.

15. In the study of nonisothermal flow of Newtonian fluid between parallel plates, the initial-value problem of the form

$$y'' = -t^2 e^y, \quad y(0) = 1, \ y'(0) = 1$$

arises. Use the Runge-Kutta method of order 4 to approximate $y(1)$ with $h = 0.05$

16. The secretion of hormones into the blood is often a periodic activity. If a hormone is secreted on a 24-hr cycle, then the rate of change of the level of the hormone in the blood may be represented by the initial-value problem

$$\frac{dy}{dt} = \alpha - \beta \cos \frac{\pi t}{12} - ky, \quad y(0) = 10,$$

where $y(t)$ is the amount of the hormone in the blood at time t, α is the average secretion rate, β is the amount of daily variation in the secretion, and k is a positive constant reflecting the rate at which the body removes the hormone from the blood. If $\alpha = \beta = 1$, and $k = 2$, use a numerical method described in the text to approximate the amount of the hormone in the blood at $t = 2$.

17. As a raindrop falls, it evaporates while retaining its spherical form. If further assumptions are made that the rate at which the raindrop evaporates is proportional to its surface area and that air resistance is negligible, then the velocity $v(t)$ of the raindrop is given by

$$\frac{dv}{dt} = -\frac{3(k/\rho)}{(k/\rho)t + 0.01}v + g.$$

Here ρ is the density of water, $k < 0$ is the constant of proportionality, and the downward direction is taken to be the positive direction. Given $k/\rho = -0.0003$ and assuming that the raindrop falls from rest, estimate $v(10)$ using a numerical method described in the text. Use $g = 32$ ft/s^2.

18. When $E = 100$ V, $R = 10\Omega$, and $L = 1$ h, the system of differential equations for the currents i_1 and i_3 in an electrical network is given by the system of differential equations

$$\frac{di_1}{dt} = -20i_1 + 10i_3 + 100$$

$$\frac{di_3}{dt} = 10i_1 + 20i_3$$

where $i_1(0) = 0$ and $i_3(0) = 0$. Use the Runge-Kutta method of order 4 to approximate $i_1(t)$ and $i_3(t)$ in the interval $[0, 0.5]$ with $h = 0.1$. Plot the graph of $i_1(t)$ and $i_3(t)$ to predict the behavior of $i_1(t)$ and $i_3(t)$ as $t \longrightarrow \infty$.

19. Circuit flowing through a resistor causes an increase in temperature, and this in turn increases its resistance R according to the formula $R = R_0 + bi^2$. When an E volt (V) battery is applied across such a resistor in a series with an L henry (h) inductor at time $t = 0$, the circuit current $i = i(t)$ in milliamperes (ma) after t milliseconds (ms) satisfies the IVP

$$L\frac{di}{dt} = E - R_0 i - bi^2, \quad i(0) = 0.$$

Suppose that $L = 4$ h, $R_0 = 1.2$ kilohms, and $b = 0.4 \times 10^{-7}$ kilohms/ma^2. Use a numerical method of your choice to approximate $i(t)$ on the interval $[0, 1]$ for (a) $E = 1.2$ V; (b) $E = 12$ V; (c) $E = 120$ V.

20. The system of ODE

$$x'(t) = -2x/\sqrt{x^2 + y^2}$$

$$y'(t) = 1 - 2y/\sqrt{x^2 + y^2}$$

describes the path of a duck attempting to swim across a river by aiming steadily at the target position T. The speed of the river is 1, and the duck speed is 2. The duck starts at S, so that $x(0) = 1$ and $y(0) = 0$ (see the figure below). Apply the Runge-Kutta method of order 4 to compute the duck's path from $t = 0$ to $t = 4$. Compare with the exact trajectory $y = (x^{1/2} - x^{3/2})/2$.

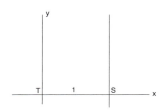

21. Consider a beam of length L subjected to a load F at its free end (see the figure below). Let x be the distance along the axis of the beam with $x = 0$ at the built-in end. The variable y measures the deflection of the beam. It can be shown that the elastic deflections satisfy the differential equation

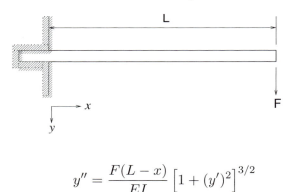

$$y'' = \frac{F(L-x)}{EI}\left[1+(y')^2\right]^{3/2}$$

where E is Young's modulus for the material and I is the moment of inertia of a cross section of the beam. Since both the deflection and its slope vanish at $x = 0$, we have the initial conditions

$$y(0) = 0, \qquad y'(0) = 0.$$

Use a numerical method of your choice to approximate $y(100)$ using $h = 2$ for the case of a beam of high-strength aluminum alloy with rectangular cross section. The beam is 100 in. long and $I = 0.128$ in.^{4}. The value of Young's modulus is $E = 10 \times 10^6$ psi. Solve the IVP for $F = 60, 80, 100, 150$, and 200 lbs to obtain some feeling about the deflection of the beam.

22. In 1926, Volterra developed a mathematical model for predator-prey systems. If r is the population density of prey (rabbits), and f is the population density of prey (foxes), then Volterra's model for the population growth is the system of ordinary differential equations

$$r'(t) = a\,r(t) - b\,r(t)f(t)$$
$$f'(t) = d\,r(t)f(t) - c\,f(t)$$

where t is time, a is the natural growth rate of rabbits, c is the natural death rate of foxes, b is the death rate of rabbits per one unit of the fox population, and d is the growth rate of foxes per one unit of the rabbit population. Use the

Runge-Kutta method for the solution of this system. Take $a = 0.03$, $b = 0.01$, $c = 0.01$, and $d = 0.01$, the interval $t \in [0, 500]$, the step size $h = 1$, and the initial values

(a) $r(0) = 1.0$, $f(0) = 2.0$,

(b) $r(0) = 1.0$, $f(0) = 4.0$.

Plot the solution functions $r(t)$ and $f(t)$.

Chapter 13

Boundary-Value Problems

All the problems discussed in Chapter 9 required conditions on $y(t)$ and its derivatives at only one point that define the initial value. For that reason, such problems are termed initial-value problems. However, in many problems, conditions are specified at more than one point. Such problems are classified as **boundary-value problems** (BVP). These problems arise in the study of beam deflection, heat flow, and various physical problems. In this chapter we shall develop two principal classes of numerical methods used for solving the following boundary-value problems:

The linear second-order boundary-value problem,

$$\begin{cases} y''(x) + p(x)y' + q(x)y = r(x) \\ y(a) = \alpha, \quad y(b) = \beta, \end{cases} \tag{13.1}$$

and the nonlinear second-order boundary-value problem,

$$\begin{cases} y''(x) = f(x, y, y') \\ y(a) = \alpha, \quad y(b) = \beta. \end{cases} \tag{13.2}$$

EXAMPLE 13.1 : Cooling Fin

Consider the steady temperature $T(x)$ in a fin of uniform cross section, which satisfies the following second order, linear, differential equation:

$$\frac{d^2 T}{dx^2} - \frac{hP}{kA}(T - T_\infty) = 0, \quad 0 < x < L.$$

Heat is conducted steadily through the fin and is lost from the sides of the fin by convection to the surrounding air with heat transfer coefficient h (W/m^2/K). Here, P is the fin perimeter (meters), A is the cross-section area of the fin (meters squared), k is the thermal conductivity (W/m/K), and T_∞ is the ambient temperature. Refer to Figure 13.1 for the geometry. The boundary conditions at the ends of the fin may be of several types; however, consider the following specific boundary conditions:

$$T(0) = T_0 \quad \text{and} \quad T(L) = T_L$$

which indicate that the temperature at both ends of the fin is fixed.

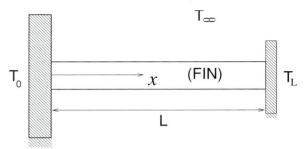

FIGURE 13.1
Geometry for the cooling fin example.

13.1   FINITE-DIFFERENCE METHODS

In this section, we assume that we have a linear differential equation. The problem we want to solve is

$$y''(x) + p(x)y' + q(x)y = r(x), \quad a \le x \le b \qquad (13.3)$$

subject to the boundary conditions

$$y(a) = \alpha \quad \text{and} \quad y(b) = \beta. \qquad (13.4)$$

To accomplish this, let us first divide the interval $[a, b]$ into N equal subintervals, each of size h, with

$$h = \frac{b - a}{N} \quad \text{and} \quad x_i = a + hi, \quad i = 0, 1, \ldots, N.$$

The **finite-difference** method consists of replacing every derivative in Eqn. (13.3) by finite-difference approximations such as the central divided-difference approximations

$$y'(x_i) \approx \frac{1}{2h} [y(x_{i+1}) - y(x_{i-1})]$$

$$(13.5)$$

$$y''(x_i) \approx \frac{1}{h^2} [y(x_{i+1}) - 2y(x_i) + y(x_{i-1})].$$

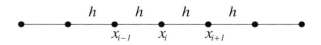

Observe that using central divided-difference approximations to derivatives will result in a finite-difference method with a truncation error of order $O(h^2)$.

Substituting (13.5) into Eqn. (13.3), and rearranging, we get the difference equation

$$\left[1 - \frac{h}{2}p_i\right] y_{i-1} + (-2 + h^2 q_i)y_i + \left[1 + \frac{h}{2}p_i\right] y_{i+1} = h^2 r_i \qquad (13.6)$$

$i = 1, 2, \ldots, N - 1$
and

$$y_0 = \alpha \quad \text{and} \quad y_N = \beta \tag{13.7}$$

where $y_i \approx y(x_i)$, $p_i = p(x_i)$, $q_i = q(x_i)$, and $r_i = r(x_i)$.

The equations given by (13.6) form a linear system in $(N-1)$ equations and $(N-1)$ unknowns. It can be represented in the tridiagonal matrix form

$$A\mathbf{y} = \mathbf{b}, \tag{13.8}$$

or

$$\begin{bmatrix} b_1 & c_1 & & & & \\ a_2 & b_2 & c_2 & & \mathbf{0} & \\ & a_3 & b_3 & c_3 & & \\ & & & \ddots & & \\ \mathbf{0} & & a_{N-2} & b_{N-2} & c_{N-2} \\ & & & a_{N-1} & b_{N-1} \end{bmatrix} \begin{bmatrix} y_1 \\ y_2 \\ y_3 \\ \vdots \\ y_{N-2} \\ y_{N-1} \end{bmatrix} = \begin{bmatrix} d_1 - a_1\alpha \\ d_2 \\ d_3 \\ \vdots \\ d_{N-2} \\ d_{N-1} - c_{N-1}\beta \end{bmatrix}$$

where a_i, b_i, c_i, and d_i are defined by

$$a_i = 1 - \frac{h}{2}p_i, \quad b_i = -2 + h^2 q_i,$$

$$c_i = 1 + \frac{h}{2}p_i, \quad d_i = h^2 r_i \tag{13.9}$$

for $i = 1, \ldots, N - 1$.

So, our problem now reduces to solving the tridiagonal system (13.8) whose solution will give approximate values for the solution of the boundary-value problem at discrete points on the interval $[a, b]$.

To obtain a better accuracy for Eqn. (13.3), we could use more accurate finite-difference approximations to the derivatives. But, the disadvantage of doing this is that the resulting system of equations would not be tridiagonal and more calculations would be required. Another way of insuring accuracy is to solve the linear system for smaller values of h, and compare the solutions at the same mesh points; the round-off error, however, will eventually increase and may become large.

EXAMPLE 13.2
Solve the following boundary-value problem

$$\begin{cases} y'' + (x + 1)y' - 2y = (1 - x^2)e^{-x}, & 0 \le x \le 1 \\ y(0) = -1, & y(1) = 0 \end{cases}$$

using the above finite-difference method with $h = 0.2$. Compare the results with the exact solution $y = (x - 1)e^{-x}$.

x	F. Diff. Method	Exact Sol.	Error
0.0	-1.00000	-1.00000	0.000000
0.2	-0.65413	-0.65498	0.000854
0.4	-0.40103	-0.40219	0.001163
0.6	-0.21848	-0.21952	0.001047
0.8	-0.08924	-0.08987	0.000624
1.0	0.00000	0.00000	0.000000

Table 13.1   The finite-difference method for Example 13.2.

In this example: $p(x) = x + 1$, $q(x) = -2$, and $r(x) = (1 - x^2)e^{-x}$; hence, Eqn. (13.6) yields

$$[1 - 0.1(x_i + 1)]y_{i-1} + (-2 - 0.08)y_i + [1 + 0.1(x_i + 1)]y_{i+1} = \\ +0.04(1 - x_i^2)e^{-x_i}$$

and

$$y_0 = -1 \quad \text{and} \quad y_5 = 0$$

where $x_i = 0.2i$, $i = 1, 2, 3, 4$.

The matrix formulation is

$$\begin{bmatrix} -2.08 & 1.12 & & \\ 0.86 & -2.08 & 1.14 & \\ & 0.84 & -2.08 & 1.16 \\ & & 0.82 & -2.08 \end{bmatrix} \begin{bmatrix} y_1 \\ y_2 \\ y_3 \\ y_4 \end{bmatrix} = \begin{bmatrix} 0.91143926 \\ 0.02252275 \\ 0.01404958 \\ 0.00647034 \end{bmatrix}.$$

The solution of this set is given in Table 13.1. For comparison, the table also gives values calculated from the analytical solution.

EXAMPLE 13.3
Solve the following boundary-value problem

$$y'' + e^x y' - xy = e^{-x}(-x^2 + 2x - 3) - x + 2$$

$$y(0) = -1 \quad \text{and} \quad y(1) = 0$$

using the finite-difference method described in Section 13.1 with $h = 0.1$. Compare the results with the exact solution $y = (x - 1)e^{-x}$.

Set

$$p(x) = e^x, \quad q(x) = -x \quad \text{and} \quad r(x) = e^{-x}(-x^2 + 2x - 3) - x + 2$$

to get the results in Table 13.2.

```
» finitediff( 'p', 'q', 'r', 0, 1, -1, 0, 10)

        Finite difference method with h = 0.1

The subdiagonal of A =

   0.9389   0.9325   0.9254   0.9176   0.9089   0.8993   0.8887   0.8770

The main diagonal of A =

   -2.0010  -2.0020  -2.0030  -2.0040  -2.0050  -2.0060  -2.0070  -2.0080  -2.0090

The superdiagonal diagonal of A =

   1.0553   1.0611   1.0675   1.0746   1.0824   1.0911   1.1007   1.1113

The Coefficient vector B' =

   0.9383  -0.0036  -0.0014   0.0002   0.0014   0.0021   0.0026   0.0028   0.0028

The solution of the BVP =
```

xi	yi	y(xi)	error
0	-1	-1	0.0E+00
0.1	-0.814197	-0.814354	1.6E-04
0.2	-0.654714	-0.654985	2.7E-04
0.3	-0.518229	-0.518573	3.4E-04
0.4	-0.401815	-0.402192	3.8E-04
0.5	-0.30289	-0.303265	3.8E-04
0.6	-0.219182	-0.219525	3.4E-04
0.7	-0.148691	-0.148976	2.8E-04
0.8	-0.089661	-0.089866	2.0E-04
0.9	-0.040549	-0.040657	1.1E-04
1	0	0	0.0E+00

Table 13.2   The finite-difference method for Example 13.3.

EXAMPLE 13.4   Cooling fin

Use the MATLAB function finited to find the temperature distribution along the fin given that the perimeter of the fin is $P = 4.5$ in., $k = 25.9$ (steel), $L = 36$ in., $h = 1.06$, $A = 1$ in.^{2}, $T_0 = 200°F$, $T_L = 70°F$, and $T_\infty = 70°F$ (see Example 13.1).

Here, $m = hP/kA = (1.06)(4.5)/(25.9 \times 1) = 0.184$. We have

$$T'' - mT = -mT_\infty.$$

So, $p = -m$, $q = 0$, and $r = -mT_\infty = -70m$. Defining p, q, and r as M-functions and calling the MATLAB function finitediff give the result shown in the following table.

x	0.00	0.25	0.5	0.75	1.00	1.25	
T	200.00	195.78	190.53	184.21	176.77	168.16	
x	1.5	1.75	2.00	2.25	2.50	2.75	3.00
T	158.32	147.19	134.71	120.82	105.45	88.54	70.00

EXERCISE SET 13.1

1. Solve the following boundary-value problems using the finite-difference method developed in Section 13.1.

   (a) $y'' + y' + xy = 0$, $y(0) = 1$, $y(1) = 0$, $\quad 0 \le x \le 1$, $\quad h = 0.2$,

   (b) $y'' + 9y = 0$, $y(0) = 1$, $y(\pi) = -1$, $\quad 0 \le x \le \pi$, $\quad h = \pi/6$,

   (c) $y'' + y = x$, $y(0) = y(\pi) = 0$, $\quad 0 \le x \le \pi$, $\quad h = \pi/4$.

   (d) $y'' + 2y = \cos x$, $y(0) = y(1) = 0$, $\quad 0 \le x \le 1$, $\quad h = 0.1$,

2. Show that $y(x) = x + 1/x$ is a solution of the boundary-value problem

$$y'' = 2y^3 - 6y - 2x^3, \quad y(1) = 2, \ y(2) = 5/2.$$

3. Apply a finite-difference method to solve the boundary-value problem

$$y'' + x^2 y' - 4xy = 0, \ y(0) = 0, \ y(1) = 5, \quad 0 \le x \le 1$$

   with $h = 0.1$. Compare with the exact solution $y = x^4 + 4x$.

4. Consider the boundary-value problem

$$y'' + xy' - x^2 y = 2x^2, \quad y(0) = 1, \ y(1) = -1.$$

   Write down the central-difference approximation to the differential equation for any x_i and h. Use $h = 0.25$ to approximate the solution of BVP.

5. Write the difference equation with $n = 10$ to approximate the solution of the boundary-value problem

$$y'' + 3y' + 2y = 4x^2, \ y(1) = 1, \ y(2) = 6, \quad 1 \le x \le 2.$$

COMPUTER PROBLEM SET 13.1

1. Consider the boundary-value problem

$$y'' = y' + 2y + \cos x, \quad y(0) = 0.3, \ y(\pi/2) = -0.1, \ 0 \le x \le \pi/2.$$

M-function 13.1

The following MATLAB function **finitediff.m** finds the solution of the linear BVP (12.1) using the finite difference method described in sec. 12.1. INPUTS are functions $p(x)$, $q(x)$, $r(x)$; the end points aa, bb; the boundary conditions y_0 and y_n; the number of steps n. The input functions p, q, and r should be defined as M-files.

```
function finitediff(p,q,r,aa,bb,y0,yn,n)
% Solve the second order linear BVP using a finite difference
% method.
fprintf('\n')
h=(bb-aa)/n;
blank='     ';
disp(['Finite difference method with h=',num2str(h)])
fprintf('\n')
for i=1:n-1
  x=aa+i*h;
  if (i~=1)
    a(i-1)=1-h/2*feval(p,x); %Compute the subdiagonal of A.
  end
  b(i)=-2+h^2*feval(q,x); %Compute the main diagonal of A.
  if (i~=n-1)
    c(i)=1+h/2*feval(p,x); %Compute the superdiagonal of A.
  end
end
disp(' The subdiagonal of A =')
disp(a)
disp(' The main diagonal of A =')
disp(b)
disp(' The superdiagonal diagonal of A =')
disp(c)
% Compute the coefficient vector B.
d(1)=h^2*feval(r,aa+h)-y0*(1-h/2*feval(p,aa+h));
d(n-1)=h^2*feval(r,bb-h)-yn*(1+h/2*feval(p,bb-h));
for i=2:n-2
  x=aa+i*h;
  d(i)=h^2*feval(r,x);
end
fprintf('\n')
disp(' The Coefficient vector B'' =')
disp(d)
disp(' The solution of the BVP =')
fprintf('\n')
```

```
disp('   xi                 yi              y(xi)            error')
fprintf('\n')
% Solve the tridiagonal system Ax=B.
for i=2:n-1
  ymult=a(i-1)/b(i-1);
  b(i)=b(i)-ymult*c(i-1);
  d(i)=d(i)-ymult*d(i-1);
end
y(n)=yn;
y(n-1)=d(n-1)/b(n-1);
for i=n-2:-1:1
  y(i)=(d(i)-c(i)*y(i+1))/b(i);
end
fprintf('%6.2f  %12.6f  %12.6f  %10.2e\n',aa,y0,y0,0)
for i=1:n
  x=aa+i*h;
%Write the exact solution if known as s=s(x) otherwise set s='n'.
  s='n';
  if (s=='n')
    fprintf('%6.2f %12.6f\n',x,y(i))
  else
    err=abs(s-y(i));
    fprintf('%6.2f  %12.6f  %12.6f  %10.2e\n',x,y(i),s,err)
  end
end
```

(a) Write a computer program in a language of your choice to solve the BVP (13.1) using the finite-difference method developed in Section 13.1 with $h = \pi/12, \pi/24$, and $h = \pi/48$.

(b) Obtain the analytical solution to the BVP by assuming the form $y(x) = A \sin x + \cos x$ and applying the given boundary conditions.

(c) Compare the numerical approximations with the exact solution by plotting the functions and computing the mean-square error in y over the numerical solution points:

$$\text{mean-square error} = \sqrt{\frac{1}{n} \sum_{i=1}^{n-1} [y_i - y(x_i)]^2}.$$

2. Use the MATLAB function finitediff to estimate the solution of the boundary-value problem

$$y'' + \sqrt{x}y' + y = e^x, \quad y(0) = 0, \ y(1) = 0, \ 0 \le x \le 1$$

on the interval $[0, 1]$ with $h = 0.1, 0.01$.

3. Use the MATLAB function finitediff to estimate the solution of the boundary-value problem

$$y'' + \frac{1}{4}y - 8 = 0, \quad y(0) = 0, \ y(10) = 0, \ 0 \le x \le 10$$

on the interval $[0, 10]$ with $h = 1$. Compare the numerical approximations with the exact solution

$$y(x) = 32 \left[\frac{\cos(5) - 1}{\sin(5)} \sin(\frac{x}{2}) - \cos(\frac{x}{2}) + 1 \right].$$

4. Solve the boundary-value problem

$$x^2 y'' - xy' + y = \ln x, \quad y(1) = 0, \ y(2) = -2, \ 1 \le x \le 2$$

using the finite-difference method with $n = 8$.

5. Use the MATLAB function finitediff to approximate the solution of the boundary-value problem

$$y'' + 2xy - y = 2(1 + x^2) \cos x, \quad y(0) = 0, \ y(\pi/2) = \pi/2.$$

Compare the approximate results with the exact solution. $y = x \sin x$.

13.2   SHOOTING METHODS

13.2.1   The nonlinear case

The finite-difference method developed in the previous section works reasonably well for linear boundary-value problems and does not present problems of instability. For a boundary-value problem involving a nonlinear differential equation, these methods run into problems, in that the resulting system is nonlinear. In such situations, it is often preferable to use the so-called **shooting** method, which we now discuss.

Consider the boundary-value problem

$$y''(x) = f(x, y, y'), \quad a \le x \le b \tag{13.10}$$

with the boundary conditions

$$y(a) = \alpha, \quad y(b) = \beta.$$

The shooting method consists of converting Problem (13.10) to a second-order initial-value problem of the form

$$y''(x) = f(x, y, y'), \quad a \le x \le b \tag{13.11}$$

subject to the initial conditions

$$y(a) = \alpha, \quad \text{and} \quad y'(a) = \gamma.$$

Since the condition $y'(a) = \gamma$ is not known, we arbitrarily choose a value γ_1, and the integration from a to b of the initial-value problem (13.11) is carried out by using any standard method for initial-value problems. Let $y(\gamma_1; b)$ be the computed solution at $x = b$. The problem is to determine γ, so that

$$y(\gamma; b) - \beta = 0. \tag{13.12}$$

Again, we arbitrarily choose another value, γ_2, and the procedure is repeated. Let $y(\gamma_2; b)$ be the computed solution at $x = b$ (see Figure 13.2). Having obtained two values of y at $x = b$, the next trial value, γ_3, for the initial solution can now be

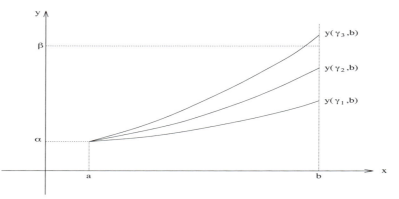

FIGURE 13.2
The shooting method and the first three approximations to its solution.

generated by using linear interpolation

$$\gamma_3 = \gamma_1 + (\gamma_2 - \gamma_1)\frac{\beta - y(\gamma_1; b)}{y(\gamma_2; b) - y(\gamma_1; b)}. \tag{13.13}$$

The general form of Eqn. (13.13) is

$$\gamma_{k+1} = \gamma_{k-1} + (\gamma_k - \gamma_{k-1})\frac{\beta - y(\gamma_{k-1}; b)}{y(\gamma_k; b) - y(\gamma_{k-1}; b)}, \quad k = 2, 3, \ldots. \tag{13.14}$$

We now repeat the procedure by using the value γ_3 to get $y(\gamma_3; b)$. This procedure is repeated until convergence has been obtained, i.e.,

$$|y(\gamma_k; b) - \beta| < \epsilon \quad \text{for a specified tolerance } \epsilon. \tag{13.15}$$

The method we have illustrated is known as the shooting method. One should note that in order for the method to converge, a good choice of γ_1 and γ_2 is necessary.

EXAMPLE 13.5

Use the nonlinear shooting method with $N = 20$ to solve the following boundary-value problem with an accuracy of 10^{-6}.

$$yy'' = -(y')^2, \quad 1 \leq x \leq 3$$

with

$$y(1) = \sqrt{2} \quad \text{and} \quad y(3) = 2.$$

Compare your results with the exact solution $y = \sqrt{1 + x}$.

For this example we have $f(x, y, y') = -\frac{(y')^2}{y}$, $a = 1$ and $b = 3$.
As a first approximation, we set $\gamma_1 = y'(1) \approx 1.4$. and we replace the boundary-value problem with the initial-value problem

$$yy'' = -(y')^2, \quad 1 \leq x \leq 3$$
$$y(1) = \sqrt{2} \quad \text{and} \quad y'(1) = \gamma_1 = 1.4$$

Using the fourth-order Runge-Kutta method with $n = 20$, we find that

$$y(\gamma_1, 3) = 3.14953128 \quad \text{and} \quad |y(\gamma_1, 3) - 2| = 1.14953128.$$

We repeat the procedure using a different estimate of $y'(1)$. For the second approximation, we set $\gamma_2 = \frac{2 - \sqrt{2}}{3 - 1} \approx 0.3$; this leads to

$$y(\gamma_2, 3) = 1.92277306 \quad \text{and} \quad |y(\gamma_2, 3) - 2| = 0.07722694.$$

For the third approximation to γ_3, we use equation (13.13) to get

$$\gamma_3 = 1.4 + (0.3 - 1.4) \frac{2 - 3.14953128}{1.92277306 - 3.14953128} \approx 0.369\,247\,25$$

which leads to

$$y(\gamma_3, 3) = 2.02207265 \quad \text{and} \quad |y(\gamma_2, 3) - 2| = 0.02207265.$$

The results of further calculations of the $y(\gamma_k, 3)$ are given in Table 13.3. After six attempts, a value $\gamma_6 = 0.35355340$ has been located for which $y(\gamma_6, 3) = 2$ to within 10 decimal places.

Remark: Equation (13.12) can also be solved using Newton's method, which requires a first guess at the solution, γ_0, and then one calculates recursively

$$\gamma_{n+1} = \gamma_n - \frac{y(\gamma_n, b) - \beta}{\frac{\partial y(\gamma_n, b)}{\partial \gamma}}, \quad n = 0, 1, 2, \ldots.$$

ti	yi	y(ti)	Error
1.0	1.41421356	1.41421356	0
1.1	1.44913767	1.44913767	1.41E-09
1.2	1.48323970	1.48323970	2.39E-09
1.3	1.51657509	1.51657509	3.05E-09
1.4	1.54919334	1.54919334	3.48E-09
1.5	1.58113883	1.58113883	3.71E-09
1.6	1.61245155	1.61245155	3.81E-09
1.7	1.64316767	1.64316767	3.78E-09
1.8	1.67332005	1.67332005	3.68E-09
1.9	1.70293863	1.70293864	3.50E-09
2.0	1.73205080	1.73205081	3.27E-09
2.1	1.76068168	1.76068169	3.00E-09
2.2	1.78885438	1.78885438	2.70E-09
2.3	1.81659021	1.81659021	2.37E-09
2.4	1.84390889	1.84390889	2.03E-09
2.5	1.87082869	1.87082869	1.67E-09
2.6	1.89736659	1.89736660	1.29E-09
2.7	1.92353841	1.92353841	9.14E-10
2.8	1.94935887	1.94935887	5.29E-10
2.9	1.97484177	1.97484177	1.40E-10
3.0	2	2	2.50E-10

Table 13.3   The nonlinear shooting method for Example 13.5.

To find $\frac{\partial}{\partial \gamma} y(\gamma_n, b)$, set $u(\gamma, t) = \frac{\partial}{\partial \gamma} y(\gamma, t)$ and differentiate the differential equation (13.11) and the initial conditions to get

$$\frac{\partial y''}{\partial \gamma} = \frac{\partial f(x, y, y')}{\partial \gamma} = \frac{\partial f}{\partial x}\frac{\partial x}{\partial \gamma} + \frac{\partial f}{\partial y}\frac{\partial y}{\partial \gamma} + \frac{\partial f}{\partial y'}\frac{\partial y'}{\partial \gamma}$$

$$= \frac{\partial f}{\partial y}u + \frac{\partial f}{\partial y'}u'.$$

$\frac{\partial f}{\partial x}\frac{\partial x}{\partial \gamma} = 0$ since x and γ are independent. For the initial conditions we obtain

$$\frac{\partial y(a)}{\partial \gamma} = \frac{\partial \alpha}{\partial \gamma} = 0 = u(\gamma, 0),$$

$$\frac{\partial y'(a)}{\partial \gamma} = \frac{\partial \gamma}{\partial \gamma} = 1 = u'(\gamma, 0).$$

So, u satisfies the initial-value problem

$$u'' = \frac{\partial f(x, y, y')}{\partial y}u + \frac{\partial f(x, y, y')}{\partial y'}u'$$

$$u(\gamma, 0) = 0, \quad u'(\gamma, 0) = 1.$$

Thus, we solve the system for $n = 0, 1, 2, \ldots$

$$y'' = f(x, y, y'), \quad y(a) = \alpha, \ y'(a) = \gamma_n$$

$$u'' = \frac{\partial f(x, y, y')}{\partial y} u + \frac{\partial f(x, y, y')}{\partial y'} u', \quad u(a) = 0, \ u'(a) = 1 \tag{13.16}$$

and then determine for $x = b$

$$\gamma_{n+1} = \gamma_n - \frac{y(\gamma_n, b) - \beta}{u(\gamma_n, b)}.$$

As in the previous method, one proceeds until the inequality (13.15) is satisfied. If the equation is linear, the system (13.16) is linear.

EXAMPLE 13.6

Solve the linear boundary-value problem

$$y'' - y = \cos x, \quad y(0) = y(1) = 0.$$

The system for y and u is

$$y'' - y = \cos x, \quad y(0) = 0, \ y'(0) = \gamma$$
$$u'' - u = 0, \quad u(0) = 0, \ u'(0) = 1.$$

In this case, we can find the solution explicitly

$$y = -\frac{1}{2} \cos x + \frac{1}{2} \cosh x + \gamma \sinh x$$
$$u = \sinh x.$$

The iteration scheme is

$$\gamma_{n+1} = \gamma_n - \frac{-\frac{1}{2} \cos 1 + \frac{1}{2} \cosh 1 + \gamma_n \sinh 1}{\sinh 1}$$
$$= \frac{\cos 1 - \cosh 1}{2 \sinh 1}$$

with γ_0 an initial guess.

Let us take for an initial guess, say, $\gamma_0 = 10$. Then

$$\gamma_1 = \frac{\cos 1 - \cosh 1}{2 \sinh 1}.$$

Observe that it is irrelevant how we choose γ_0; γ_1 is already the correct value. This is typical for linear problems when one shoots using Newton's method. The solution is

$$y = -\frac{1}{2} \cos x + \frac{1}{2} \cosh x + \frac{\cos 1 - \cosh 1}{2 \sinh 1} \sinh x.$$

EXAMPLE 13.7

Set the system of differential equations that need to be solved using nonlinear shooting with Newton's method for the BVP

$$y'' = -\frac{(y')^2}{y}, \quad y(0) = 1, \ y(1) = 2.$$

We have $f(x, y, y') = -\frac{(y')^2}{y}$ and

$$\frac{\partial f}{\partial y} = \frac{(y')^2}{y^2} \text{ and } \frac{\partial f}{\partial y'} = \frac{-2y'}{y}.$$

So, we need to solve the system of differential equations

$$\begin{cases} y'' = -\frac{(y')^2}{y}, & y(0) = 1, \ y'(0) = 1 = \gamma_n, \\ u'' = \frac{(y')^2}{y^2}u - \frac{2y'}{y}u', & u(0) = 0, \ u'(0) = 1. \end{cases}$$

Update γ by Newton's method

$$\gamma_{n+1} = \gamma_n - \frac{y(\gamma_n, 1) - 2}{u(\gamma_n, 1)}, \quad n = 0, 1, 2, \ldots.$$

13.2.2   The linear case

Consider the linear second-order boundary-value problem,

$$\begin{cases} y'' + p(x)\, y' + q(x)\, y = r(x) \\ \quad y(a) = \alpha, \quad y(b) = \beta. \end{cases} \tag{13.17}$$

The solution to this problem can be obtained by forming a linear combination of the solution to the following two second-order initial-value problems

$$u'' + p(x)u' + q(x)u = r(x), \quad u(a) = \alpha, \quad u'(a) = 0 \tag{13.18}$$

and

$$v'' + p(x)v' + q(x)v = 0, \quad v(a) = 0, \quad v'(a) = 1. \tag{13.19}$$

We claim that the linear combination

$$y(x) = u(x) + Av(x) \tag{13.20}$$

is a solution to the differential equation $y'' + p(x)y' + q(x)y = r(x)$ in (13.17).

To show that, we substitute Eqn. (13.20) into the differential equation in (13.17) to get

$$u''(x) + Av''(x) + p(x)[u'(x) + Av'(x)] + q(x)[u(x) + Av(x)] = r(x)$$

$$\underbrace{u''(x) + p(x)u'(x) + q(x)u(x)}_{r(x)} + A\underbrace{[v''(x) + p(x)v'(x) + q(x)v(x)]}_{0} = r(x).$$

To find the constant A, we use the boundary conditions in (13.17) to get

$$y(a) = u(a) + Av(a) = \alpha + 0 = \alpha$$
$$y(b) = u(b) + Av(b) = \beta$$

from which

$$A = \frac{\beta - u(b)}{v(b)}. \tag{13.21}$$

Therefore, if $v(b) \neq 0$, the solution to boundary-value problem (13.17) is

$$y(x) = u(x) + \frac{\beta - u(b)}{v(b)} v(x). \tag{13.22}$$

The solutions to the second-order initial-value problems (13.18) and (13.19) were described in Section 12.10 and can be obtained by using the fourth-order Runge-Kutta method.

EXAMPLE 13.8
Solve the boundary-value problem

$$y'' + (x+1) y' - 2 y = (1 - x^2) e^{-x}, \quad 0 \le x \le 1, \quad y(0) = y(1) = 0$$

using the linear shooting method with $h = 0.2$.

The functions p, q, and r are $p(x) = x + 1$, $q(x) = -2$, and $r(x) = (1 - x^2) e^{-x}$, respectively. The numerical solution to this problem is

$$y(x) = u(x) - \frac{u(1)}{v(1)} v(x) \tag{13.23}$$

where $u(x)$ and $v(x)$ are the solutions of the initial-value problems

$$u'' + (x+1)u' - 2u = (1 - x^2) e^{-x}, \quad u(0) = 0, \quad u'(0) = 0,$$
$$v'' + (x+1)v' - 2v = 0, \quad v(0) = 0, \quad v'(0) = 1,$$

respectively. They can be expressed in the form

$$\begin{array}{ll} u' = w, & u(0) = 0, \\ w' = 2u - (x+1)w + (1 - x^2) e^{-x}, & w(0) = 0, \quad 0 \le x \le 1 \end{array} \tag{13.24}$$

and

$$\begin{array}{ll} v' = z, & v(0) = 0, \\ z' = 2v - (x+1) z, & z(0) = 1, \quad 0 \le x \le 1. \end{array} \tag{13.25}$$

The Runge-Kutta method of order 4 is used to construct numerical solutions u_i and v_i to the linear systems (13.24) and (13.25). The approximations of u_i and v_i are given in Table 13.4.

x	ui	vi
0	0	0
0.2	0.01743428	0.18251267
0.4	0.06017154	0.33919207
0.6	0.11620204	0.48140428
0.8	0.17679396	0.61776325
1	0.23633393	0.75454368

Table 13.4   Numerical solutions of systems (13.24) and (13.25).

For example, using (13.23) and the Table 13.4 the approximate solution y_2 at $x = 0.4$ is

$$y_2 = u(0.4) - \frac{u(1)}{v(1)} v(0.4)$$

$$= 0.06017154 - \frac{0.23633393}{0.75454368} (0.33919207)$$

$$= -0.046068294.$$

EXAMPLE 13.9

Solve the boundary-value problem

$$y'' - \frac{3}{x} y' + \frac{3}{x^2} y = 2x^2 e^x, \quad 1 \le x \le 2, \quad y(1) = 0, \quad y(2) = 4e^2$$

using the linear shooting method with $h = 0.1$. Compare the results with the exact solution $y = 2xe^x(x - 1)$.

The functions p, q, and r are $p(x) = -3/x$, $q(x) = 3/x^2$, and $r(x) = 2x^2 e^x$, respectively. By defining the MATLAB functions

```
function f=f1(x,u,v);
f=3/x*u-3/x.^2*v+2*x.^2*exp(x);
```

```
function f=f2(x,u,v);
f=3/x*u-3/x.^2*v;
```

and then calling the MATLAB function lshoot, we obtain the results shown in Table 13.5.

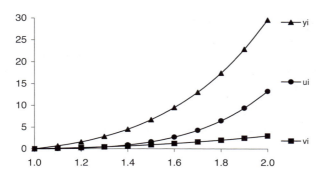

FIGURE 13.3

Numerical solutions ui and vi used to form $y(x) = u(x) + \frac{4e^2 - u(2)}{v(2)} v(x)$ **to**

solve $y'' - \frac{3}{x} y' + \frac{3}{x^2} y = 2x^2 e^x$, $n = 10$.

EXERCISE SET 13.2

1. Approximate the solution of the following boundary-value problems using the linear shooting method:

   (a) $y'' + (1 - x)y' + xy = x$, $y(0) = 0$, $y(1) = 2$, $h = 0.2$,

   (b) $y'' - \frac{1}{x}y' + \frac{1}{x^2}y = \frac{\ln x}{x^2}$, $y(1) = 0$, $y(2) = -2$, $h = 0.2$,

   (c) $y'' + xy' + y = x$, $y(0) = 1$, $y(1) = 0$, $h = 0.2$.

2. Use the shooting method to solve the boundary-value problem

$$y'' = y - xy' + 2x + \frac{2}{x}, \quad y(1) = 0, \quad y(2) = 4 \ln 2.$$

   Compare with the exact solution $y = 2x \ln x$.

3. The solution to the linear boundary-value problem

$$y'' - y = -2e^{-x}, \quad y(0) = -1, \ y(1) = 0$$

   is $y = (x - 1)e^{-x}$. Solve this problem by both shooting methods and compare the results.

4. Show that in the case of the linear boundary-value problem

$$y'' + A(x)y' + B(x)y = F(x), \quad y(a) = \alpha, \ y(b) = \beta$$

M-function 13.2

The following MATLAB function **lshoot.m** finds the solution of the BVP (12.17) using the linear shooting method. INPUTS are functions $f1(x,u,v) = -p(x)*u-q(x)*v+r(x)$ and $f2(x,u,v) = -p(x)*u-q(x)*v$; the end points a, b; the boundary conditions α and β; the number of steps n. The input functions $f1$ and $f2$ should be defined as M-files.

```
function lshoot(f1,f2,a,b,alfa,beta,n)
% Solve the 2nd order BVP using the linear shooting method.
% f1= -p(x)*u-q(x)*v+r(x),  f2= -p(x)*u-q(x)*v.
h=(b-a)/n;
y1=alfa;
y2=0;
u=0;
v=alfa;
for i=1:n
% RK of order 4
   x=a+(i-1)*h;
   k1=feval(f1,x,u,v);
   c1=u;
   k2=feval(f1,x+h/2,u+h*k1/2,v+h*c1/2);
   c2=u+h/2*k1;
   k3=feval(f1,x+h/2,u+h*k2/2,v+h*c2/2);
   c3=u+h/2*k2;
   k4=feval(f1,x+h,u+h*k3,v+h*c3);
   c4=u+h*k3;
   u=u+h*(k1+2*k2+2*k3+k4)/6;
   v=v+h*(c1+2*c2+2*c3+c4)/6;
   y1(i+1)=v;
end
u=1;
v=0;
for i=1:n
% RK of order 4
   x=a+(i-1)*h;
   k1=feval(f2,x,u,v);
   c1=u;
   k2=feval(f2,x+h/2,u+h*k1/2,v+h*c1/2);
   c2=u+h/2*k1;
   k3=feval(f2,x+h/2,u+h*k2/2,v+h*c2/2);
   c3=u+h/2*k2;
```

```
      k+4=feval(f2,x+h,u+h*k3,v+h*c3);
      c4=u+h*k3;
      u=u+h*(k1+2*k2+2*k3+k4)/6;
      v=v+h*(c1+2*c2+2*c3+c4)/6;
      y2(i+1)=v;
   end
fprintf('\n')
disp('          Linear shooting method')
fprintf('\n')
disp([' u(b) = ',num2str(y1(n+1)),' '])
disp([' v(b) = ',num2str(y2(n+1)),' '])
fprintf('\n')
disp('_____')
disp('          xi          ui          vi          yi        ')
disp('_____')
for i=1:n+1
   x=a+(i-1)*h;
   w=y1(i)+(beta-y1(n+1))/y2(n+1)*y2(i);
   %Write the exact solution if known as s=s(x) otherwise set s='n'.
   s='n';
   if (s=='n')
      fprintf('%6.2f %12.6f %12.6f %12.6f\n',x,y1(i),y2(i),w)
   else
      err=abs(w-s);
      fprintf('%6.2f %12.6f %12.6f %12.6f %12.6f %10.2e\n', x, y1(i),
y2(i), w, s, err)
   end
end
```

the shooting method based on Newton's method converges in one iteration. Assume $A(x)$, $B(x)$, and $F(x)$ are smooth functions.

Hint. Recall that there are two linearly independent solutions to $y'' + A(x)y' + B(x)y = 0$. Call them $\varphi(x)$ and $\phi(x)$ and choose them so that $\varphi(a) = 1$, $\varphi'(a) = 0$ and $\phi(a) = 0$, $\phi'(a) = 1$. Let $y_p'' + A(x)y_p' + B(x)y_p = F(x)$, $y_p(0) = y_p'(0) = 0$. Now set up Newton's iteration scheme.

5. Use the nonlinear shooting method to approximate the solution of the following boundary-value problems with $TOL = 10^{-4}$.

(a) $y'' = xy' + 2y$, $y(0) = -1$, $y(1) = 0$, $h = 0.2$,

(b) $y'' = -\frac{y'^2 + 2y^2}{y}$, $y(0) = 1$, $y(\pi/6) = \sqrt{2}$, $h = \pi/30$,

(c) $y'' = \frac{1 - y'^2}{y}$, $y(0) = 1$, $y(2) = 2$, $h = 0.4$.

6. The solution to the nonlinear boundary-value problem

$$y'' - 2y' - 2x^2 y = -2x^2, \quad y(1) = 0, \ y(2) = 3$$

is $y = x^2 - 1$. Solve this problem numerically by both shooting methods and compare the results. Check for accuracy.

```
lshoot('f1','f2',1,2,0,4*exp(2),10)

        Linear shooting method

u(b) = 13.24634
v(b) = 3.000015
```

xi	ui	vi	yi	y(xi)	Error
1.0	0.00000	0.00000	0.00000	0.00000	0
1.1	0.03298	0.11550	0.66091	0.66092	3.75e-006
1.2	0.15838	0.26400	1.59365	1.59366	6.91e-006
1.3	0.42372	0.44850	2.86204	2.86205	9.38e-006
1.4	0.88840	0.67200	4.54181	4.54182	1.11e-005
1.5	1.62568	0.93750	6.72252	6.72253	1.18e-005
1.6	2.72490	1.24801	9.50981	9.50982	1.17e-005
1.7	4.29404	1.60651	13.02798	13.02799	1.04e-005
1.8	6.46274	2.01601	17.42298	17.42298	8.12e-006
1.9	9.38564	2.47951	22.86575	22.86576	4.65e-006
2.0	13.24634	3.00002	29.55622	29.55622	0.00e+000

Table 13.5 The linear shooting method for Example 13.9.

7. The solution to the nonlinear boundary-value problem

$$y'' = -2yy', \quad y(0) = 1, \ y(1) = 1/2, \ 0 \le x \le 1$$

is $y = \frac{1}{x+1}$ Solve this problem numerically by the nonlinear shooting method and compare with the exact solution.

8. In using the shooting method on the problem

$$y'' = \frac{1}{2}y - \frac{2(y')^2}{y}, \quad y(0) = 1, \quad y(1) = 1.5$$

the following results were obtained

$$y'(0) = 0 \Rightarrow y(1) = 1.54308, \ y'(0) = -0.05 \Rightarrow y(1) = 1.42556.$$

Which value of $y'(0)$ should be used in the next "shot"?

COMPUTER PROBLEM SET 13.2

1. Write a computer program in a language of your choice to solve the following boundary-value problem

$$y'' = f(x, y, y'), \quad y(a) = \alpha, \quad y(b) = \beta, \quad b \le x \le b$$

using the nonlinear shooting method.

Test your program to solve the boundary-value problem

$$y'' = x(y')^2, \quad y(0) = \pi/2, \quad y(2) = \pi/4, \quad 0 \le x \le 2.$$

Compare your answer to the actual solution $y = \cot^{-1} \frac{x}{2}$.

2. Use the MATLAB function lshoot to compute the solution of the boundary-value problem

$$y'' + (1 - x)y' + xy = x, \quad y(0) = 0, \; y(1) = 2$$

on the interval $[0, 1]$ with $n = 10$.

3. Using a shooting routine from a program library solve the following BVP in $[0, 1]$ with $n = 20$:

$$y'' = -\frac{1}{(1 + y)^2}, \quad y(0) = y(1) = 0.$$

4. Consider the boundary-value problem

$$y'' = y' - \sin(xy), \quad y(0) = 1, \; y(1) = 1.5.$$

Use the shooting method to approximate the solution of this problem.

5. Use MATLAB function lshoot to approximate the solution of the the boundary-value problem in $[0, 1]$ with $n = 20$.

$$y'' - y\cos^2 x + \sin x e^{\sin x}, \quad y(0) = 1, \; y(\pi) = 1$$

with $h = \pi/10$. Compare with the exact solution $y = e^{\sin x}$.

APPLIED PROBLEMS FOR CHAPTER 13

1. Under a uniform load, small deflections y of a simple supported beam are given by

$$EI(d^2y/dx^2) = qx(x - L)/2, \quad y(0) = Y(L) = 0$$

where $L = 10$ ft, $EI = 1900$ kip·ft^2, and $q = 0.6$ kip/ft. The beam extends from $(x = 0)$ to $(x = L)$. Find y at every 2 ft by the shooting method.

2. Suppose an 8-lb weight stretches a spring 2 ft beyond its natural length. The weight is then pulled down another $\frac{1}{2}$ ft and released with an initial velocity of 6 ft/sec. The displacement y of the weight at any time t is given by the differential equation

$$\frac{d^2y}{dt^2} + 16y = 0$$

with $y(0) = \frac{1}{2}$ and $y(\pi/2) = \frac{1}{2}$. Approximate the solution of the differential equation in the time interval $[0, \pi/2]$ using the finite difference method described in Section 13.1.

3. Suppose a chain hangs between two points, (a, α) and (b, β). See the figure below. The curve describing the line of the chain is given by $y(x)$ and satisfies

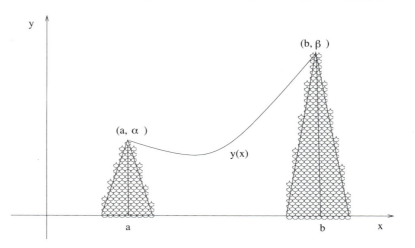

FIGURE 13.4
Chain between two poles with different heights.

the boundary-value problem

$$y'' = k[1 + (y')^2]^{1/2}$$
$$y(a) = \alpha, \; y(b) = \beta.$$

Here k is a constant. In this problem we shall take it to be known. Suppose $(a, \alpha) = (1, 5)$ and $(b, \beta) = (4, 15)$. For $k = 0.5, 2, 4$ calculate the length, l, of the chain, which is given by

$$l = \int_1^4 [1 + (y')^2]^{1/2} dx.$$

Find the point where the chain is lowest, that is find the values of x and y where $y(x)$ is a minimum.

Now suppose the chain of length l hangs between two points $(a, \alpha) = (1, 4)$ and $(b, \beta) = (3, 4)$. Solve

$$y'' = k[1 + (y')^2]^{1/2}$$
$$y(1) = 4, \; y(3) = 4$$
$$l = \int_1^3 [1 + (y')^2]^{1/2} dx.$$

Note: y is a function of x and the parameter k. In the shooting method one solves the differential equation with initial conditions $y(1) = 4$, $y'(1) = \gamma$ so

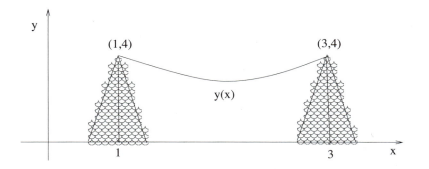

FIGURE 13.5
Chain between two poles with equal heights.

that $y = y(x, \gamma, k)$. γ and k must be chosen so that

$$y(3, \gamma, k) = 4 \text{ and } \int_1^3 \sqrt{1 + [y'(x, \gamma, k)]^2} dx = l$$

where l is given.

(a) Develop a general Newton method for solving the problem.

(b) Find and graph $y(x)$ where $l = 3$.

(c) Find and graph the solution when $l = 4$. Find the lowest point of the chain.

(d) If $l = 3.9$, what is the lowest point of the chain? If $l = 3.8$, what is the lowest point of the chain? How does shortening the length of the chain affect the position of the lowest point of the chain?

4. The electrostatic potential u between two concentric spheres of radius $r = 1$ and $r = 4$ is determined from the BVP

$$\frac{d^2u}{dr^2} + \frac{2}{r}\frac{du}{dr} = 0, \quad u(0) = 50, \; u(4) = 100.$$

Use the shooting method to approximate the solution of the BVP.

5. In the study of a vibrating spring with damping, we are led to the boundary-value problem of the form

$$mx''(t) + bx'(t) + kx(t) = 0, \quad x(0) = x_0, \; x(a) = x_1$$

where m is the mass of the spring system, h is the damping constant, k is the spring constant, x_0 is the initial displacement, $x(t)$ is the displacement from equilibrium of the spring system at time t (see the figure below). Determine the displacement of this spring system in the time interval $[0, 10]$ when $m = 36$ kg, $b = 12$ kg/sec, $k = 37$ kg/sec^2, $x_0 = 70.0$ cm, and $x_1 = -13.32$ cm. Use a finite-difference method. Plot the displacement over the interval $[0, 10]$.

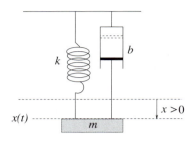

6. In the design of a sewage treatment plant, the following boundary-value problem arose:

$$60 - h = (77.7)\frac{d^2h}{dt^2} + (19.42)\left(\frac{dh}{dt}\right)^2, \quad h(0) = 0, \ h(5) = 5.956$$

where h is the level of the fluid in an ejection chamber and t is the time in seconds. Use the shooting method to determine the level of the fluid in the time interval $[0, 5]$. Plot the level of the fluid over the interval $[0, 5]$.

Chapter 14

Eigenvalues and Eigenvectors

The calculation of eigenvalues and eigenvectors is a problem that plays an important part in a large number of applications, both theoretical and practical. They touch most areas in science, engineering, and economics. Some examples are the solution of the Schrödinger equation in quantum mechanics, the various eigenvalues representing the energy levels of the resulting orbital, the solution of ordinary equations, space dynamics, elasticity, fluid mechanics, and many others.

In this chapter we introduce a new method called the **quadratic method** (see [15]) for computing the eigenvalues and eigenvectors of a symmetric matrix A. The method is simple and has the advantage of computing the eigenpairs in parallel.

14.1 BASIC THEORY

Le A be a real square, $n \times n$ matrix and let $\mathbf{x}$ be a vector of dimension n. We want to find scalars λ for which there exists a nonzero vector $\mathbf{x}$ such that

$$A\mathbf{x} = \lambda\mathbf{x}. \tag{14.1}$$

When this occurs, we call λ an eigenvalue and $\mathbf{x}$ an eigenvector that corresponds to λ. Together they form an eigenpair $(\lambda, \mathbf{x})$ of A. Note that Eqn. (14.1) will have a nontrivial solution only if

$$p(\lambda) = \det(A - \lambda I) = 0. \tag{14.2}$$

The function $p(\lambda)$ is a polynomial of degree n and is known as the characteristic polynomial. The determinant in Eqn. (14.2) can be written in the form

$$\begin{vmatrix} a_{11} - \lambda & a_{12} & \cdots & a_{1n} \\ a_{21} & a_{22} - \lambda & \cdots & a_{2n} \\ \vdots & \vdots & \ddots & \vdots \\ a_{n1} & a_{n2} & \cdots & a_{nn} - \lambda \end{vmatrix} = 0. \tag{14.3}$$

It is known that p is an nth degree polynomial with real coefficients and has at most n distinct zeros not necessarily real. Each root λ can be substituted into

Eqn. (14.1) to obtain a system of equations that has a nontrivial solution vector **x**. We now state the following definitions and theorems necessary for the study of eigenvalues. Proofs can be found in any standard texts on linear algebra. See, for example, [7].

DEFINITION 14.1   *The spectral radius $\rho(A)$ of an $n \times n$ matrix A is defined by*

$$\rho(A) = max_{1 \le i \le n}|\lambda_i|$$

where λ_i are the eigenvalues of A.

THEOREM 14.1
The eigenvalues of a symmetric matrix are all real numbers.

THEOREM 14.2
*For distinct eigenvalues λ there exists at least one eigenvector **v** corresponding to λ.*

THEOREM 14.3
If the eigenvalues of an $n \times n$ matrix A are all distinct, then there exists n eigenvectors $\mathbf{v}_j$, for $j = 1, 2, ..., n$.

THEOREM 14.4 (Gerschgorin's Circle Theorem)
Let A be an $n \times n$ matrix and let C_i be the disc in the complex plane with center a_{ii} and radius

$$r_i = \sum_{\substack{j=1 \\ j \neq i}}^{n} |a_{ij}|$$

that is, C_i consists of all points z such that

$$|z - a_{ii}| \le \sum_{\substack{j=1 \\ j \neq i}}^{n} |a_{ij}|.$$

    Let D be the union of all the disc C_i for $i = 1, 2, ..., n$; then all the eigenvalues of A lie within D.

As an example consider the matrix

$$A = \begin{bmatrix} 2 & 3 & 0 \\ 3 & 8 & -2 \\ 0 & -2 & -4 \end{bmatrix}.$$

All the eigenvalues of this matrix are real since A is symmetric. We have (see Figure 14.1):

C_1 is the disc with center $(2,0)$ and radius $= 3 + 0 = 3$.
C_2 is the disc with center $(8,0)$ and radius $= 3 + |-2| = 5$.
C_3 is the disc with center $(-4,0)$ and radius $= 0 + |-2| = 2$.

The union of these discs is

$$D = [-1, 5] \cup [3, 13] \cup [-6, -2] = [-6, -2] \cup [-1, 13]$$

Thus, the eigenvalues of A must lie within D. In fact, they are

$$\lambda_i = -4.3653, \ 0.8684, \ \text{and } 9.4969.$$

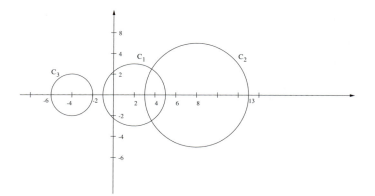

FIGURE 14.1
Gerschgorin's Circle for the matrix A.

EXAMPLE 14.1
Find the eigenpairs for the matrix

$$\begin{bmatrix} 2 & -3 & 6 \\ 0 & 3 & -4 \\ 0 & 2 & -3 \end{bmatrix}.$$

The characteristic equation $\det(A - \lambda I) = 0$ is

$$-\lambda^3 + 2\lambda^2 + \lambda - 2 = 0.$$

The roots of the equation are the three eigenvalues $\lambda_1 = 1$, $\lambda_2 = 2$, and $\lambda_3 = -1$.

To find the eigenvector $\mathbf{x}_1$ corresponding to λ_1, we substitute $\lambda_1 = 1$ to Eqn. (14.1) to get the system of equations

$$x_1 - 3x_2 + 6x_3 = 0$$
$$2x_2 - 4x_3 = 0$$
$$2x_2 - 4x_3 = 0.$$

Since the last two equations are identical, the system is reduced to two equations in three unknowns. Set $x_3 = \alpha$, where α is an arbitrary constant, to get $x_2 = 2\alpha$ and $x_1 = 0$. Hence, by setting $\alpha = 1$, the first eigenpair is

$$\lambda_1 = 1 \quad \text{and} \quad \mathbf{x}_1 = (0, 2, 1)^T.$$

To find $\mathbf{x}_2$, substitute $\lambda_2 = 2$ to Eqn. (14.1) to get the system of equations

$$-3x_2 + 6x_3 = 0$$
$$x_2 - 4x_3 = 0$$
$$2x_2 - 5x_3 = 0.$$

The solution of this system is $x_1 = \alpha$, $x_2 = x_3 = 0$. Hence, by setting $\alpha = 1$, the second eigenpair is

$$\lambda_2 = 2 \quad \text{and} \quad \mathbf{x}_2 = (1, 0, 0)^T.$$

Finally, to find $\mathbf{x}_3$ substitute $\lambda_3 = -1$ to Eqn. (14.1) to get the system of equations

$$3x_1 - 3x_2 + 6x_3 = 0$$
$$4x_2 - 4x_3 = 0$$
$$2x_2 - 2x_3 = 0.$$

The solution of this system is $x_1 = -\alpha$, $x_2 = x_3 = \alpha$. Hence, by setting $\alpha = 1$ the third eigenpair is

$$\lambda_3 = -1 \quad \text{and} \quad \mathbf{x}_3 = (-1, 1, 1)^T.$$

As one can see from this example, the method is practical when the dimension n is small but for a large n it is difficult to determine the zeros of $p(\lambda)$ and also to find the nonzero solutions of the homogeneous linear system $(A - \lambda I)\mathbf{x} = 0$. There are many techniques for approximating the eigenpairs of a matrix. The most popular ones are the QR and QL algorithms (see [7] for more details). In Section 14.3 we shall introduce a new iterative technique, called the quadratic method, for approximating the eigenpairs of a symmetric matrix, but before that we give a brief description of the classical power method.

EXERCISE SET 14.1

1. Find the characteristic polynomials $\rho(\lambda)$ and eigenpairs for the following matrices.

(a) $A = \begin{bmatrix} 3 & 8 \\ 1 & 1 \end{bmatrix}$, (b) $B = \begin{bmatrix} -2 & 6 \\ 3 & 0 \end{bmatrix}$,   (c) $C = \begin{bmatrix} -1 & 3 \\ -2 & 4 \end{bmatrix}$.

2. Let $A = \begin{bmatrix} 1 & a+2 \\ a-1 & 4 \end{bmatrix}$,

    (a) Find the characteristic polynomial $\rho(\lambda)$ of A.

    (b) Show that the eigenvalues of A are $\lambda_1 = a+3$ and $\lambda_2 = 2 - a$.

    (c) Show that the eigenvectors of A are $V_1 = (1,1)^T$ and $V_2 = (-\frac{a+2}{a-1}, 1)^T$.

3. Let $A = \begin{bmatrix} 3 & 5 & -1 \\ 1 & 4 & -2 \\ 4 & 2 & 1 \end{bmatrix}$.

    (a)   Use the Gerschgorin Circle Theorem to determine a region containing all the eigenvalues of A.

    (b) Find an interval that contains the spectral radius of A.

4. Let $A = \begin{bmatrix} -2 & 3+m & -2 \\ 1 & 6 & -1 \\ 2+m & 2 & 14 \end{bmatrix}$. If $m > 0$, find all values of m such that the circles in the Gerschgorin Theorem are all disjoint.

5. Find the characteristic polynomial $\rho(\lambda)$ and eigenpairs for the following matrices:

    (a) $A = \begin{bmatrix} 1 & 2 & -1 \\ 1 & 0 & 1 \\ 4 & -4 & 5 \end{bmatrix}$,   (b) $B = \begin{bmatrix} 1 & -2 & 2 \\ -2 & 1 & 2 \\ 2 & 2 & 1 \end{bmatrix}$.

6. Let $A = \begin{bmatrix} -2 & 1 & -2 \\ 1 & 4 & a+2 \\ b+1 & 1 & 12 \end{bmatrix}$. If $a,b \geq 0$, find the values of a and b such that the center circle in the Gerschgorin Theorem will be tangent to the two others.

14.2   THE POWER METHOD

The **power method** is a classical method of use mainly to determine the largest eigenvalue in magnitude, called the **dominant eigenvalue,** and the corresponding eigenvector of the system

$$A\mathbf{x} = \lambda\mathbf{x}.$$

Let A be a real $n \times n$ matrix. Let $\lambda_1, \lambda_2, ..., \lambda_n$ be the distinct eigenvalues of A such that

$$|\lambda_1| > |\lambda_2| > ... > |\lambda_n| > 0$$

and $\mathbf{x}_1, \mathbf{x}_2, ..., \mathbf{x}_n$ be the corresponding eigenvectors. The procedure is applicable if a complete system of n independent eigenvectors exists, even though some of eigenvalues $\lambda_2, ..., \lambda_n$ may not be distinct. Then any eigenvector $\mathbf{x}$ in the space of eigenvectors $\mathbf{x}_1, \mathbf{x}_2, ..., \mathbf{x}_n$ can be written as

$$\mathbf{x} = c_1\mathbf{x}_1 + c_2\mathbf{x}_2 + ... + c_n\mathbf{x}_n. \tag{14.4}$$

Premultiplying by A and substituting $A\mathbf{x}_1 = \lambda_1\mathbf{x}_1$, $A\mathbf{x}_2 = \lambda_2\mathbf{x}_2$, etc., Eqn. (14.4) becomes

$$A\mathbf{x} = \lambda_1\, c_1\mathbf{x}_1 + \lambda_2 c_2\mathbf{x}_2 + ... + \lambda_n c_n\mathbf{x}_n$$
$$= \lambda_1 \left[c_1\mathbf{x}_1 + c_2 \left(\frac{\lambda_2}{\lambda_1}\right) \mathbf{x}_2 + ... + c_n \left(\frac{\lambda_n}{\lambda_1}\right) \mathbf{x}_n \right]. \tag{14.5}$$

Premultiplying by A again and simplifying, Eqn. (14.5) becomes

$$A^2\mathbf{x} = \lambda_1^2 \left[c_1\mathbf{x}_1 + c_2 \left(\frac{\lambda_2}{\lambda_1}\right)^2 \mathbf{x}_2 + ... + c_n \left(\frac{\lambda_n}{\lambda_1}\right)^2 \mathbf{x}_n \right]$$
$$...$$
$$A^k\mathbf{x} = \lambda_1^k \left[c_1\mathbf{x}_1 + c_2 \left(\frac{\lambda_2}{\lambda_1}\right)^k \mathbf{x}_2 + ... + c_n \left(\frac{\lambda_n}{\lambda_1}\right)^k \mathbf{x}_n \right] \tag{14.6}$$

$$A^{k+1}\mathbf{x} = \lambda_1^{k+1} \left[c_1\mathbf{x}_1 + c_2 \left(\frac{\lambda_2}{\lambda_1}\right)^{k+1} \mathbf{x}_2 + ... + c_n \left(\frac{\lambda_n}{\lambda_1}\right)^{k+1} \mathbf{x}_n \right]. \tag{14.7}$$

As $k \to \infty$, the right-hand sides of (14.6) and (14.7) tend to $\lambda_1^k c_1\mathbf{x}_1$ and $\lambda_1^{k+1} c_1\mathbf{x}_1$, respectively, since $|\lambda_i/\lambda_1| < 1$, $i = 2, ..., n$. The vector $c_1\mathbf{x}_1 + c_2 \left(\lambda_2/\lambda_1\right)^k \mathbf{x}_2 + ... + c_n \left(\lambda_n/\lambda_1\right)^k \mathbf{x}_n$ tends to $c_1\mathbf{x}_1$ which is the eigenvector corresponding to λ_1. The eigenvalue λ_1 is obtained as the ratio of the corresponding components of $A^k\mathbf{x}$ and $A^{k+1}\mathbf{x}$

$$\lambda_1 = \lim_{k\to\infty} \frac{(A^{k+1}\mathbf{x})_r}{(A^k\mathbf{x})_r}, \quad r = 1, 2, ..., n$$

where r denotes the rth component of the vector.

The power method is usually applied as follows. Start with an initial estimate $\mathbf{x}_0$ of the eigenvector $\mathbf{x}$. This estimate is usually chosen to be

$$\mathbf{x}_0 = (1, 1, ..., 1)^T$$

so that the infinity norm $\|\mathbf{x}\|_\infty = \max_{1 \le i \le n} \{\, |x_i| \,\}$ of the vector is 1. We then generate the sequence $\{\mathbf{x}_k\}$ recursively, using

$$\mathbf{z}_{k-1} = A\mathbf{x}_{k-1},$$
$$\mathbf{x}_k = \frac{1}{m_k}\mathbf{z}_{k-1}$$

where m_k is the component of $\mathbf{z}_{k-1}$ with the largest magnitude. If the method converges, the final value of m_k is the required eigenvalue and the final value of $\mathbf{x}_k$ is the corresponding eigenvector. That is,

$$\lim_{k \to \infty} \mathbf{x}_k = \mathbf{x} \quad \text{and} \quad \lim_{k \to \infty} m_k = \lambda.$$

The iteration process is terminated when the infinity norm $\|\mathbf{x}_k - \mathbf{x}_{k-1}\|_\infty$ is less than a specified tolerance.

The method is best illustrated by an example.

EXAMPLE 14.2

Consider the eigenvalue problem

$$\begin{bmatrix} -9 & 14 & 4 \\ -7 & 12 & 4 \\ 0 & 0 & 1 \end{bmatrix} \begin{bmatrix} x_1 \\ x_2 \\ x_3 \end{bmatrix} = \lambda \begin{bmatrix} x_1 \\ x_2 \\ x_3 \end{bmatrix}$$

where the eigenvalues of A are $\lambda_1 = 5$, $\lambda_2 = 1$, and $\lambda_3 = -2$.

As a first guess, we choose $\mathbf{x}_0 = (1, 1, 1)^T$. Now compute

$$\begin{bmatrix} -9 & 14 & 4 \\ -7 & 12 & 4 \\ 0 & 0 & 1 \end{bmatrix} \begin{bmatrix} 1 \\ 1 \\ 1 \end{bmatrix} = \begin{bmatrix} 9 \\ 9 \\ 1 \end{bmatrix} = 9 \begin{bmatrix} 1 \\ 1 \\ \frac{1}{9} \end{bmatrix} = m_1 \mathbf{x}_1.$$

We have normalized the vector by dividing through by its largest element. The next iteration yields

$$\begin{bmatrix} -9 & 14 & 4 \\ -7 & 12 & 4 \\ 0 & 0 & 1 \end{bmatrix} \begin{bmatrix} 1 \\ 1 \\ \frac{1}{9} \end{bmatrix} = \frac{49}{9} \begin{bmatrix} 1 \\ 1 \\ \frac{1}{49} \end{bmatrix} = m_2 \mathbf{x}_2.$$

After 10 iterations, the sequence of vectors converges to

$$\mathbf{x} = \left(1, 1, 1.02 \times 10^{-8} \right)^T,$$

and the sequence m_k of constants converges to

$$\lambda = 5.000\,000\,205.$$

A summary of the calculations is given in Table 14.1.

The major disadvantage of the power method is that its rate of convergence is slow when the *dominance ratio*

$$r = \left| \frac{\lambda_2}{\lambda_1} \right|$$

of the eigenvalues with the two largest magnitudes is close to one. However, the convergence rate can be accelerated by various strategies. The simplest of these is

i	$m_i \mathbf{x}_i$
1	$9[1, 1, \frac{1}{9}]' = 9\,[1.0, 1.0, 1.11 \times 10^{-1}]'$
2	$\frac{49}{9}[1, 1, \frac{1}{49}]' = 5.444\,444\,[1.0, 1.0, 2.04 \times 10^{-2}]'$
3	$\frac{249}{49}[1, 1, \frac{1}{249}]' = 5.081\,633\,[1.0, 1.0, 4.02 \times 10^{-3}]'$
4	$\frac{1249}{249}[1, 1, \frac{1}{1249}]' = 5.016\,064\,[1.0, 1.0, 8.01 \times 10^{-4}]'$
5	$\frac{6249}{1249}[1, 1, \frac{1}{6249}]' = 5.003\,2023\,[1.0, 1.0, 1.60 \times 10^{-4}]'$
...	...
10	$\frac{97\,656\,249}{19\,531\,249}[1, 1, \frac{1}{97656249}]' = 5.000\,000\,205\,[1.0, 1.0, 1.02 \times 10^{-8}]'$

Table 14.1 Power method for Example 14.2.

done by using the power method with **shift**. We know that A and $A - kI$ have the same set of eigenvectors and for each eigenvalue λ_i of A we have, for $A - kI$, the eigenvalue $\lambda_i - k$. That is

$$(A - kI)\mathbf{x} = A\mathbf{x} - k\mathbf{x} = \lambda\mathbf{x} - k\mathbf{x} = (\lambda - k)\mathbf{x}.$$

Therefore, if we subtract k from the diagonal elements of A, then each eigenvalue is reduced by the same factor and the eigenvectors are not changed. The power method can now be used as

$$\mathbf{z}_k = (A - kI)\,\mathbf{x}_k,$$
$$\mathbf{x}_{k+1} = \frac{1}{m_{k+1}}\mathbf{z}_k.$$

For example, if 20 and -18 are the largest eigenvalues in magnitude, the dominance ratio is $r = 0.9$, which is close to 1 and convergence will be relatively slow. However, if we choose $k = \frac{1}{2}(10 + 18) = 14$, then the new eigenvalues are 6 and -32, yielding the dominance ratio of $r = 0.1875$ and a faster rate of convergence. Of course, the choice of k is difficult unless we know, a priori, an estimate of the eigenvalues. One way of obtaining an estimate of the eigenvalues is by using the Gerschgorin Circle Theorem.

Finally, the eigenvalue with the smallest magnitude can be obtained in a similar way using the **inverse power method**. If λ is an eigenvalue of A and $\mathbf{x}$ is the corresponding eigenvector, then $1/\lambda$ is an eigenvalue of A^{-1} corresponding to the same eigenvector $\mathbf{x}$. Applying the power method to A^{-1} gives an approximation to the eigenvalue of A with the smallest magnitude. The power method can also be used to determine eigenvalues other than the dominant one using *deflation techniques*. For a complete discussion see [21].

EXERCISE SET 14.2

1. Find the dominant eigenpairs of the given matrices

(a) $A = \begin{bmatrix} -4 & 1 & 1 \\ 1 & 5 & -1 \\ 0 & 1 & -3 \end{bmatrix}$,

(b) $A = \begin{bmatrix} 2 & 1 & -1 & 3 \\ 1 & 7 & 0 & -1 \\ -1 & 0 & 4 & -2 \\ 3 & -1 & -2 & 1 \end{bmatrix}$,　(c) $A = \begin{bmatrix} 1 & -3 & 1 & 0 \\ -1 & 2 & 0 & 3 \\ 4 & -1 & 4 & 1 \\ 2 & 1 & -1 & 8 \end{bmatrix}$. (d)

$A = \begin{bmatrix} -17 & 36 & -16 \\ -12 & 25 & -10 \\ 0 & 0 & 3 \end{bmatrix}$,

2. Let $(\lambda, \mathbf{x})$ be an eigenpair of A. If $\lambda \neq 0$, show that $(1/\lambda, \mathbf{x})$ is an eigenpair of A^{-1}.

14.3　THE QUADRATIC METHOD

We now describe the quadratic method for computing the eigenpairs of a real symmetric matrix. It is an iterative method that is based on solving quadratic systems.

Let A be an $n \times n$ symmetric matrix. Let $\mathbf{x_i}$ be an eigenvector of A, which has 1 in the ith position and λ_i as its corresponding eigenvalue. Then, the algebraic eigenvalue problem

$$A\mathbf{x_i} = \lambda_i \mathbf{x_i} \tag{14.8}$$

is a nonlinear system of n equations in n unknowns: $\lambda_i, x_i, i = 1, 2, ..., i-1, i+1, ..., n$.

The ith Eqn. of (14.8) is given by

$$a_{ii} + \sum_{p=1}^{i-1} a_{ip}x_p + \sum_{p=i+1}^{n} a_{ip}x_p = \lambda_i, \tag{14.9}$$

and if $j \neq i$ the jth Eqn. of (14.8) is

$$a_{ji} + \sum_{p=1}^{i-1} a_{jp} x_p + \sum_{p=i+1}^{n} a_{jp} x_p = \lambda_i x_j. \tag{14.10}$$

Using (14.9) and (14.10) we obtain, for $j = 1, ..., i - 1, i + 1, ..., n$

$$a_{ji} + \sum_{p=1}^{i-1} a_{jp}x_p + \sum_{p=i+1}^{n} a_{jp}x_p = \left[a_{ii} + \sum_{p=1}^{i-1} a_{ip}x_p + \sum_{p=i+1}^{n} a_{ip}x_p \right] x_j. \tag{14.11}$$

For $j = 1, ..., i - 1$, (14.11) takes the form

$$f_j = a_{ij}x_j^2 + \left[a_{ii} - a_{jj} + \sum_{p=1}^{j-1} a_{ip}x_p + \sum_{p=j+1}^{i-1} a_{ip}x_p + \sum_{p=i+1}^{n} a_{ip}x_p \right] x_j$$

$$- \left[a_{ji} + \sum_{p=1}^{j-1} a_{jp}x_p + \sum_{p=j+1}^{i-1} a_{jp}x_p + \sum_{p=i+1}^{n} a_{jp}x_p \right] = 0, \tag{14.12}$$

and for $j = i + 1, ..., n$ (14.11) takes the form

$$f_j = a_{ij}x_j^2 + \left[a_{ii} - a_{jj} + \sum_{p=1}^{i-1} a_{ip}x_p + \sum_{p=i+1}^{j-1} a_{ip}x_p + \sum_{p=j+1}^{n} a_{ip}x_p \right] x_j$$

$$- \left[a_{ji} + \sum_{p=1}^{i-1} a_{jp}x_p + \sum_{p=i+1}^{j-1} a_{jp}x_p + \sum_{p=j+1}^{n} a_{jp}x_p \right] = 0. \tag{14.13}$$

For a fixed i and for $j = 1, 2, ..., i - 1, i + 1, ..., n$, we define $A_j = a_{ij}$, B_j the coefficient of x_j, and C_j the constant term in Eqns. (14.12) and (14.13).

Hence, finding the eigenvectors of A is reduced to the solution of the quadratic system given by

$$A_j x_j^2 + B_j x_j + Cj = 0, \quad j = 1, 2, ..., i - 1, i + 1, ..., n. \tag{14.14}$$

We now apply Newton's method to approximate the solution of the system. The $(n - 1) \times (n - 1)$ Jacobian matrix $J = (q_{jl})$ of the system is defined by

$$(q_{jl}) = \begin{cases} 2A_j x_j + B_j & \text{if } j = l \\ a_{il}x_j - a_{jl} & \text{if } j \neq l \end{cases}. \tag{14.15}$$

Assuming that the $det(J) \neq 0$, the functional iteration procedure evolves from selecting $\mathbf{x}^{(0)}$ and generating $\mathbf{x}^{(k)}$ for $k \geq 1$, as follows:

- Calculate $\mathbf{F}(\mathbf{x}^{(k-1)})$ and $J(\mathbf{x}^{(k-1)})$

- Solve the $(n - 1) \times (n - 1)$ linear system $J(\mathbf{x}^{(k-1)}) \mathbf{z} = -\mathbf{F}(\mathbf{x}^{(k-1)})$

- Generate $\mathbf{x}^{(k)} = \mathbf{x}^{(k-1)} + \mathbf{z}^{(k-1)}$

where $\mathbf{F} = -(f_1, ..., f_{i-1}, f_{i+1}, ..., f_n)$.

Once the eigenvectors are obtained, the corresponding eigenvalues are readily obtained from (14.9).

It is well known that Newton's method is generally expected to give a quadratic convergence, provided that a sufficiently accurate starting value is known and $J^{-1}(\mathbf{x})$ exists. To insure an accurate starting value, we introduce a continuation method (see [50]) defined by the family of matrices $H(t)$ such that

$$H(t) = D + tP, \quad 0 \leq t \leq 1 \tag{14.16}$$

where t is a parameter and

$$D = \begin{bmatrix} a_{11} & 0 & \cdots & 0 \\ 0 & a_{22} & \ddots & \vdots \\ \vdots & \ddots & \ddots & 0 \\ 0 & \cdots & 0 & a_{nn} \end{bmatrix}, \quad tP = \begin{bmatrix} 0 & ta_{12} & \cdots & ta_{1n} \\ ta_{21} & 0 & \ddots & \vdots \\ \vdots & \ddots & \ddots & ta_{n-1,n} \\ ta_{n1} & \cdots & ta_{n,n-1} & 0 \end{bmatrix}.$$

Note that the eigenvalues of $H(0) = D$ are given by the diagonal elements of A and that $H(1) = A$. We now consider the problem of finding the eigenpairs of the family of matrices $H(t)$ with $t \in [0,1]$. Suppose that for each $t \in [0,1]$ and a given $i = 1,2,\ldots,n$, $H(t)$ has an eigenvalue $\lambda_i(t)$, which depends continuously on t. Then $\lambda_i(t)$ describes a curve (solution curve) in R^2 with one endpoint at the ith diagonal element $\lambda_i(0) = a_{ii}$ of A and the other endpoint at the ith eigenvalue $\lambda_i(1)$ of A. To obtain $\lambda_i(1)$, we need to define first a partition of the interval $[0,1]$ by

$$t_k = kh, \ k = 0, 1, .., M, \quad M = 1/h. \tag{14.17}$$

One way of selecting h is by using the Gerschgorin circle theorem. Choose $h = 2^{-s}$ so that all the Gerschgorin circles of the matrix $H(t_1)$ are disjoint (see Figure 14.2), with s given by

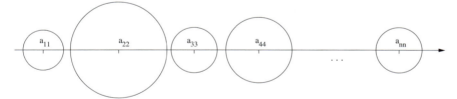

FIGURE 14.2

Gerschgorin circles $\left(a_{ii}, h \sum_{\substack{j=1 \\ i \neq j}}^{n} |a_{ij}| \right)$ **of** $H(t_1)$.

$$INT\left(\frac{\ln(r/d)}{\ln 2} + 2\right), \quad r = \max_i \left(\sum_{\substack{j=1 \\ i \neq j}}^{n} |a_{ij}| \right), \text{ and } d = \min_{\substack{1 \leq i,j \leq n \\ i \neq j}} |a_{ii} - a_{jj}| \tag{14.18}$$

provided that $a_{ii} \neq a_{jj}$. Here $INT(x)$ represents the value of x truncated to an integer. Thus, given h, the method consists of finding the sequence of eigenpairs of $H(t_0), H(t_1), ..., H(t_M)$ obtained by solving, for each t_k, the nonlinear system (14.12) and (14.13) using Newton's method.

The steps for finding the starting value for the application of Newton's method are as follows:

- At $t_0 = 0$, the eigenvalues of $H(0)$ are known and given by the diagonal elements of A and the corresponding eigenvectors are

$$\mathbf{x}^{(0)} = \left(x_1^{(0)}, x_2^{(0)}, ..., x_n^{(0)} \right) = \left(\begin{bmatrix} 1 \\ 0 \\ \vdots \\ 0 \end{bmatrix}, \begin{bmatrix} 0 \\ 1 \\ \vdots \\ 0 \end{bmatrix}, ..., \begin{bmatrix} 0 \\ 0 \\ \vdots \\ 1 \end{bmatrix} \right).$$

Here, $x_i^{(k)}$ denotes the ith eigenvector of the matrix $H(t_k)$.

- At $t = t_1$, Newton's method is applied to the matrix $H(t_1)$ using $\mathbf{x}^{(0)}$ as a starting value to get $\mathbf{x}^{(1)}$.

- At $t = t_2$, the two-point interpolation formula

$$\mathbf{y}^{(2)} = 2\mathbf{x}^{(1)} - \mathbf{x}^{(0)} \tag{14.19}$$

is used to find the starting value $\mathbf{y}^{(2)}$. Here $\mathbf{x}^{(0)}$ and $\mathbf{x}^{(1)}$ are the eigenvectors of $H(t_0)$ and $H(t_1)$, respectively.

- For $t_k > t_2$, the previous three eigenvectors, i.e., $\mathbf{x}^{(k-1)}$, $\mathbf{x}^{(k-2)}$, $\mathbf{x}^{(k-3)}$ of $H(t_{k-1})$, $H(t_{k-2})$, $H(t_{k-3})$, respectively, are used together with the following three-point interpolation formula

$$\mathbf{y}^{(k)} = 3\mathbf{x}^{(k-1)} - 3\mathbf{x}^{(k-2)} + \mathbf{x}^{(k-3)}, \quad k \geq 3 \tag{14.20}$$

to find the starting value $\mathbf{y}^{(k)}$ for Newton's method that is needed to compute the eigenpairs of $H(t_k)$.

Hence, the method consists of finding the eigenpairs of the sequence of matrices $H(t_0), H(t_1), ..., H(t_M)$ by solving at each step quadratic nonlinear systems using formula (14.20) to compute the starting values.

EXAMPLE 14.3
Use the quadratic method to find the eigenpairs for the matrix

$$A = \begin{bmatrix} 3 & 2 & -2 \\ 2 & -4 & 2 \\ -2 & 2 & 6 \end{bmatrix}.$$

For this matrix we have $h = 0.25$ obtained using formula 14.18. We begin by considering the following sequence of matrices

$$\tilde{A}_k = D + t_k P = \begin{bmatrix} 3 & 2t_k & -2t_k \\ 2t_k & -4 & 2t_k \\ -2t_k & 2t_k & 6 \end{bmatrix}$$

where $t_k = kh = 0.25k, \quad k = 0, 1, ..., 4$.

The algorithm of the quadratic method for finding the first eigenpair $(\lambda_1, \mathbf{x}_1)$ where $\mathbf{x}_1 = [1, x_2, x_3]^T$ proceeds as follows:

From Eqn. (14.1) we have

$$
\begin{bmatrix}
3 & 2t_k & -2t_k \\
2t_k & -4 & 2t_k \\
-2t_k & 2t_k & 6
\end{bmatrix}
\begin{bmatrix}
1 \\ x_2 \\ x_3
\end{bmatrix}
= \lambda_1
\begin{bmatrix}
1 \\ x_2 \\ x_3
\end{bmatrix}
$$

which can be written as

$$3 + 2x_2 t_k - 2x_3 t_k = \lambda_1 \tag{14.21}$$
$$2t_k - 4x_2 + 2x_3 t_k = \lambda_1 x_2$$
$$-2t_k + 2x_2 t_k + 6\, x_3 = \lambda_1 x_3.$$

Substitute λ_1 into the last two equations to get the following nonlinear system in two unknowns x_2 and x_3

$$-2x_2^2 t_k + (2x_3 t_k - 7)x_2 + 2x_3 t_k + 2t_k = 0 \tag{14.22}$$
$$2x_3^2 t_k - (2x_2 t_k - 3)x_3 + 2x_2 t_k - 2t_k = 0.$$

We now solve system (14.22) for different values of t_k, using Newton's method with a tolerance of 10^{-6}.

Step 1: To find the eigenpair of $\tilde{A}_0$, set $t_0 = 0$ in (14.22) to get the following eigenvector

$$\mathbf{x}_1^{(0)} = [1, 0, 0]^T$$

and then its corresponding eigenvalue

$$\lambda_1^{(0)} = a_{11} = 3.$$

Step 2: To find the eigenpair of $\tilde{A}_1$, set $t_1 = 0.25$ in (14.22) to get the following system

$$-0.5x_2^2 + (0.5x_3 - 7)x_2 + 0.5x_3 + 0.5 = 0$$
$$0.5x_3^2 - (0.5x_2 - 3)x_3 + 0.5x_2 - 0.5 = 0.$$

By applying Newton's method with

$$\mathbf{y}_1^{(1)} = \mathbf{x}_1^{(0)} = [1, 0, 0]^T$$

as the starting value, we obtain after 3 iterations the following eigenvector:

$$\mathbf{x}_1^{(1)} = [1, 0.0826308129, 0.15116808965]^T.$$

The corresponding eigenvalue is obtained by substituting $\mathbf{x}_1^{(1)}$ into Eqn. (14.21), that is:

$$\lambda_1^{(1)} = 3 + 0.5(0.0826308129) - 0.5\,(0.15116808965) = 2.96573136162.$$

Step 3: To find the eigenpair of $\tilde{A}_2$, set $t_2 = 0.5$ in (14.22) to get the following system

$$-x_2^2 + (x_3 - 7)x_2 + x_3 + 1 = 0$$
$$x_3^2 - (x_2 - 3)x_3 + x_2 - 1 = 0.$$

By applying Newton's method with

$$\mathbf{y}_1^{(2)} = 2\mathbf{x}_1^{(1)} - \mathbf{x}_1^{(0)} = [1, -.834\,738\,374\,2, -.697\,663\,820\,6]^T$$

as the starting value, we obtain, after 3 iterations, the following eigenvector

$$\mathbf{x}_1^{(2)} = [1,\; 0.1828779200, 0.2651074706]^T.$$

The corresponding eigenvalue is obtained by substituting $\mathbf{x}_1^{(2)}$ into Eqn. (14.21), that is:

$$\lambda_1^{(2)} = 3 + (0.1828779200) - (0.2651074706) = 2.9177704494.$$

Step 4: To find the eigenpair of $\tilde{A}_3$, set $t_3 = 0.75$ in (14.22) to get the following system

$$-1.5x_2^2 + (1.5x_3 - 7)x_2 + 1.5x_3 + 1.5 = 0$$
$$1.5x_3^2 - (1.5x_2 - 3)x_3 + 1.5x_2 - 1.5 = 0.$$

By applying Newton's method with

$$\mathbf{y}_1^{(3)} = 3\mathbf{x}_1^{(2)} - 3\mathbf{x}_1^{(1)} + \mathbf{x}_1^{(0)} = [1, 1.\,300\,741\,321, 1.\,341\,818\,143]^T$$

as the starting value, we obtain, after 3 iterations, the following eigenvector

$$\mathbf{x}_1^{(3)} = [1, 0.2915541714, 0.3450029259]^T.$$

The corresponding eigenvalue is obtained by substituting $\mathbf{x}_1^{(3)}$ into Eqn. (14.21), that is:

$$\lambda_1^{(3)} = 3 + 1.5(0.2915541714) - 1.5\,(0.3450029259) = 2.919\,826\,868.$$

Step 5: Finally, to find the eigenpair of $\tilde{A}_4 = A$, set $t_4 = 1$ in (14.22) to get the following system

$$-2x_2^2 + (2x_3 - 7)x_2 + 2x_3 + 2 = 0$$
$$2x_3^2 - (2x_2 - 3)x_3 + 2x_2 - 2 = 0.$$

By applying Newton's method with

$$\mathbf{y}_1^{(4)} = 3\mathbf{x}_1^{(3)} - 3\mathbf{x}_1^{(2)} + \mathbf{x}_1^{(1)} = [1, 0.\,408\,659\,567\,1, 0.\,390\,854\,455\,9]^T$$

as the starting value, we obtain, after 3 iterations, the following eigenvector

$$\mathbf{x}_1^{(4)} = \mathbf{x}_1 = [1, 0.4, 0.4]^T.$$

The corresponding eigenvalue is obtained by substituting $\mathbf{x}_1^{(4)}$ into Eqn. (14.21), that is

$$\lambda_1^{(4)} = \lambda_1 = 3 + 2(0.4) - 2\,(0.4) = 3.0.$$

The two other eigenpairs of A can be obtained in a similar manner. Figure (14.3) shows the solution curves of A.

It is important to note the unique feature of this method, which is that all the eigenpairs can be found independently of each other, hence, the suitability of this procedure for parallel processing. The main disadvantage of the method is that it requires the solution of an $(n-1) \times (n-1)$ quadratic system at each step.

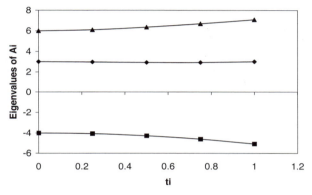

FIGURE 14.3
Solution curves obtained from the eigenvalues of A_i at $t_i, i = 0, ..., 4.$ for Example 14.3.

EXAMPLE 14.4
Use the quadratic method to find the eigenpairs for the matrix

$$A = \begin{bmatrix} 3 & 1 & 0 \\ 1 & 4 & 2 \\ 0 & 2 & 3 \end{bmatrix}.$$

Note that in this example $a_{11} = a_{33}$, which means that the formula in (14.18) cannot be applied. The following modifications must be made to account for this case:

We apply a "relative perturbation" say $\varepsilon = 10^{-3}$, to the identical diagonal elements of matrix A to get the following new matrix:

$$B = \begin{bmatrix} 3 & 1 & 0 \\ 1 & 4 & 2 \\ 0 & 2 & 3(1+\varepsilon) \end{bmatrix}$$

whose diagonal elements are distinct. We now apply the quadratic method to B as in Example 14.3 and use the resulting eigenvectors of B as the starting values for Newton's method with $h = 1$ to find the eigenpairs of A. The result is:

By applying the quadratic method we obtain the following eigenvectors of B:

$$\mathbf{x}_B = \left[\begin{pmatrix} 1.0 \\ -1.7898267750 \\ 1.9966533298 \end{pmatrix}, \begin{pmatrix} 0.3581371370 \\ 1.0 \\ 0.7170446750 \end{pmatrix}, \begin{pmatrix} -1.9988008642 \\ -.0011998560 \\ 1.0 \end{pmatrix} \right].$$

Using $\mathbf{x}_B$ as the starting values for Newton's method, we apply the quadratic method with $h = 1$ to get the following eigenpairs of A:

$$\lambda_1 = 1.2087121525 \text{ and } \mathbf{x}_1 = [1.0, -1.7912878475, 2.0]^T,$$
$$\lambda_2 = 5.7912878475 \text{ and } \mathbf{x}_2 = [0.3582575695, 1.0, 0.7165151390]^T,$$
$$\lambda_3 = 3.0000000000 \text{ and } \mathbf{x}_3 = [-2.0, 0.0, 1.0]^T.$$

MATLAB's Methods

The MATLAB function for finding the eigenvalues and eigenvectors of a given square matrix A is

```
>> [V,D]=eig(A)
```

When called, the function returns a diagonal matrix D of eigenvalues and a full matrix V whose columns are the corresponding eigenvectors. For example, if we call this function to find the eigenpairs of the matrix given in Example 14.4, we get the output

```
>> A=[3  1  0 ;  1  4  2 ;  0  2  3];
>> [V,D]=eig(A)
```

V =

$$\begin{matrix} 0.3490 & 0.8944 & 0.2796 \\ -0.6252 & -0.0000 & 0.7805 \\ 0.6981 & -0.4472 & 0.5592 \end{matrix}$$

D =

$$\begin{matrix} 1.2087 & 0 & 0 \\ 0 & 3.0000 & 0 \\ 0 & 0 & 5.7913 \end{matrix}$$

These results are in complete agreement with the ones obtained by using the quadratic method.

EXERCISE SET 14.3

1. Use the quadratic method to find the eigenpairs of the following matrices.

   (a) $A = \begin{bmatrix} 3 & -4 & 6 \\ -4 & 9 & 2 \\ 6 & 2 & 4 \end{bmatrix}$,   (b) $A = \begin{bmatrix} 1 & 2 & 2 \\ 2 & 0 & 3 \\ 2 & 3 & 0 \end{bmatrix}$,

   (c) $A = \begin{bmatrix} -7 & 0 & 6 \\ 0 & 5 & 0 \\ 6 & 0 & 2 \end{bmatrix}$,   (d) $A = \begin{bmatrix} -1 & 1 & 2 \\ 1 & 2 & 1 \\ 2 & 1 & -1 \end{bmatrix}$,

   (e) $A = \begin{bmatrix} -5 & 1 & 1 & 0 \\ 1 & -8 & 2 & 0 \\ 1 & 2 & -6 & 1 \\ 0 & 0 & 1 & -3 \end{bmatrix}$.

2. Consider the system of differential equations

$$X'(t) = \begin{bmatrix} 1 & 2 & 2 \\ 2 & 0.1 & 3 \\ 2 & 3 & 0.2 \end{bmatrix} X(t).$$

   The general solution to this problem is

$$X(t) = Ae^{\lambda_1 t} X_1 + Be^{\lambda_2 t} X_2 + Ce^{\lambda_3 t} X_3$$

   where (λ_1, X_1), (λ_2, X_2), and (λ_3, X_3) are the eigenpairs of the coefficient matrix of the system.

   Find the general solution of the system using the quadratic method.

3. Use the quadratic method to find the eigenpairs of the following tridiagonal matrix:

$$A = \begin{bmatrix} 5 & -2 & 0 & 0 & 0 \\ -2 & 5 & -2 & 0 & 0 \\ 0 & -2 & 5 & -2 & 0 \\ 0 & 0 & -2 & 5 & -2 \\ 0 & 0 & 0 & -2 & 5 \end{bmatrix}.$$

4. The Hilbert matrix A of order n is defined by

$$a_{ij} = \frac{1}{i+j-1}, \quad i,j = 1,2,\ldots,n.$$

   Use the quadratic method to find the eigenvalues of A for $n = 4, 8$.

5. Find the largest eigenvalue of the matrix $A = \begin{bmatrix} -1 & 0 & 3 \\ 0 & 3 & 0 \\ 3 & 0 & -1 \end{bmatrix}$.

14.4   EIGENVALUES FOR BOUNDARY-VALUE PROBLEMS

Eigenvalue problems arise in many branches of applied mathematics, and, in particular, in the solutions of differential equations such as problems in the fields of elasticity and vibration. Consider, for example, the homogeneous second-order differential equation

$$\frac{d^2y}{dx^2} + k^2\, y = 0, \quad 0 \le x \le L \tag{14.23}$$

subject to the boundary conditions

$$y(0) = 0 \quad \text{and} \quad y(L) = 0 \tag{14.24}$$

where k^2 is a parameter. Here, $y(x)$ denotes the deflection of a thin elastic column of length L when a constant vertical compressive force P is applied to its top, as shown in Figure 14.4. The parameter k is given by $k = P/EI$, where E is Young's modulus of elasticity and I is the moment of inertia of a cross section about a vertical line through its centroid.

Eqn. (14.23), along with the boundary conditions (14.24), falls into a special class of boundary-value problems known as eigenvalue problems. The values of k^2 for which the boundary-value problem has nontrivial solutions are called the **eigenvalues**. The general solution of (14.23) is

$$y(x) = c_1 \cos(kx) + c_2 \sin(kx). \tag{14.25}$$

If we apply the boundary conditions (14.24) to the general solution we get:

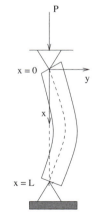

FIGURE 14.4
Bucking of a thin vertical column.

The condition $y(0) = 0$ yields $c_1 = 0$, but $y(L) = 0$ implies

$$c_2 \sin(kL) = 0.$$

If $c_2 = 0$, then necessarily $y = 0$ which is called the **trivial solution**. However, if $c_2 \neq 0$, then $\sin(kL) = 0$. The last condition is true only for the certain values of k:

$$k_n = \pm \frac{n\pi}{L}, \quad n = 1, 2, 3, \dots.$$

These are the eigenvalues for the equation, and the corresponding eigenfunctions are

$$y_n = c_2 \sin(\frac{n\pi}{L} x), \quad n = 1, 2, 3, \dots.$$

These eigenvalues are important for the vibration problem because they give the natural frequency of the system. That is, if the external load P is applied very near to these frequencies, resonance causes an amplification of the motion.

We now describe a numerical solution to the eigenvalue problems (14.23) and (14.24).

Let us divide the interval $[0, L]$ into n equal subintervals defined by

$$x_i = hi, \quad h = L/n, \quad i = 0, 1, \dots, n.$$

By replacing the second derivative in (14.23) with a central difference approximation, we obtain the difference equation

$$\frac{y_{i-1} - 2y_i + y_{i+1}}{h^2} + k^2 y_i = 0$$

or

$$-y_{i-1} + 2y_i - y_{i+1} = h^2 k^2 y_i, \quad i = 1, 2, \dots, n-1. \tag{14.26}$$

In matrix form Eqn. (14.26) can be written as

$$A\mathbf{y} = \lambda \mathbf{y}$$

where

$$A = \begin{bmatrix} 2 & -1 & 0 & \cdots & & 0 \\ -1 & 2 & -1 & \ddots & & \vdots \\ 0 & -1 & \ddots & \ddots & & 0 \\ \vdots & \ddots & \ddots & & 2 & -1 \\ 0 & \cdots & & 0 & -1 & 2 \end{bmatrix}, \quad \mathbf{y} = \begin{bmatrix} y_1 \\ y_2 \\ \vdots \\ y \\ y_{n-1} \end{bmatrix} \quad \text{and} \quad \lambda = h^2 k^2. \tag{14.27}$$

The solutions of the system of equations (14.26) are given by the eigenpairs of A, which can be obtained by using the quadratic method described in the previous section. For example, let $L = 1$ and $n = 6$ or $h = 1/6$. The eigenvalues of the 5×5 matrix A are

$$\lambda_1 = 0.2679, \ \lambda_2 = 1.0, \ \lambda_3 = 2.0, \ \lambda_4 = 3.0, \ \text{and} \ \lambda_5 = 3.7321$$

which give these values for k

$$k_1 = \pm 3.1058, \ k_2 = \pm 6.0, \ k_3 = \pm 8.4853,$$
$$k_4 = \pm 10.3923, \ k_5 = \pm 11.5911.$$

The analytical values for k are given by

$$k_1 = \pm 3.1415, \ \ k_2 = \pm 6.2832, \ \ k_3 = \pm 9.4248,$$
$$k_4 = \pm 12.5664, \ \text{ and } k_5 = \pm 15.7080.$$

We did not get a good approximation of the values of k due to the small number n of the subintervals. To get a better approximation, one needs to choose a larger number of subintervals.

EXERCISE SET 14.4

1. Estimate the eigenvalues of the boundary-value problem

$$y'' - 2y' + 4k^2 y = 0, \quad y(0) = 0, \quad y(1) = 0$$

   using $h = 0.25$.

14.5 BIFURCATIONS IN DIFFERENTIAL EQUATIONS

Systems of ordinary differential equations, modeling physical or biological problems, typically contain parameters, which in a given situation must be assigned definite numerical values. The behavior of the solutions may, however, be radically affected by the values of the parameters chosen. Consider for example the logistic or Verhulst population model

$$\frac{dN}{dt} = rN - aN^2,$$

which is of interest only for $N \geq 0$. We assume $a > 0$ has been chosen, and consider the dependence of the solution on the parameter r. The fixed points or equilibrium solutions are given by

$$rN - aN^2 = 0.$$

If $r \leq 0$, the only fixed point is $N = 0$, and it is stable. In fact, for every initial condition

$$N(0) = N_0 > 0,$$

the solution satisfies

$$\lim_{t \to \infty} N(t) = 0.$$

If $r > 0$, there are two fixed points, $N = 0$ and $N = r/a$. The fixed point $N = 0$ becomes unstable and the fixed point $N = r/a$ is stable. The fixed point is said to bifurcate at $r = 0$. If we think of the fixed point, N, as a function of r so that $N = N(r)$, then $(0,0)$ is called a **bifurcation point**. We illustrate the behavior described with the diagram shown in Figure 14.5.

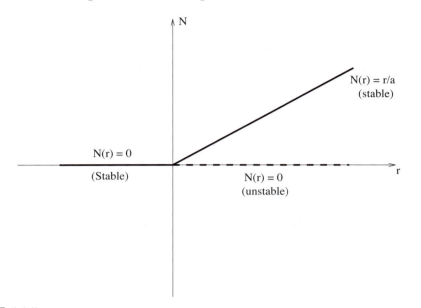

FIGURE 14.5
Bifurcation point.

The philosophy for the general case is quite similar. For the sake of simplicity in presentation, we shall restrict ourselves to the case of two differential equations in two unknown functions, $x(t)$ and $y(t)$, and consider the autonomous system

$$\frac{dx}{dt} = f(x, y, r) \quad \text{and} \quad \frac{dy}{dt} = g(x, y, r), \tag{14.28}$$

where r is a parameter. However, the discussion works for any number of differential equations. Moreover, the second-order differential equation

$$\frac{d^2u}{dt^2} = F\left(u, \frac{du}{dt}, r\right)$$

can be reduced to a form equivalent to (14.28) by setting $x = u$ and $y = du/dt$ to obtain

$$\frac{dx}{dt} = y \quad \text{and} \quad \frac{dy}{dt} = F(x, y, r).$$

The fixed points of (14.28) are obtained by solving the nonlinear system

$$f(x, y, r) = 0 \quad \text{and} \quad g(x, y, r) = 0 \tag{14.29}$$

to obtain the solutions $(x^*(r), y^*(r))$, which depend on r. Intuitively, for a fixed r, (x^*, y^*) is stable if a solution that starts sufficiently close to (x^*, y^*) remains close to

(x^*, y^*) and it is unstable if it is forced to leave the given neighborhood no matter how close one starts to (x^*, y^*). Now form the matrix

$$J(x^*, y^*, r) = \begin{bmatrix} f_x(x^*, y^*, r) & f_y(x^*, y^*, r) \\ g_x(x^*, y^*, r) & g_y(x^*, y^*, r) \end{bmatrix}. \tag{14.30}$$

Let us note that for a given r, there may be several distinct values of x and y that satisfy (14.29). Let us suppose that $\bar{r}$, $f(\bar{x}, \bar{y}, \bar{r}) = 0$ and $g(\bar{x}, \bar{y}, \bar{r}) = 0$. If

$$\det[J(\bar{x}, \bar{y}, \bar{r})] \neq 0,$$

then in a neighborhood of $\bar{r}$, there are functions $x^*(r)$ and $y^*(r)$ such that

$$x^*(\bar{r}) = \bar{x} \quad \text{and} \quad y^*(\bar{r}) = \bar{y}$$

by the implicit function theorem. Now we are seeking values of r where branching can occur, and at this branching there is often a change in the stability of the fixed point. A necessary condition for branching to occur is, therefore,

$$\det[J(x^*, y^*, r)] = 0. \tag{14.31}$$

EXAMPLE 14.5

Consider the case of a frictionless bead of mass m moving on a hoop of radius r that is, itself, twirled at a rate of constant circular frequency $\omega \geq 0$. See Figure 14.6.

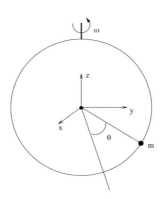

FIGURE 14.6
Frictionless bead.

The position of m at any time, t, is

$$x = R \cos \omega t \sin \theta$$
$$y = R \sin \omega t \sin \theta$$
$$z = -R \cos \theta.$$

Its kinetic energy, K, is

$$K = \frac{mR^2}{2} \left(\omega^2 \sin^2 \theta + \theta'^2 \right)$$

and its potential energy, V, is

$$V = mgR \left(1 - \cos\theta\right).$$

The equation of motion is

$$mR^2 \frac{d^2\theta}{dt^2} = mR^2\omega^2 \sin\theta\cos\theta - mgR\sin\theta.$$

Divide the equation by mR^2 and let $u = \theta$ and $v = d\theta/dt$ to obtain the system

$$\frac{du}{dt} = v \quad \text{and} \quad \frac{dv}{dt} = \omega^2 \sin u \cos u - \frac{g}{R}\sin u. \tag{14.32}$$

The equilibrium points in the uv-phase plane satisfy $v = 0$ and

$$\omega^2 \sin u \cos u - \frac{g}{R}\sin u = \omega^2 \sin u \left(\cos u - \frac{g}{\omega^2 R}\right) = 0.$$

Now $\sin u = 0$ implies $u = 0$ or $u = \pi$. One can, of course, add multiples of 2π, but that is irrelevant. The point $(0, 0)$ corresponds to the point on the very bottom of the hoop and the point $u = \pi$ corresponds to the case where the bead sits at the very top of the hoop. Whether or not there are other fixed points depends upon the existence of solutions to $\cos u - g/\omega^2 R = 0$ and this in turn depends upon how fast the hoop spins. If the hoop spins slowly, then ω is small and $g/\omega^2 R > 1$ so there are no more fixed points. While if ω is large and $g/\omega^2 R < 1$ then an additional fixed point corresponds to $\theta = \arccos\left(g/\omega^2 R\right)$. The Jacobian matrix of the system (14.32) is

$$A\left(u, v, \omega\right) = \begin{bmatrix} 0 & 1 \\ \omega^2 \left\{(\cos^2 u - \sin^2 u) - \frac{g}{\omega^2 R}\cos u\right\} & 0 \end{bmatrix}$$

$$A\left(0, 0, \omega\right) = \begin{bmatrix} 0 & 1 \\ \omega^2 - g/R & 0 \end{bmatrix}$$

$$A\left(\pi, 0, \omega\right) = \begin{bmatrix} 0 & 1 \\ \omega^2 + g/R & 0 \end{bmatrix}$$

$$A\left(\cos^{-1}\frac{g}{\omega^2 R}, 0, \omega\right) = \begin{bmatrix} 0 & 1 \\ (g^2/\omega^2 R^2) - \omega^2 & 0 \end{bmatrix}.$$

$\det[A\left(\pi, 0, \omega\right)] = -\left(\omega^2 + g/R\right)$ is never zero, so this fixed point will have no bifurcations from it. When $\omega = \sqrt{g/R}$ then $\det\left[A\left(0, 0, \omega\right)\right] = -\left(\omega^2 - g/R\right) = 0$ and the bottom can bifurcate. The bifurcation diagram for the fixed point θ as a function of ω is shown in Figure 14.7.

The investigation of stability will take us too far afield in this short introduction, and we shall limit ourselves to determining stability experimentally by taking many values of ω (see the exercises below).

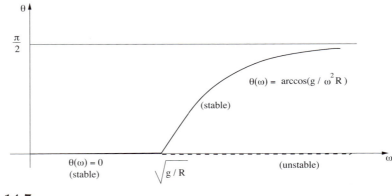

FIGURE 14.7
Bifurcation point.

EXERCISE SET 14.5

1. Find the fixed points and draw bifurcation diagrams for the following one-dimensional problems. Stability is not an issue here.

   (a) $rx - x^3 = 0$,

   (b) $r^2x - x^3 = 0$,

   (c) $r^2x + x^3 = 0$.

   (d) $x^3 + 2x^2 + rx = 0$,

2. Consider the problem of the bead on the hoop of Example 14.5. Let $g = 1$ and $R = 1$. In this problem we will take $\theta'(0) = 0$.

   (a) Let $\omega = 0.5$. Make phase plane plots corresponding to the initial values $\theta(0) = 0.1$ and $\theta(0) = 3.14$. On the basis of these plots, what can you say about the stability of the point $(0, 0)$ and $(\pi, 0)$?

   (b) Let $\omega = 1.5$. Make phase plane plots corresponding to the initial values $\theta(0) = 0.1$ and $\theta(0) = 3.14$. Compare these plots with those of part (a).

   (c) Let $\omega = 10$. Make phase plane plots corresponding to the initial values $\theta(0) = 0.1$ and $\theta(0) = 3.14$. Compare these plots with those of part (b).

3. Consider, again, the problem of the bead on the hoop of Example 14.5. Assume that friction forces act and that the equation of motion is

$$mR^2\frac{d^2\theta}{dt^2} = -mR^2c\frac{d\theta}{dt} + mR^2\omega^2\sin\theta\cos\theta - mgR\sin\theta.$$

Find the fixed points and determine if they bifurcate with ω. Let $c = 2$ and answer parts (b) and (c) of Problem 2 for this case.

4. Consider the system of differential equations

$$\frac{dx}{dt} = rx + y$$

$$\frac{dy}{dt} = - \left(x + ry + rx^3 \right)$$

with $r > 0$. Show that the origin is the only fixed point where $r < 1$. Find the fixed points for $r > 1$. Show that $r = 1$ is a critical point, that is, a possible bifurcation point. Choose a number of initial values $(x(0), y(0))$ for different values of r and determine the stability of the origin and the other fixed points experimentally. What happens when $r = 0$?

APPLIED PROBLEMS FOR CHAPTER 14

1. Consider the Lorentz system of differential equations

$$\frac{dx}{dt} = \sigma (y - x)$$

$$\frac{dy}{dt} = rx - y - xz$$

$$\frac{dz}{dt} = -\beta + xy$$

where σ, β, r are positive constants. In this problem, σ and β will be regarded as fixed but we will assign different values to r, although, in any given example, r will be fixed. In the numerical examples, we will take $\sigma = 10$ and $\beta = 8/3$, which are traditional values.

(a) Find the fixed points for the system.

Note : $(0, 0, 0)$ will always be a fixed point. You will find that it is the only fixed point when $r < 1$ but there will be others when $r > 1$.

(b) Show experimentally that the origin is stable when $r < 0$ and unstable when $r > 1$.

(c) Show that $r = 1$ is a bifurcation point.

(d) Make three-dimensional plots in the phase plane of the solutions to the Lorentz system for $r = 0.5, 2, 10, 28, 56, 100$. Choose initial conditions close to the origin.

Chapter 15

Partial Differential Equations

This chapter presents a brief introduction to some techniques available for approximating the solution to partial differential equations (PDEs) of the form

$$A\frac{\partial^2}{\partial x^2}u(x,y) + B\frac{\partial^2}{\partial x \partial y}u(x,y) + C\frac{\partial^2}{\partial y^2}u(x,y) = f\left(x,y,u,\frac{\partial u}{\partial x},\frac{\partial u}{\partial y}\right). \qquad (15.1)$$

Eqns. (15.1) are classified into three categories depending on the values of the coefficients A, B, and C:

$$\begin{aligned}
&\text{If} \quad B^2 - 4AC > 0, \qquad \text{the equation is } \textbf{hyperbolic.} \\
&\text{If} \quad B^2 - 4AC = 0, \qquad \text{the equation is } \textbf{parabolic.} \\
&\text{If} \quad B^2 - 4AC < 0, \qquad \text{the equation is } \textbf{elliptic.}
\end{aligned}$$

In many applications, the y variable refers to the time and when that is the case, we write t instead of y.

In the next sections, we will discuss solution methods for the parabolic partial differential equation given by

$$\alpha\frac{\partial^2 u}{\partial x^2} = \frac{\partial u}{\partial t}, \qquad \alpha > 0, \qquad (15.2)$$

and known as the **heat** or **diffusion equation**.

Another equation for which we discuss solution methods is the hyperbolic partial differential equation given by

$$\alpha^2\frac{\partial^2 u}{\partial x^2} = \frac{\partial^2 u}{\partial t^2}, \qquad (15.3)$$

and known as the **wave equation**.

Finally, we will discuss solution methods for the elliptic partial differential equation given by

$$\frac{\partial^2 u}{\partial x^2} + \frac{\partial^2 u}{\partial y^2} = 0 \qquad (15.4)$$

and known as **Laplace's equation**.

15.1   PARABOLIC EQUATIONS

We begin our discussion with the one-dimensional PDEs of parabolic type. Consider Eqn. (15.2)

$$\alpha \frac{\partial^2 u}{\partial x^2} = \frac{\partial u}{\partial t}, \quad 0 < x < L, \quad 0 < t \le T \tag{15.5}$$

subject to the boundary conditions

$$u(0,t) = 0, \quad u(L,t) = 0, \quad 0 \le t \le T, \tag{15.6}$$

and the initial conditions

$$u(x,0) = f(x), \quad 0 \le x \le L. \tag{15.7}$$

Here, $u(x,t)$ denotes the temperature at any time t along a thin, long rod of length L in which heat is flowing as illustrated in Figure 15.1. We assume that the rod is of homogeneous material and has a cross-sectional area A that is constant throughout the length of the rod. The rod is laterally insulated along its entire length. The constant α in Eqn. (15.5) is determined by the thermal properties of the material and is a measure of its ability to conduct heat.

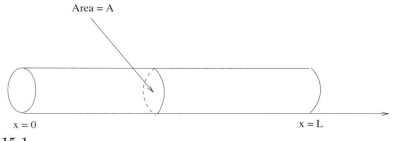

FIGURE 15.1
One-dimensional rod of length L.

15.1.1   Explicit methods

One important approach to approximate the solution to Eqns. (15.5)-(15.7) is the finite-difference method. In order to approximate the solution for our problem, a network of grid points is first established throughout the rectangular region

$$R = \{(x,t) \,|\, 0 \le x \le L, 0 \le t \le T\}$$

as shown in Figure 15.2. We partition R by dividing the interval $[0, L]$ into n equal subintervals, each of length $h = L/n$ and the interval $[0, T]$ into m equal subintervals, each of length $k = T/m$. The corresponding points of the intervals $[0, L]$ and $[0, T]$ are denoted by x_i, for $i = 0, ..., n$ and t_j, for $j = 0, ..., m$, respectively. The points (x_i, t_j) are called the **mesh** or **grid** points and are defined by

$$x_i = hi, \quad t_j = kj \quad \text{for } i = 0, ..., n, \text{ and } j = 0, ..., m.$$

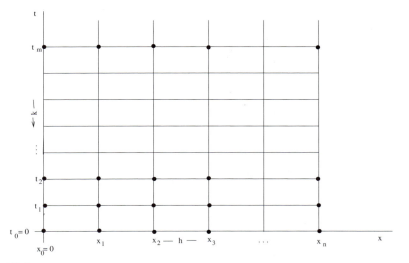

FIGURE 15.2
The region R and the mesh points.

 The approximate solution of $u(x,t)$ at the mesh point (x_i, t_j) is denoted by u_i^j and the true solution is denoted by $u(x_i, t_j)$.

 Two formulas from Section 9.1 are used in this context: The central-difference formula for approximating $u_{xx}(x_i, t_j) = \frac{\partial^2 u}{\partial x^2}(x_i, t_j)$

$$u_{xx}(x_i, t_j) = \frac{u(x_{i+1}, t_j) - 2u(x_i, t_j) + u(x_{i-1}, t_j)}{h^2} + O(h^2) \tag{15.8}$$

and the forward-difference formula for approximating

$$\frac{\partial u}{\partial t}(x_i, t_j) = u_t(x_i, t_j) = \frac{u(x_i, t_{j+1}) - u(x_i, t_j)}{k} + O(k). \tag{15.9}$$

 By substituting Eqns. (15.8) and (15.9) into Eqn. (15.5) and neglecting error terms $O(h^2)$ and $O(k)$, we obtain

$$\frac{u_i^{j+1} - u_i^j}{k} = \alpha \frac{u_{i+1}^j - 2u_i^j + u_{i-1}^j}{h^2}. \tag{15.10}$$

 If we set $\lambda = \alpha\left(k/h^2\right)$ and solve for u_i^{j+1} in Eqn. (15.10), we obtain the explicit difference formula

$$u_i^{j+1} = (1 - 2\lambda)\, u_i^j + \lambda\left(u_{i+1}^j + u_{i-1}^j\right), \tag{15.11}$$

for $i = 1, ..., n-1$, and $j = 0, ..., m-1$, known as the **Forward-Difference** or **classical explicit method**.

 In schematic form, Eqn. (15.11) is shown in Figure 15.3. The solution at every point $(i, j+1)$ on the $(j+1)$th time level is expressed in terms of the solution values at the points $(i-1, j)$, (i, j), and $(i+1, j)$ of the previous time level. Such a method is called an **explicit** method. It can be shown that the Forward-Difference method has an accuracy of the order $O(k + h^2)$ (see [32]).

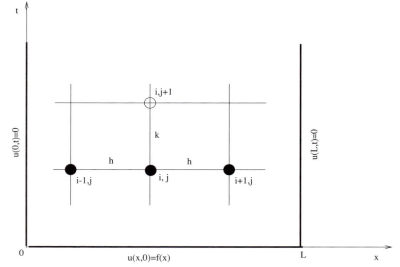

FIGURE 15.3
Schematic form of the Forward-Difference method.

The values of the initial condition $u(x_i, 0) = f(x_i)$, for $i = 0, ..., n$, are used in Eqn. (15.11) to find the values of u_i^1, for $i = 1, ..., n - 1$. The boundary conditions, $u(0, t_j) = u(L, t_j) = 0$, imply that $u_0^j = u_n^j = 0$, for $j = 0, ..., m$. Once the approximations u_i^1, $i = 1, \ldots, n - 1$ are known, the values of $u_i^2, u_i^3, ..., u_i^m$ can be obtained in a similar manner.

EXAMPLE 15.1
We illustrate the forward-difference method by solving the heat equation

$$\frac{\partial^2 u}{\partial x^2} = \frac{\partial u}{\partial t}, \quad 0 < x < 1, \quad t > 0$$

subject to the initial and boundary conditions

$$u(x, 0) = \sin \pi x, \quad 0 \leq x \leq 1 \quad \text{and} \quad u(0, t) = u(1, t) = 0, \quad t \geq 0.$$

The exact solution to this problem is

$$u(x, t) = e^{-\pi^2 t} \sin \pi x.$$

The solution will be approximated first with $T = 0.025$, $h = 0.1$, and $k = 0.0025$ so that $\lambda = k/h^2 = 0.25$, and then with $T = 0.05$, $h = 0.1$, and $k = 0.01$ so that $\lambda = k/h^2 = 1.0$.

With $h = 0.1$ and $k = 0.0025$, the approximations of u at $t = 0.0025$ for $i = 1, 2$ are

$$u_1^1 = \frac{1}{4} \left(u_0^0 + 2u_1^0 + u_2^0 \right) = \frac{1}{4} \left[\sin(0.0) + 2 \sin(\pi 0.1) + \sin(\pi 0.2) \right]$$

$$= 0.301\,455,$$

$$u_2^1 = \frac{1}{4}\left(u_1^0 + 2u_2^0 + u_3^0\right) = \frac{1}{4}\left[\sin\left(\pi 0.1\right) + 2\sin\left(\pi 0.2\right) + \sin\left(\pi 0.3\right)\right]$$

$$= 0.573\,401.$$

With $h = 0.1$ and $k = 0.01$, the approximations of u at $t = 0.01$ for $i = 1, 2$ are

$$u_1^1 = u_0^0 - u_1^0 + u_2^0 = \sin\left(0.0\right) - \sin\left(\pi 0.1\right) + \sin\left(\pi 0.2\right) = 0.278\,768,$$

$$u_2^1 = u_1^0 - u_2^0 + u_3^0 = \sin\left(\pi 0.1\right) - \sin\left(\pi 0.2\right) + \sin\left(\pi 0.3\right) = 0.530\,249.$$

If we continue in this manner, we get the results shown in Tables 15.1 and 15.2 obtained from using the MATLAB function heat. A three-dimensional representation of Table 15.1 is shown in Figure 15.4.

```
» heat('f1',0,0,1,.025,.1,.0025,1)

lambda =
         0.25
```

t \ x	0	0.1	0.2	0.3	0.4	0.5 ...	0.9	1
0.0000	0.000000	0.309017	0.587785	0.809017	0.951057	1.000000	0.309017	0.000000
0.0025	0.000000	0.301455	0.573401	0.789219	0.927783	0.975528	0.301455	0.000000
0.0050	0.000000	0.294078	0.559369	0.769905	0.905078	0.951655	0.294078	0.000000
0.0075	0.000000	0.286881	0.545680	0.751064	0.882929	0.928367	0.286881	0.000000
0.0100	0.000000	0.279861	0.532327	0.732685	0.861322	0.905648	0.279861	0.000000
0.0125	0.000000	0.273012	0.519300	0.714755	0.840244	0.883485	0.273012	0.000000
0.0150	0.000000	0.266331	0.506591	0.697263	0.819682	0.861865	0.266331	0.000000
0.0175	0.000000	0.259813	0.494194	0.680200	0.799623	0.840773	0.259813	0.000000
0.0200	0.000000	0.253455	0.482100	0.663554	0.780055	0.820198	0.253455	0.000000
0.0225	0.000000	0.247253	0.470303	0.647316	0.760966	0.800127	0.247253	0.000000
0.0250	0.000000	0.241202	0.458793	0.631475	0.742343	0.780546	0.241202	0.000000
Error at t = 0.025	0	0.0002465	0.0004689	0.0006453	0.0007586	0.0007977	0.000247	0

Table 15.1 Forward-Difference method for Example 15.1 with $\lambda = 0.25$.

```
» heat('f1',0,0,1,.05,.1,.01,1)

lambda =
         1
```

t \ x	0	0.1	0.2	0.3	0.4	0.5 ...	0.9	1
0.00	0.000000	0.309017	0.587785	0.809017	0.951057	1.000000	0.309017	0.000000
0.01	0.000000	0.278768	0.530249	0.729825	0.857960	0.902113	0.278768	0.000000
0.02	0.000000	0.251480	0.478344	0.658384	0.773977	0.813808	0.251480	0.000000
0.03	0.000000	0.226864	0.431521	0.593937	0.698215	0.734147	0.226864	0.000000
0.04	0.000000	0.204657	0.389280	0.535798	0.629869	0.662283	0.204657	0.000000
0.05	0.000000	0.184624	0.351175	0.483351	0.568213	0.597454	0.184624	0.000000
Error at t = 0.05	0	0.004	0.0077	0.0106	0.0124	0.013	0.004	0

Table 15.2 Forward-Difference method for Example 15.1 with $\lambda = 1.0$.

This explicit method, just described, does not necessarily produce good results. One can see from Table 15.2 that the results obtained for the case when $\lambda = 1.0$

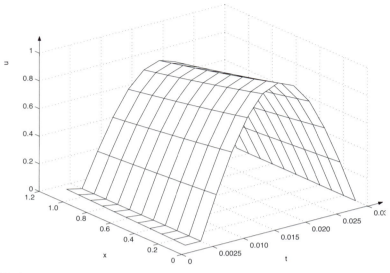

FIGURE 15.4
$u = u_{ij}$ **for Example 15.1 with** $\lambda = 0.25$.

are poor approximations of the solution and do not reflect an accuracy of the order $O(k+h^2)$. This is due to the stability condition of the explicit method. Appropriate choices have to be made for the step sizes, h and k, that determine the values of λ. It can be shown (see [32], for further discussion) that the explicit method is stable if the **mesh ratio** λ satisfies

$$0 < \lambda \le \frac{1}{2}. \tag{15.12}$$

This means that the error made at one stage of calculation does not cause increasingly large errors as the computations are continued, but rather will eventually damp out.

15.1.2   Implicit methods

In the explicit method previously described, the approximate solution u_i^{j+1} depends on the values $u_{i-1}^j, u_i^j \; u_{i+1}^j$ of u at the previous time level. Furthermore, the requirement $0 < \lambda \le \frac{1}{2}$ places an undesirable restriction on the time step k that can be used.

The implicit method, now to be described, overcomes the stability requirement by being unconditionally stable. The finite-difference equation of this method is obtained by replacing $u_{xx}(x,y)$ in Eqn. (15.5) with the average of the centered difference at the time steps $j+1$ and j and $u_t(x,y)$ with the forward difference. The result is

$$\frac{u_i^{j+1} - u_i^j}{k} = \frac{\alpha}{2}\left[\frac{u_{i+1}^j - 2u_i^j + u_{i-1}^j}{h^2} + \frac{u_{i+1}^{j+1} - 2u_i^{j+1} + u_{i-1}^{j+1}}{h^2}\right]. \tag{15.13}$$

By setting $\lambda = \alpha k/h^2$ as before, Eqn. (15.13) can be written as

$$-\lambda u_{i-1}^{j+1} + 2(1+\lambda) u_i^{j+1} - \lambda u_{i+1}^{j+1} = \lambda u_{i-1}^j + 2(1-\lambda) u_i^j + \lambda u_{i+1}^j \qquad (15.14)$$

for $i = 1, 2, ..., n-1$. This method is called the **Crank-Nicolson** method. In schematic form, Eqn. (15.14) is shown in Figure 15.5. The solution value at any point $(i, j+1)$ on the $(j+1)$th time level is dependent on the solution values at the neighboring points on the same level and three points on the jth time level. Since values at the $(j+1)$th time level are obtained implicitly, the method is called an **implicit** method. It can be shown that the Crank-Nicolson method has an accuracy of the order $O(k^2 + h^2)$ and is unconditionally stable (see [32]).

If, in Eqn. (15.5) we replace the time derivative by a forward difference and the

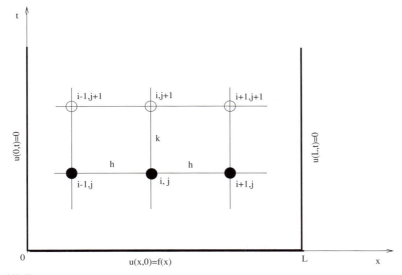

FIGURE 15.5
Schematic form of the Crank-Nicolson method.

space derivative by a centered difference at the forward time step $j+1$, we get the **classical implicit** method

$$u_i^j = -\lambda u_{i-1}^{j+1} + (1+2\lambda)u_i^{j+1} - \lambda u_{i+1}^{j+1}$$

where $\lambda = \alpha k/h^2$. This method is unconditionally stable.

The matrix form of the Crank-Nicolson method is

$$A\,\mathbf{u}^{(j+1)} = B\,\mathbf{u}^{(j)}, \quad j = 0, 1, 2, ... \qquad (15.15)$$

where $\mathbf{u}^{(j+1)} = \left[u_1^{j+1}, u_2^{j+1}, ..., u_{n-1}^{j+1}\right]^t$,

$$
A = \begin{bmatrix}
2(1+\lambda) & -\lambda & 0 & \cdots & & 0 \\
-\lambda & 2(1+\lambda) & -\lambda & \ddots & & \vdots \\
0 & -\lambda & \ddots & \ddots & & 0 \\
\vdots & & \ddots & \ddots & \ddots & -\lambda \\
0 & & \cdots & 0 & -\lambda & 2(1+\lambda)
\end{bmatrix},
$$

and

$$
B = \begin{bmatrix}
2(1-\lambda) & \lambda & 0 & \cdots & & 0 \\
\lambda & 2(1-\lambda) & \lambda & \ddots & & \vdots \\
0 & \lambda & \ddots & \ddots & & 0 \\
\vdots & & \ddots & \ddots & \ddots & \lambda \\
0 & & \cdots & 0 & \lambda & 2(1-\lambda)
\end{bmatrix}.
$$

The tridiagonal matrix A is positive definite and strictly diagonally dominant. Therefore, A is nonsingular and the system of equation (15.14) can be solved by any method described in Chapter 4.

```
» heat_crank('f1',0,0,1,0.5,0.2,0.05,1)

lambda =
            1.25

   t \ x        0        0.2        0.4        0.6        0.8        1

   0.00    0.000000   0.587785   0.951057   0.951057   0.587785   0.000000
   0.05    0.000000   0.361228   0.584480   0.584480   0.361228   0.000000
   0.10    0.000000   0.221996   0.359197   0.359197   0.221996   0.000000
   0.15    0.000000   0.136430   0.220748   0.220748   0.136430   0.000000
   0.20    0.000000   0.083844   0.135662   0.135662   0.083844   0.000000
   0.25    0.000000   0.051527   0.083372   0.083372   0.051527   0.000000
   0.30    0.000000   0.031666   0.051237   0.051237   0.031666   0.000000
   0.35    0.000000   0.019461   0.031488   0.031488   0.019461   0.000000
   0.40    0.000000   0.011960   0.019351   0.019351   0.011960   0.000000
   0.45    0.000000   0.007350   0.011893   0.011893   0.007350   0.000000
   0.50    0.000000   0.004517   0.007309   0.007309   0.004517   0.000000
----------------------------------------------------------------------------
Error at        0      0.00029   0.000469   0.000469    0.00029      0
t = 0.5
```

Table 15.3   Crank-Nicolson method for Example 15.2 with $\lambda = 1.25$.

EXAMPLE 15.2

Solve the heat equation

$$
\frac{\partial^2 u}{\partial x^2} = \frac{\partial u}{\partial t}, \quad 0 < x < 1, \quad 0 < t < 0.5
$$

subject to the initial and boundary conditions

$$u(x,0) = \sin \pi x, \quad 0 \le x \le 1$$
$$u(0,t) = u(1,t) = 0, \quad 0 \le t \le 0.5$$

using the Crank-Nicolson method.

We choose $h = 0.2$, $k = 0.05$, so that $\lambda = 1.25$, $n = 5$, and $m = 10$. By replacing the value of λ into (15.14), we get the simplified difference equation

$$-1.25\, u_{i-1}^{j+1} + 4.5 u_i^{j+1} - 1.25\, u_{i+1}^{j+1} = 1.25\, u_{i-1}^j - 0.5\, u_i^j + 1.25\, u_{i+1}^j$$

for $i = 1, ..., 4$. At the first time step $t = k$, u_i^1 is given by the solution of the tridiagonal system

$$\begin{bmatrix} 4.5 & -1.25 & 0 & 0 \\ -1.25 & 4.5 & -1.25 & 0 \\ 0 & -1.25 & 4.5 & -1.25 \\ 0 & 0 & -1.25 & 4.5 \end{bmatrix} \begin{bmatrix} u_1^1 \\ u_2^1 \\ u_3^1 \\ u_4^1 \end{bmatrix} = \begin{bmatrix} -0.5\, u_1^0 + 1.25\, u_2^0 \\ 1.25\, u_1^0 - 0.5\, u_2^0 + 1.25\, u_3^0 \\ 1.25\, u_2^0 - 0.5\, u_3^0 + 1.25\, u_4^0 \\ 1.25\, u_3^0 - 0.5\, u_4^0 \end{bmatrix}$$

$$= \begin{bmatrix} 0.\,894\,928 \\ 1.\,448\,024 \\ 1.\,448\,024 \\ 0.\,894\,928 \end{bmatrix}$$

where $u_i^0 = \sin(\pi i h)$. The solution to the tridiagonal system is

$$u_1^1 = 0.361228, \quad u_2^1 = 0.584480 \quad u_3^1 = 0.584480, \quad u_4^1 = 0.361228.$$

Using the MATLAB function heat_crank gives the solution shown in Table 15.3.

EXERCISE SET 15.1

1. In which part of the xy-plane is the following equation parabolic?

$$u_{xx} + 4u_{xy} + (x^2 + y^2)u_{yy} = \sin(xy)$$

2. Verify that $u(x,t) = e^{-t} \cos \pi(x - \frac{1}{2})$ is a solution to the heat equation $\frac{1}{\pi^2} u_{xx} = u_t$.

3. Approximate the solution of the following parabolic partial differential equations using the Forward Difference:

m-function 15.1a

The following MATLAB function **heat.m** finds the solution of the heat equation using the Forward-Difference method. INPUTS are a function f; the boundary condition $c1$, $c2$; the endpoint L; the maximum time T; the step sizes h and k; the constant α. The input function $f(x)$ should be defined as an M-file.

```
function heat(f,c1,c2,L,T,h,k,alpha)
% Solve the heat equation with I.C. u(x,0)=f(x) and B.C. u(0,t)=c1
%  and u(L,t)=c2  using the Forward-Difference method.
n=L/h; m=T/k;
lambda=alpha^2*k/(h^2)
z=0:h:L;
disp('_____')
fprintf('    t       x =                                 ')
fprintf('%4.2f        ',z)
fprintf('\n')
disp('_____')
fprintf('% 5.4f  ',0)
for i=1:n+1
   u(i)=feval(f,(i-1)*h);
fprintf('%10.6f   ',u(i))
end
fprintf('\n')
for j=1:m
  t=j*k;
  fprintf('% 5.4f  ',t)
  for i=1:n+1
    if (i==1)
      y(i)=c1;
    elseif (i==n+1)
       y(i)=c2;
    else
       y(i)=(1-2*lambda)*u(i)+lambda*(u(i+1)+u(i-1));
    end;
  fprintf('%10.6f   ',y(i))
  end;
  fprintf('\n')
  u=y;
end
```

M-function 15.1b

The following MATLAB function **heat_crank.m** finds the solution of the heat equation using the Crank-Nicolson method. INPUTS are a function f; the boundary condition $c1,\ c2$; the endpoint L; the maximum time T; the step sizes h and k; the constant α. The input function $f(x)$ should be defined as an M-file.

```
function heat_crank(f,c1,c2,L,T,h,k,alpha)
% Solve the heat equation with I.C. u(x,0)=f(x) and B.C. u(0,t)=c1,
% and u(L,t)=c2 using Crank-Nicolson method.
n=L/h; m=T/k; lambda=alpha^2*k/(h^2)
z=0:h:L;
disp('_____')
fprintf('        t         x = ')
fprintf('%4.2f      ',z)
fprintf('\n')
disp('_____')
fprintf('% 4.2f ',0)
for i=1:n+1
  u(i)=feval(f,(i-1)*h);
fprintf('%10.6f   ',u(i))
end
fprintf('\n')
for i=2:n
  if (i~=n)
     a(i)=-lambda;
  end
  b(i)=2*(1+lambda);
  if (i~=n)
     c(i)=-lambda;
  end
end
bb=b;
for j=1:m
  t=j*k;
  fprintf('% 4.2f ',t)
  for i=2:n
    d(i)=lambda*u(i-1)+2*(1-lambda)*u(i)+lambda*u(i+1);
  end
  y(n+1)=c2;
  y(1)=c1;
  for i=3:n
     ymult=a(i-1)/bb(i-1);
```

```
      bb(i)=bb(i)-ymult*c(i-1);
      d(i)=d(i)-ymult*d(i-1);
    end
    y(n)=d(n)/bb(n);
    for i=n-1:-1:2
      y(i)=(d(i)-c(i)*y(i+1))/bb(i);
    end
      for i=1:n+1
        fprintf('%10.6f   ',y(i))
      end
    fprintf('\n')
    u=y;
    bb=b;
  end
```

(a) $u_{xx} = u_t$, $0 < x < 1$, $0 < t < 0.2$;

  $u(0,t) = u(1,t) = 0$, $0 \le t \le 0.2$;

  $u(x,0) = 4x - 4x^2$, $0 \le x \le 1$.

  Use $h = 0.2$, $k = 0.02$.

(b) $10\, u_{xx} = u_t$, $0 < x < 100$, $0 < t < 0.05$;

  $u(0,t) = u(100,t) = 0$, $0 \le t \le 0.05$;

  $$u(x,0) = \begin{cases} 0 & 0 \le x < 25, \\ 50 & 25 \le x \le 75, \\ 0 & 75 < x \le 100. \end{cases}$$

  Use $h = 10$, $k = 0.01$.

(c) $u_{xx} = u_t$, $0 < x < 1$, $0 < t < 0.1$;

  $u(0,t) = u(1,t) = 0$, $0 \le t \le 0.1$;

  $$u(x,0) = \begin{cases} 1, & 0 \le x < 0.5, \\ 0, & 0.5 < x \le 1. \end{cases}$$

  Use $h = 0.2$, $k = 0.01$.

4. Use the Crank-Nicolson method to approximate the solution to the following parabolic partial differential equations:

(a) $\frac{1}{\pi^2}\, u_{xx} = u_t$, $0 < x < 1$, $0 < t < 0.1$;

  $u(0,t) = u(1,t) = 0$, $0 \le t \le 0.1$;

  $u(x,0) = \sin \pi x$, $0 \le x \le 1$.

  Use $h = 0.1$, $k = 0.01$. Compare your results with the exact solution $u(x,t) = e^{-t} \sin \pi x$.

(b) $u_{xx} = u_t$, $0 < x < 1$, $0 < t < 0.05$;

$$u(0,t) = u(\pi,t) = 0, \quad 0 \le t \le 0.05;$$

$$u(x,0) = \sin \pi x + \sin 2\pi x, \quad 0 \le x \le 1.$$

Use $h = 0.2$, $k = 0.01$.

(c) $3\,u_{xx} = u_t, \quad 0 < x < \pi,\ 0 < t < 0.1;$

$$u(0,t) = u(\pi,t) = 0, \quad 0 \le t \le 0.1;$$

$$u(x,0) = \sin x, \quad 0 \le x \le \pi.$$

Use $h = 0.52$, $k = 0.01$.

5. Derive a finite-difference method for solving the nonlinear parabolic equation

$$u_{xx} = uu_t.$$

15.2 HYPERBOLIC EQUATIONS

In this section, we discuss methods for the numerical solution of the one-dimensional PDEs of hyperbolic type. Consider the wave equation given by Eqn. (15.3)

$$\alpha^2 \frac{\partial^2 u}{\partial x^2} = \frac{\partial^2 u}{\partial t^2}, \quad 0 < x < L, \quad t > 0 \tag{15.16}$$

subject to the boundary conditions

$$u(0,t) = 0, \quad u(L,t) = 0, \quad t > 0, \tag{15.17}$$

and the initial conditions

$$u(x,0) = f(x), \quad 0 \le x \le L,$$
$$u_t(x,0) = g(x), \quad 0 \le x \le L. \tag{15.18}$$

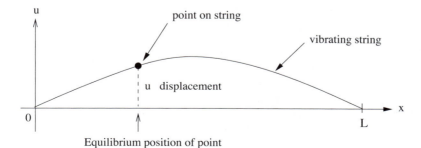

FIGURE 15.6
Vibrating string.

Here, $u(x,t)$ denotes the vertical displacement of a uniform, perfectly flexible string of constant density that is tightly stretched between two fixed points, 0 and L. We assume that the equilibrium position of the string is horizontal, with the string aligned along the x-axis as shown in Figure 15.6. Suppose the string is plucked at time $t = 0$ causing the string to vibrate. Our problem is to determine the vertical displacement $u(x,t)$ of a point x at time t. We assume that the horizontal displacement is so small relative to the vertical displacement as to be negligible. We also assume that the maximum displacement of each point on the string is small in comparison with the length L of the string.

We assume that the region $R = \{(x,t) \,|\, 0 \le x \le L, 0 \le t \le T\}$ to be subdivided into rectangles as shown in Figure 15.2. To derive a difference equation for the solution, we start in the usual way by replacing the space derivative by the central-difference formula

$$u_{xx}(x_i, t_j) = \frac{u(x_{i+1}, t_j) - 2u(x_i, t_j) + u(x_{i-1}, t_j)}{h^2} + O(h^2)$$

and the time derivative by the central-difference formula

$$u_{tt}(x_i, t_j) = \frac{u(x_i, t_{j+1}) - 2u(x_i, t_j) + u(x_i, t_{j-1})}{k^2} + O(k^2).$$

Substituting these central-difference formulas into (15.16) and dropping the terms $O(h^2)$ and $O(k^2)$ gives the difference equation

$$\alpha^2 \frac{u_{i+1}^j - 2u_i^j + u_{i-1}^j}{h^2} = \frac{u_i^{j-1} - 2u_i^j + u_i^{j+1}}{k^2}.$$

If we set $\lambda = \alpha\,(k/h)$ and rearrange the order of the terms, we obtain the **explicit three-level difference** formula

$$u_i^{j+1} = 2(1 - \lambda^2)\, u_i^j + \lambda^2 \left(u_{i+1}^j + u_{i-1}^j \right) - u_i^{j-1}, \tag{15.19}$$

for $i = 1, ..., n-1$ and $j = 1, ..., m-1$. In schematic form, Eqn. (15.19) is shown in Figure 15.7. The solution at every point $(i, j+1)$ on the $(j+1)$th time level is expressed in terms of the solution values at the points $(i-1, j)$, (i, j), $(i+1, j)$, and $(i, j-1)$, of the two previous time levels. The explicit formula (15.19) has a stability problem and it can be shown that the method is stable if

$$0 < \lambda \le 1$$

(see [32] for proofs and detailed discussion).

The right-hand side of Eqn. (15.19) shows that to calculate the entry u_i^{j+1} on the $(j+1)$th time level, we must use entries from the jth and $(j-1)$th time levels. This requirement creates a problem at the beginning, because we only know the very first row from the initial condition $u_i^0 = f(x_i)$. To obtain the second row corresponding to u_i^1, the second initial condition $u_t(x, 0) = g(x)$ may be used as follows:

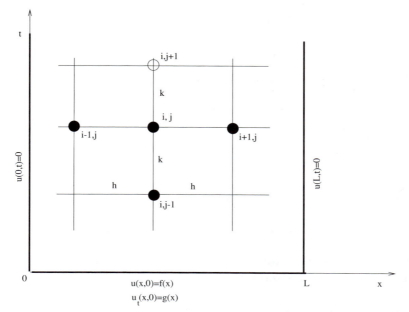

FIGURE 15.7
Schematic form of the three-level difference method.

If we use the forward-difference approximation

$$u_t(x_i, 0) = \frac{u(x_i, t_1) - u(x_i, 0)}{k} + O(k),$$

we get a finite-difference equation that gives an approximation for the second row with a local truncation error of only $O(k)$. To get a better approximation, we consider the Taylor series expansion of $u(x, t)$ about the point $(x_i, 0)$

$$u(x_i, k) = u(x_i, 0) + k\, u_t(x_i, 0) + \frac{k^2}{2}\, u_{tt}(x_i, 0) + O(k^3). \qquad (15.20)$$

Assuming that the second derivative of $f(x)$ exists, we have the result obtained from Eqns. (15.16) and (15.18)

$$u_{tt}(x_i, 0) = \alpha^2 u_{xx}(x_i, 0) = \alpha^2 f''(x_i). \qquad (15.21)$$

Substituting (15.21) into Eqn. (15.20) and using the initial conditions $u_t(x_i, 0) = g(x_i)$ and $u(x_i, 0) = f(x_i)$ gives

$$u(x_i, k) = f(x_i) + k\, g(x_i) + \frac{k^2}{2}\, \alpha^2\, f''(x_i) + O(k^3).$$

Finally, by replacing $f''(x_i)$, in the last equation, by the central-difference formula, we get a difference formula for the numerical approximation in the second row

$$u_i^1 = f(x_i) + k\, g(x_i) + \frac{k^2 \alpha^2}{2h^2} \left[f(x_{i-1}) - 2\, f(x_i) + f(x_{i+1}) \right]$$

$$= (1 - \lambda^2)\, f(x_i) + \frac{\lambda^2}{2} \left[f(x_{i+1}) + f(x_{i-1}) \right] + k\, g(x_i), \ i = 1, ..., n - 1$$

which has an accuracy of the order $O(k^3 + h^2 k^2)$.

EXAMPLE 15.3

Approximate the solution of the wave equation

$$16 u_{xx} = u_{tt}, \quad 0 < x < 1, \quad t > 0,$$

subject to the conditions

$$u(0, t) = u(1, t) = 0, \quad t > 0,$$
$$u(x, 0) = \sin \pi x, \quad 0 \le x \le 1,$$
$$u_t(x, 0) = 0, \quad 0 \le x \le 1.$$

The solution will be approximated with $T = 0.5$, $h = 0.2$, and $k = 0.05$ so that $\lambda = 4 \, (0.05) \, /0.2 = 1.0$. The approximations of u at $t = 0.05$ for $i = 1, 2, 3, 4$ are as follows:

The boundary conditions give

$$u_5^j = u_0^j = 0, \quad j = 1, ..., 10,$$

and the initial conditions give

$$u_i^0 = \sin(\pi 0.2 i), \quad i = 0, ..., 5,$$
$$u_i^1 = \frac{1}{2} \left[f(0.2(i + 1)) + f(0.2(i - 1)) \right] + 0.05 \, g(0.2 i)$$
$$= 0.5 \left\{ \sin[\pi 0.2(i + 1)] + \sin[\pi 0.2(i - 1)] \right\}, \quad i = 1, ..., 4.$$

Hence,

$$i = 1, \; u_1^1 = 0.475\,528\,26$$
$$i = 2, \; u_2^1 = 0.769\,420\,88$$
$$i = 3, \; u_3^1 = 0.769\,420\,88$$
$$i = 4, \; u_4^1 = 0.475\,528\,26.$$

For $t = 2k = 0.1$, we get the difference equation

$$u_i^2 = u_{i+1}^1 + u_{i-1}^1 - u_i^0, \quad i = 1, 2, 3, 4$$

which implies that

$$i = 1, \; u_1^2 = \; u_2^1 - u_1^0 = 0.181\,635\,63$$
$$i = 2, \; u_2^2 = \; u_3^1 + u_1^1 - u_2^0 = 0.29389263$$
$$i = 3, \; u_2^2 = \; u_4^1 + u_2^1 - u_3^0 = 0.29389263$$
$$i = 4, \; u_3^2 = \; u_3^1 - u_4^0 = 018163563.$$

Using the MATLAB function hyperbolic with $T = 0.5$, $h = 0.2$, and $k = 0.05$ gives the solution shown in Table 15.4. A three-dimensional representation of Table 15.4 is shown in Figure 15.8.

```
» hyperbolic('f1','g1',0,0,1,0.5,0.2,0.05,4)

lambda =
        1.0

  t \ x        0         0.2        0.4        0.6        0.8        1.0

  0.00     0.000000   0.587785   0.951057   0.951057   0.587785  0.000000
  0.05     0.000000   0.475528   0.769421   0.769421   0.475528  0.000000
  0.10     0.000000   0.181636   0.293893   0.293893   0.181636  0.000000
  0.15     0.000000  -0.181636  -0.293893  -0.293893  -0.181636  0.000000
  0.20     0.000000  -0.475528  -0.769421  -0.769421  -0.475528  0.000000
  0.25     0.000000  -0.587785  -0.951057  -0.951057  -0.587785  0.000000
  0.30     0.000000  -0.475528  -0.769421  -0.769421  -0.475528  0.000000
  0.35     0.000000  -0.181636  -0.293893  -0.293893  -0.181636  0.000000
  0.40     0.000000   0.181636   0.293893   0.293893   0.181636  0.000000
  0.45     0.000000   0.475528   0.769421   0.769421   0.475528  0.000000
  0.50     0.000000   0.587785   0.951057   0.951057   0.587785  0.000000

Error at     0          0       1.1e-16    1.1e-16    1.1e-16      0
t = 0.5
```

Table 15.4 Explicit three-level difference method for Example 15.3.

EXERCISE SET 15.2

1. Approximate the solution of the following hyperbolic partial differential equations using the three-level difference method

   (a) $u_{xx} = u_{tt}$, $0 < x < 1$, $0 < t < 0.5$;

   $u(0,t) = u(1,t) = 0$, $0 \le t \le 0.1$;

   $u_t(x,0) = x^2$, $u(x,0) = x$, $0 \le x \le 1$.
   Use $h = 0.1$, $k = 0.05$.

   (b) $u_{xx} = u_{tt}$, $0 < x < \pi$, $0 < t < 1.0$;

   $u(0,t) = u(\pi,t) = 0$, $0 \le t \le 1.0$;

   $u_t(x,0) = 0$, $u(x,0) = 2x/\pi(\pi - x)$, $0 \le x \le \pi$.
   Use $h = \pi/5$, $k = 0.1$.

   (c) $u_{xx} = u_{tt}$, $0 < x < \pi$, $0 < t < 1.0$;

   $u(0,t) = u(\pi,t) = 0$, $0 \le t \le 1.0$;

   $u_t(x,0) = \sin x$, $u(x,0) = \sin x$, $0 \le x \le \pi$.
   Use $h = \pi/10$, $k = 0.25$.

   (d) $u_{xx} = u_{tt}$, $0 < x < 4$, $t > 0$;

   $u(0,t) = u(4,t) = 0$, $0 \le t \le 1.0$;

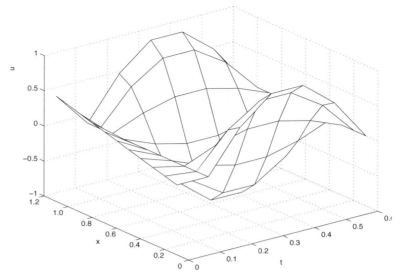

FIGURE 15.8
Explicit three-level difference method for Example 15.3.

$$u(x,0) = \begin{cases} 16x^2(1-x)^2, & 0 \le x \le 1, \\ 0, & 1 \le x \le 4. \end{cases}$$

$$u_t(x,0) = 0, \quad 0 \le x \le 4.$$

Use $h = 0.4$, $k = 0.25$.

(e) $u_{xx} = u_{tt}, \quad 0 < x < 1, 0 < t < 1;$

$u(0,t) = u(1,t) = 0, \quad 0 \le t \le 1;$

$u_t(x,0) = \sin \pi x, \quad u(x,0) = \sin 2\pi x, \quad 0 \le x \le 1.$

Use $h = 0.25$, $k = 0.25$.

2. In which part of the xy-plane is the following equation hyperbolic?

$$u_{xx} + 4u_{xy} + (x^2 + y^2)u_{yy} = \sin(xy).$$

15.3   ELLIPTIC EQUATIONS

Our discussion on elliptic equations will focus on the formulation of the finite-difference equation for the two-dimensional Laplace equation

$$\frac{\partial^2 u}{\partial x^2} + \frac{\partial^2 u}{\partial y^2} = 0 \tag{15.22}$$

M-function 15.2

The following MATLAB function **hyperbolic.m** finds the solution of the hyperbolic equation using a three level explicit method. INPUTS are functions *f* and *g*; the boundary condition *c1, c2*; the endpoint *L*; the maximum time *T*; the step sizes *h* and *k*; the constant α. The input functions *f(x)* and *g(x)* should be defined as M-files.

```
function hyperbolic(f,g,c1,c2,L,T,h,k,alpha)
% Solve the hyperbolic equation with u(x,0)=f(x), ut(x,0)=g(x) and
% u(0,t)=c1, u(L,t)=c2 using a three level explicit method.
n=L/h; m=T/k;
lambda=alpha*k/h;
z=0:h:L;
disp('_____')
fprintf('        t              x =                    ')
fprintf('%4.2f      ',z)
fprintf('\n')
disp('_____')
fprintf('% 4.2f ',0)
for i=1:n+1
  u0(i)=feval(f,(i-1)*h);
  fprintf('%10.6f   ',u0(i))
end
fprintf('\n')
fprintf('% 4.2f %10.6f   ',k,c1)
for i=1:n-1
  u1(i+1)=(1-lambda^2)*feval(f,i*h)+lambda^2/2*(feval(f,(i+1)*h)...
      +feval(f,(i-1)*h))+k*feval(g,i*h);
  fprintf('%10.6f   ',u1(i+1))
end
fprintf('%10.6f   ',c2)
fprintf('\n')
for j=2:m
  t=j*k;
  fprintf('% 4.2f ',t)
  u1(1)=c1; ui(1)=c1; u1(n+1)=c2; ui(n+1)=c2;
  fprintf('%10.6f   ',ui(1))
  for i=2:n
    ui(i)=2*(1-lambda^2)*u1(i)+lambda^2*(u1(i+1)+u1(i-1))-u0(i);
    fprintf('%10.6f   ',ui(i))
  end
  fprintf('%10.6f   ',ui(n+1))
```

```
    fprintf('\n')
    u0=u1;
    u1=ui;
 end
```

on a rectangular domain $R = \{(x, y) \mid a < x < b, c < y < d\}$ and subject to the Dirichlet boundary conditions

$$u(x, c) = f_1(x), \quad u(x, d) = f_2(x), \quad a \leq x \leq b$$
$$u(a, y) = g_1(y), \quad u(b, y) = g_2(y), \quad c \leq y \leq d. \tag{15.23}$$

Eqn. (15.22) arises in the study of steady-state or time-independent solutions of heat equations. Because these solutions do not depend on time, initial conditions are irrelevant and only boundary conditions are specified. Other applications include the static displacement $u(x, y)$ of a stretched membrane fastened in space along the boundary of a region; the electrostatic and gravitational potentials in certain force fields; and, in fluid mechanics for an ideal fluid. A problem of particular interest is in studying a steady-state temperature distribution in a rectangular region R, as illustrated in Figure 15.9. This situation is modeled with Laplace's equation in R with the boundary conditions (15.23).

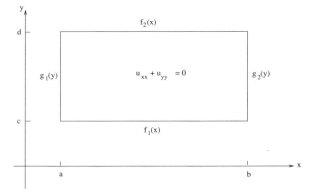

FIGURE 15.9
Steady-state heat flow.

To solve Laplace's equation by difference methods, the region R is partitioned into a grid consisting of $n \times m$ rectangles with sides h and k. The mesh points are given by

$$x_i = a + ih, \quad y_j = c + jk, \quad i = 0, 1, ..., n, \quad j = 0, 1, ..., m.$$

By using central-difference approximations for the spacial derivatives, the finite-difference equation for Eqn. (15.22) is

$$\frac{u_{i+1,j} - 2u_{i,j} + u_{i-1,j}}{h^2} + \frac{u_{i,j+1} - 2u_{i,j} + u_{i,j-1}}{k^2} = 0$$

or

$$2\left(\frac{h^2}{k^2}+1\right)u_{i,j}-(u_{i-1,j}+u_{i+1,j})-\frac{h^2}{k^2}(u_{i,j-1}+u_{i,j+1})=0 \tag{15.24}$$

for $i=1,2,...,n-1$ and $j=1,2,...,m-1$, and the boundary conditions are

$$\begin{aligned}
u(x_i,y_0)&=f_1(x_i), & i&=1,...,n-1,\\
u(x_i,y_m)&=f_2(x_i), & i&=1,...,n-1,\\
u(x_0,y_j)&=g_1(y_j), & j&=1,...,m-1,\\
u(x_n,y_j)&=g_2(y_j), & j&=1,...,m-1.
\end{aligned} \tag{15.25}$$

Often it is desirable to set $h=k$, and in this case (15.24) becomes

$$-4u_{i,j}+u_{i,j-1}+u_{i,j+1}+u_{i-1,j}+u_{i+1,j}=0. \tag{15.26}$$

The computational molecule for Eqn. (15.24) is shown in Figure 15.10. Eqn. (15.24) is referred to as the **five-point difference formula** and has an accuracy of the order $O(h^2+k^2)$.

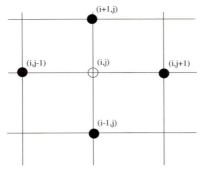

FIGURE 15.10
The computational molecule for Laplace's equation.

Eqn. (15.24) with the boundary conditions (15.25) form an $(n-1)(m-1)$ by $(n-1)(m-1)$ linear system of equations. The coefficient matrix of the system is sparse, but is not banded in quite the same way as we have come to expect of sparse matrices. Instead, the matrix is "striped" with a tridiagonal band along the main diagonal, but with two additional bands displaced from the main diagonal by a considerable amount.

For general use, iterative techniques often represent the best approach to the solution of such systems of equations. However, if the number of equations is not too large, a direct solution of such systems is practical. We now illustrate the solution of the Laplace's equation when $n=m=4$.

EXAMPLE 15.4

Find an approximate solution to the steady-state heat flow equation,

$$u_{xx}+u_{yy}=0, \quad 0<x<2, 0<y<2$$

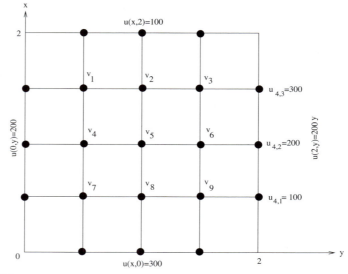

FIGURE 15.11
Numbering system for internal grid points.

with

$$u(x,0) = 300, \ u(x,2) = 100, \quad 0 \le x \le 2,$$
$$u(0,y) = 200, \ u(2,y) = 200\,y, \quad 0 < y < 2,$$

using $h = k = 0.5$.

To set up the linear system, the nine interior grid points are labeled row-by-row from v_1 to v_9 starting from the left-top corner point as shown in Figure 15.11. The resulting system is

$$\begin{bmatrix} -4 & 1 & 0 & 1 & 0 & 0 & 0 & 0 & 0 \\ 1 & -4 & 1 & 0 & 1 & 0 & 0 & 0 & 0 \\ 0 & 1 & -4 & 0 & 0 & 1 & 0 & 0 & 0 \\ 1 & 0 & 0 & -4 & 1 & 0 & 1 & 0 & 0 \\ 0 & 1 & 0 & 1 & -4 & 1 & 0 & 1 & 0 \\ 0 & 0 & 1 & 0 & 1 & -4 & 0 & 0 & 1 \\ 0 & 0 & 0 & 1 & 0 & 0 & -4 & 1 & 0 \\ 0 & 0 & 0 & 0 & 1 & 0 & 1 & -4 & 1 \\ 0 & 0 & 0 & 0 & 0 & 1 & 0 & 1 & -4 \end{bmatrix} \begin{bmatrix} v_1 \\ v_2 \\ v_3 \\ v_4 \\ v_5 \\ v_6 \\ v_7 \\ v_8 \\ v_9 \end{bmatrix} = \begin{bmatrix} -u_{0,3} - u_{1,4} = -300 \\ -u_{2,4} = -100 \\ -u_{4,3} - u_{3,4} = -400 \\ -u_{0,2} = -200 \\ 0 \\ -u_{4,2} = -200 \\ -u_{0,1} - u_{1,0} = -500 \\ -u_{2,0} = -300 \\ -u_{3,0} - u_{4,1} = -400 \end{bmatrix}.$$

Using Gaussian elimination, the temperature at the interior grid points is

$$v_1 = 166.071, \quad v_2 = 164.286, \quad v_3 = 191.071,$$
$$v_4 = 200.000, \quad v_5 = 200.000, \quad v_6 = 200.000,$$
$$v_7 = 233.929, \quad v_8 = 235.714, \quad v_9 = 208.929.$$

Of course, with such a small number of grid points, we do not expect high accuracy. If we choose a smaller value of h, the accuracy should improve.

As stated above, direct solution of the finite-difference equations for elliptic PDEs is expensive. The alternative to direct solution is to use an iterative method. One such method applied to Eqn. (15.26) with Dirichlet boundary conditions is a procedure of the type

$$(L + D)\psi^{(n+1)} + U\psi^{(n)} = 0, \quad n = 0, 1, 2, ...,$$

where L and U are respectively lower and upper triangular matrices with zero diagonal entries and D is the diagonal matrix. This leads to replacing Eqn. (15.26) by

$$u_{i,j}^{(n+1)} = \frac{1}{4}\left(u_{i,j-1}^{(n+1)} + u_{i,j+1}^{(n)} + u_{i-1,j}^{(n+1)} + u_{i+1,j}^{(n)}\right), \quad n = 0, 1, 2, ...,$$

which is the basis of the **Gauss-Seidel** method. This method is also known as the **method of successive displacements**; the elements of $\psi^{(n+1)}$ replace those of $\psi^{(n)}$ in the calculation as soon as they have been computed. The iterative process is set to stop when

$$\left|u_{i,j}^{(n+1)} - u_{i,j}^{(n)}\right| < \epsilon$$

where ϵ is a specified tolerance. For relevant expositions on iterative methods for the solution of large systems of linear equations, we refer to [60], Chapter 8.

EXAMPLE 15.5

Use the MATLAB function laplace with $n = 9$ to approximate the solution of the Laplace's equation in the square region $0 < x < \pi$ and $0 < y < \pi$ subject to the boundary conditions

$$u(x,0) = 0, \ u(x,\pi) = \sin x, \quad 0 \leq x \leq \pi,$$
$$u(0,y) = 0, \ u(\pi,y) = 0, \quad 0 \leq y \leq \pi$$

We set the tolerance value to 10^{-6} and the maximum number of iterations to 120 to get the result shown in Table 15.5.

EXERCISE SET 15.3

1. Solve Laplace's equation for a square $0 < x < a$ and $0 < y < a$, subject to the specified boundary conditions.

   (a) $u(x,0) = u(x,\pi) = u(\pi,y) = 0$, $u(0,y) = \sin^2 y$.
   (b) $u(x,0) = u(0,y) = u(x,1) = 0$, $u(1,y) = -y(y - 1)$.
   (c) $u(x,0) = u(0,y) = u(x,2) = 0$, $u(2,y) = 10$.
   (d) $u(x,0) = u(0,y) = u(x,\pi) = 0$, $u(\pi,y) = \sin y$.

M-function 15.3

The following MATLAB function **laplace.m** finds the solution of the Laplace's equation using a five-point difference formula. INPUTS are functions *f1, f2, g1,* and *g2*; the endpoint *a*; the number of subintervals *n*; the maximum number of iterations *itmax*; and a tolerance *tol*. The input functions should be defined as M-files.

```
function laplace(f1,f2,g1,g2,a,n,itmax,tol)
% Solve the laplace equation in a square with B.C.  u(x,0)=f₁(x)
% u(x,a)=f₂(x)  and u(0,y)=g₁(y) and u(a,y)=g₂(y) using a
% five-point difference method.
h=a/n;
z=0:h:a;
h
disp('_____')
fprintf('         u= x\\y                              ')
fprintf('%4.2f     ',z)
fprintf('\n')
disp('_____')
for i=1:n+1
  u(i,1)=feval(f1,(i-1)*h);
  u(i,n+1)=feval(f2,(i-1)*h);
  u(1,i)=feval(g1,(i-1)*h);
  u(n+1,i)=feval(g2,(i-1)*h);
end
iter=0;
err=tol+1;
while (err>tol)&(iter<itmax)
   err=0;
  for i=2:n
    for j=2:n
      oldu=u(i,j);
      u(i,j)=(u(i+1,j)+u(i-1,j)+u(i,j+1)+u(i,j-1))/4;
      res=abs(u(i,j)-oldu);
      if (err<res);
        err=res;
      end
    end
  end
  iter=iter+1;
end
for i=1:n+1
  fprintf('   %4.2f',z(i))
```

```
laplace('f1','f2','g1','g2', pi,9,120,10^(-6))
```

```
h =
            0.3489
```

u= x\y	0.35	0.70	1.05	1.40	1.75	...	2.79	3.14
0.00	0	0	0	0	0		0	0
0.35	0.0108	0.0228	0.0377	0.0570	0.0833		0.2416	0.3420
0.70	0.0202	0.0429	0.0708	0.1072	0.1566		0.4541	0.6428
1.05	0.0273	0.0578	0.0954	0.1444	0.2109		0.6118	0.8660
1.40	0.0310	0.0658	0.1085	0.1643	0.2399		0.6957	0.9848
1.75	0.0310	0.0658	0.1085	0.1643	0.2399		0.6957	0.9848
2.09	0.0273	0.0578	0.0954	0.1444	0.2109		0.6118	0.8660
2.44	0.0202	0.0429	0.0708	0.1072	0.1566		0.4541	0.6428
2.79	0.0108	0.0228	0.0377	0.0570	0.0833		0.2416	0.3420
3.14	0	0	0	0	0		0	0

Table 15.5 Approximate solution for Example 15.5.

```
    for j=1:n+1
        fprintf('%10.4f ',u(i,j))
    end
    fprintf('\n')
  end
  iter
```

15.4 NONLINEAR PARTIAL DIFFERENTIAL EQUATIONS

Nonlinear partial differential equation problems are encountered in many fields of engineering and sciences and have applications to many physical systems including fluid dynamics, porous media, gas dynamics, traffic flow, shock waves, and many others. In this section we will give a brief description of the numerical solution to some nonlinear physical models arising in engineering and sciences.

15.4.1 Burger's equation

Consider the one-dimensional quasi-linear parabolic partial differential equation

$$\frac{\partial u}{\partial t} + u\frac{\partial u}{\partial x} = \frac{1}{Re}\frac{\partial^2 u}{\partial x^2} \tag{15.27}$$

with the initial and boundary conditions

$$u(x,0) = f(x), \quad 0 < x < 1$$
$$u(0,t) = g_1(t), \ u(1,t) = g_2(t), \quad 0 < t \leq T$$

where Re is the Reynolds number characterizing the size of viscosity and $f(x)$, $g_1(x)$, and $g_2(x)$ are sufficiently smooth given functions.

Equation (15.27) is known as Burger's equation from fluid mechanics and has been widely used for various applications, such as modeling of gas dynamics, traffic flow, shock waves, etc. It is considered as a good model for the numerical solution of nonlinear PDEs.

To solve equation (15.27) using an explicit finite-difference scheme, we approximate the time derivative by using a forward difference and the space derivatives using a central-difference scheme to get

$$\frac{u_j^{n+1} - u_j^n}{\Delta t} + u_j^n \frac{u_{j+1}^n - u_{j-1}^n}{2\Delta x} = \frac{1}{Re} \frac{u_{j+1}^n - 2u_j^n + u_{j-1}^n}{\Delta x^2}$$

Solving for u_j^{n+1} gives

$$u_j^{n+1} = (1 - 2\frac{\lambda}{Re})u_j^n + \left(\frac{\lambda}{Re} - \frac{\Delta t}{2\Delta x}u_j^n\right)u_{j+1}^n + \left(\frac{\lambda}{Re} + \frac{\Delta t}{2\Delta x}u_j^n\right)u_{j-1}^n \quad (15.28)$$

$$\lambda = \frac{\Delta t}{\Delta x^2}, \ j = 2, 3, ..., j_{\max} - 1, \ n = 1, 2, ...$$

Given that at the boundary conditions $u_1^n = g_1(n\Delta t)$ and $u_{j_{max}}^n = g_2(n\Delta t)$, $n = 1, 2, ...$ and the initial condition $u_j^1 = f(j\Delta x)$, $j = 1, 2, ..., j_{\max}$.

Stability analysis show that the explicit scheme is stable for $\left|\frac{\lambda}{Re}\right| \leq 0.5$.

EXAMPLE 15.6

Figure 15.12 shows the numerical solution of (15.27) at different time levels for $f(x) = \sin(\pi x)$ and the homogeneous boundary conditions $g_1(t) = g_2(t) = 0$.

An implicit finite-difference scheme is obtained as follows:

$$\frac{u_j^{n+1} - u_j^n}{\Delta t} + u_j^n \frac{u_{j+1}^{n+1} - u_{j-1}^{n+1}}{2\Delta x} = \frac{1}{Re} \frac{u_{j+1}^{n+1} - 2u_j^{n+1} + u_{j-1}^{n+1}}{\Delta x^2}$$

which can be written as

$$\left(\frac{\Delta t}{2\Delta x}u_j^n - \frac{\lambda}{Re}\right)u_{j+1}^{n+1} + (1 + 2\frac{\lambda}{Re})u_j^{n+1} - \left(\frac{\Delta t}{2\Delta x}u_j^n + \frac{\lambda}{Re}\right)u_{j-1}^{n+1} = u_j^n \quad (15.29)$$

$$\lambda = \frac{\Delta t}{\Delta x^2}, \ j = 2, 3, ..., j_{\max} - 1.$$

Using the boundary conditions, the values of u at $j = 1$ and $j = j_{max}$ are known. That is, $u_1^n = g_1(n\Delta t)$ and $u_{j_{max}}^n = g_2(n\Delta t)$, $n = 1, 2, ...$

The advantage of the implicit schemed is that it is unconditionally stable but at each time level a system of equations needs to be solved for the unknowns u_j^{n+1}, $j = 2, 3, ..., j_{\max} - 1$.

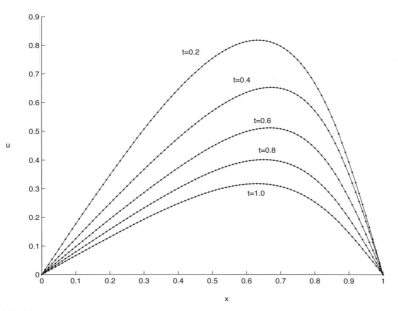

FIGURE 15.12
Numerical solutions of Burger's equation (15.27) at different times for
$Re = 10$, $\Delta x = 0.01$ **and** $\Delta t = 0.0005$.

15.4.2 Reaction-diffusion equation

Consider the one-dimensional reaction-diffusion equation that has the nonlinear term $u(1 - u)(u - \mu)$

$$\frac{\partial u}{\partial t} = \frac{\partial^2 u}{\partial x^2} + u(1 - u)(u - \mu) \tag{15.30}$$

with the initial and boundary conditions

$$u(x, 0) = f(x), \quad 0 < x < 10$$
$$u(0, t) = c_1, \ u(10, t) = c_2, \quad 0 < t \leq T$$

where μ is a parameter.
This equation arises in combustion theory and is well known as the Zeldovich equation.

An explicit scheme is obtained by using a forward-difference scheme for the time derivative and central scheme for the spatial derivative. That is,

$$\frac{u_j^{n+1} - u_j^n}{\Delta t} = \frac{u_{j+1}^n - 2u_j^n + u_{j-1}^n}{\Delta x^2} + u_j^n(1 - u_j^n)(u_j^n - \mu)$$

Solve for u_j^{n+1} to get

$$u_j^{n+1} = (1 - 2\gamma)u_j^n + \gamma\left(u_{j+1}^n + u_{j-1}^n\right) + \Delta t\, u_j^n(1 - u_j^n)(u_j^n - \mu) \tag{15.31}$$

$$\lambda = \frac{\Delta t}{\Delta x^2}, \ j = 2, 3, ..., j_{\max} - 1, \ n = 1, 2, ...$$

EXAMPLE 15.7

Figure 15.13 and 15.14 show the numerical solution of (15.30) at different time levels for

$$f(x) = \begin{cases} 1, & x < 5 \\ 0, & x \geq 5 \end{cases} \text{ and } u(0,t) = 1 \text{ and } u(10,t) = 0.$$

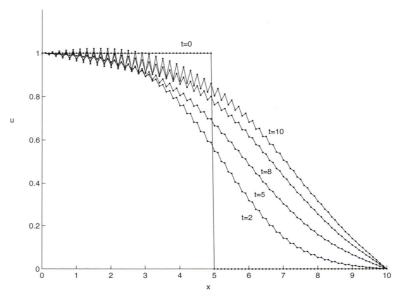

FIGURE 15.13
Numerical solutions of the reaction-diffusion equation (15.30) at different

times for $\mu = 0.3$, $\Delta x = 0.1$ and $\Delta t = 0.005$.

An example of an implicit finite-difference scheme for equation (15.30) is obtained by using a forward-difference scheme for the time derivative and Crank-Nicolson scheme for the spatial derivative. That is

$$\frac{u_j^{n+1} - u_j^n}{\Delta t} = \frac{u_{j+1}^n - 2u_j^n + u_{j-1}^n}{2\Delta x^2} + \frac{u_{j+1}^{n+1} - 2u_j^{n+1} + u_{j-1}^{n+1}}{2\Delta x^2} + u_j^n(1 - u_j^n)(u_j^n - \mu),$$

which can be written as

$$-\gamma u_{j-1}^{n+1} + 2(1 + \gamma)u_j^{n+1} - \gamma u_{j+1}^{n+1} = 2(1 - \gamma)u_j^n + \gamma u_{j-1}^n + \gamma u_{j+1}^n - 2\Delta t \, u_j^n(1 - u_j^n)(u_j^n - \mu),$$

$$\lambda = \frac{\Delta t}{\Delta x^2}, \; j = 2, 3, ..., j_{\max} - 1 \qquad (15.$$

Eqn. (15.32) is a system of equations for the unknowns u_j^{n+1}, $j = 2, 3, ..., j_{\max} - 1$.

Using the boundary conditions, the values of u at $j = 1$ and $j = j_{max}$ are known. That is $u_1^n = c_1$ and $u_{j_{\max}}^n = c_2$, $n = 1, 2, ...$

Stability analysis show that the implicit scheme is unconditionally stable.

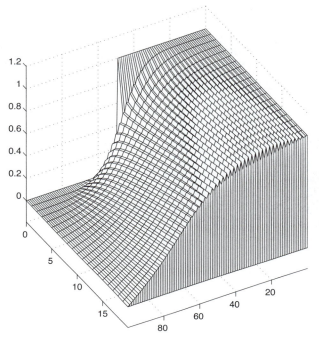

FIGURE 15.14
Solution of the reaction-diffusion equation for Example 15.7.

15.4.3 Porous media equation

Consider the equation

$$u_t = (u_x)^2 + uu_{xx} \tag{15.33}$$

with the initial and boundary conditions

$$u(x,0) = x, \quad 0 < x < 1$$
$$u(0,t) = t, \ u(1,t) = 1 + t, \quad 0 < t \leq T$$

known as the porous media equation. It is encountered in nonlinear problems of heat and mass transfer, combustion theory, and flows in porous media. It has also applications to many physical systems including the fluid dynamics of thin films.

An explicit scheme for the solution of this problem can be obtained by using a forward-difference scheme for the time derivative and central scheme for the spatial derivative. That is

$$\frac{u_j^{n+1} - u_j^n}{\Delta t} = \left(\frac{u_{j+1}^n - u_{j-1}^n}{2\Delta x} \right)^2 + u_j^n \frac{u_{j+1}^n - 2u_j^n + u_{j-1}^n}{\Delta x^2}.$$

Solve for u_j^{n+1} to get

$$u_j^{n+1} = u_j^n + \frac{\gamma}{4} \left(u_{j+1}^n - u_{j-1}^n \right)^2 + \gamma u_j^n \left(u_{j+1}^n - 2u_j^n + u_{j-1}^n \right) \tag{15.34}$$

$$\lambda = \frac{\Delta t}{\Delta x^2}, \ j = 2, 3, ..., j_{\max} - 1, \ n = 1, 2, ...$$

EXAMPLE 15.8

Figure 15.15 shows the numerical solution of (15.33) at different time levels for $f(x) = x$ and $u(0, t) = t$ and $u(1, t) = t + 1$. The exact solution is $u(x, t) = x + t$.

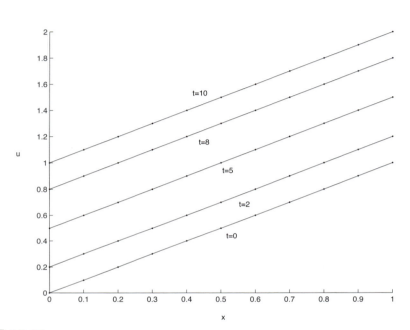

FIGURE 15.15
Numerical solutions of the porous media equation at different times for $\Delta x = 0.1$ and $\Delta t = 0.0001$.

15.4.4 Hamilton-Jacobi-Bellman equation

The Hamilton-Jacobi-Bellman equation is a parabolic partial differential equation of the form

$$u_t + F(t, x, u, u_x, u_{xx}) = 0$$

An interesting example of financial model that results in a nonlinear Hamilton-Jacobi-Bellman (HJB), is the passport option. We do not present the underlying model here and refer to [64] for more details. The reduced one-dimensional equation is

$$u_t + \gamma u - \sup_{|q| \le 1} \left\{ ((r - \gamma - r_c) q - (r - \gamma - r_t) x) u_x + \frac{1}{2}\sigma^2(x - q)^2 u_{xx} \right\} = 0 \quad (15.35)$$

with the initial condition $u(0, x) = \max(x, 0)$, $x_{\min} < x < x_{\max}$.

Here t is the time variable, x is a real number representing the wealth in the trading account per unit of underlying stock, r is the risk-free interest rate, γ is the dividend rate, r_c is a cost of carry rate, and r_t is an interest rate for the trading account.

We consider the case when q, the number of shares of the underlying holder holds at time t, is either $+1$ or -1, $r - \gamma - r_t = 0$ and $r - \gamma - r_c < 0$. (15.35) reduces to

$$u_t + \gamma u - \max\left\{(r - \gamma - rc)\,u_x + \frac{1}{2}\sigma^2(x-1)^2 u_{xx}, -(r - \gamma - r_c)\,u_x + \frac{1}{2}\sigma^2(x-1)^2 u_{xx}\right\} = 0$$
(15.36)

An explicit scheme of Eqn. (15.36) is obtained by using a forward-difference scheme for the time derivative, an appropriate forward- or backward-finite difference for the first partial derivative and a central scheme for the second spatial derivative. The result is

$$\frac{u_j^{n+1} - u_j^n}{\Delta t} + \gamma u_j^n - \max\left\{\begin{array}{l}(r - \gamma - r_c)\frac{u_j^n + u_{j-1}^n}{\Delta x} + \frac{1}{2}\sigma^2(x_j - 1)^2\frac{u_{j+1}^n - 2u_j^n + u_{j-1}^n}{\Delta x^2}, \\ -(r - \gamma - r_c)\frac{u_{j+1}^n + u_j^n}{\Delta x} + \frac{1}{2}\sigma^2(x_j - 1)^2\frac{u_{j+1}^n - 2u_j^n + u_{j-1}^n}{\Delta x^2}\end{array}\right\} = 0$$

Solve for u_j^{n+1} to get

$$u_j^{n+1} = u_j^n(1 - \gamma\Delta t) + \Delta t\max\left\{\begin{array}{l}(r - \gamma - r_c)\frac{u_j^n + u_{j-1}^n}{\Delta x} + \frac{1}{2}\sigma^2(x_j - 1)^2\frac{u_{j+1}^n - 2u_j^n + u_{j-1}^n}{\Delta x^2}, \\ -(r - \gamma - r_c)\frac{u_{j+1}^n + u_j^n}{\Delta x} + \frac{1}{2}\sigma^2(x_j - 1)^2\frac{u_{j+1}^n - 2u_j^n + u_{j-1}^n}{\Delta x^2}\end{array}\right\}$$
(15.37)

$j = 2, 3, ..., j_{\max} - 1$, $n = 1, 2, ...$

It can be shown that this scheme is stable if

$$\frac{\Delta t}{(\Delta x)^2} \leq \frac{1}{(\Delta x)^2\,\gamma + (r - \gamma - r_c)\,\Delta x + \sigma^2\max\left\{\max_j\{j\Delta x - 1\}^2, \max_j\{j\Delta x + 1\}^2\right\}}$$
(15.38)

EXAMPLE 15.9

Figure 15.16 shows the numerical solution of (15.36) at different time levels using the boundary conditions $u(-3, t) = 0$ and $u(4, t) = 4$, $-3 < x < 4$.

An implicit scheme is obtained by replacing the explicit scheme by a fully implicit upwind scheme that is unconditionally stable, that is

$$\frac{u_j^{n+1} - u_j^n}{\Delta t} + \gamma u_j^{n+1} - \max\left\{\begin{array}{l}(r - \gamma - r_c)\frac{u_j^{n+1} + u_{j-1}^{n+1}}{\Delta x} + \frac{1}{2}\sigma^2(x_j - 1)^2\frac{u_{j+1}^{n+1} - 2u_j^{n+1} + u_{j-1}^{n+1}}{\Delta x^2}, \\ -(r - \gamma - r_c)\frac{u_{j+1}^{n+1} + u_j^{n+1}}{\Delta x} + \frac{1}{2}\sigma^2(x_j - 1)^2\frac{u_{j+1}^{n+1} - 2u_j^{n+1} + u_{j-1}^{n+1}}{\Delta x^2}\end{array}\right\} = 0.$$

$j = 2, 3, ..., j_{\max} - 1$, $n = 1, 2, ...$

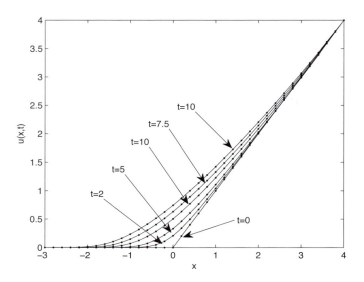

FIGURE 15.16
Numerical solutions of the Hamilton-Jacobi-Bellman equation at differ-ent times for $\Delta x = 0.2$, $\Delta t = 0.0025$, and $T_{\max} = 10$.

EXERCISE SET 15.4

1. Solve Burger's equation in Example 15.6 using the implicit scheme (15.29) described above.

2. Solve the reaction-diffusion equation in Example 15.7 using the implicit scheme (15.32) described above.

3. Solve Hamilton-Jacobi-Bellman's equation in Example 15.9 using the implicit scheme (15.37) described above.

15.5   INTRODUCTION TO THE FINITE-ELEMENT METHOD

Numerical methods typically involve approximations in a finite-dimensional set-ting. There are many techniques available for deriving these approximations. The method to be described below involves the approximation of functions such as the solution to differential equations by finite-linear combinations of simple, linearly in-dependent functions. These linear combinations of the simple or basis functions are thought of as generating a space of functions and the methods that we now describe belong to a family of so-called **projection methods**. There are several names associated with these methods including those of Ritz, Galerkin, finite elements, and others. They are ultimately based on the following theorem and corollaries thereof.

15.5.1 Theory

THEOREM 15.1 (Fundamental Lemma of the Calculus of Variations)
Suppose $M(x)$ is a continuous function defined on the interval $a \le x \le b$. Suppose further that for every continuous function, $\zeta(x)$,

$$\int_a^b M(x)\zeta(x)dx = 0.$$

Then

$$M(x) = 0 \ \ for \ all \ \ x \in [a, b].$$

Proof Suppose $M(x)$ is not zero at some point $x_0 \in (a, b)$. Suppose for definiteness that $M(x_0) > 0$. Then by continuity there is a $\delta > 0$ such that

$$\frac{M(x_0)}{2} < M(x) - M(x_0) < \frac{M(x_0)}{2} \ \ \text{for} \ \ |x - x_0| < \delta \ \ \text{with} \ \ x \in [a, b].$$

Thus, $M(x) > M(x_0)/2$ in that interval. Now choose $\zeta(x)$ such that

$$\zeta(x) = \begin{cases} 0 & \text{if} \ a \le x \le a_1 = \max(x_0 - \delta, a) \\ > 0 & \text{if} \ |x - x_0| < \delta, \ x \in [a, b] \\ 0 & \text{if} \ \min(x_0 + \delta, b) = b_1 \le x \le b. \end{cases}$$

See the Figure 15.17. Then

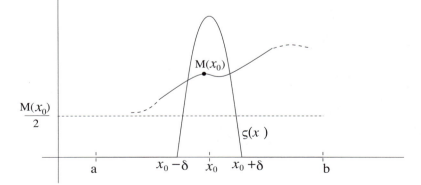

FIGURE 15.17

$$0 = \int_a^b M(x)\zeta(x)dx = \int_{a_1}^{b_1} M(x)\zeta(x)dx > \frac{1}{2}M(x_0) \int_{a_1}^{b_1} \zeta(x)dx > 0,$$

a contradiction.

If $M(x_0) < 0$, redo the argument with $-M(x)$ instead of $M(x)$. The cases where $x_0 = a$ or $x_0 = b$ follow in the same way with only minor modifications. This proves the theorem.

COROLLARY 15.1

The result of Theorem 15.1 holds if one requires in addition that

$$\zeta(a) = \zeta(b) = 0.$$

Theorem 15.1 and the corollary admit the following generalization:

Suppose $I = [a, b]$ is a bounded interval. In the sequel we shall only deal with bounded intervals. Suppose further that $\varphi_1(x), \varphi_2(x), \ldots$ is a sequence of linearly independent functions such that every continuous function can be approximated arbitrarily closely by linear combinations of the functions $\{\varphi_n\}$. More precisely, if $f(x)$ is a continuous function on I and $\epsilon > 0$ is given, then there are constants $a_1, a_2, \ldots$ and an N such that

$$\left| f(x) - \sum_{j=1}^{N} a_j \varphi_j(x) \right| < \epsilon \tag{15.39}$$

for all $x \in I$. For our purposes, it is not necessary to know how to do this but only that it can in principle be done. Such a set $\{\varphi_n\}$ will be called a **basis set**. An example of such a basis set is the set of powers of x, that is, $\varphi_n(x) = x^n$, $n = 0, 1, \ldots$. The assertion (15.39) is then called the Weierstrass approximation theorem.

COROLLARY 15.2

Suppose $M(x)$ is continuous on the interval $I = [a, b]$ and suppose $\{\varphi_n(x)\}_{n=1}^{\infty}$ is a set of basis functions. Suppose, moreover, that

$$\int_a^b M(x) \varphi_n(x) dx = 0 \ \ for \ \ n = 1, 2, \ldots.$$

Then $M(x) = 0$ for all $x \in [a, b]$.

Proof   Let K be a constant such that $|M(x)| \leq K$ for all $x \in [a, b]$ and let $\zeta(x)$ be an arbitrary continuous function. Let $\epsilon > 0$ be given and let us approximate ζ by

$$\zeta(x) = \sum_{k=1}^{N} \alpha_k \varphi_k(x) + r_N(x)$$

where $|r_N(x)| < \epsilon$. Then

$$\int_a^b M(x)\zeta(x)dx = \int_a^b M(x) \sum_{k=1}^{N} \alpha_k \varphi_k(x)dx + \int_a^b M(x)r_N(x)dx$$

$$= \int_a^b M(x)r_N(x)dx.$$

Hence

$$\left| \int_a^b M(x)\zeta(x) \right| \le \int_a^b |M(x)||r_N(x)|dx < K(b-a)\epsilon$$

for any $\epsilon > 0$. Thus,

$$\int_a^b M(x)\zeta(x)dx = 0$$

and by Theorem 15.1, $M(x) = 0$ for all $x \in [a,b]$, which proves the corollary.

Finally, let us remark that the continuity requirement on the ζ's in these results can be replaced by more stringent differentiability requirements. The theorem and corollaries are still true if we require ζ to be piecewise differentiable, or once differentiable, or even infinitely often differentiable. Thus, we have great flexibility in our choice of basis functions. An additional remark here is also in order. We usually only choose a finite number of basis functions. Any such set can be extended to a set of basis functions, but again we do not have to do that but only to realize that it can in principle be done. To get an idea how we can implement these ideas and obtain an algorithm, let us approximate the solution to the boundary problem

$$y'' - y = 0, \quad y(0) = 0, \ y(1) = 1. \tag{15.40}$$

The solution to the problem is, of course, $y = \sinh x / \sinh 1$.

EXAMPLE 15.10
Chose the basis functions $1, x, x^2, x^3, \dots$.

For the purpose of the calculation, take

$$y_3 = a_0 + a_1 x + a_2 x^2 + a_3 x^3.$$

y_3 must satisfy the boundary conditions

$$y_3(0) = a_0 = 0 \quad \text{and} \quad y_3(1) = a_1 + a_2 + a_3 = 1$$

and

$$y_3'' - y_3 = 2a_2 - a_0 + (6a_3 - a_1)x - a_2 x^2 - a_3 x^3.$$

There are four coefficients to be chosen. Two of them are chosen to satisfy the boundary conditions. Choose the next two by demanding that

$$\int_0^1 (y_3'' - y_3)x^k dx = \int_0^1 [2a_2 + (6a_3 - a_1)x - a_2 x^2 - a_3 x^3]x^k dx = 0$$

for $k = 0, 1$. We find the system

$$a_0 = 0$$
$$a_1 + a_2 + a_3 = 1$$
$$-6a_1 + 20a_2 + 33a_3 = 0$$
$$-20a_1 + 45a_2 + 108a_3 = 0$$

with the solution $a_0 = 0.0$, $a_1 = 0.8511980$, $a_2 = -0.0151324$, $a_3 = 0.1639344$ so that the approximation is

$$y_3(x) = 0.851198x - 0.0151324x^2 + 0.1639344x^3.$$

Comparing the results with the time solution at the points $x = 0.25$, $x = 0.5$, $x = 0.75$, we find

x	$y_3(x)$	$y(x)$	% error
0.25	0.2144152	0.2149524	0.25
0.50	0.4423077	0.4434094	0.25
0.75	0.6990464	0.6997241	0.10

EXAMPLE 15.11

Let us try to approximate the solution to (15.40) with the same philosophy but another set of basis functions.

Observe also that the function $u(x) = x$ satisfies the boundary conditions so let us try the approximation

$$y_3(x) = x + a_1 \sin(\pi x) + a_2 \sin(2\pi x) + a_3 \sin(3\pi x).$$

$y_3(x)$ satisfies the boundary conditions $y_3(0) = 0$, $y_3(1) = 1$. The basis functions are $\{\sin n\pi x\}$, $n = 1, 2, \ldots$, of which we take just three. Then

$$y_3'' - y_3 = -x - (\pi^2 + 1)a_1 \sin(\pi x) - (4\pi^2 + 1)a_2 \sin(2\pi x)$$
$$- (9\pi^2 + 1)a_3 \sin(3\pi x).$$

Choose a_1, a_2, a_3 so that

$$\int_0^1 \left[(y_3'' - y_3) \sin(n\pi x) \right] dx = 0, \quad n = 1, 2, 3.$$

We find

$$a_1 = -\frac{2}{\pi(\pi^2 + 1)} = -0.0585688$$

$$a_2 = \frac{1}{\pi(4\pi^2 + 1)} = 0.0078637$$

$$a_3 = -\frac{2}{3\pi(9\pi^2 + 1)} = -0.0023624$$

so the approximation is

$$y_3(x) = x - 0.0585688 \sin(\pi x) + 0.0078637 \sin(2\pi x) - 0.0023624 \sin(3\pi x).$$

Again, the comparison with the time solution is

x	$y_3(x)$	$y(x)$	% error
0.25	0.2147788	0.2149524	0.08
0.50	0.4437936	0.4434094	0.09
0.75	0.6990514	0.6997241	0.10

These approximations are quite crude and yet the numerical values are fairly accurate. They have, moreover, the feature that they automatically approximate the solution over the whole interval in question so that no interpolation between individual computer points is necessary as is the case for methods based on finite difference methods. The values given by the two approximations are slightly but not appreciably different. That is actually to be expected.

Now let us consider the general procedure. Consider the differential equation

$$Ly = (p(x)y')' + q(x)y' + r(x)y = f(x) \tag{15.41}$$

together with the boundary conditions

$$y(a) = A, \quad y(b) = B. \tag{15.42}$$

Let

$$\ell(x) = A\frac{b-x}{b-a} + B\frac{x-a}{b-a}, \quad a \le x \le b. \tag{15.43}$$

Reduce the boundary conditions to zero by setting

$$y(x) = u(x) + \ell(x) \tag{15.44}$$

so that $u(x)$ satisfies the differential equation

$$Lu = f(x) - L\ell(x) \equiv F(x) \tag{15.45}$$

and the boundaries

$$u(a) = 0, \quad u(b) = 0. \tag{15.46}$$

Note that if u satisfies (15.45), it also satisfies for any ζ with $\zeta(a) = \zeta(b) = 0$ the identity

$$0 = \int_a^b \zeta[Lu - F(x)]dx = \int_a^b [-p(x)u'\zeta' + q(x)u'\zeta + r(x)u\zeta - F(x)\zeta]dx.$$

The converse is also under mild, natural assumptions which is what we need for developing our algorithm. Let $\varphi_1(x), \varphi_2(x), \ldots$ be a basis with $\varphi_j(a) = \varphi_j(b) = 0, \quad j = 1, 2, \ldots$ and let

$$u_n(x) = \sum_{j=1}^n a_j \varphi_j(x) \tag{15.47}$$

be the approximation to the solution, $u(x)$. The approximate solution to $y(x)$ will be

$$y_n(x) = u_n(x) + \ell(x). \tag{15.48}$$

The general algorithm has the form

$$\sum_{j=1}^{n} \left\{ \int_a^b [p(x)\varphi_j'(x)\varphi_i'(x) - q(x)\varphi_j'(x)\varphi_i'(x) - r(x)\varphi_j(x)\varphi_i(x)]dx \right\} a_j$$

$$= - \int_a^b F(x)\varphi_i(x)dx, \quad i = 1, \dots, n. \tag{15.49}$$

The system (15.49) is a linear system of equations for the n unknowns, $a_1, \dots, a_n$.

The disadvantage of the above method is that finding good basis functions $\{\varphi_n(x)\}$ is not easy, and polynomials which, are the usual choice, may interpolate poorly. In the next section we shall give special basis functions that lead to efficient and popular algorithms for solving such systems.

EXERCISE SET 15.5.1

1. Use the methods of this section to approximate the solution to

$$y'' + y = 3x^2, \quad y(0) = 0, \ y(2) = 3.5.$$

   For basis functions, take $n = 2$ and $\varphi_1(x) = x(x-2)$, $\varphi_2(x) = x^2(x-2)$.(Note that $u(x) = 7x/4$ satisfies the boundary conditions)

2. Do Exercise 1 using a shooting method. Compare the results. The exact solution is $6\cos x + 3(x^2 - 2)$.

3. Use the methods of this section to approximate the solution to

$$y'' + y^2 = x, \quad y(0) = y(1) = 0.$$

   Let $n = 1$ and for a basis function take $\varphi_1(x) = x\sin\pi x$. This is a nonlinear problem but the technique still works.

4. Use the methods of this section to approximate the solution to

$$y'' + y^3 = \cos(8\pi x), \quad y(0) = y(1) = 0.$$

   Let $n = 3$ and for basis functions let

$$\varphi_1(x) = \sin\pi x, \ \varphi_2(x) = \sin 2\pi x, \ \varphi_3(x) = \sin 3\pi x.$$

   Hint: You will have to use the Newton-Raphson method to solve the nonlinear system.

5. The solution to

$$((x+1)y')' - (x+2)y = \frac{x+2}{e-1}, \quad y(0) = 0, \ y(1) = 1$$

is $y = \dfrac{e^x - 1}{e - 1}$.

(a) Use the methods of this section with $n = 5$, $\varphi_j(x) = \sin(j\pi x)$, $j = 1, \ldots, 5$ to find an approximation, $y_5(x)$, to the solution $y(x)$.

(b) Use the methods of this section with $n = 10$ and again $\varphi_j(x) = \sin(j\pi x)$, $j = 1, \ldots, 10$, to find an approximation, $y_{10}(x)$, to the solution.

(c) Use the shooting methods based on Newton's method to calculate an approximate solution, $y_5(x)$. Divide the interval up into twenty subintervals so that $h = 1/20$.

(d) Compare the values of the approximate solutions with the true solution in the points $x_1 = .25$, $x_2 = .5$, $x_3 = .75$, i.e., calculate $y(x_j)$, $y_5(x_i)$, $y_{10}(x_j)$, $y_{15}(x_j)$, $j = 1, 2, 3$ and the quantities.

$$|y(x_j) - y_5(x_j)|, (|y(x_j) - y_5(x_j)|/|y(x_j)|) \times 100, \text{ etc.}$$

6. The solution to

$$y'' - 8(e^x + x^2)y = x, \quad y(0) = y(1) = 0$$

is unknown. Again take as basis function $\varphi_k(x) = \sin(k\pi x)$, $k = 1, 2, \ldots$. Use the methods of this section to construct a sequence, $\{y_n(x)\}$, of approximations to the solution $y(x)$. Take n sufficiently large that you are convinced the approximation is good. How large is large? Why do you think you are right? Graph the solution you think is a good approximation. Note: You will have to give a criterion telling what you mean by a good approximation.

7. (Theoretical Exercise). Suppose $M(x)$ is defined and continuous on an infinite interval, I, (i.e., on $(-\infty, a]$ or $[b, \infty)$ or $(-\infty, \infty)$). Suppose further that $\displaystyle\int_I M(x)\zeta(x)\,dx = 0$ for all continuous functions, $\zeta(x)$, such that $\zeta(x) = 0$ for all x outside a bounded interval J in I. J depends on the particular ζ and will in general be different for different functions. Then $M(x) = 0$ for all $x \in I$.

15.5.2   The Finite-Element Method

The theoretical Ritz-Galerkin method is a powerful one. The approximations are very good and the method is easy to implement. Unfortunately, most basis functions, lead to full matrices, that must be inverted to obtain the approximation. Moreover, to obtain better and better approximations, one is forced to choose more basis functions which in turn leads to even larger matrices. The question, therefore, arises whether there are not basis functions which lead to sparse matrices, that is matrices that are made up of a great many zeros such as band matrices. The answer to that question is yes and we now take up such methods.

The finite-element method is simply the Ritz-Galerkin method, where the finite set of basis functions $\varphi_1(x), \ldots, \varphi_n(x)$ are splines. We shall restrict ourselves to the splines of the following form: Graphically, for $h > 0$, or more explicitly

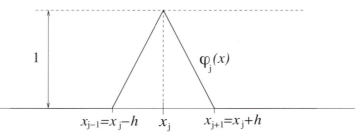

FIGURE 15.18
Hat function.

$$
\varphi_j(x) = \begin{cases}
0, & x \leq x_{j-1} \\
\dfrac{x - x_{j-1}}{h}, & x_{j-1} \leq x \leq x_j \\
-\dfrac{x - x_{j+1}}{h}, & x_j \leq x \leq x_{j+1} \\
0, & x \geq x_{j+1}.
\end{cases} \tag{15.50}
$$

These continuous piecewise linear functions, which vanish identically outside the interval $[x_{j-1}, x_{j+1}]$, are often referred to as **hat functions** or **chapeau functions**. In using them in the Ritz-Galerkin scheme, we must calculate the derivative of $\varphi_j(x)$. This derivative does not exist at the points x_{j-1}, x_j, x_{j+1} but it does exist everywhere else, namely

$$
\varphi_j'(x) = \begin{cases}
0, & x < x_{j-1} \\
\dfrac{1}{h}, & x_{j-1} < x < x_j \\
-\dfrac{1}{h}, & x_j < x < x_{j+1} \\
0, & x_{j+1}
\end{cases} \tag{15.51}
$$

and $\varphi_j'(x)$ is integrable.

To see how the method is implemented, let us consider the following example.

EXAMPLE 15.12
Find an approximation to the solution to

$$
y'' - y = -x, \quad 0 < x < 1, \; y(0) = y(1) = 0
$$

using the method of finite elements with the hat functions as basis functions.

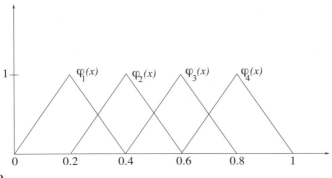

FIGURE 15.19
Hat functions for Example 15.12.

Subdivide the interval into five equal subintervals so $h = 0.2$. In this case $n = 4$. The graphs of the hat functions are
We have

$$y_4(x) = \sum_{j=1}^{4} a_j \varphi_j(x).$$

The system (15.49) has the form

$$\int_0^1 (\varphi_i' y_4' + \varphi_i y_4) dx = \int_0^1 x \varphi_i(x), \quad i = 1, 2, 3, 4,$$

or written out

$$\sum_{j=1}^{4} \left(\int_0^1 [\varphi_i'(x)\varphi_j'(x) + \varphi_i(x)\varphi_j(x)] dx \right) a_j = \int_0^1 x \varphi_i(x) dx, \quad i = 1, 2, 3, 4.$$

To set up the system, we must calculate the integrals. Note that $x_i = ih$, $i = 0, \ldots, 5$ so that first

$$\int_0^1 x \varphi_i(x) dx = \int_{x_{i-1}}^{x_i} x \varphi_i(x) dx + \int_{x_i}^{x_{i+1}} x \varphi_i(x) dx$$

$$= \int_{x_{i-1}}^{x_i} \frac{x(x - x_{i-1})}{h} dx + \int_{x_i}^{x_{i+1}} \frac{x(x_{i+1} - x)}{h} dx$$

$$= \frac{(3i - 1)h^2}{6} + \frac{(3i + 1)h^2}{6} = ih^2 = 0.04i,$$

$i = 1, 2, 3, 4$. The matrix entries require more work. We calculate only the entries for the first row.

$$\int_0^1 [\varphi_1'^2(x) + \varphi_1^2(x)] dx = \int_0^{x_1} [\varphi_1'^2(x) + \varphi_1^2(x)] dx + \int_{x_1}^{x_2} [\varphi_1'^2(x) + \varphi_1^2(x)] dx$$

$$= \int_0^{x_1} \left[\frac{1}{h^2} + \left(\frac{x}{h} \right)^2 \right] dx + \int_{x_1}^{x_2} \left[\frac{1}{h^2} + \frac{(x - x_2)^2}{h^2} \right] dx$$

$$= \frac{2h}{h^2} + \frac{2x_1^3}{3h^2} = \frac{2}{h} + \frac{2h}{3} = 10.1333333$$

to seven decimal places. Next, from the figure

$$\int_0^1 [\varphi_1'(x)\varphi_2'(x) + \varphi_1(x)\varphi_2(x)]dx = \int_{x_1}^{x_2} [\varphi_1'(x)\varphi_2'(x) + \varphi_1(x)\varphi_2(x)]dx$$

$$= \int_{x_1}^{x_2} \left[-\frac{1}{h} \cdot \frac{1}{h} - \frac{(x-x_2)^2}{h^2} \right] dx$$

$$= -\frac{1}{h} - \frac{h}{3} = -5.066667.$$

Obviously,

$$\int_0^1 [\varphi_1'(x)\varphi_3'(x) + \varphi_1(x)\varphi_3(x)]dx = \int_0^1 [\varphi_1'(x)\varphi_4'(x) + \varphi_1(x)\varphi_4(x)]dx = 0.$$

The other integrals are calculated in the same way. We therefore arrive at the system

$$\begin{bmatrix} 10.1333333 & -5.0666667 & 0 & 0 \\ -5.066667 & 10.1333333 & -5.0666667 & 0 \\ 0 & -5.0666667 & 10.1333333 & -5.0666667 \\ 0 & 0 & -5.06666667 & 10.1333333 \end{bmatrix} \begin{bmatrix} a_1 \\ a_2 \\ a_3 \\ a_4 \end{bmatrix} = \begin{bmatrix} .04 \\ .08 \\ .12 \\ .16 \end{bmatrix}.$$

We solve this by Gauss elimination to obtain

$$a_1 = .0315790 \quad a_2 = .0526318 \quad a_3 = .0631579 \quad a_4 = .0473684.$$

The true solution is $y(x) = x - \sinh x / \sinh 1$. Let us compare the approximate solution with the values of the true solution at three interior points of the interval.

x	$y_4(x)$	$y(x)$	absolute error	%error
0.2	0.0315790	0.0286795	0.0028995	10.11
0.5	0.0578949	0.0565906	0.0013043	2.30
0.8	0.0473684	0.0442945	0.0030739	6.94

This example points up several advantages of the finite element method. First, it leads to sparse matrices. In the case of the hat functions, these matrices are tridiagonal. One can continually refine the size of the mesh spacing and get better and better approximations. Setting up the coefficient matrix will, of course, take more time, but the resulting system remains easy to solve.

We can also take more general splines and obtain a finite element method corresponding to that set of basis functions. However, we shall be satisfied with the hat functions. If more accuracy is needed, refine the mesh.

In summary, we proceed as follows. To solve the boundary-value problem

$$(p(x)y')' + q(x)y' + r(x)y = f(x), \quad y(a) = A, \ y(b) = B, \tag{15.52}$$

let first

$$\ell(x) = B\frac{x-a}{b-a} + A\frac{b-x}{b-a}$$

and set $y(x) = u(x) + \ell(x)$. Then $u(x)$ satisfies the problem

$$(p(x)u')' + q(x)u' + r(x)u = F(x), \quad u(a) = 0, \ u(b) = 0$$

where

$$F(x) = f(x) - \left(p(x)\frac{B-A}{b-a}\right)' - q(x)\frac{B-A}{b-a} - r(x)\left[\frac{B(x-a)}{b-a} + \frac{A(b-x)}{b-a}\right].$$

Divide the interval $[a, b]$ up into $(n+1)$ subintervals so there will be n interior grid points given by $x_j = a + jh, \quad j = 1, \ldots, n$ where $h = (b-a)/(n+1)$. Set $x_0 = a$ and $x_{n+1} = b$. Construct the hat functions, $\varphi_j(x), \quad j = 1, \ldots, n$ and set

$$u_n(x) = \sum_{j=1}^{n} a_j \varphi_j(x), \tag{15.53}$$

where the a_j are determined by solving the linear system of equations

$$\sum_{j=1}^{n} \left\{ \int_a^b (-p(x)\varphi_i'(x)\varphi_j'(x) + q(x)\varphi_i(x)\varphi_j'(x) + r(x)\varphi_i(x)\varphi_j(x))dx \right\} a_j$$

$$= \int_a^b F(x)\varphi_i(x)dx, \tag{15.54}$$

$i = 1, \ldots, n$. Once the a_j have been determined from (15.54), the approximation, $y_n(x)$, to the solution, $y(x)$ of (15.52), is given by

$$y_n(x) = u_n(x) + \ell(x). \tag{15.55}$$

Setting up the coefficient matrix for the a_j can be time-consuming if the differential equation is complicated, but that procedure should be automated.

EXERCISE SET 15.5.2

In the following exercises, use hat functions for the approximations.

1. The solution to

$$y'' + xy = x^3 - 4/x^3, \quad y(1) = -1, \ y(2) = 3$$

is $y = x^2 - 2/x$.

(a) Let $n = 3$ and find $y_3(x)$.

(b) Graph $y_3(x)$ and $y(x)$ on the same axes.

(c) Compare the approximations with the true solution at the points $x = 1.25$, $x = 1.5$, and $x = 1.75$ by computing both the absolute errors and percent errors at those points.

2. Use the shooting method to solve Exercise 1 with $h = .25$ and compare these results with those obtained in Exercise 1 with finite elements.

3. Solve the problem

$$y'' - x^2 y = e^{2x}, \quad y(0) = y(1) = 0,$$

by both the finite-element method and the shooting method. Let first $n+1 = 10$ so $h = 0.1$ and then let $n+1 = 20$ so $h = 0.05$. Which method do you think gives the best answers? Why?

4. Throughout these last two sections, we have tacitly assumed there exists a unique solution. That is not always the case. Consider the problem

$$y'' + y = 0, \quad y(0) = y(\pi) = 0.$$

There are infinitely many solutions, all of the form: $y = C \sin x$, where C is an arbitrary constant. Nevertheless, use the finite-element method to solve this problem. Take larger and larger values of n. Recall the theory of eigenvalues for matrices.

5. Use the finite-element method to find an approximation for the smallest eigenvalue and an approximation to the corresponding eigenfunction of

$$y'' - xy + \lambda y = 0, \quad y(0) = y(1) = 0.$$

APPLIED PROBLEMS FOR CHAPTER 15

1. Consider a homogeneous iron rod 100 cm long that is heated to the initial temperature distribution $f(x) = 100 - x$, $0 \le x \le 100$. At time $t = 0$ the lateral surface is insulated, and the left end of the rod is plunged into a reservoir of oil kept at 30^0C, while the right end remains at the constant temperature of 70^0C. The thermal diffusivity α of iron is $0.15 \ cm^2/\text{sec}$. Use $h = 10$ in. and an appropriate value of k so that $\lambda = \alpha 2k/h^2 = 0.5$. Carry on the solution for five time steps.

2. A large steel plate 2 cm thick is heated according to the initial temperature distribution $f(x) = x$ if $0 \le x \le 1$ and $f(x) = 2 - x$ if $1 \le x \le 2$. Both faces are maintained at 0^0C. The thermal diffusivity α of the steel is $0.15 \ cm^2/\text{sec}$. Assume that the lateral flow of heat relative to the flow perpendicular to the faces is neglected. Use the heat equation with $h = 0.25$ in. and an appropriate value of k so that $\lambda = \alpha 2k/h^2 = 0.5$ to carry on the solution for five time steps.

3. Solve the wave equation

$$u_{xx} = u_{tt}, \quad 0 < x < 4, \quad t > 0;$$

Subject to the boundary conditions

$$u(0, t) = 0, \quad u(4, t) = 0 \quad t > 0;$$

and the initial conditions

$$u(x, 0) = 0, \quad 0 \le x \le 4$$

$$\frac{\partial u(0, 0)}{\partial t} = \begin{cases} 0 & 0 \le x < 1 \\ 4 & 1 \le x \le 1.5 \\ 0 & 1.5 < x < 3 \\ -4 & 3 \le x \le 3.5 \\ 0 & 3.5 < x \le 4 \end{cases}.$$

The total energy in the string with pinned ends at $x = 0$ and $x = 4$ is made up of the kinetic plus the potential energy and is at any time t given by

$$E(t) = \frac{1}{2} \int_0^4 \left[\left(\frac{\partial u(x, t)}{\partial t} \right)^2 + \left(\frac{\partial u(x, t)}{\partial x} \right)^2 \right] dx.$$

If initially $u(x, 0) = f(x)$ and $\partial u(x, 0)/\partial t = g(x)$, the initial energy is

$$E(0) = \frac{1}{2} \int_0^4 \left[(g(x))^2 + (f'(x))^2 \right] dx.$$

The energy is conserved if $E(t) = E(0)$ for all t.

Bibliography and References

[1] AKAI, T.J., *Applied Numerical Methods for Engineers*, John Wiley & Sons, New York, 1993.

[2] ALEFRED, G., and J. HERZBERGER, *Introduction to Interval Computations*, Academic Press, New York, 1983.

[3] AYYUB, B.M., and R.H. McCUEN, *Numerical Methods for Engineers*, Prentice-Hall, Upper Saddle River, NJ, 1996.

[4] BIRKHOFF, G., and G.C. ROTA, *Ordinary Differential Equations*, John Wiley & Sons, New York, 1978.

[5] BRACEWELL, R.N., *The Fourier Transform and Its Applications*, McGraw-Hill, New York, 1978.

[6] BRENT, R.P., *Algorithms for Minimization without Derivatives*, Prentice-Hall, Englewood Cliffs, NJ, 1973.

[7] BURDEN, R.L., and J.D. FAIRS, *Numerical Analysis*, Fourth Ed. PWS-KENT Publishing Comp., Boston, MA., 1988.

[8] CARNAHAN, B., H.A. LUTHER, and J.O. WILKES, *Applied Numerical Methods*, John Wiley & Sons, New York, 1969.

[9] CHAPRA, S.C., *Numerical Methods for Engineers: With Personal Computer Applications*, McGraw-Hill, New York, 1985.

[10] CHENEY, W., and D. KINCAID, *Numerical Mathematics and Computing*, 4th ed. Brooks/Cole Publishing Comp., New York, 1999.

[11] DE BOOR, C., and S.D. CONTE, *Elementary Numerical Analysis*, McGraw-Hill, New York, 1980.

[12] DAHLQUIST, G., and A. BJORCK, *Numerical Methods*, Prentice-Hall, Englewood Cliffs, NJ, 1974.

[13] DATTA, B.N., *Numerical Linear Algebra and Applications*, Brooks/Cole, Pacific Grove, CA, 1995.

[14] DAVIS, P.J., and P. RABINOWITZ, *Methods of Numerical Integration*, Academic Press, New York, 2nd ed., 1984.

[15] ELJINDI, M., and A. KHARAB, *The Quadratic Method for Computing the Eigenpairs of a Matrix*, IJCM 73, 530, 2000.

[16] EVANS, G., *Practical Numerical Analysis*, John Wiley & Sons, Chichester, England, 1995.

[17] FAUSETT, L.V., *Applied Numerical Analysis Using MATLAB*, Prentice-Hall, Upper Saddle River, NJ, 1999.

[18] FORSYTHE, G.E., M.A. MALCOLM, and C.B. MOLER, *Computer Methods for Mathematical Computations*, Prentice-Hall, Englewood Cliffs, NJ, 1977.

[19] GERALD, C.F., and P.O. WHEATLEY, *Applied Numerical Analysis*, Addison-Wesley, Reading, MA, 1989.

[20] GILL, P.E., W. MURRAY, and M.H. WRIGHT, *Numerical Linear Algebra and Optimization*, volume 1, Addison-Wesley, Redwood City, CA, 1991.

[21] GOLUB, G., and C. VAN LOAN, *Matrix Computations*, Johns Hopkins Press, Baltimore, 1983.

[22] GOLUB, G., and J. M. ORTEGA, *Scientific Computing and Differential Equations: An Introduction to Numerical Methods*, Academic Press, Inc., Boston, 1992.

[23] GREENSPAN, D., and V. CASULLI, *Numerical Analysis for Applied Mathematics, Science and Engineering*, Addison-Wesley, New York, 1988.

[24] HAGEMAN, L.A., and D.M. YOUNG, *Applied Iterative Methods*, Academic Press, New York, 1981.

[25] HAGER, W.W., *Applied Numerical Linear Algebra Methods*, Prentice-Hall, Englewood Cliffs, NJ, 1988.

[26] HANSELMAN, D., and B. Littlefield, *Mastering MATLAB 5: A Comprehensive Tutorial and Reference*, Prentice-Hall, Upper Saddle River, NJ, 1998.

[27] HOFFMAN, J.D., *Numerical Methods for Engineering and Scientists*, McGraw-Hill, New York, 1992.

[28] HEATH, M.T., *Scientific Computing: An Introductory Survey*, McGraw-Hill, New York, 1997.

[29] HORN, R.A., and C.R. JOHNSON, *Matrix Analysis*, Cambridge University Press, Cambridge, 1985.

[30] HULTQUIST, P.F., *Numerical Methods for Engineers and Computer Scientists*, Benjamin/Cummings Publishing Comp., CA, 1988.

[31] INCROPERA, F.P., and D.P. DeWITT, *Introduction to Heat Transfer*, Wiley, New York, 2nd ed., 1990.

[32] ISAACSON, E., and H.B. KELLER, *Analysis of Numerical Methods*, Wiley, New York, 2nd ed., 1990.

[33] JOHNSTON, R.L., *Numerical Methods: A Software Approach*, John Wiley & Sons, New York, 1982.

[34] KAHANER, D., C. MOLER, and S. NASH, *Numerical Methods and Software*, Prentice Hall, Englewood Cliffs, NJ, 1989.

[35] KERNIGHAN, B.W., and R. PIKE, *The Practice of Programming*, Addison-Wesley, Reading, MA, 1999.

[36] KINGS, J.T., *Introduction to Numerical Computation*, McGraw-Hill, New York, 1984.

[37] LAMBERT, J.D., *Numerical Methods for Ordinary Differential Systems*, John Wiley & Sons, Chichester, 1991.

[38] LAMBERT, J.D., *The Initial Value Problem for Ordinary Differential Equations*, The state of the art in numerical analysis, D. Jacobs, editor. Academic Press, New York, 1977.

[39] LAWSON, C. L., and R.J. HANSON, *Solving Leclsi-Squares Problems*, Prentice Hall, Englewood Cliffs, NJ, 1974.

[40] LINDFIELD, G.R., and J.E.T. PENNY, *Microcomputers in Numerical Analysis*, Ellis Horwood, Chichester, 1989.

[41] MARCHAND, P., *Graphics and GUIs with MATLAB*, CRC Press, Boca Raton, FL, 3rd ed., 2008.

[42] MARON, M.J., and R.J. LOPEZ, *Numerical Analysis: A Practical Approach*, 3rd ed. Wadsworth, Belmont, CA, 1991.

[43] MATHEWS, H., and K. D. FINK, *Numerical Methods Using MATLAB*, Prentice Hall, Upper Saddle River, NJ, 3rd ed., 1999.

[44] MATHEWS, J.H., *Numerical Methods for Mathematics, Science, and Engineering*, 2nd ed. Prentice-Hall International, Englewood Cliffs NJ, 1992.

[45] McNEARY, S.S., *Introduction to Computational Methods for Students of Calculus*, Prentice-Hall, Englewood Cliffs. NJ, 1973.

[46] MILLER, W., *The Engineering of Numerical Software*, Prentice Hall, Englewood Cliffs, NJ, 1984.

[47] MILLER, W., *A Software Tools Sampler*, Prentice-Hall, Englewood Cliffs, NJ, 1987.

[48] MOOR, R.V., *Interval Analysis*, Prentice-Hall, Englewood Cliffs, NJ, 1966.

[49] MORRIS, J.L., *Computational Methods in Elementary Theory and Application of Numerical Analysis*, John Wiley & Sons, New York, 1983.

[50] ORTEGA, J.M., and W.G. POOLE, Jr., *An Introduction to Numerical Methods for Differential Equations*, Pitman Press, Marshfield, MA, 1981.

[51] POLKING, J.C., *Ordinary Differential Equations Using MATLAB*, Prentice-Hall, Englewood Cliffs, NJ, 1995.

[52] RAMIREZ, R.W., *The FFT, Fundamentals and Concepts*, Prentice-Hall, Englewood Cliffs, NJ, 1985.

[53] RATSCHEK, H., *Uber Einige Intervallarithmetische Grundbegriffe*, Computing 4, 43, 1969.

[54] RICE, J.R., *Numerical Methods, Software, and Analysis*, IMSL Reference Edition, McGraw-Hill, New York, 1983.

[55] SCHIAVONE P., C. COUTANDA, and A. MIODUCHAWSKI, *Integral Methods in Science and Engineering*, Birkhauser, Boston, 2002.

[56] SCHWARTZ, H.R., *Numerical Andlysis: A Comprehensive Introduction*, Wiley, New York, 1989.

[57] SHAMPINE, L.F., R.C. ALLEN, and S. PRUESS Jr., *Fundamentals of Numerical Computing*, John Wiley & Sons, New York, 1997.

[58] SHAMPINE, L., and M.W. REICHELT, *The MATLAB ODE suite*, SIAM Journal on Scientific Computing 18(1), pp. 1-22, Jan 1997.

[59] SHINGAREVA, I., and C. LIZARRAGA-CELAYA, *Solving Nonlinear Partial Differential Equations with Maple and Mathematica*, Springer, Wien, New York, 2011.

[60] STOER, J., and R. BULIRSCH, *Introduction to Numerical Analysis*, Springer-Verlag, New York, 1993.

[61] The MATHWORKS, Inc., *Using MATLAB*, The Mathworks, Inc., Natick, MA, 1996.

[62] The MATHWORKS, Inc., *Using MATLAB Graphics*, The Mathworks, Inc., Natick, MA, 1996.

[63] The MATHWORKS, Inc., *MATLAB Language Reference Manual*, The Mathworks, Inc., Natick, MA, 1996.

[64] TOURIN, A., *An Introduction to Finite Difference Methods for PDEs in Finance*, The Fields Institute, Toronto, 2010.

[65] TRICOMI, F.G., and C.H.H. BAKER, *Treatment of Inegral Equations by Numerical Methods*, Birkhauser, Boston, 2002.

[66] VARGA, R.S., *Matrix Iterative Analysis*, Prentice-Hall, Englewood Cliffs, NJ, 1962.

Appendix A

Calculus Review

We assume that the reader is familiar with topics normally covered in the under-graduate calculus sequence. For purpose of reference this appendix gives a summary of the relevant results from calculus on limits, continuity, differentiation, and integration. The proof of these results can be found in almost any calculus book.

A.1   Limits and continuity

DEFINITION A.1   *Let a function f be defined in an open interval and L be a real number. We write*

$$\lim_{x \to a} f(x) = L$$

if for every $\epsilon > 0$, there is a $\delta > 0$ such that

$$\text{if} \quad 0 < |x - a| < \delta, \quad \text{then} \quad 0 < |f(x) - L| < \epsilon.$$

DEFINITION A.2   *A function f is continuous at $x = a$ if it satisfies the following three conditions:*

(a) f *is defined at* $x = a$.

(b) $\lim_{x \to a} f(x)$ *exists.*

(c) $\lim_{x \to a} f(x) = f(a)$.

THEOREM A.1
A polynomial function f is continuous at each point of the real line.

THEOREM A.2
Let f be a continuous function on the interval $[a, b]$. Then $f(x)$ assumes its maximum and minimum values on $[a, b]$; that is, there are numbers $x_1, x_2 \in [a, b]$ such that

$$f(x_1) \leq f(x) \leq f(x_2)$$

for all $x \in [a, b]$.

A.2 Differentiation

DEFINITION A.3   *The derivative of a function f is the function f' defined by*

$$f'(x) = \lim_{h \to 0} \frac{f(x+h) - f(x)}{h}$$

provided the limit exists.

For example, let $f(x) = \sqrt{x}$, then if $x > 0$

$$
\begin{aligned}
f'(x) &= \lim_{h \to 0} \frac{\sqrt{x+h} - \sqrt{x}}{h} \\
&= \lim_{h \to 0} \frac{\sqrt{x+h} - \sqrt{x}}{h} \cdot \frac{\sqrt{x+h} + \sqrt{x}}{\sqrt{x+h} + \sqrt{x}} \\
&= \lim_{h \to 0} \frac{1}{\sqrt{x+h} + \sqrt{x}} = \frac{1}{2\sqrt{x}}.
\end{aligned}
$$

THEOREM A.3
A polynomial function f is differentiable at each point of the real line.

THEOREM A.4
If a function f is differentiable at $x = a$, then f is continuous at a.

A.3 Integration

THEOREM A.5 (Fundamental Theorem for Calculus)
Suppose that f is continuous on the closed interval $[a, b]$.

Part I: If the function F is defined on $[a, b]$ by

$$F(x) = \int_a^x f(t) \, dt,$$

then F is an antiderivative of f for x in $[a, b]$.

Part II: If G is any antiderivative of f on $[a, b]$, then

$$\int_a^b f(x) \, dx = G(b) - G(a).$$

For example,

$$\int_1^4 \sqrt{x}\, dx = \left[\frac{2}{3} x^{3/2}\right]_1^4 = \frac{2}{3}\,(4)^{3/2} - \frac{2}{3}\,(1)^{3/2} = \frac{14}{3}.$$

THEOREM A.6 (First Mean Value Theorem for Integrals)
If f is continuous on $[a, b]$, then

$$\int_a^b f(x)\, dx = f(\xi)(b - a)$$

for at least one number ξ in the open interval (a, b).

For example,

$$\int_0^3 x^2\, dx = 9 = f(\xi)(3 - 0) = 3\,\xi^2.$$

This implies that $\xi = \sqrt{3} \in (0, 3)$ satisfies the conclusion of the theorem.

THEOREM A.7 (Second Mean Value Theorem for Integrals)
If f is continuous on $[a, b]$ and if the function $g(x)$ is of constant sign in the interval (a, b), then

$$\int_a^b f(x)\, g(x)\, dx = f(\xi) \int_a^b g(x)\, dx$$

for at least one number ξ in the open interval (a, b).

THEOREM A.8
Let f be continuous on $[a, b]$ and $a \le c \le b$, then

$$\frac{d}{dx} \int_c^x f(t)\, dt = f(x)$$

for every x in $[a, b]$.

THEOREM A.9 (Integration by Parts)
Let $u(x)$ and $v(x)$ be real-valued functions with continuous derivatives. Then

$$\int u'(x)\, v(x)\, dx = u(x)v(x) - \int u(x)\, v'(x)\, dx.$$

For example,

$$\int x e^x\, dx = x e^x - \int e^x\, dx = x e^x - e^x + C.$$

Appendix B

MATLAB Built-in Functions

Listed below are some of the MATLAB built-in functions grouped by subject areas.

Special Matrices

diag	diagonal
eye	identity
hadamard	Hadamard
hankel	Hankel
hilb	Hilbert
invhilb	inverse Hilbert
linspace	linearly spaced vectors
logspace	logarithmically spaced vectors
magic	magic square
meshdom	domain for mesh points
ones	constant
pascal	Pascal
rand	random elements
toeplitz	Toeplitz
vander	Vandermonde
zeros	zero

Roots of Functions

fzero	zeros of a function
roots	roots of a polynomial

Interpolation

spline	cubic spline
table1	1-D table look-up
table2	2-D table look-up
interp1	interpolation 1-D

Numerical Integration

quad	numerical function integration
quadl	numerical function integration

Differential Equation Solution

ode23	2nd and 3rd order Runge-Kutta method
ode45	4th and 5th order Runge-Kutta-Fehlberg method
ode15s	Numerical function of ODE
ode113	Numerical function of ODE
ode23s	Numerical function of ODE
ode23t	Numerical function of ODE

Graph Paper

plot	linear X-Y plot
loglog	loglog X-Y plot
semilogx	semi-log X-Y plot
semilogy	semi-log X-Y plot
polar	polar plot
mesh	3-dimensional mesh surface
contour	contour plot
meshdom	domain for mesh plots
bar	bar charts
stairs	stairstep graph
errorbar	add error bars

Elementary Math Functions

abs	absolute value or complex magnitude
angle	phase angle
sqrt	square root
real	real part
imag	imaginary part
conj	complex conjugate
round	round to nearest integer
fix	round toward zero
floor	round toward $-\infty$
ceil	round toward ∞
sign	signum function
rem	remainder
exp	exponential base e
log	natural logarithm
log10	log base 10

Text and Strings

abs	convert string to ASCII values
eval	evaluate text macro
num2str	convert number to string
int2str	convert integer to string
setstr	set flag indicating matrix is a string
sprintf	convert number to string
isstr	detect string variables
strcomp	compare string variables
hex2num	convert hex string to number

Control Flow

if	conditionally execute statements
elseif	used with if
else	used with if
end	terminate if, for, while
for	repeat statements a number of times
while	do while
switch	switch expression
break	break out of for and while loops
return	return from functions
pause	pause until key pressed

Programming and M-files

input	get numbers from keyboard
keyboard	call keyboard as M-file
error	display error message
function	define function
eval	interpret text in variables
feval	evaluate function given by string
echo	enable command echoing
exist	check if variables exist
casesen	set case sensitivity
global	define global variables
startup	startup M-file
getenv	get environment string
menu	select item from menu
etime	elapsed time

Decompositions and Factorizations

backsub	back substitution
chol	Cholesky factorization
eig	eigenvalues and eigenvectors
inv	inverse
lu	factors from Gaussian elimination
nnls	nonnegative least squares
orth	orthogonalization
qr	orthogonal-triangular decomposition
rref	reduced row echelon form
svd	singular value decomposition

Appendix C

Text MATLAB Functions

In this appendix we list all the MATLAB functions that are supplied in this text. These functions are contained in a CD-ROM attached to the cover of the book.

MATLAB Functions	Definition	Section
abash	Adams-Bashforth of orders 2, 3, and 4	12.6
amoulton	Adams-Moulton method	12.8
bisect	Bisection method	3.1
derive	Differentiation	9.2
euler	Euler's method	12.1
explsqr	Least-squares fit, exponential	7.3.1
falsep	False position method	3.2
fibonacci	Fibonacci search method	8.2.3
finitediff	Finite-difference methods	13.1
fixed	Fixed-point iteration	3.6
fredholmTrapez	Integral equations	11.2
fredholmGN45	Integral equations	11.2b
gaussel	Gaussian elimination, scaled pivoting .	4.3
gauss_quad	Gaussian quadrature	10.4
golden	Golden section search	8.2.2
heat	Parabolic D.E., Explicit method	15.1.1
heat_crank	Parabolic D.E., Implicit method	15.1.2
hyperbolic	Hyperbolic differential equations	15.2
hyplsqr	Least-squares fit, hyperbolic	7.3.2
jacobi	Jacobi iteration	4.5.1
lagrange	Lagrange interpolation polynomial ...	5.4
laplace	Laplace's equation	15.3
linlsqr	Least-squares fit, linear	7.1
lshoot	Linear shooting method	13.2
lufact	LU decomposition	4.4

MATLAB Functions	Definition	Section
midpoint	Midpoint method	12.4
newton	Newton's method	3.4
newton_sys	Newton's method for systems	3.8
newton2	Newton's method, multiple roots	3.7
newtondd	Interpolation, Newton divided differences	5.2
ngaussel	Gaussian elimination, naive	4.2
parabint	Successive parabolic interpolation	8.2.4
polylsqr	Least-squares fit, polynomial	7.2
rk2_4	Runge-Kutta method, orders 2 and 4	12.4
romberg	Romberg integration	10.3
secant	Secant method	3.5
seidel	Gauss-Seidel iteration	4.5.2
simpson	Simpson's rule	10.2
spl1	Spline, linear	6.1
spl2	Spline, quadratic	6.2
spl3	Spline, cubic	6.3
sys_rk4	System of differential equations	12.11
trapez	Trapezoidal rule	10.1
VolterraEuler	Integral equations	11.5a
VolterraHeun	Integral equations	11.5b

Appendix D

MATLAB GUI

In this appendix we will show the use of MATLAB Graphical User Interface (GUI) to run some of the MATLAB functions of the book. The main reason we introduced GUI in this edition is make it easier for the students to run the MATLAB functions of the book. A readme file called "SUN Package readme" located in the directory "NMETH" of the CD of the book gives instructions with more details on the use of the M-functions of the book using MATLAB GUI.

D.1   Roots of Equations

We start by solving the following example:

EXAMPLE D.1
The function $f(x) = x^3 - x^2 - 1$ has exactly one zero in $[1, 2]$. Use the bisection algorithm to approximate the zero of f to within 10^{-4}.

Open up MATLAB and make sure that your current directory is NMETH (located in CD of the book). At the command window type

$>>$ sun

to bring the window shown in Figure D.1 (a). To get the results shown in Figure D.1 (b) follow the following steps:

* Select from the window menu tool.

* From the list of Chapters

> Root Finding
> Linear Equations
> Interpolation
> Integration

Ordinary Differential Equations (IVP)
Ordinary Differential Equations (BVP)
Partial Differential Equations (PDE)
Least-Squares Method
Differentiation

select Root Finding.

* Choose bisection from the pop up menu beside Algorithm.

* In the text box beside Name enter any name, for example bis1.

* Now enter your data: Function x^3-x^2-1, Region 1 to 2, and a Tolerance say
  10^(-4).

Finally, click on the button Add Set once and then choose bis1 from the pop up
menu Data set to get the results in Figure D.1 (b) and the output of the MATLAB
function bisect.m in the main window of MATLAB.

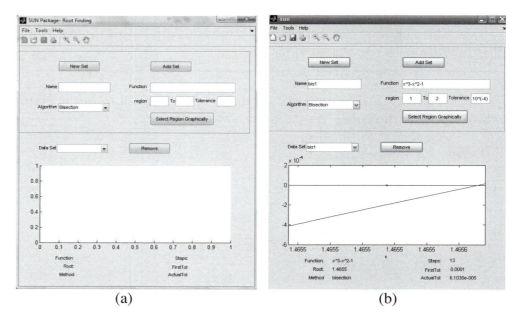

(a)   (b)

FIGURE D.1
MATLAB GUI: Bisection method.

D.2 System of Linear Equations

EXAMPLE D.2

Solve the system of equations using both Naive Gauss elimination and Gauss-Seidel method

$$
\begin{bmatrix}
4 & 1 & 1 & 1 \\
2 & 8 & 1 & 5 \\
-1 & 1 & -5 & 3 \\
3 & 1 & 2 & -7
\end{bmatrix}
\begin{bmatrix}
x_1 \\
x_2 \\
x_3 \\
x_4
\end{bmatrix}
=
\begin{bmatrix}
4 \\
3 \\
-2 \\
1
\end{bmatrix}.
$$

The results obtained by using GUI are shown in Figure D.2.

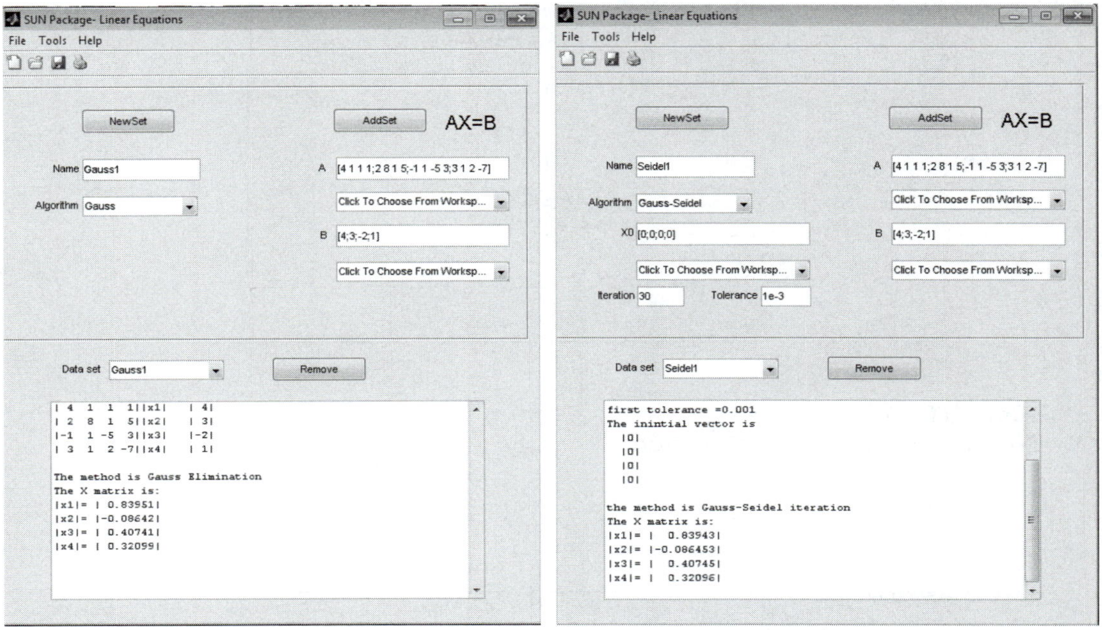

(a)   (b)

FIGURE D.2
(a) Naive Gauss elimination; (b) Gauss-Seidel method.

D.3 Interpolation

EXAMPLE D.3

Find Newton's interpolating polynomial for the following data.

x	1	2	3	4	5	6	7	8	9	10
y	3	10	21	36	55	78	105	136	171	210

The results obtained by using GUI are shown in Figure D.3.

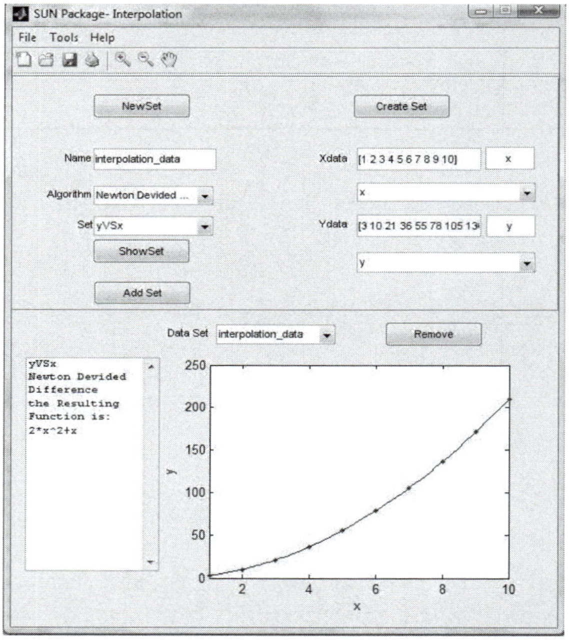

FIGURE D.3
Newton's divided difference.

D.4 The Method of Least Squares

EXAMPLE D.4

Find the least-squares polynomial of degree two that fit the following table of values.

x	1	2	3	4	5	6	7	8	9	10
y	6	9	14	21	30	41	54	69	86	105

The results obtained by using GUI are shown in Figure D.4.

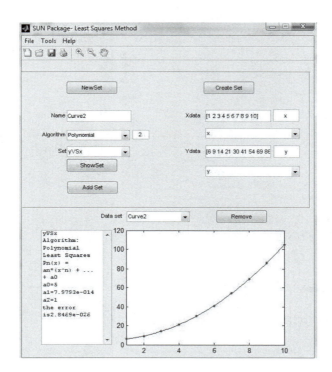

FIGURE D.4
Least-Squares Method.

D.5 Integration

EXAMPLE D.5
Use Simpson's method to approximate $\int_0^1 \sin x^2 dx$.

The results obtained by using GUI are shown in Figure D.5.

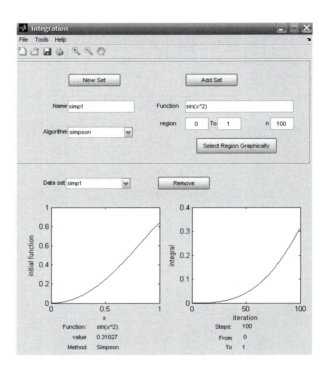

FIGURE D.5
Simpson's method.

D.6   Differentiation

EXAMPLE D.6
Given $f(x) = e^x$ and $h = 0.25$, compute $D_{6,6}$ to approximate $f'(1)$.

The results obtained by using GUI are shown in Figure D.6.

D.7   Numerical Methods for Differential Equations

EXAMPLE D.7
Solve the initial-value problem

$$\frac{dy}{dt} = 2t - y, \quad y(0) = -1$$

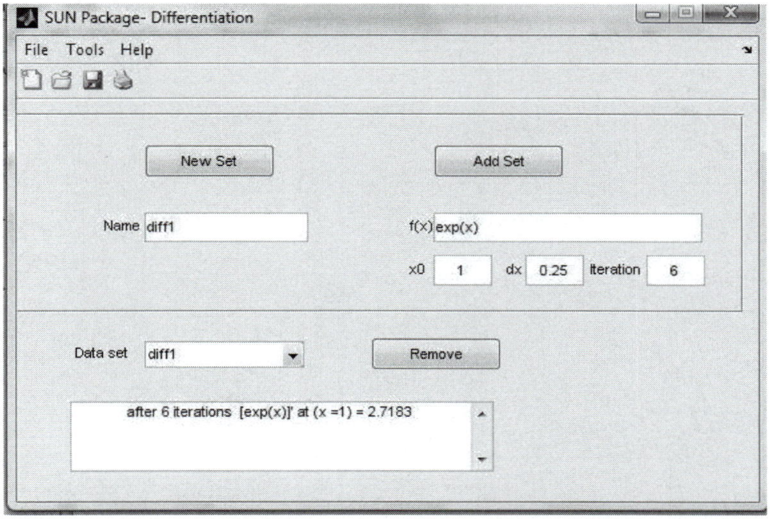

FIGURE D.6
Romberg method.

with $N = 10$ to approximate the value of y at $t = 1$ using the midpoint method.

The results obtained by using GUI are shown in Figure D.7 (a).

D.8 Boundary-Value Problems

EXAMPLE D.8

Solve the following boundary-value problem

$$y'' - \frac{3}{x}y' + \frac{3}{x^2}y = 2x^2 \exp(x)$$

$$y(0) = 0 \quad \text{and} \quad y(1) = 4\exp(2)$$

using the finite-difference method with $h = 0.1$.

The results obtained by using GUI are shown in Figure D.7 (b).

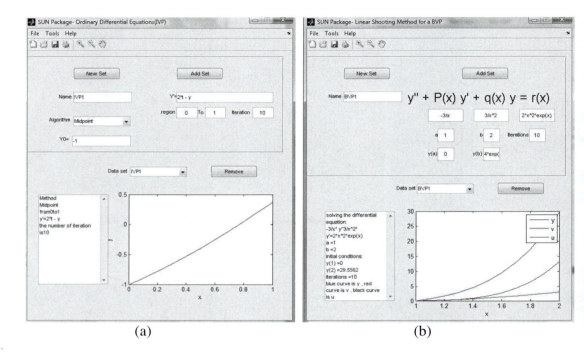

(a)   (b)

FIGURE D.7
(a) Midpoint method; (b) Finite-difference method.

D.9   Numerical Methods for PDEs

EXAMPLE D.9
Solve the heat equation

$$\frac{\partial^2 u}{\partial x^2} = \frac{\partial u}{\partial t}, \quad 0 < x < 1, \quad 0 < t < 0.5$$

subject to the initial and boundary conditions

$$u(x,0) = \sin \pi x, \quad 0 \le x \le 1$$
$$u(0,t) = u(1,t) = 0, \quad 0 \le t \le 0.5$$

using the Crank-Nicolson method.

The results obtained by using GUI are shown in Figure D.8 (a).

EXAMPLE D.10

Approximate the solution of the wave equation

$$16u_{xx} = u_{tt}, \quad 0 < x < 1, \quad t > 0,$$

subject to the conditions

$$u(0,t) = u(1,t) = 0, \quad t > 0,$$
$$u(x,0) = \sin \pi x, \quad 0 \le x \le 1,$$
$$u_t(x,0) = 0, \quad 0 \le x \le 1.$$

using an explicit method.

The results obtained by using GUI are shown in Figure D.8 (b).

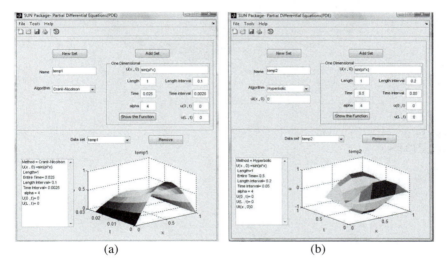

(a)   (b)

FIGURE D.8
(a) Crank-Nicolson method; (b) Explicit three-level difference method.

EXAMPLE D.11

Use the MATLAB function laplace with $n = 9$ to approximate the solution of the Laplace's equation in the square region $0 < x < \pi$ and $0 < y < \pi$ subject to the boundary conditions

$$u(x,0) = 100, \ u(x,10) = 50, \quad 0 \le x \le 10,$$
$$u(0,y) = 0, \ u(10,y) = 10 \sin x, \quad 0 \le y \le 10$$

The results obtained by using GUI are shown in Figure D.9.

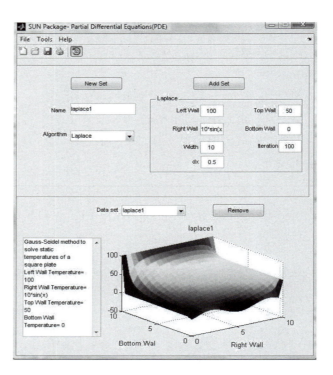

FIGURE D.9
 Elliptic PDEs.

Answers to Selected Exercises

CHAPTER 1

Section 1.3

2. (a) No   (b) $c = 2/(3\sqrt{3})$ (c) No.

3. $\{\sqrt{3}\}$.

4. No.

5. $f(-0.88) \approx 2.381$.

7. $f(0.2) \approx 0.2426667$ for $n = 3$.

10. $\sqrt{e} \approx 1.6483475$.

14. $n = 7$, $n = 14$.

19. $\sin^2 x = x^2 - x^4/3 + 2x^6/45 - x^8/315 + O(x^{10})$.

20. (a) 0.946082766440

(b) 0.239814814815

(c) 1.317901815240

(d) 3.059116423148

CHAPTER 2

Section 2.1

1. (a) $(0.110111) \times 2^2$ (b) $(0.1100111) \times 2^4$.

5. (a) -848 (b) 34.5   (c) -13824.

6. (a) $2^{-1022} \approx 2.2 \times 10^{-308}$,

(b) $-(2 - 2^{-52})2^{1023} \approx -1.8 \times 10^{308}$,

(c) $2 \cdot 2046 \cdot 2^{52} + 1 \approx 1.8 \times 10^{19}$.

Section 2.2

1. (a) (i) 2.083   (ii) 2.083,

    (b) (i) -0.1617 (ii) -0.1618,

    (c) (i) 1.143   (ii) 1.142.

3. (a) $7.571450291E - 5$ (b) $1.051116167E - 5$.

5. 10.51921875, 8.60.

8. (a) $149.85 < p < 150.15$,

    (b) $1498.5 < p < 1501.5$.

11. $a_2 = 50,000 + a_0$ and for all values of a_0 the result will always be $a_2 = 50,000$. The final value is independent of a_0 because of the large multiplier on a_1 compared to the precision of the computation.

Section 2.3

1. $e^x = 1 + x + \frac{x^2}{2!} + \frac{x^3}{3!} + \cdots$
$e^{0.5} \approx 1.625$, error $= 0.0237213$.

2. -0.666667.

Section 2.4

2. (a) $[1, 4]$

case	condition	result
1	$a > 0, c > 0$	$[ac, bd]$
2	$a > 0, d < 0$	$[bc, ad]$
3	$b < 0, c > 0$	$[ad, bc]$
4	$b < 0, d < 0$	$[bd, ac]$
5	$a < 0 < b, c > 0$	$[ad, bd]$
6	$a < 0 < b, d < 0$	$[bc, ac]$
7	$a > 0, c < 0 < d$	$[bc, bd]$
8	$b < 0, c < 0 < d$	$[ad, ac]$
9	$a < 0 < b, c < 0 < d$	$[ad, bc, ac, bd]$

3.

4. (a) $[(1)(3), (2)(4)] = [3, 8]$,

    (b) $[(2)(-4), (-1)(-4)] = [-8, 4]$.

5. (c) $[-20, 35]$,

    (d) $[-4.4, 8.8]$,

    (e) $[-2, 2]$,

    (f) $[-6, 12]$,

    (g) $[-1, 0] + [-10, 15] = [-11, 15]$.

CHAPTER 3

Section 3.1

1. $x = 1.037055$, $n = 17$.

2. $x = 1.000012$ with $TOL = 10^{-4}$, iter. $= 13$.

4. $x = 1.388130$.

5. $x = 0.223632$ in $[0, 0.5]$ with $TOL = 10^{-3}$.

11. 50 or more.

12. Smallest integer $> \ln[(b-a)/\epsilon]/\ln 2$.

Section 3.2

1. $x = 1.037013$, iter. $= 4$.

2. $x = 0.999994668$ with $TOL = 10^{-4}$, iter. $= 7$.

3. $x = 0.2234256$.

Section 3.3

2. (a) $x = 3.000001$ (b) $x = -0.99999$ (c) no convergence.

3. (a) $x = 4.669565$ (c) $x = -2.7183971$.

4. (a) $g(x) = x^5 - 0.25$, $x = -0.25099$,
5. (b) $g(x) = 2\sin x$, $x = 1.895490$,
   (c) $g(x) = \sqrt{3x+1}$, $x = 3.30277$,
   (d) $g(x) = (2 - e^x + x^2)/3$, $x = 0.257531$.

9. (a) Diverge,
   (b) Converge.

14. For $p_0 = 1.5$ the sequence diverge, $p_0 = 2.5$ is slowly convergent to the unique fixed point 2. The Theorem does not hold because $g'(2) = 1$.

Section 3.4

1. $x = 0.11378$.

2. (a) $x = 0.260805$ (b) $x = 0.934929$ with $TOL = 10^{-6}$.

4. x_2 does not exist because $f(x_1) - f(x_2) = 0$.

5. $x = 3.999999$.

9. (a) $x_2 = 3.32258064516129032$, $x_3 = 3.30160226201696513$,

   (b) $x_2 = 1.32791327913279133$, $x_3 = 1.324643265171403$,

   (c) $x_2 = 0.614678899082568808$, $x_3 = 0.61738958600928762$.

Section 3.5

4. $x = 0.655318$.

5. $x = 0.444130$ with $x_0 = 0.5$.

6. $x = 3.715365$ with $x_0 = 4$ and $x = 2.630872$ with $x_0 = 3$.

9. $x = 0.73908511$ for the secant method and $x = 0.73908513$ for Newton's method.

11. $x = 0.2$.

14. $f(x) = 1/a$

20. Yes.

21. No, $f(x)$ oscillates with many points x_0 where $f'(x_0) = 0$.

Section 3.6

1. $x = 1.380278$,
   $0.269723, 1.014931$ for $n = 2$,
   $0.073836, 1.339098$ for $n = 3$,
   $0.007300, 1.508588$ for $n = 4$.

2. $x = 1.465571$,
   $0.215571, 0.772065$ for $n = 2$,
   $0.088947, 1.211811$ for $n = 3$,
   $0.023235, 1.010581$ for $n = 4$.

7. $x_{n+1} = x_n - (x_n^3 - 36)/3x_n^2$ is quadratically convergent to $\alpha = x_2 = 743/225 = 3.3022$.

8. Yes since $x_{n+1} = x_n^2$.

9. For $x = 1$, converges to $x = 0$ but convergence is slow since it is a triple root. For $x = 2$ convergence is quadratic; 5 iterations to converge to $x = 2.327807$.

Section 3.7

1. No convergence. The tangent line to the graph of f at $x = 0$ is horizontal.

5. $x = 2.0$.

6. $x = 1.000000$ after 5 iterations, yes the quadratic convergence is recovered.

Section 3.8

5. (a) $x = 2.4997$, $y = 4.4990$,

   (b) $x = 2.2817$, $y = 1.6427$.

7. $(x_1, y_1) = (21/25, 22/25)$.

8. Starting with $(1, 1)$ we get $(0.72591, 0.502947)$ after 5 iterations. Starting with $(-1, 0)$ we get $(1.670093, 0.345134)$ after 8 iterations.

CHAPTER 4

Section 4.1

2. (a) $AB = \begin{bmatrix} 29 & 9 & -7 & -30 \\ 6 & 12 & 32 & 11 \\ -15 & -1 & -5 & 15 \end{bmatrix}$, $\quad BA$ does not exist,

   (b) $A^2 = \begin{bmatrix} 23 & -18 & -18 \\ -24 & 41 & 24 \\ 1 & -3 & 10 \end{bmatrix}$, $\quad B^2$ does not exist,

   (c) $B^T A = \begin{bmatrix} -1 & 12 & -18 \\ -1 & 14 & 14 \\ -17 & 34 & 14 \\ -5 & 6 & 34 \end{bmatrix}$.

Section 4.2

1. Naive Gaussian elimination algorithm fails.

3. (a) $x_1 \approx 0.3333333$, $x_2 \approx 0.6666667$ for $n = 4$.
   $x_1 \approx 0.3333333461$, $x_2 \approx 0.6666666667$ for $n = 8$.

   (b) $\bar{r} = \begin{bmatrix} 0 \\ -3.7E - 14 \end{bmatrix}$, $\bar{e} = \begin{bmatrix} 3.7E - 14 \\ -7.4E - 17 \end{bmatrix}$ with $n = 4$.

   $\bar{r} = \begin{bmatrix} 0 \\ 1.3E - 08 \end{bmatrix}$, $\bar{e} = \begin{bmatrix} -1.3E - 08 \\ 1.5E - 16 \end{bmatrix}$ with $n = 8$.

4. (a) $\begin{bmatrix} 1 & 1/2 & 1/3 & 1/4 & 1/5 \\ 1/2 & 1/3 & 1/4 & 1/5 & 1/6 \\ 1/3 & 1/4 & 1/5 & 1/6 & 1/7 \\ 1/4 & 1/5 & 1/6 & 1/7 & 1/8 \\ 1/5 & 1/6 & 1/7 & 1/8 & 1/9 \end{bmatrix}$.

5. (a) $x = 2$, $y = 0$, $z = 0$, $v = -1$, $w = 0$.

   (b) $x \approx -16.981$, $y \approx 11.231$, $z \approx 1.954$, $w \approx 13.306$.

   (c) $x = 1$, $y = 7$, $z = -2$.

Section 4.3

1. (a) $x \approx 0.3143$, $y \approx 1.1143$, $z \approx -2.2571$.

    (b) $x_1 \approx -1.889$, $x_2 = 1.0$, $x_3 \approx 3.370$, $x_4 \approx -1.074$.

3. Algorithm fails.

5. $x_1 \approx 2.886$, $x_2 = -1.348$, $x_3 = -2.625$, $x_4 \approx 0.603$, $x_5 \approx 0.989$.
    scale vector $= [4, 3, 3, 3, 2]^T$.
    index vectors at steps 1, 2, 3, and 4 are
    $$\begin{bmatrix} 5 \\ 2 \\ 3 \\ 4 \\ 1 \end{bmatrix}, \begin{bmatrix} 5 \\ 2 \\ 3 \\ 4 \\ 1 \end{bmatrix}, \begin{bmatrix} 5 \\ 2 \\ 4 \\ 3 \\ 1 \end{bmatrix}, \begin{bmatrix} 5 \\ 2 \\ 4 \\ 3 \\ 1 \end{bmatrix}.$$

10. (a) $x = [9, -36, 30]'$,

    (b) $[9.671, -39.508, 33.284]'$.

Section 4.4

2. (a) $L = \begin{bmatrix} 4 & & \\ 3 & 3.75 & \\ 3 & 3.75 & -6 \end{bmatrix}$, $U = \begin{bmatrix} 1 & -0.25 & 0.25 \\ & 1.0 & 1.93 \\ & & 1.0 \end{bmatrix}$,

    (c) $L = \begin{bmatrix} 2 & & & \\ 1 & 2 & & \\ 0 & -3 & 1 & \\ 2 & -2 & 1 & 1 \end{bmatrix}$, $U = \begin{bmatrix} 1 & 0 & 0 & 0 \\ & 1 & 0 & 0 \\ & & 1 & 0 \\ & & & 1 \end{bmatrix}$.

3. $x = [15, 14, 12, 9, 5]^T$.

5. (a) $x_1 \approx -5.5428$, $x_2 \approx -5.0571$, $x_3 \approx 3.5143$, $x_4 \approx -7.8571$,

    (b) $x = [2.4444, 3, 1.2222]^T$.

6. $A = \begin{bmatrix} 1 & -3 & -3 & 4 \\ -3 & 13 & 7 & -8 \\ -3 & 7 & 19 & 4 \\ 4 & -8 & 4 & 72 \end{bmatrix}$.

Section 4.5

2. Both diverge with $\mathbf{x} = [0.1, 1.1]^T$. Gauss-Seidel diverges more rapidly.

5. When converging: $x_1 \approx 1.000005$, $x_2 \approx -4.00001$.

7. (a) The method diverges,

    (b) $x \approx 0.000393$, $y \approx 0.631038$, $z \approx 1.210084$,
        $TOL = 10^{-3}$.

8. $x = [-1.07688, 1.990028, 1.474477, -1.906078]'$ converge after 35 iterations
    with tol $= 0.0001$ and $x_0 = [1, 1, 1, 1]$.

CHAPTER 5

Section 5.1

1. $p(x) = 18.37 - 30.69x + 14.66x^2 - 1.13x^3$.

2. $p(x) = -1.04 + 0.35x + 0.5x^2$.

6. (a) At least 5 points should be used,

   (b) For $n = 5$ the coefficients are closest to correct, $n = 10$ coefficients looks a little bit different. This example demonstrates a poor performance of the Vandermonde approach.

Section 5.2

2. $p(x) = 2 - x + 2x^2$.

4. $p(x) = 14.5 + 5(x - 1) + 3(x - 1)(x - 2) + (x - 1)(x - 2)(x - 3)$.
   $p(4.5) = 71.275$.

5. $p(x) = -41 + 24(x + 3) - 6(x + 3)(x + 2) + (x + 3)(x + 2)(x + 1) - 0.167(x + 3)(x + 2)(x + 1)x$.

6.

x_i	$f[.]$	$f[.,.]$	$f[.,.,.]$	$f[.,.,.,.]$
1.1	2.45			
		0.609		
2.2	3.12		0.079	
		0.782		0.024
3.3	3.98		0.157	
		1.127		
4.4	5.22			

8. (a) $p(x) = 22 + (x - 1) + \frac{1}{2}(x - 1)(x - 2) + \frac{1}{3}(x - 1)(x - 2)(x - 3) - \frac{1}{2}(x - 1)(x - 2)(x - 3)(x - 4)$.

   (b) 23.47 pounds.

11. $p(x) = (x - 1) - (1/6)(x - 1)(x - 2)$.

13. $p = 2 + 2(x - 1) - 2(x - 1)(x - 2)$.

Section 5.3

7. $h = 0.04472135955$, $n = 12$.

8. (a) $f^{(n)}(x) = (-2)^n e^{-2x}$, Error $\leq 2^n/n!$,

   (b) $h^{(n)}(x) = \frac{1+(-1)^n}{2} \sin x + \frac{1+(-1)^n}{2} \cos x$, Error $\leq 1/n!$,

   (c) $g^{(n)}(x) = (-1)^n (n!)(x + 1)^{-(n+1)}$, Error $\leq n!/n! = 1$,

(d) $k^{(n)}(x) = \begin{cases} \frac{(-1)^n \Pi_{i=1}^n (2i-3)}{2^n}(x-1)^{-n+1/2}, & x > 1 \\ \frac{\Pi_{i=1}^n (2i-3)}{2^n}(1-x)^{-n+1/2}, & x < 1 \end{cases}$

Error $= \infty$, unbounded.

All are smooth except $k^{(n)}(x)$ does not exist at $x = 1$.

f and h will eventually converge. g will converge very slowly or unlikely. Convergence for k is unlikely.

9. Error $= \frac{f^{(3)}(c)}{6}(x-1)(x-2)(x-3)$. We choose the interval $[1,3]$, thus the max of $f^{(3)} = 8e^6$. Then error $= \frac{8e^6}{6}(.5)(.5)(1.5) = e^6/2 \approx 202$.

10. $h = 0.02260678$.

Section 5.4

1. $p(x) = -\frac{1}{48}(x-2)(x-4)(x-6) - \frac{1}{16}x(x-4)(x-6) - \frac{3}{16}x(x-2)(x-6) + \frac{1}{12}x(x-2)(x-4)$.

2. $L_1(x) = \frac{1}{3}(x-0.5)(x-1)$,

   $L_2(x) = -\frac{3}{4}(x+1)(x-1)$,

   $L_3(x) = (x+1)(x-0.5)$.

6. $p_2(x) = \frac{(x-3)(x-6)}{(2-3)(2-6)}0.6932 + \frac{(x-2)(x-6)}{(3-2)(3-6)}1.0986 + \frac{(x-2)(x-3)}{(6-2)(6-3)}1.7918$.

8. $p = 3L_1 - L_3$, $\quad L_1 = \frac{(x-2)(x-3)}{(1-2)(1-3)}$, $\quad L_3 = \frac{(x-1)(x-2)}{(3-1)(3-2)}$.

CHAPTER 6

Section 6.1

1. (a) No,   (b) Yes.

2. $S(2.3) = 3.7$.

3. $S(0.4) = 0.614658$, $e = S(0.4) - f(0.4) = 0.017929$.

5. $S_1(x) = 0.6711x + 1$, $S_2 = 0.6801x + 0.99982$.

   $\int_0^{0.04} s(x)dx = 0.04053868$, $e = 5.7E - 07$.

Section 6.2

2. $a = -0.5$, $b = 0.5$.

3. $a = 0$, $b = 2$, $c = 0$, $d = -2$.

4. No.

Section 6.3

1. $S(1.6) = 3.04$.

2. 1.876.

3. $a = 6$, $b = -18$, $c = -36$, $d = -24$.

6. No.

7. $S(2.05) = 1.6075$, $S(2.15) = 2.84$, $S(2.25) = 4.5325$.

CHAPTER 7

Section 7.1

1. $y = 1.97x + 0.08$, $E(a, b) = 1.059$.

3. $y = -1.77x + 80$, $E(a, b) = 1.86E + 3$, homework #13 = 57.

5. $y = 3.14x + 23$, $E(a, b) = 41.71$.

Section 7.2

2. (a) $p(x) = 26.72 - 52.34x + 29.92x^2$, $E = 67.97$,

   (b) $p(x) = 115.26 - 211.51x + 118.64x^2 - 15.51x^3$, $E = 31.73$.

3. The polynomial of degree two is
   $p(x) = 7 + 3x + x^2$, $E = 1.2E - 27$.

5. $p(x) = 12.59 - 9.99x + 1.95x^2$, $f(6.0) \approx 22.70$.

Section 7.3

1. $y = 1.99e^{.48x}$, $E = 0.002677$, $y(5.0) \approx 14.67$.

2. $y = 2.57 + 1.21/x$, $y(3.0) \approx 2.97$.

8. $a = 210.882776$, $b = 0.098784$, $c = 69.621755$.

10. $a = 0.6853$, $b = 0.3058$.

12. $y = ax^b$, $a = 2.30$, $b = 1.39$.

CHAPTER 8

Section 8.1

1. (a) No relative extrema,

   (b) Relative minimum at $x = -1/\sqrt{27}$, relative maximum at $x = 1/\sqrt{27}$.

2. (a) Increasing: $[-1/\sqrt{2}, 0] \cup [1/\sqrt{2}, \infty)$
      Decreasing: $(-\infty, -1/\sqrt{2}] \cup [0, 1/\sqrt{2}]$,

   (c) Increasing: $(-\infty, 1]$, decreasing: $[1, \infty)$.

Section 8.2

2. One relative maximum located at $x = n$.

3. $(1, 0)$.

4. $(-\sqrt{2}, 1)$, $(\sqrt{2}, 1)$.

5. $(1, 2)$.

Section 8.3

iteration	x_n	$f(x_n)$	$|x_n - x_{n-1}|$
1	-1	2.62091	
2	-2.03822	-0.35178	1.03822
3	-1.90688	0.01063	0.13135
4	-1.91063	0.00001	0.00375
5	-1.91063	0	0

1.

iteration	x_n	$f(x_n)$	$|x_n - x_{n-1}|$
1	-2	-0.24844	
2	-1	2.62091	1
3	-1.91342	-0.00787	0.91342
4	-1.91068	-0.00014	0.00273
5	-1.91063	0	0.00005
6	-1.91063	0	0

2.

CHAPTER 9

Section 9.1

5. $f'(1) \approx 2.722814$.

10. It is a scheme for $f'(x)$, error$= (5h/2)f''(\xi)$.

11. Set $f''(x_0) = Af(x_0) + Bf(x_0 + h) + Cf(x_0 + 3h)$ to get $A = 2/(3h^2)$, $B = -1/h^2$, $C = 1/(3h^2)$; Error $= (4/3)hf'''(\xi)$.

Section 9.2

1. 1.003353 for $h = 0.1$; 1.00003 for $h = 0.01$.

2. $\ln 6 \approx 1.79175945$ for $n = 3$ and $h = 0.1$.

3. $\ln 3 \approx 1.09861195$ for $n = 2$ and $h = 0.1$.

9. (a) $g''(2) = [g(4) - 2g(2) + g(0)]/2^2 = 5/4$,
    (b) $g''(2) = [(4)(1) - 5/4]/3 = 11/12$.

13. $f'(x) = 0.70710678071652773$ using 4 rows.

CHAPTER 10

Section 10.1

2. 3.1415965.

3. 0.5381721 for $h = 0.04$, $E = 17(0.2)(0.04)^2 e^{1.25}/12 \approx 0.001505$.

5. (a) -0.433155,   (c) 0.292290.

7. 1.436004, $n = 5$,
   1.429270, $n = 7$,
   1.426503, $n = 9$.

8. $I = 1/2[f(2) + 2f(3) + 2f(4) + 2f(5) + f(6)] = 359/420$.

10. When $h = 0.000488281$, $T = 0.666664434$, error $= 0.000002233$. The error decreases by about a factor of 3. The error analysis does not apply.

Section 10.2

1. 1.5801713.

2. $h \approx 0.15331$.

3. 0.000002.

4. $0.940012 < 1 < 1.061267$.

12. $\int_0^\pi \sqrt{1 + \cos^2 x}\, dx = 3.820197789$ with $h = 0.09817477$.

14. $S = \frac{h}{3}[f(0) + 4f(3) + f(6)] = 0$ with $h = 3$. Error is of the order $O(h^4)$.

21. $I = 2.00421$, exact $= 2.0$, error $= 0.00421$.

22. (a) $n = 8$, (b) 1.111443.

23. Error ≈ 0.000115.

Section 10.3

2. 82/15.

3. 418.408987.

4. 0.8591451666.

6. 1.09863055, $E = 0.000018$.

7.

$I(h)$	1.570796	2.004560	1.999984	1.999999
$I(h/2)$	1.896119	2.000270	1.999999	
$I(/h/4)$	1.974232	2.000016		
$I(h/8)$	1.993570.			

8. 20271 only at integer points.

Section 10.4

1. 0.627393.

2. -0.9059848, $n = 5$.

4. 5.300966.

5. (a) -0.43989 (b) 1.238583   (c) 0.292893   (d) 1.656840.

7. 0.786885, $n = 2$,
   0.785267, $n = 3$,
   0.785402, $n = 4$.

12. 20.0473.

13. $w_0 = -2/3$, $w_1 = -2/3$, $w_2 = 8/3$.

16. $4/3[f(\sqrt{(0.4)}) + f(-\sqrt{(0.4)})]$, $I = 32/15 = 2.1333333333$.

CHAPTER 11

Section 11.1

1. $x + \int_0^x (t - x)\sin(t)dt = x + \sin x + x(\cos x - 1) - x\cos x = \sin x$.

2. $1 + \int_0^x e^t dt = 1 + e^x - 1 = e^x$.

3. $1 + x\int_0^1 t\left(1 + \frac{3t}{4}\right)dt = 1 + x\left[\frac{1}{4}t^2(t + 2)\right]\Big|_0^1 = 1 + 3x/4$.

4. $x + \int_0^1 (1 + xt^2)(-3)dt = x - x - 3 = -3$.

5. $y(x) + \int_0^x k(x,t)y(t)dt = F(x)$, where
   $k(x,t) = (x - t)[p(t) - q'(t)] + q(t)$ and
   $F(x) = \int_0^x (x - t)y(t)dt + a + bx + a\,q(0)\,x$

6. $u(x) + \int_0^x (t - x)tu(t)dt = x + e^x$.

7. $u(x) + \int_0^1 k(x,t)t^2 u(t)dt = -\int_0^x t(1 - x)\sin t\,dt - x\int_x^1 (1 - t)\sin t\,dt$
   or $u(x) + \int_0^1 k(x,t)t^2 u(t)dt = x\sin 1 - \sin x$,
   where $k(x,t) = \begin{cases} t(1 - x), & 0 \le t \le x \\ x(1 - t), & x \le t \le 1 \end{cases}$

Section 11.2

1.

	x_i	u_i	$u(x_i)$	Error
	0	0	0	0
	0.25	0.2516	0.25	0.0016
	0.50	0.5031	0.50	0.0031
	0.75	0.7547	0.75	0.0047
(a)	1.00	1.0063	1.00	0.0063

	x_i	u_i	$u(x_i)$	Error
	0	0	0	0
	0.250	0.3067	0.3125	0.0058
	0.500	0.7150	0.7500	0.0350
	0.750	1.2247	1.3125	0.0878
(b)	1.000	1.8359	2.0000	0.1641

	x_i	u_i	$u(x_i)$	Error
	0	1.0591	1.0000	0.0591
	0.250	1.0987	1.0396	0.0591
	0.500	1.2307	1.1716	0.0591
	0.750	1.5056	1.4465	0.0591
(c)	1.000	2.0591	2.0000	0.0591

	x_i	u_i	$u(x_i)$	Error
	0	2.0052	2.0000	0.0052
	0.250	2.2892	2.2840	0.0052
	0.500	2.6539	2.6487	0.0052
	0.750	3.1222	3.1170	0.0052
(d)	1.000	3.7235	3.7183	0.0052

2.

	x_i	u_i
	0.33001	0.4590
	0.06943	0.0744
	0.66999	1.3093
(a)	0.93057	2.3599

	x_i	u_i
	0.25919	1.07028
	0.05453	1.00295
	0.52621	1.33734
(b)	0.73087	1.80365

x_i	u_i
0.33001	0.44889
0.06943	0.86596
0.66999	0.10891
0.93057	0.00482

(c)

x_i	u_i
2.07351	1.92497
0.43625	2.06690
4.20967	−2.88858
5.84693	−0.25422

(d)

x_i	u_i
0.51838	0.25684
0.10906	0.01187
1.05242	0.91416
1.46173	1.45305

(e)

x_i	u_i
0.33001	−0.30383
0.06943	−0.41901
0.66999	−0.22051
0.93057	−0.22501

(f)

x_i	u_i
0.33001	−0.39832
0.06943	−0.34317
0.66999	−0.26611
0.93057	−0.00828

(g)

x_i	u_i
2.07351	−157.67021
0.43625	69.16195
4.20967	40.60670
5.84693	169.15287

(h)

3.

(a) Trapezoidal rule

x_i	u_i	$u(x_i)$	Error
0	0	0	0
0.393	0.3846	0.3827	0.001906
0.785	0.7109	0.7071	0.003813
1.178	0.9296	0.9239	0.005719
1.571	1.0076	1.0000	0.007626

(b) Gauss-Nyström method

x_i	u_i	$u(x_i)$	Error
0.51838	0.4955	0.4955	0.000000
0.10906	0.1088	0.1088	0.000000
1.05242	0.8686	0.8686	0.000000
1.46173	0.9941	0.9941	0.000000

Section 11.3

1. $u(x) = x$

2. $u(x) = x^2 + x$

3. $u(x) = \frac{2}{e^2-1}e^x$

4. $u(x) = \sec^2 x$

5. $u(x) = e^x + 1$

Section 11.4

1. $u(x) \approx \cos x + \frac{2\pi}{\pi^2-4}\sin x - \frac{\pi^2}{\pi^2-4}\cos x$

2. $u(x) = -\frac{59}{560} - \frac{33}{280}x^2 + x^3$

3. $u(x) \approx \frac{27}{76} - \frac{15}{152}x^2$

4. $u(x) \approx -\frac{375}{2017}x^2 + \frac{558}{2017}x + \frac{1863}{4034}$

5. $\cos(xt) \approx 1 - \frac{1}{2}(xt)^2 + \frac{1}{24}(xt)^4$

   Solution using Schmidt's method: $u(x) = 0.12331x^4 - 1.3890x^2 + x + 0.45237$

   Schmidt and Trapezoidal rule with with $n = 10$

x_i	Schmidt	T(n=10)	Error
0.0	0.45237	0.0926	0.3598
0.62832	0.55155	0.4108	0.1408
1.25664	−0.17692	0.1174	0.2943
2.19911	−1.18190	−0.6544	0.5275
3.14159	1.89660	0.9567	0.9399

   Schmidt and Gauss-Nyström ($n = 5$) methods

x_i	Schmidt	G-N(n=5)	Error
0.72497	0.48137	0.41213	0.06924
0.14737	0.56963	0.27519	0.29444
1.57080	−0.65334	−0.27026	0.38308
2.41662	−1.03720	−0.70683	0.33037
2.99422	0.90507	0.57740	0.32767

6. $\exp(xt) \approx 1 + xt + \frac{1}{2}(xt)^2$

   Solution using Schmidt's method: $u(x) = \frac{3}{5}x$.

   Gauss-Nyström ($n = 5$) and Schmidt methods

x_i	GN(n=5)	Schmidt
-0.5385	-0.322106	-0.323082
-0.9062	-0.522669	-0.543708
0.0000	0.000000	0.000000
0.5385	0.322106	0.323082
0.9062	0.522669	0.543708

   Trapezoidal rule ($n = 10$) and Schmidt's method

x_i	T(n=10)	Schmidt
-1.0	-0.563530	-0.600000
-0.6	-0.354608	-0.360000
-0.2	-0.120853	-0.120000
0.2	0.120853	0.120000
0.6	0.354608	0.360000
1.0	0.563530	0.600000

Section 11.5

1.

   Euler's method with $h = 0.2$

x_i	u_i	$u(x_i)$	Error
0.0	1.0	1.0	0.0
0.20	0.840000	0.856192	0.016192
0.40	0.792000	0.810960	0.018960
0.60	0.833600	0.846435	0.012835
0.80	0.946880	0.947987	0.001107
1.00	1.117504	1.103638	0.013866

   Heun's method with $h = 0.2$

x_i	u_i	$u(x_i)$	Error
0.0	1.0	1.0	0.0
0.20	0.854545	0.856192	0.001647
0.40	0.808264	0.810960	0.002696
0.60	0.843125	0.846435	0.003309
0.80	0.944375	0.947987	0.003612
1.00	1.099943	1.103638	0.003695

Euler's method with $h = 0.1$

x_i	u_i	$u(x_i)$	Error
0.0	1.0	1.0	0.0
0.20	0.849000	0.856192	0.007192
0.40	0.802690	0.810960	0.008270
0.60	0.841179	0.846435	0.005256
0.80	0.948355	0.947987	0.000368
1.00	1.111167	1.103638	0.007529

Heun's method with $h = 0.1$

x_i	u_i	$u(x_i)$	Error
0.0	1.0	1.0	0.0
0.20	0.855782	0.856192	0.000410
0.40	0.810289	0.810960	0.000671
0.60	0.845611	0.846435	0.000824
0.80	0.947087	0.947987	0.000900
1.00	1.102718	1.103638	0.000921

2.

$u(x) = 1 + \frac{1}{12}x^4 + \int_0^x \frac{x-t}{1+t}u(t)dt$

Heun's method with $n = 10$

x_i	u_i
0.0	1.0
0.40	1.076138
0.80	1.306879
1.20	1.761592
1.60	2.580244
2.00	3.972943

3.

$u(x) = x + \int_0^x (t-x)u(t)dt$

Heun's method with $n = 10$

x_i	u_i	$u(x_i)$	Error
0.0	0.0	0.0	00.0
0.26	1.008587	0.951057	0.057530
2.51	0.581670	0.587785	0.006115
3.77	-0.673127	-0.587785	0.085342
5.03	-0.969875	-0.951057	0.018818
6.28	0.113783	-0.000000	0.113783

4.

Modified Heun's method with $n = 5$

x_i	u_i
0.0	0.0
0.20	0.202591
0.40	0.404607
0.60	0.640514
0.80	0.900876
1.00	1.164941

CHAPTER 12

Section 12.1

2.

t_i	$y_i(h = 0.1)$
1.2	0.504545
1.4	0.524791
1.6	0.557778
1.8	0.602524
2.0	0.659112

$f(t) = 0.655812 - 0.31609t + 0.159036t^2$.

(i) $y(1.02) \approx f(1.02) = 0.49885892$, $E = 0.0012$,
(ii) $y(1.67) \approx f(1.67) = 0.57147362$, $E = 0.0136$,
(iii) $y(1.98) \approx f(1.98) = 0.65343451$, $E = 0.0194$.

4.

t_i	$y_i(h = 0.1)$	Error
0.2	0.01	0.00960
0.4	0.057824	0.01610
0.6	0.133878	0.01728
0.8	0.222919	0.01343
1.0	0.309146	0.00691

5.

t_i	$y_i(h = 0.1)$
0.2	1.98
0.4	1.882188
0.6	1.716555
0.8	1.500612
1.0	1.256313

(a)

(b)

t_i	$y_i(h = 0.2)$
0.2	1.0
0.4	0.992053
0.6	0.961147
0.8	0.896023
1.0	0.793180

(c)

t_i	$y_i(h = 0.2)$
1.2	1.2
1.4	1.4
1.6	1.6
1.8	1.8
2.0	2.0

(e)

t_i	$y_i(h = 0.1)$
2.2	3.332497
2.4	3.662115
2.6	3.989580
2.8	4.315387
3.0	4.639891

(f)

t_i	$y_i(h = 0.5)$
1.0	2.5641
2.0	7.0835
3.0	17.6232
4.0	44.1189
5.0	115.5593

6.

(a)

t_i	$y_i(h = 0.1)$
0.2	2.401603
0.4	1.823684
0.6	1.295803
0.8	0.846296
1.0	0.496120

(b)

t_i	$y_i(h = 0.1)$
1.9	-0.000290
2.0	0.000514

t_i	$y_i(h = 0.1)$
0.2	2.56
0.4	1.6384
0.6	1.048576
0.8	0.671088
1.0	0.429496

(c)

t_i	$y_i(h = 0.1)$
1.1	0.765902
1.2	0.731161

(d)

Section 12.2

1. $0.1(e - 1)$.

4. $h = 0.0116$.

Section 12.3

4. $y(0.1) = 1.11583, \quad y(0.5) = 1.97916$.

5. (a) $y_{i+1} = y_i + y_i^2[h^2(4t_i^2 y_i - 1) - 2ht_i]$,
   (b) $y_{i+1} = y_i + 3h(t_i - 1)(t_i + h - 1)$.

7. $y(0.5) = 1.7974313$, Error $= 0.00001$.

Section 12.4

1.

t_i	$y_i(h = 0.1)$
1.2	0.508918
1.4	0.532794
1.6	0.569412
1.8	0.618206
2.0	0.679570

$p(x) = 0.64 - 0.30t + 0.16t^2$.

(i) $y(1.02) \approx 4.99625, \ E = 4.7E - 4$,
(ii) $y(1.67) \approx 0.58470, \ E = 4.0E - 4$,
(iii) $y(1.98) \approx 0.67314, \ E = 3.0E - 4$.

2.

t_i	$y_i(h = 0.1)$	Error
1.2	1.764317	5.7E-3
1.4	2.551362	1.6E-2
1.6	4.079304	3.7E-2
1.8	7.177027	8.4E-2
2.0	13.850020	2.0E-1

3. (a) $y(1.25) \approx 2.07$, $E = 0.16$,

    (b) $y(1.65) \approx 6.44$, $E = 1.8$.

4.

t_i	$y_i(h = 0.1)$
0.2	2.40485
0.4	1.836221
0.6	1.321514
0.8	0.885249
1.0	0.543803

(a)

t_i	$y_i(h = 0.1)$
0.2	2.6896
0.4	1.808487
0.6	1.216026
0.8	0.817656
1.0	0.549792

(c)

5.

t_i	$y_i(h = 0.1)$
0.2	2.681297
0.4	1.797338
0.6	1.204800
0.8	0.807606
1.0	0.541358

(c)

t_i	$y_i(h = 0.1)$
0.2	1.639642
0.4	1.354902
0.6	1.137127
0.8	0.976082
1.0	0.862152
1.2	0.787500

(d)

6.

t_i	$y_i(h = 0.1)$
0.2	1.960397
0.4	1.846232
0.6	1.670540
0.8	1.452298
1.0	1.213061

(a)

(c)

t_i	$y_i(h = 0.2)$
1.4	1.4
1.8	1.8
3.0	3.0

(e)

t_i	$y_i(h = 0.5)$
1.0	3.17803
2.0	8.80537
3.0	22.07598
4.0	56.25298
5.0	149.13270

8. $y(0.5) \approx 1.648720$, $E = 6.3E - 7$,
   $y(1.0) \approx 2.718281$, $E = 2.1E - 6$.

Section 12.6

1.

(b)

t_i	$y_i(h = 0.1)$
0.2	1.960397
0.4	1.846252
0.6	1.670616
0.8	1.452436
1.0	1.213246

(c)

t_i	$y_i(h = 0.2)$
0.4	0.979225
0.8	0.850831
1.2	0.607661
1.6	0.353770
2.0	0.177025

(d)

t_i	$y_i(h = 0.2)$
0.2	2.681600
0.4	1.797745
0.8	0.814563
1.2	0.372990
1.6	0.172457
2.0	0.082279

(f)

t_i	$y_i(h = 0.2)$
0.2	2.404192
0.6	1.318572
1.2	0.297438
1.6	0.059185
2.0	0.021230

2.

t_i	$y_i(h = 0.1)$	Error
0.4	0.268114	1.4E-5
0.6	0.329255	3.2E-5
0.8	0.359421	4.2E-5
1.0	0.367832	4.7E-5

Section 12.8

4.

t_i	$y_i(h = 0.1)$	Error
1.4	0.532793	1.2E-6
1.6	0.569410	2.4E-6
1.8	0.618203	3.0E-6
2.0	0.679567	3.5E-6

5.

(a)

t_i	$y_i(h = 0.1)$
0.4	1.834524
0.6	1.318481
0.8	0.880768
1.0	0.538093

(b)

t_i	$y_i(h = 0.1)$
1.4	0.060102
1.6	0.011540
1.8	-0.003703
2.0	-0.088238

(c)

t_i	$y_i(h = 0.1)$
0.4	1.797290
0.6	1.204704
0.8	0.807500
1.0	0.541258

(d)

t_i	$y_i(h = 0.1)$
0.4	1.354907
0.6	1.137137
0.8	0.976094
1.0	0.862164
1.2	0.787512

Section 12.9

2. $a = 1$, $b = 1/3$. The scheme is weakly stable.

5. The method is strongly stable.

7. $|1 - h + h^2/2| \leq 1$.

8. The method is stable.

Section 12.10

1.

(a)

$t_i(h = 0.2)$	x_i	y_i
0.2	1.216733	0.270133
0.4	1.447219	0.728703
0.6	1.643529	1.466837
0.8	1.723895	2.605075
1.0	1.558194	4.297922

(c)

$t_i(h = 0.1)$	x_i	y_i
0.2	0.180796	0.873687
0.4	0.331208	0.872484
0.6	0.467719	0.969720
0.8	0.603843	1.146398
1.0	0.745363	1.388707

2.

(a)

t_i	$y_i(h = 0.1)$
0.2	1.001403
0.4	1.011876
0.6	1.042759
0.8	1.109481
1.0	1.235013

(c)

t_i	$y_i(h = 0.1)$
0.2	-0.761336
0.4	-0.450754
0.6	-0.076667
0.8	0.351802
1.0	0.824356

Section 12.11

1. (b)

$t_i(h = 0.1)$	x_i	y_i	$|x_i - x(t_i)|$	$|y_i - y(t_i)|$
1.0	-34.08	-3.71	5.517	0.605

Section 12.12

1. $\frac{dx}{dt} = y$, $\frac{dy}{dt} = \sin(x + y)$.

2. $\frac{dx}{dt} = y$, $\frac{dy}{dt} = -ty + x^2 \cos t$.

3. $\frac{dx}{dt} = y$, $\frac{dy}{dt} = -y + \sin x$.

4. $\frac{dx}{dt} = y$, $\frac{dy}{dt} = (1 - x^2)y + x - x^3$.

6. $(0, 0)$.

7. $(0, 0)$, $(1, 1)$.

8. $y = \frac{dx}{dt} = 0$, $x = 0$, $x = 1$, $x = -1$ so the fixed points are $(0, 0)$, $(0, 1)$, and $(0, -1)$.

CHAPTER 13

Section 13.1

1.

x_i	$y_i(h = 0.2)$
0.0	1.0
0.2	0.727356
0.4	0.498994
0.6	0.304895
0.8	0.139434
1.0	0.0

(a)

x_i	$y_i(h = \pi/6)$
0	1.0
$\pi/6$	0.59805
$\pi/3$	-1.27953
$\pi/2$	-1.1E-16
$2\pi/3$	1.27953
$5\pi/6$	-0.59805
π	-1.0

(b)

x_i	$y_i(h = \pi/4)$
0	0.0
$\pi/4$	26.92366
$\pi/2$	37.72392
$3\pi/4$	26.22312
π	0.0

(c)

x_i	$y_i(h = 0.1)$
0.1	-0.05000
0.3	-0.11653
0.5	-0.13585
0.7	-0.10951
0.9	-0.04410
1.0	0.0

(d)

3.

x_i	$y_i(h = 0.1)$	Exact	Error
0.1	0.401034	0.4001	.0009
0.3	1.210314	1.2081	.0022
0.5	2.065237	2.0625	.0027
0.7	3.042561	3.0401	.0025
0.9	4.257267	4.2561	.0012
1.0	5.0	5.0	.0

Section 13.2

1.

x_i	$y_i(h = 0.2)$
1.1	-0.157272
1.3	-0.508495
1.5	-0.898971
1.7	-1.320660
1.9	-1.768007

(b)

x_i	$y_i(h = 0.2)$
0.2	0.778899
0.4	0.543991
0.6	0.321652
0.8	0.134837

(c)

2.

x_i	$y_i(h = 0.1)$	Error
1.2	0.4375723075	5.7E-07
1.4	0.9421229809	7.2E-07
1.6	1.5040122213	6.1E-07
1.8	2.1160323388	3.5E-07

5.

x_i	$y_i(h = 0.2)$
0.2	-0.74220
0.4	-0.53524
0.6	-0.35616
0.8	-0.18486

(a)

x_i	$y_i(h = 0.4)$
0.4	0.979910
0.8	1.113563
1.2	1.356478
1.6	1.661331

(c)

CHAPTER 14

Section 14.1

1. (a) $\lambda^2 - 4\lambda - 5$
   $\lambda_1 = -1; v^{(1)} = (-2, 1)^T$
   $\lambda_2 = 5; v^{(2)} = (4, 1)^T$.

   (c) $\lambda^2 - 3\lambda + 2$
   $\lambda_1 = 1; v^{(1)} = (3/2, 1)^T$
   $\lambda_2 = 2; v^{(2)} = (1, 1)^T$.

2. (a) $\lambda^2 - 5\lambda + 6 - a^2 - a$.

5. (a) $\lambda^3 - 6\lambda^2 + 11\lambda - 6$
   $\lambda_1 = 2; v^{(1)} = (-2, 1, 4)^T$
   $\lambda_2 = 3; v^{(2)} = (-1, 1, 4)^T$
   $\lambda_3 = 1; v^{(3)} = (-1, 1, 2)^T$.

   (b) $\lambda^3 - 3\lambda^2 - 9\lambda + 27$
   $\lambda_1 = 3; v^{(1)} = (1, 0, 1)^T$
   $\lambda_2 = 3; v^{(2)} = (-1, 1, 0)^T$
   $\lambda_3 = -3; v^{(3)} = (-1, -1, 1)^T$.

Section 14.2

1. (a) $\lambda = 5$, $\mathbf{x} = [0.121826, 1, 0.121826]^T$,

   (b) $\lambda = 7.23117887$, $\mathbf{x} = [0.173101, 1, 0.035389, -0.058309]^T$,

   (c) $\lambda = 8.16062223$, $\mathbf{x} = [-0.234068, 0.524851, -0.106132, 1]^T$.

   (d) $\lambda = 7.00251001$, $\mathbf{x} = [1, 0.666726, 5.97 \times 10^{-5}]^T$,

Section 14.3

1. (b) $\lambda_1 = -1; v^{(1)} = (-2, 1, 1)^T$
   $\lambda_2 = 5; v^{(2)} = (1, 1, 1)^T$
   $\lambda_3 = -3; v^{(3)} = (0, -1, 1)^T$.

(c) $\lambda_1 = 5;\ v^{(1)} = (0, 1, 0)^T$
$\lambda_2 = 5;\ v^{(2)} = (1, 0, 2)^T$
$\lambda_3 = -10;\ v^{(3)} = (-2, 0, 1)^T.$

(d) $\lambda_1 = 3;\ v^{(1)} = (1, 2, 1)^T$
$\lambda_2 = 0;\ v^{(2)} = (-1, 1, -1)^T$
$\lambda_3 = -3;\ v^{(3)} = (-1, 0, 1)^T.$

2. $X(t) = Ae^{5.10t}[1, 1.019, 1.032]^T + Be^{-2.85t}[-0.013, 1, -0.975]^T +$

$$Ce^{-0.95t}[-1.998, 0.949, 1]^T.$$

3. $\lambda_1 = 3;\ v^{(1)} = (1, 1, 0, -1, -1)^T$
$\lambda_2 = 1.5359;\ v^{(2)} = (.5774, 1, 1.1547, 1, .5774)^T$
$\lambda_3 = 5;\ v^{(3)} = (-1, 0, 1, 0, -1)^T$
$\lambda_4 = 8.4641;\ v^{(4)} = (-.5774, 1, -1.1547, 1, -.5774)^T$
$\lambda_5 = 7;\ v^{(5)} = (-1, 1, 0, -1, 1)^T.$

5. $\lambda = 3.$

Section 14.4

1. $k_1 = \pm 1.5883,\ k_2 = \pm 2.8284,\ k_3 = \pm 3.6711.$

Section 14.5

1. (c) $x(r + x^2) = 0$ so $x = 0$ is the only fixed point. There is no bifurcation.
   (d) $x(x^2 + 2x + r) = 0$ so, $x = 0$ and $x = -1 \pm \sqrt{(1 + r)}.$

2. For small values of ω, $(0, 0)$ is stable while for larger it is unstable.

CHAPTER 15

Section 15.1

3.

(a)

$t\backslash x$	0.0	0.2	0.4	0.6	0.8	1.0
0.20	0.00	0.196328	0.262656	0.317734	0.162344	0.00

(b)

$t\backslash x$	0.0	20	40	60	80	100
0.05	0.00	2.35492	49.95245	49.95245	2.35492	0.00

(c)

$t\backslash x$	0.0	0.2	0.4	0.6	0.8	1.0
0.10	0.00	0.141066	0.219796	0.209357	0.124175	0.00

4.

(a)

$t \backslash x$	0.0	0.2	0.4	0.6	0.8	1.0
0.01	0.00	0.532286	0.861257	0.861257	0.532286	0.00
Error	0.00	0.0004	0.0007	0.0007	0.0004	0.00

5. $\dfrac{u_{i+1}^{k+1} - 2u_{i+1}^{k} + u_{i+1}^{k-1}}{(\Delta x)^2} = u_i^k \left[\dfrac{u_{i+1}^k - u_i^k}{\Delta y} \right].$

Section 15.2

1.

(a)

$t \backslash x$	0.0	0.25	0.5	0.75	1.0
0.25	0.00	0.176777	0.250000	0.176777	0.00
1.00	0.000	1.000000	-0.000000	-1.000000	0.00

(b)

$t \backslash x$	0.0	0.2	0.4	0.6	0.8	1.0
0.05	0.00	0.202000	0.408000	0.618000	0.832000	0.00
0.50	0.00	0.252190	0.473242	-0.056159	-0.119455	0.00

(e)

$t \backslash x$	0.0	0.63	1.26	1.88	2.51	3.14
0.10	0.00	0.998943	1.501598	1.501598	0.998943	0.00
1.00	0.00	0.482248	0.880878	0.880878	0.482248	0.00

Section 15.3

1.

(a)

$y \backslash x$	0.0	0.25	0.50	0.75	1.00
0.79	0.5000	0.2633	0.1249	0.0491	0.0000
1.57	1.0000	0.4285	0.1874	0.0714	0.0000
2.36	0.5000	0.2634	0.1250	0.0491	0.0000
3.14	0.0000	0.0000	0.0000	0.0000	0.0000

(b)

$y \backslash x$	0.0	0.25	0.50	0.75	1.00
0.25	0.0000	0.0150	0.0390	0.0865	0.1875
0.50	0.0000	0.0211	0.0546	0.1194	0.2500
0.75	0.0000	0.0150	0.0390	0.0865	0.1875
1.00	0.0000	0.0000	0.0000	0.0000	0.0000

(c)

$y \backslash x$	0.0	0.50	1.00	1.50	2.00
0.50	0.0000	0.7142	1.8749	4.2857	10.0000
1.00	0.0000	0.9820	2.4999	5.2678	10.0000
1.50	0.0000	0.7142	1.8750	4.2857	10.0000
2.00	0.0000	0.0000	0.0000	0.0000	10.0000

$y\backslash x$	0.0	0.79	1.57	2.36	3.14
0.79	0.0000	0.0583	0.1508	0.3318	0.7071
1.57	0.0000	0.0825	0.2133	0.4692	1.0000
2.36	0.0000	0.0583	0.1509	0.3318	0.7071
3.14	0.0000	0.0000	0.0000	0.0000	0.0000

(d)

Section 15.4

1. $u(0.5, 0.05) = 0.0550$, $u(0.6, 0.05) = 0.0489$, $u(0.8, 0.05) = 0.0398$.

2. $u(2, 5) = 0.9431$, $u(4, 5) = 0.8053$, $u(6, 5) = 0.4828$, $u(8, 5) = 0.1581$.

Section 15.5.1

1.

x	0.25	0.5	0.75	1.0	1.25	1.5	1.75
y	-0.01981	-0.03379	0.02035	0.20491	0.58217	1. 21443	2. 16399

2.

x	0.25	0.5	0.75	1.0	1.25	1.5	1.75
y	0.00011	0.01383	0.075278	0.238921	0.576185	1.171017	2.114664

x	0.25	0.5	0.75	1.0	1.25	1.5	1.75
Err. Exer. 1	0.021	0.049	0.057	0.037	0.003	0.040	0.046
Err. Exer. 2	0.001	0.002	0.002	0.003	0.003	0.003	0.003

Section 15.5.2

2.

x	1.25	1.5	1.75
Finite E.	-0.0381	0.9174	1.9197
Shooting M.	-0.0375	0.9167	1.9196
Exact	-0.0375	0.9167	1.9196

Index